10,001 Titillating Tidbits of Avian Trivia

by

Frank S. Todd

Ibis Publishing Company

10,001 Titillating Tidbits of Avian Trivia

ISBN Number 0-934797-08-0

Ibis Publishing Company
3420 Fredas Hill Road
Vista, CA 92084

Printed in the United States of America

To my best friend, life-long companion and most ardent supporter, my wife Sherlyn. She, if only via osmosis, has become far more enlightened about the fascinating world of birds than she ever thought she wanted to be. This volume is also dedicated to those fortunate souls with ever-inquiring minds, especially those with a passionate thirst for information about our feathered friends. At the same time, I feel compelled to remind readers of the sage advice of the beloved American poet Walt Whitman, who wisely cautioned against an overabundance of knowledge: *You must not know too much, or be too precise or scientific about birds and trees and flowers. A certain free margin . . . helps your enjoyment of these things.*

CONTENTS

INTRODUCTION

The first time I set eyes on Frank Todd it appeared as though a California Condor was attempting to carry him away. That was in December 1971, and Frank was Curator of birds at the Los Angeles Zoo. *Topatopa* was the name of the California Condor perched on his right forearm, and Frank was attempting to feed him (although at the time, the great vulture seemed to be feeding on Frank). I had come to see him about birding in Ecuador, knowing he had some good contacts there for seeing umbrellabirds and cocks-of-the-rock. He didn't disappoint me, and thereafter our friendship blossomed.

I soon came to appreciate that he was a fountain of knowledge, avian and otherwise, but waterfowl and penguins were his real love, especially penguins. It seemed as though tidbits of avian "trivia" crept into our every conversation. We kept in touch when he joined the staff at Sea World in San Diego, and eventually was instrumental in the development and stocking of their world-famous Penguin Encounter exhibit.

Our first mutual "penguin encounter" occurred aboard the Antarctic cruise ship *World Discoverer* in 1979 where Frank was the on-board ornithologist and dispenser of avian trivia. Because of him, I soon joined the lecture team on those voyages, and during the ensuing years we made numerous journeys together to both polar regions of the world.

Together with like-minded people aboard we spent many delightful hours exchanging and challenging bits of avian information; sort of like a bird-watcher's "can-you-top this?" It was then that I began to appreciate what a great fund of knowledge lay behind that red (now gray) mustache. It was on one of our later cruises to the European Arctic that he introduced me to an unfinished manuscript which he had titled *10,001 Tantalizing Tidbits of Avian Trivia.*

Apparently, during all these years he had been carefully collating and researching these gems of bird lore gleaned from his own readings, experiences and conversations with friends, colleagues and other ornithologists. The amount of information he had amassed was staggering, most of which was certainly not trivial.

After reading it, I realized that what he had done was to create a compendium, or in truth, an encyclopedia, of thousands of bits of hard information about birds. There seemed to be almost no fact pertaining to birds–their lives, habits, behavior, dimensions, morphology, anatomy, etc., that was omitted.

If you needed to know how many feathers clothe a hummingbird, which bird wears the most feathers, what is the largest egg produced by birds, which birds construct nests that are edible, which birds eat their own feathers, which bird has a laterally deflected bill, and more, much more, it was all here. Much of what I read, I had never known before, nor did I know how else I could have obtained that knowledge. The manuscript covered such a broad range of topics, much of which was information I thought I would never need to know. I confess to the profuse use of it ever since.

Frank had vague hopes of one day having it published, and I urged him to pursue that dream. Eventually I received in the mail a more polished copy and knew that this book had no peer in ornithological literature. It is a book that dispenses pleasure as well as information. It is a book one can read anywhere and at any time. Just open it to any page, begin reading, and the facts fly like Peregrine Falcons. You will never know just how much you did or didn't know about birds until then. However, if you seek particular information about this or that in the bird world, the index will direct you to the right section.

All through those years Frank Todd was carefully sweeping up and saving all those cast-off bits of avian "trivia", and produced a work that will pleasure and enrich us all.

Arnold Small
Beverly Hills, California
December, 1993

FOREWORD

As often happens, this volume started out with relatively humble beginnings, but gradually evolved into a major project over the years. During countless nocturnal cocktail sessions with ornithological, avicultural and birding associates, bird trivia questions were continually tossed back and forth, and these informal gatherings typically extended well into the wee hours of the morning, and sometimes until dawn. These thought-provoking, brain-racking sessions were mentally stimulating (and often exhausting), and even people who were not especially bird-oriented initially found themselves caught up in the ensuing mania, and became active participants. Ultimately, it was the surprisingly keen interest of these *outsiders* that convinced me that a comprehensive book on avian trivia might be of general interest.

It was decided that no subject was too trivial for inclusion, including questions that might not be directly related to birds, such as individuals, books, or even movies named after birds. The nagging problem of how best to organize the book was troublesome from the very beginning. A number of my colleagues strongly urged that topics be neatly organized according to subject or listed taxonomically. This sage advice obviously has considerable merit, but such a structured organization was not really consistent with the purpose of the book. A shotgun kaleidoscopic approach was taken instead, but in doing so, I fully recognize that some readers may become frustrated at not being able to locate specific items easily, although the comprehensive index will resolve most of these difficulties. Nevertheless, this compendium is not intended to be a standard reference volume, although it hopefully will be instructive as well as fun.

It is highly unlikely that anyone, ornithologist or otherwise, will be able to correctly answer all the questions. It is anticipated that some answers will be challenged because the opinions of varying authorities often differ, sometimes considerably so. I was astounded at the variety of expert opinions on a number of subjects. Even apparent cut-and-dried issues are not resolved; for example, to this day there is major disagreement on when the last Labrador Duck was taken, and amazingly, even when the last *captive* Carolina Parakeet died. The numerous, and sometimes drastic, revisions in avian taxonomy in recent years will also no doubt fuel a multitude of arguments.

I have discovered (often the hard way) during the course of compiling this book, that trivia fanatics can be especially vocal and passionate. Some answers will undoubtedly be controversial in light of the fact that new and updated information is being generated at an astounding rate. And some data will admittedly already be yesterday's news the very day the book goes to press. I expect to receive some flack, but if readers feel obligated to look things up, the book will have served one of its purposes.

A number of my colleagues graciously reviewed portions of the manuscript and offered invaluable constructive advice. I am especially indebted to Arnold Small, Joe Jehl, Jr., Richard Brooke, Bob Clements, W. Roy Siegfried, Paul Johnsgard, Jim Clements, Bill Everett and Wayne and Sue Trivelpiece. My long-time friend and fellow world traveler, Arnold Small, was most encouraging when the book was in its infancy, and I can think of no one more qualified to write the introduction. Norman Lasca and Cory Drieschman were particularly helpful with the thankless, and seemingly endless, task of editing.

My publisher, Jim Clements, was also an early and enthusiastic fan of the project. An ornithologist himself, Jim is a recognized world-class birder who contributed significantly to the organization of the book and quietly, but persuasively, put forth numerous suggestions that were ultimately incorporated. Jim assumed responsibility for compiling the extensive index, a labor of love that will greatly benefit all readers. I owe a great deal to all of these individuals who gave graciously of their time and considerable expertise, and to many others too numerous to list. However, any shortcomings, omissions or errors fall entirely on my shoulders.

Frank S. Todd
January 1994

Questions

1. Which is the only vulture that is not primarily a carnivore, and on what does it mainly feed?
2. Which birds echo-locate in the manner of bats?
3. Excluding northern swans, which is the only *wild* waterfowl with completely white plumage?
4. Which shorebird has the longest bill, and how long is it?
5. Which is the largest woodpecker, and what is its length?
6. What is the popular nickname of the first individual to fly solo across the Atlantic Ocean?
7. Which birds are collectively known as raptors?
8. Which bird declined to the lowest population level and subsequently recovered?
9. Which is the rarest North American songbird?
10. Which is the only two-toed bird, and why are two toes advantageous?
11. What was the origin of the expression *a quill of gold*?
12. Which bird laid the largest egg, and how large was its egg?
13. Which bird endures the coldest temperature?
14. Which surviving bird has the greatest *documented* wingspan, and how great is it?
15. What do chick-rearing Common Moorhens, Red-cockaded Woodpeckers, Florida Scrub Jays, Chimney Swifts and Brown-headed Nuthatches have in common?
16. Which is the only Antarctic bird that lacks webbed feet?
17. To which bird does the expression *the ugly duckling* refer?
18. Which bird has the longest bill, and how long is it?
19. Which *group* of birds is generally credited with laying the most beautiful eggs?
20. Which bird breeds the farthest north?
21. Which birds construct the delicate nests highly sought after for bird's-nest soup?
22. Which was the last *species* of waterfowl (ducks, geese and swans) to become extinct, and when did it disappear?
23. Which is the fastest running bird, and how swift is it?
24. Which is the most numerous gull?
25. Which of Darwin's finches does not inhabit the Galapagos Islands, and where does it occur?
26. What is an alcid?
27. For what is Hinckly, Ohio, best known?
28. Which is the heaviest raptor, and how much does it weigh?
29. Does the colorful bill of an adult puffin fade during the winter?
30. Which birds are the closest relatives of the penguins?
31. Which is the deepest diving waterfowl, and how deep can it dive?
32. What are the two *primary* functions of a pelican's pouch?
33. How many flightless Lake Atitlan Grebes remain?
34. What is the status of the Mexican Imperial Woodpecker?
35. Which eagle has the greatest wingspan, and how great is it?
36. Where is a bird from if it has a neotropical distribution?
37. A few birds are noted for *regular* tool use, but which two are generally regarded as the most notable?

Questions

38. How did the Sacred Ibis obtain its name?
39. Which is the only vertebrate that breeds in Antarctica during the dark winter?
40. Where does the Spectacled Eider winter?
41. Which is the rarest bird?
42. What is a tiercel?
43. Which is the only *species* of waterfowl that does not go through a flightless stage during the molt?
44. Which is the only bird with a sideways curving bill?
45. Is the Madagascar Elephant-bird the heaviest known bird?
46. Which is the smallest seabird, and what is its weight?
47. Why were the riflebirds (birds-of-paradise) so named?
48. Based on actual contour feather counts, which bird has the most feathers, and how many feathers were counted?
49. Based on actual contour feather counts, which bird has the least number of feathers, and how many feathers were counted?
50. Which bird lays the largest egg relative to its size?
51. What is a furcula?
52. What is an obligate brood parasitic-nesting bird?
53. Which bird has the longest feathers, and how long are they?
54. Which are the only penguins that construct no nests?
55. Asian cave swiftlet nests consist of what material?
56. Which two bird groups have the shortest bills, and which bird has the shortest bill of all?
57. Which three species of birds breed the farthest south?
58. Do birds have skeletons that weigh less than their feathers?
59. In terms of total mass and weight, which North American bird lays the largest clutch of eggs relative to its size?
60. What is the largest single cell?
61. How did the South American Oilbird obtain its name?
62. What is the carpal joint?
63. Which are the only three groups of birds that swim underwater using their wings, rather than their feet?
64. Which surviving alcids are flightless?
65. What was one of the initial tasks assigned to John Gould (1804-1881), the famous 19th-century ornithologist and bird artist, when he commenced work for the Zoological Society of London at the age of 20?
66. Which is the largest North American bird if *both* weight and wingspan are considered?
67. Which raptor has the most highly developed sense of smell, and why?
68. Which is the only bird with a *solid* ivory-like casque?
69. Which North American bird has the most restricted *breeding* range, how large is its range, and where is it located?
70. How did the southern whalebirds (prions) obtain their name?
71. After which bird is the currency of Guatemala named?
72. How did the booby obtain its name?

Questions

73. Which bird lays the smallest egg relative to its weight?
74. Which is the only bird known to fly over the South Pole?
75. What was the origin of the term *guano?*
76. Which was the first bird to be called a penguin?
77. How many hummingbird species have been recorded north of the Mexican border?
78. Which is the longest-lived wild bird, and how old is (or was) it?
79. Which birds have the longest carpal spurs, and how long are the spurs?
80. What is a Cape Pigeon?
81. Which bird constructs the largest tree nest, and how large was the largest recorded nest?
82. How large are the eggs laid by the largest living bird?
83. Which birds are noted for *distinct* eyelashes?
84. Which animal has the longest annual migration?
85. Which is the rarest American shorebird?
86. Which species of tanagers occur naturally in North America?
87. To which group of birds is the Dodo most closely related?
88. Are *adult* Ostriches capable of swimming?
89. Which bird is credited with having the most luxurious down?
90. Which bird was completely eliminated by a single cat in less than a year?
91. Which is the smallest North American bird of prey, and what does it weigh?
92. Which is the only seabird in which the male alone incubates?
93. Do birds prey on mollusks in the genus *Tridacna?*
94. Why were the big three guano producers of offshore Peruvian islands formerly called *billion dollar birds?*
95. Which are the only birds restricted to California?
96. Which birds construct the largest ground nest, and how large are the nests?
97. Why do swallows, nightjars, flycatchers and some other birds that capture flying insects on the wing, have hair-like feathers surrounding the mouth known as rictal bristles?
98. Which arboreal bird has two carpal *claws* on each wing, and what function do they serve?
99. With respect to foraging, what do baleen whales and flamingos have in common?
100. Why was Ben Franklin not in favor of the Bald Eagle as the national bird, and which bird did he propose as an alternative?
101. Which bird has the longest bill relative to its body length, and how long is the bill?
102. How many birds were named after the leaders of the Lewis and Clark expedition, and which ones are they?
103. Which is the only trogon that *regularly* occurs in the U.S.?
104. Which is the smallest bird, and what is its length and weight?
105. What is a *swan song*?
106. Which order contains the most species of birds, and how many species are in the order?
107. Which Cuban bird has the unusual distinction of having a scientific name bearing both the first name and the surname of a naturalist?

Questions

108. Which is the largest bird species named by the scientific community since 1970?
109. Where have the greatest number of exotic species of birds been introduced, and how many species were introduced?
110. Which birds are collectively called mollymauks, and what is the origin of the name?
111. On what unusual substance do Old World honeyguides feed?
112. Which bird nests at the highest altitude, and how high does it breed?
113. What was the greatest *documented* age for any bird, and what bird was it?
114. Why were the incubator-birds (megapodes) so named?
115. Excluding introduced species, which is the most widespread passerine?
116. The contents of how many hummingbird eggs are required to fill an egg of the extinct Elephant-bird of Madagascar?
117. Which endangered bird is dependent on burned jack pine forests as breeding habitat?
118. Does the Sacred Ibis inhabit Egypt?
119. Although symbolic of Antarctica, which are the only two of the 17 species of penguins essentially restricted to the frozen southern continent?
120. Which birds have the thinnest, most delicate skin?
121. When and where was the last Great Auk killed?
122. Which bird is known as a "Sore-eyed Pigeon," and why?
123. Which animal encounters the most daylight?
124. What is a murre called in the Old World?
125. Which bird is known as a Crocodile-bird, and why?
126. Which is the only obligate brood parasitic-nesting waterfowl?
127. How many species of Hawaiian birds have become extinct in historical times?
128. Which two geese are the ancestors of all domestic geese?
129. Which birds have the longest incubation periods, and how long are they?
130. Which bird returns annually to breed at the famous Spanish mission at San Juan Capistrano, California?
131. Which North American bird is the most specialized feeder, and on what does it feed?
132. How do nestling petrels and albatrosses defend themselves?
133. Which is the only penguin not found in association with the Kelp Gull?
134. Which bird lays the smallest egg, and how much does its egg weigh?
135. Excluding honeyguides, which birds feed on wax?
136. Which is the only endemic passerine of the subantarctic?
137. Which birds have the shortest incubation periods, and how long are they?
138. How did the megapodes obtain their name?
139. How much more acute is the eyesight of a diurnal raptor than that of a human?
140. Which is the heaviest passerine, and how much does it weigh?
141. Which are the three most prevalent avian pests introduced into America, and why were the birds originally introduced?
142. Which is the only carnivorous parrot?
143. Which bird is called a Laughing Jackass, and what type of bird is it?
144. Which flying bird is credited with being the fastest runner, and how fast can it run?
145. What is the origin of the name Oldsquaw duck?
146. Which is usually regarded as the most numerous bird, and what is its population size?

Questions

147. What single physical feature distinguishes birds from all other animals?
148. What is powder-down, and which type of birds have it?
149. Which nightjar relative does not prey on insects or small vertebrates, and on what does it feed?
150. What is the Bearded Vulture generally called?
151. Which small non-passerines of Africa and Asia have a particularly strong sense of smell?
152. How much heavier is the largest petrel than the smallest?
153. What is the "knee" of a flamingo?
154. Which *terrestrial* bird winters the farthest north?
155. Why do vultures have naked heads and necks?
156. What is *caviar of the East* or *white gold?*
157. Which is the only flightless, nocturnal parrot?
158. Which is the largest grouse, and what is its length and weight?
159. Do any birds truly hibernate?
160. Which two passerines attain the greatest length, and what is their length?
161. Which are the only birds that are *completely* independent at hatching?
162. In what direction does the deflected bill of a New Zealand Wrybill *always* curve?
163. Which is the only waterfowl that has three distinct annual plumages?
164. Which bird has the longest fledging period, and how long is it?
165. Which seabird has a distinct citrus-like odor, and what causes the odor?
166. Which country has the most breeding bird species?
167. Which birds have the greatest number of cervical (neck) vertebrae, and how many do they possess?
168. Why did feathers initially evolve?
169. Which American bird has the shortest migration?
170. How can the flightless Ostrich survive in the presence of fearsome terrestrial predators, such as lions, leopards, hyenas, cheetahs and Cape hunting dogs?
171. How many penguins have regal names, and which ones are they?
172. Is the Mute Swan really mute?
173. What is the function of the bizarre bill of the African Shoebill stork?
174. What is a stoop?
175. Which bird is popularly known as the Mexican eagle?
176. Which cave-dwelling bird has a keen sense of smell?
177. Which is the smallest flightless bird, and what is its length and weight?
178. Which bird has undergone the most dramatic *natural* range expansion?
179. On what do Lammergeiers (Bearded Vultures) feed?
180. What is the function of a pelican's over-sized subcutaneous air-sacs?
181. Which penguin has the most southerly breeding distribution?
182. Which birds have the longest toes relative to their size?
183. Which is the rarest New Zealand parrot, and what is its current status?
184. What is the origin of the vernacular name "goatsucker?"
185. The skull of a male Helmeted Hornbill accounts for how much of the bird's overall weight?

Questions

186. What is a wryneck, and why was it so named?
187. Which birds are credited with being the swiftest swimmers, and at what speed to they swim?
188. How many species of birds are officially named?
189. How did the honeyguides obtain their name?
190. Which two parrots ranged north of the Rio Grande River?
191. What is a gular flutter?
192. Which birds have the most rapid wingbeats, and what is the wingbeat per second?
193. Which is the fastest flying bird, and what speed can it obtain?
194. What is the origin of the term rookery?
195. How does the chick care of the Sungrebe differ from all other birds?
196. Which birds become torpid?
197. Which are the only birds with nostrils located at the tip of their bill?
198. Which bird has the longest legs relative to its size?
199. Which group of birds can rotate their heads to the greatest extreme, and to what extreme can their heads be rotated?
200. Why were storks said to bring human babies?
201. Which birds have the most-vestigial wings, and how long are the wings?
202. Which is the only bird that normally never sets foot on solid ground?
203. On what does the Limpkin feed almost exclusively?
204. Which birds have the slowest wingbeat?
205. Why do toucans have such over-sized bills?
206. How do Lammergeiers (Bearded Vultures) obtain bone marrow?
207. Which birds have functional tongues?
208. What is pigeon's milk?
209. Which is often cited as the most aerial bird?
210. Which American woodpecker migrates the farthest?
211. Which bird consumes windshield wipers?
212. What is the range of avian body temperatures?
213. To which group of birds are sandgrouse believed to be most closely related?
214. Which birds have the shortest legs relative to their size?
215. Can owls see in *total* darkness?
216. Name at least five *groups* of birds in which the females are larger than the males.
217. What is the uropygial gland, and what is its function?
218. How many museum specimens exist of the extinct Great Auk?
219. Which bird digs the longest burrow, and what is the length of the longest burrow on record?
220. Which two birds are the most commonly asked for in crossword puzzles?
221. Which was the most abundant bird of historical times, and how numerous was it?
222. The calls of which two exotic birds most often embellish the soundtracks of many Hollywood jungle movies?
223. Which land bird inhabiting North America has the longest tail feathers, and how long is its tail?
224. Which American bird is called "golden slippers," and why?

Questions

225. Which New World bird was deliberately exterminated?
226. Are birds the only vertebrates with a nictitating membrane, and what is its purpose?
227. Which penguin occupies the largest rookery, and where is it located?
228. Which *native* North American breeding bird has the longest tail feathers relative to its size, and how long is its tail?
229. Which birds have the poorest vision?
230. What is a polyandrous bird, and name at least five of them?
231. Which bird is popularly known as a gooney bird?
232. Which is the heaviest waterfowl, and how much does it weigh?
233. In historical times, have California Condors always graced the skies of North America?
234. Which birds are pictured on the New Zealand one, two, five, 10, 20, 50 and 100 dollar banknotes?
235. Why were Oilbirds valued by South American Indians?
236. Which bird had the longest continually used tree nest?
237. Which two North American birds have the longest incubation periods, and how long are they?
238. In which family of birds does the male work the hardest while the female alone incubates?
239. Why do most incubating birds turn their eggs, and which birds do not turn their eggs?
240. What are altricial, psilopaedic and nidicolous birds?
241. Do cassowaries have flight feathers?
242. Does the male Indian Peacock have the longest tail of any bird?
243. What is a sulid?
244. What is the height and weight of the tallest bird that ever lived?
245. What is the estimate of the total number of birds inhabiting the planet?
246. Which bird has the longest tail, and how long is its tail?
247. How do megapodes (incubator-birds) regulate the temperature in their incubation mounds?
248. Which seabirds are called "Jesus Christ birds," and why?
249. What are loons called in Europe?
250. Which bird seeks blood on which to feed, and how does it obtain the blood?
251. Which aquatic bird in America is known as the "snake bird," and why?
252. Which is the only North American stork?
253. Which birds are commonly called lily-trotters?
254. Which were the first domesticated birds, and when did domestication occur?
255. Excluding some owls, which is the most terrestrial raptor?
256. Which is the largest endangered endemic bird of New Caledonia, and how many still survive?
257. Which is the heaviest non-domesticated flying bird, and how much does it weigh?
258. What is the origin of the name "jaeger?"
259. What pigment causes the bright red wing coloration of many African turacos?
260. What type of bird is the Greater Roadrunner?
261. Which marine reptile is named after a bird?

Questions

262. Why were the African mousebirds so named?
263. What is the most striking physical feature of the motmots?
264. Which Old World birds are the hummingbird counterparts?
265. What type of bird is the American Robin?
266. Where do most Elegant Terns nest?
267. Which birds consume the most food daily relative to their weight, and how much food is required daily?
268. Which is the smallest waterfowl, and how much does it weigh?
269. Why do cassowaries have large casques?
270. Why have relatively few avian fossils been discovered?
271. What Lammergeier body part is modified, enabling it to feed on bone marrow?
272. Which is the smallest North American passerine, and how much does it weigh?
273. Which bird is known as Pharaoh's chicken?
274. Why do Harlequin Ducks and Torrent Ducks have such high-pitched whistles?
275. What is a ratite, and what is the origin of the name?
276. What is a shag, and what is the meaning of the name?
277. Which seabirds are known as stinkers, and why?
278. The inhabitants of which country are commonly known by the name of their national bird?
279. Why are New World and Old World vultures separated taxonomically?
280. Which mammal has the most bird-like characteristics?
281. Which bird has traditionally been called a duck hawk?
282. Which land bird spends the most time on the wing?
283. How did Ian Fleming come up with the name for his well-known character *007*?
284. Which flightless seabird is sometimes run over by cars?
285. Which enormous land mass has the least number of bird species, and how many breed there?
286. How do pelicans and gannets incubate?
287. Which bird undergoes the longest fast, and how long is the bird without food?
288. Which two parrots have the largest, most powerful bills?
289. How did the Secretary-bird obtain its name?
290. Which is the tallest American bird, and what height does it attain?
291. Why do hornbills have casques?
292. Which waterfowl are collectively known as stiff-tails?
293. What happened to the San Benedicto Rock Wren?
294. What is the greatest single cause for the demise of the Ivory-billed Woodpecker?
295. What is the incubation period of the neotropical Sungrebe?
296. Why is the flight of an owl so silent?
297. For which group of birds is New Gúinea most famous?
298. What is the more accepted name for the tree-ducks?
299. The chicks of which birds never see their parents?
300. How much did the extinct Dodo weigh?
301. Which bird is the most social nester?
302. What is the *sport of kings*?

Questions

303. Which is the heaviest parrot, and how much does it weigh?
304. What is considered to be the earliest proto-bird fossil, and how ancient is it?
305. Which bird in America is known as the fish hawk?
306. At what age does a greater albatross commence breeding?
307. Which are the only birds with a lower mandible longer than the upper mandible?
308. Which is the only tree-nesting alcid?
309. How many *full* species of birds became extinct between 1600 and 1980?
310. Three of the largest, most powerful raptors inhabit the forests of the Neotropics, Asia and Africa, with the Harpy Eagle occupying the New World and the Monkey-eating Eagle restricted to the Philippines, but which eagle is the African counterpart?
311. Which pigeons are the largest, and where do they live?
312. What is an onomatopoetic name?
313. Which bird is popularly known as a chaparral cock?
314. What are Old World buzzards?
315. Which North American waterfowl has the brightest-colored bill?
316. The vocalizations of which waterfowl resulted in both its scientific and common name?
317. Which is the only jay common to both the Old and New World?
318. Which penguin is named after an equine, and why?
319. How did the neotropical puffbirds obtain their name?
320. Which bird is believed to be the mythical Roc of Arabian Nights fame?
321. Which is the only parrot that constructs a communal nest?
322. How did the Old World rollers obtain their name?
323. Which duck is the most highly prized by American hunters?
324. What type of bird is the Nightingale?
325. Which is the only nocturnal fruit-eating bird?
326. Which waterfowl has been persecuted as much as the American coyote and, like the coyote, has persisted, even thrived, in spite of it?
327. What type of raptor is the American Bald Eagle?
328. Why are molting penguins obligated to fast?
329. What are the three primary physical differences between the superficially similar swifts and swallows?
330. The seven New World vultures may be related to which group of non-raptor birds?
331. Why do Black-necked Swans carry young on their backs?
332. What was the wingspan of the largest extinct flying bird?
333. What is anting, and what is its function?
334. Which birds are the least capable of walking?
335. What is the height and weight of the largest surviving bird?
336. Which bird is known as a sprig?
337. How many species of turkeys are described?
338. How did the name "limpkin" originate?
339. Which birds are collectively known as peeps?
340. When did the California Condor disappear in the wild?
341. What is the most distinctive shrike habit?

Questions

342. Which northern hummingbird barely ranges west of the Mississippi River?
343. Which birds are known as muttonbirds?
344. Which is the only albatross that nests near the equator?
345. Which is the heaviest booby, and how much does it weigh?
346. Which is the most numerous native North American bird?
347. Which penguin breeds the farthest north?
348. Where do the Alaska-nesting Asian Golden-Plovers winter?
349. Other than various species of waterfowl, which birds does the obligate brood parasitic-nesting South American Black-headed Duck select as host species?
350. On what does the Laughing Falcon primarily feed?
351. Are honeyguides related to toucans?
352. Which bird has the longest continuous incubation stretch, and how long is it?
353. When was the Ring-necked (Common) Pheasant first introduced into North America, and from where did the stock originate?
354. What tool does an oystercatcher's bill resemble?
355. Which seabird is called a "stonecracker," and why?
356. What type of birds are lapwings?
357. What unusual feeding technique is used by phalaropes?
358. Which bird is known as *the eagle of the Antarctic*?
359. What is a noddy?
360. Which sex of the Australian Malleefowl incubates?
361. The parasitic-nesting behavior of some Old World cuckoos is such that it resulted in the introduction of what word into the English language?
362. Which caprimulgids do not capture prey on the wing?
363. At what angle does a perched Sword-billed Hummingbird hold its bill, and why?
364. How far south did the extinct Great Auk winter?
365. What are marginal pectinations, which birds have them, and what purpose do they serve?
366. Which Old World birds are known as dollarbirds, and why?
367. Which birds are commonly called man-o'-war birds?
368. Which of the three puffins is most inclined to nest in a cliff face crevice, rather than dig its own burrow?
369. Which bird is the deepest diver, how deep can it dive, and how long can it remain submerged?
370. Which bird is known as a "water turkey," and why?
371. What is the third eyelid?
372. How did the shearwaters obtain their collective name?
373. How did the expression *crazy as a loon* originate?
374. Which of the six crested penguins is the most widespread?
375. How do the nostrils of an albatross differ from those of its close relative, the petrel?
376. Which are the only birds with neither wing nor tail plumes?
377. To which birds are the neotropical tinamous most closely related?
378. How many Little Blue (Fairy) Penguins are required to equal the weight of a large Emperor Penguin?

Questions

379. Which is the second largest bird, and how much does it weigh?
380. Which are the only truly aquatic passerines?
381. Which bird is the state bird of no less than seven states, and name the states?
382. Which are the only birds that have occurred naturally on all continents?
383. Which large, long-legged birds are superior dancers?
384. What is a lek, and name at least three birds that use leks?
385. Why are the neotropical umbrellabirds so named?
386. Which is the only alcid that does not always nest on the ground?
387. What is a carinate bird, and what is the origin of the name?
388. Do any cotingas range north of the Rio Grande River?
389. Which is the most recently described waterfowl species, and when was it described?
390. Where and when did falconry originate?
391. Birds-of-paradise were formerly believed to lack what?
392. What is a Blue Goose?
393. Which two American birds are known as "camp robbers?"
394. What is the greatest threat facing the endangered Kokako of New Zealand?
395. Why are Lesser Flamingos more inclined to feed at night than Greater Flamingos?
396. What is a cygnet?
397. Which bird spanned the longest interval from "extinct" status to ultimate rediscovery?
398. What is a Goosander, and how did the name originate?
399. What is a granivorous bird?
400. What is an eyrie?
401. Do megapodes use the same nesting mound year after year?
402. Which South American bird is called a "guacharo," and why?
403. How many species of hummingbirds nest in Canada, and which ones are they?
404. Which goose is the most abundant, and how numerous is it?
405. Birds in which two *families* are noted for large casques?
406. Which American bird is known as a "muffle-jawed egret?"
407. What is a cere?
408. Which American birds are known as "hell-divers," and why?
409. Which waterfowl are the most colonial nesters?
410. How many species of kittiwakes are described?
411. Why are many captive birds subjected to laparotomies?
412. What is a crèche, and name three birds that use crèches?
413. Which is the only swan that regularly lines its nest?
414. How long is the incubation period of the Malleefowl?
415. What is a loomery?
416. Which birds are known as dabchicks?
417. Which American bird is called a "darter," and why?
418. The sexes of which parrot are so unlike in appearance that males and females were formerly considered separate species?
419. What is eclipse plumage?
420. As of early 1994, what was the size of the California Condor population?
421. How have sandgrouse adapted to desert nesting?

Questions

422. When and where were Ross' Goose breeding grounds discovered?
423. How large is the Cahow (Bermuda Petrel) population?
424. Why do jacanas have such long toes and nails?
425. How do Mountain Quail undertake their seasonal altitudinal movements?
426. Which was the first hibernating bird to be discovered, when was it discovered, and what was the bird's temperature?
427. How many breeding pairs of the recently described Amsterdam Albatross exist?
428. Which is the smallest owl?
429. How many species of seabirds are known in which the female alone incubates?
430. How many states have selected the Northern Mockingbird as the state bird, and which ones are they?
431. The range of which bird essentially corresponds to the range of the Galapagos Penguin?
432. When were tropicbirds first documented nesting in Chile?
433. Who was the main character in James Fenimore Cooper's classic *The Last of the Mohicans*?
434. Are petrels good underwater swimmers?
435. Do Brown Pelicans generally forage within sight of land?
436. If a King Penguin chick were fed daily, how long would it take to fledge?
437. Do Wood Storks occur as far south as Argentina?
438. Do Barn Owls prey on bats?
439. Does the South American Oilbird have a short tail?
440. Which was the first bird protected by humans?
441. What is a sapsucker, and what is the origin of the name?
442. Can a Common (Ring-necked Pheasant) elevate its ear (auricular) feathers?
443. Which birds may copulate on the wing?
444. What alternate name is used for the large shorebirds known as thick-knees?
445. Which raptor has the largest bill?
446. Which birds are sometimes called sea swallows, and why?
447. How much weight do fasting incubating male Emperor Penguins lose?
448. Which birds are called "Christbirds" or "lotus birds?"
449. Which bird is known as a bonxie?
450. What is a speculum?
451. Which American bird is known as a butcher-bird?
452. Why do kiwis have nostrils located near the tip of their bills?
453. Which South American birds are known as ñandu, and why?
454. Which island is called Wideawake Island, and why?
455. What is a wreck?
456. By what name do Europeans know the American Wood Duck?
457. Why is the Helmeted Hornbill prized by Malayan and Indonesian natives?
458. What is mobbing?
459. Do cormorants sight their prey while in flight?
460. What is hawking?
461. What are oxpeckers?

Questions

462. Which birds are considered the New World counterparts of the hornbills?
463. What is a crepuscular bird?
464. What is the resting heartbeat of a hummingbird?
465. Why is the Canvasback duck so tasty?
466. Which are the only two birds that nest well into the *interior* of Antarctica?
467. Which are the only two northern sea-ducks with speculums?
468. What is allopreening?
469. What are fish-ducks?
470. Which land mass is called *the continent of parrots?*
471. Is an ani an aberrant neotropical potoo?
472. Which birds are known as Mother Carey's Chickens, and why were they so named?
473. What do perched motmots often do with their long racket tails?
474. Why are laparotomies always performed on the left side of a bird?
475. Which bird was formerly called the "woolly penguin?"
476. What is an aracari?
477. What is the hunter's name for the American Wigeon, and how did the name originate?
478. Which two birds lay eggs resembling avocados, both in size and color?
479. How can ant-eating woodpeckers consume large quantities of ants without being harmed by the ant's formic acid?
480. What is a fossorial bird?
481. What racetrack is famous for a large flock of breeding Caribbean flamingos?
482. Which Old World parrot had a bounty placed on it, and why?
483. What well-known brands of whiskey might be consumed by an American ornithologist or birder?
484. Which bird is known as a cricket teal, and why?
485. How does the foraging behavior of brown pelicans and white pelicans differ?
486. What is the national bird of Chile?
487. Which common North American native bird causes humans considerable loss of sleep in the spring?
488. What is the mascot of the U.S. Air Force Academy?
489. Why does the inner border of the third claw of herons and bitterns have small serrations, and which other birds have claws of this type?
490. The Acorn Woodpecker has what unusual foraging habit?
491. Which birds have the most dexterous feet?
492. What is a "Tasmanian squab?"
493. Which two birds have the largest horns?
494. Which is the only spoonbill that is not a colonial nester?
495. Which is the only waterfowl that occurs naturally in all hemispheres without any subspecific variation?
496. Which raptor has the widest distribution?
497. How did the South American ovenbirds obtain their name?
498. Who was the *Birdman of Alcatraz,* and what actor portrayed him in the 1962 movie of the same name?
499. Why are there no penguins in the Arctic?

Questions

500. Which brightly-colored duck has such a high-pitched call that groups of the ducks are called "sea mice?"
501. With the exception of hovering flight, what is the minimal speed required for birds to remain airborne?
502. To which birds are flamingos probably most closely related?
503. Which North American bird has the slowest reproductive rate?
504. Male and female swans are known by what special names?
505. How many of the 18 species of flightless rails still exist?
506. How did the Graylag Goose obtain its name?
507. What is a greenshank?
508. Which two northern galliform birds have circumpolar distributions?
509. What type of bird is a hobby?
510. What avian color is not caused by any pigment?
511. What is bird banding called in the Old World?
512. What is unusual about the incubation habits of the Congo Peacock?
513. When and where did the Labrador Duck disappear?
514. Which birds are collectively known as psittacines?
515. Which family of small, red-throated, non-passerine birds is restricted to the West Indies?
516. Which is the only North American-breeding tern that nests above ground level?
517. What are tragopans?
518. What distinct physical feature typifies the adults of all three tropicbirds?
519. Which birds are collectively known as the tubenoses?
520. Do tropicbirds feed at night?
521. Which group of birds are the northern ecological counterparts of the penguins?
522. What is the European White-winged Scoter commonly called?
523. Why are Snowy Owls rare in Iceland?
524. Which is the smallest North American-breeding bird?
525. Which American forest-dwelling shorebird is managed as a gamebird?
526. The entire breeding population of Royal Penguins is essentially restricted to what island?
527. Which is the largest New World bird, and how much does it weigh?
528. What was the first U.S. National Wildlife Refuge, and when and where was it established?
529. Which North American waterfowl is the most vocal?
530. What is the state bird of Louisiana?
531. What is a brood-patch?
532. What well-known bird activity in the U.S. occurs around Christmas?
533. Which is the only goose named after a vegetable, and why was it so named?
534. Excluding poaching, what illegal bird "sport" occurs in some parts of southern U.S.?
535. What is an egg tooth?
536. The myna bird is a member of which large family?
537. What is the dorsal orange-gold "sail" of a breeding-plumaged drake Mandarin Duck?
538. What is a blood feather?

Questions

539. Which two ducks are the ancestors of *all* domestic ducks?
540. When chicks start to break out of the egg, it is called pipping in America, but what is this called in Europe?
541. What percentage of all bird species are regarded as seabirds?
542. What is a dump nest?
543. During the egg-collector era, the egg of which bird was occasionally sold as that of a California Condor?
544. What is a pochard, and how many species breed in North America?
545. Which group of birds is noted for the most extreme bone pneumaticity (hollowness)?
546. What is candling?
547. What is a cursorial bird?
548. What is a zygodactyl foot, and name at least two groups of birds with feet of this type?
549. What is a feral bird?
550. What is a Hamerkop?
551. What is the bastard wing?
552. Which birds are known as tinkerbirds?
553. What is treading?
554. How did the Cape Pigeon (Painted Petrel) obtain its name?
555. What are shorebirds collectively called in Europe?
556. Which is the smallest North American waterfowl, and how much does it weigh?
557. Is the olfactory sense highly developed in birds?
558. Why do Emperor Penguins breed during the middle of the Antarctic winter?
559. Which was the most recent North American bird declared extinct, and when did this occur?
560. What is a quiet conversation sometimes called?
561. Why do the Chinese believe that bird's-nest soup has aphrodisiac qualities?
562. Which birds porpoise?
563. Why can some birds drink salt water without ill effects?
564. Which currently numerous American duck was literally pulled back from the brink of extinction?
565. Are tattlers extremely vocal thrushes?
566. In the higher northern latitudes, which gull is most likely to forage out of sight of land?
567. When and where did the last Passenger Pigeon perish?
568. Which birds are best known for distinct facial discs?
569. To which group of birds are swifts most closely related?
570. How did the Honey-buzzard obtain its name?
571. Into what *country* have the most species of birds been introduced, and how many were introduced?
572. Why were the neotropical tapaculos so named?
573. Which is the most familiar American thrush?
574. Which goose is so inedible that it has been described as *fit only for a starving man*?
575. The overall population of which flightless rail has not been significantly reduced by humans?

576. At its lowest level in 1941, what was the entire Whooping Crane population?
577. Which American upland gamebird has a drumming courtship display?
578. The Lammergeier is considered to be what type of raptor?
579. Which seabirds have the smallest feet relative to their size?
580. When did Cliff Swallows first begin nesting at the San Juan Capistrano mission?
581. Who was known as the *father of Indian ornithology?*
582. Which birds dive underwater carrying chicks on their backs?
583. Why do the Japanese value the cormorant?
584. On what do openbill storks primarily feed?
585. Which are the smallest diurnal birds of prey, and how much do they weigh?
586. When was the Congo Peacock discovered?
587. What is a group of courting greater albatrosses called?
588. Which are the only shorebirds with lobed toes?
589. Which two penguins have musical calls?
590. What color is a breeding Ross' Gull?
591. Where do harriers generally nest?
592. The American Robin was named after which bird, and why?
593. Which birds are sometimes called "sea pigeons?"
594. Which seabird is a tree nester, but builds no nest?
595. Which birds can drink without raising their heads?
596. Which large terrestrial parrot dwells above the snow line?
597. Which is the only North American tern with a yellow-tipped black bill?
598. Which captive-reared endangered bird has been released in large cities, and why?
599. Which waterfowl has the longest incubation period, and how long is it?
600. Why are caracaras, many of which are scavengers, considered closely related to falcons, the most skilled of aerial hunters?
601. What is a bird strike?
602. Which is the smallest gull, and how much does it weigh?
603. What was the name of the FBI trainee played by Jodie Foster in the 1990 hit movie *The Silence of the Lambs*?
604. Which group of birds do the neotropical tinamous most closely resemble?
605. What mechanical device is known as a rainbird?
606. Which is the smallest bird that migrates across the Gulf of Mexico?
607. What do birds, squid and octopus have in common?
608. In the game of golf, what is a birdie?
609. Which is the most valuable bird in the world?
610. How did the Barnacle Goose obtain its name?
611. Which bird, previously known only from a single specimen collected as a fledgling in 1855 on Gua Island (Fiji Islands), was rediscovered in 1983, and what were the circumstances surrounding its rediscovery?
612. Which birds construct the smallest nests?
613. Which sulid is noted for obligate siblicide?
614. What was the approximate height and weight of the largest known extinct penguin, and where was the fossil discovered?

Questions

615. Which bird has killed a number of New Guinea natives?
616. Which bird was named after Madame de Pompadour, the famous French courtesan and mistress of the French King, Louie XV?
617. Where is the largest bird egg collection in the New World housed, and how large is it?
618. How many species of gulls inhabit Antarctica, and name them?
619. Which is the smallest of the four eiders, and how much does it weigh?
620. Do flamingos breed in western Europe?
621. Which birds have the longest tongues relative to their size?
622. Which extinct birds, or those on the verge of extinction, did Audubon probably see?
623. What are piscivorous birds?
624. Why do many white birds, such as some storks, cranes, pelicans, gulls and Snow Geese have dark-colored wingtips?
625. What is a mist net?
626. Which is the rarest northern albatross?
627. The young of which two waterfowl are fed by their parents?
628. The Reddish Egret is known for two morphs, one of which is reddish or dark blue, but what color is the other?
629. Which bird is called a South American Ostrich?
630. What is a pullet?
631. Which birds regularly toboggan as a means of locomotion?
632. What is a stint?
633. How did the name "murre" originate?
634. Which large New Zealand bird, known from only four specimens and believed to be extinct, was rediscovered in 1948, and what is its status?
635. Which albatrosses have long, pointed tails?
636. When did the Passenger Pigeon disappear in the wild?
637. How did the lapwings obtain their collective name?
638. What is the more accurate name for the disease psittacosis?
639. What sound is produced by large hornbills in flight, and what causes the sound?
640. All six crested penguins lay a two-egg clutch, but how often do the penguins rear two chicks?
641. Which American bird uses a *bubbling* display?
642. Which birds do the French know as *manchot*?
643. In the Old World, which bird is called Brunnich's Guillemot?
644. What is a lappet?
645. How many species of kingfishers inhabit North America, and which ones are they?
646. Is the neotropical Jabiru a heron?
647. What is a group of snipe called?
648. Do storks in flight catch flying insects?
649. What is an addled egg?
650. In which U.S national wildlife refuge have the most species of birds been recorded, and how many have been seen there?
651. Is the Osprey a common bird in Japan?
652. To which birds are stilts most closely related?

Questions

653. Which bird of the Bible was called an "arrowsnake?"
654. What is a Brolga?
655. What is a club?
656. Which are the most arboreal of the galliform birds?
657. Which American bird is called a "bog pumper," or "stake driver," and why?
658. What mammal book did John James Audubon illustrate?
659. What was the *Emu War*?
660. How many North American birds have been reported as regularly eating snakes?
661. Why do Carmine Bee-eaters ride the backs of Ostriches, storks and Kori Bustards?
662. What is a male goose commonly called?
663. Why are the Ravens of the Tower of London protected?
664. Which two flamingos are restricted to the Andean highlands?
665. Which bird was known as a garefowl?
666. Which North American gull migrates *north* for the winter?
667. On what does the Yellow-crowned Night-Heron primarily feed?
668. Which bird is known as a moorfowl?
669. Which long-legged, aberrant African bird constructs a huge, bulky, enclosed nest, often requiring months to construct?
670. What is imping?
671. Which waterfowl uses the communal stick-nest of the Monk Parakeet?
672. Which two diseases constitute the greatest threat to captive penguins?
673. Which alcid is called an auklet, but in reality is not, and what is it?
674. What is an accipiter?
675. Why were the Central and South American bellbirds so named?
676. Which bird uses a *booming ground?*
677. What behavior is known as *busking?*
678. What is a capon?
679. Which crane is the smallest, and how much does it weigh?
680. Which bird is known as a blackcock?
681. What is an ornitholite?
682. What causes the pink coloring of flamingos?
683. Which is the smallest of the surviving 22 alcids, and how much does it weigh?
684. Which is the most familiar and widely distributed North American shorebird?
685. Which American bird is called a "levee walker," and why?
686. How many of the more than 90 species of birds that became extinct since 1600 were insular endemics?
687. Which is the only tropical cormorant of South America?
688. With respect to birds, what did U.S. presidents Theodore Roosevelt and Franklin D. Roosevelt have in common?
689. Why do woodpeckers have rigid tails?
690. Which American gamebirds have feathered toes?
691. Which continents are not inhabited by the Black-crowned Night-Heron?
692. The eggs of which birds are shaped to prevent them from rolling off a ledge?
693. The extinct Heath Hen was a subspecies of which bird?

Questions

694. After whom was Eleonora's Falcon named, and why?
695. What was the origin of the expression *as thin as a rail*?
696. Which four groups of birds have distinct lobate webbing on their toes?
697. Which are the only birds with legs encased within the body down to the ankle joint?
698. Why were the tropicbirds so named?
699. On which day are Cliff Swallows supposed to reappear at San Juan Capistrano Mission in southern California?
700. Which birds have the longest incubation periods relative to their small size?
701. Which of the seven guineafowl is the largest, and how much does it weigh?
702. What is a Whimbrel?
703. Why do frigatebirds seldom land on water?
704. Which bird has the greatest global distribution?
705. What happened to the Wake Island Rail?
706. Which group of flying birds has the least wing surface relative to their weight?
707. Which group of Southern Hemisphere seabirds most closely resembles the non-crested auklets of the north, such as a Dovekie or Cassin's Auklet?
708. How large is the communal twig nest of the South American Monk Parakeet?
709. Which swan is the smallest, and how much does it weigh?
710. Name four of the six families of birds where all four toes are joined by webbing, and what are these birds collectively called?
711. Which are the smallest of the true quails?
712. Which is the only Antarctic-nesting bird that hatches naked, blind chicks?
713. Which upland gamebirds have two distinct plumages?
714. Which is the smallest North American heron, and how much does it weigh?
715. Which Asian waterfowl is known from only four specimens?
716. Which bird does the little Green Heron resemble in flight?
717. Why do woodpeckers have tough, thick skins?
718. Which is the largest North American owl?
719. How did the South American steamer-ducks obtain their name?
720. Which is the only African pheasant?
721. Which is the largest American grouse, and how much does it weigh?
722. What is the avian vocal apparatus called?
723. Which American shorebird is known as a "teeter-tail?"
724. What did Edgar Allan Poe's raven say?
725. Which American gamebird has the widest distribution?
726. What famous American institution was named for the pelican?
727. Which is the only inland-nesting North American cormorant?
728. Which American gamebird typically uses a drumming log?
729. What is the population size of the Hyacinth Macaw?
730. Why do most waterfowl thickly line their nests with down?
731. Budgerigars (Budgies) are native to what country?
732. In the game of golf, what is an eagle?
733. How many families of birds are restricted exclusively to North America?
734. Which American bird is sometimes called a "timber doodle?"

735. Are there any birds that breed in both the Arctic and Antarctic?
736. What was the Merlin formerly called?
737. What caused the near extinction of the Short-tailed Albatross?
738. What is a gannet colony called?
739. What color are newly-hatched Brown Pelican chicks?
740. The Black Swan is native to what country?
741. Which group of seabirds is sexually dimorphic?
742. Which cranes perch in trees?
743. Which is the largest tern, and how much does it weigh?
744. Are Kelp Gulls recent arrivals to the Antarctic?
745. Which bird is symbolic of the National Audubon Society?
746. Which ratites have long, cat-like whiskers?
747. Which large, extremely long-lived, mythical bird burned to death and rose from the ashes to live again?
748. What is mantling?
749. What type of bird is a bluebird?
750. Which American raptor has spicules on its feet, and why?
751. How do the sexes of the Belted Kingfisher differ?
752. Which bird is called a "monkey-faced owl?"
753. Is the Ladder-backed Woodpecker the smallest North American woodpecker?
754. What is a huge, wheeling flock of broad-winged hawks called?
755. What was "Peale's Egret?"
756. Which North American corvid builds a massive domed nest of sticks, the base of which may be cemented together with mud?
757. What are the two main species on which the Snowy Owl preys?
758. Why do ptarmigan have feathered toes?
759. Which is the only North American gull that, as an adult, has a forked tail?
760. Which American bird is locally called a "lizard bird?"
761. Considering both weight and wingspan, which is the largest flying bird?
762. What is a gathering of larks called?
763. What must an American hunter do to his/her duck stamp to make it legal?
764. Which is the only North American gull that is uniformly dark ventrally?
765. Which goose is greatly dependent on eel-grass beds?
766. Which is the only quail hunted in the eastern U.S.?
767. Which is the largest kingfisher, and how much does it weigh?
768. Which bird is sometimes known as a "canvasback" by the inhabitants (kelpers) of the Falkland Islands?
769. Which three groups of pelecaniform birds lay a single-egg clutch?
770. Which group of passerines spends the most time on the wing?
771. What are "Cuddy Ducks?"
772. Is the Crab Plover a South American bird?
773. Which are the only two gulls with strongly forked tails?
774. What is the Louisiana Heron commonly called?
775. Which is the only tern that *consistently* nests under cover?

776. Why did the Dodo ultimately vanish?
777. Which are the only pelecaniform birds that lay speckled or blotched eggs?
778. Do Sooty Terns nest in the continental U.S.?
779. Which of the two murres is most inclined to nest on a narrow rocky ledge, rather than a broad ledge?
780. At what time of day do skimmers generally feed, and why?
781. Which parrot is the best mimic?
782. Which birds are collectively called plantain-eaters?
783. Which bird is known as "Grayson's Dove?"
784. Which bird most closely resembles the extinct Great Auk?
785. Were Carolina Parakeets ever raised in captivity?
786. Which bird is the most flexible of the American obligate brood parasitic nesters, and how many species of birds has it been known to parasitize?
787. The extinction of which bird gave rise to a world-famous expression describing finality?
788. Why do Common Terns cover their eggs with sand?
789. How does the Palm Cockatoo differ physically from other cockatoos?
790. Which group of birds is most closely related to the owls?
791. What is a squab?
792. Which alcids are the most accomplished at terrestrial locomotion?
793. What type of birds are kingbirds?
794. Which birds are known as "moth owls?"
795. Which group of cuckoos is the most gregarious?
796. Which bird is called a "sea quail?"
797. What is the difference between a pigeon and a dove?
798. How did the African Hamerkop stork obtain its name?
799. Which is the brightest-colored North American flycatcher?
800. Which American bird is called a "fool hen," and why?
801. Which pigeons lack gall bladders and oil glands?
802. What is the favored nesting site of the Elf Owl?
803. Which is the only gamebird to *nest* in every state except Alaska?
804. How do the feeding habits of a kiskadee sometimes differ from other tyrant-flycatchers?
805. What does the name "phalarope" mean?
806. Alcids are primarily birds of the Arctic, but which three species breed as far south as Mexico?
807. What type of nest site is favored by a potoo?
808. Besides preying on fish, what do the Osprey and most fish-eating owls have in common?
809. Do swans occur in the Antarctic?
810. In what type of tree does a Short-eared Owl typically nest?
811. Why were the three South American plantcutters so named?
812. What do nesting Great Crested Flycatchers of eastern North America and the Paradise Riflebird of New Guinea have in common?

Questions

813. Which bird has the most restricted range?
814. Why were the Australian lyrebirds so named?
815. Which alcids have the longest, sharpest claws?
816. What type of bird is a martin?
817. Do the seven species of Caribbean and tropical American potoos have distinct facial bristles?
818. Ducks, geese and swans are known as waterfowl in America, but what are these birds more typically called in Europe?
819. How many species of birds breed north of the Rio Grande?
820. Where do most Great Shearwaters nest, and how many breed there?
821. Which of the five loons is the largest, and how much does it weigh?
822. Aside from the Least Auklet, which is the smallest alcid?
823. How can the neotropical chachalacas be sexed?
824. In America, which goose is essentially restricted to Alaska?
825. Other than the two sooty albatrosses, are there any other primarily dark albatrosses?
826. Which American waterfowl is particularly fond of acorns?
827. Which is the heaviest North American quail, and how much does it weigh?
828. Do all loons have straight bills?
829. Where else besides Alaska do Red-faced Cormorants breed?
830. How many warblers are named after states, and name the warblers?
831. How can adult Great Hornbills be sexed?
832. Which is the smallest North American accipiter?
833. Which is the only cracid that ranges north of Mexico?
834. How many birds are estimated to inhabit North America?
835. What do foraging flickers do that is unusual for northern woodpeckers?
836. How do male Western Bluebirds differ in appearance from bluebirds of the east?
837. What is a "calico heron?"
838. What is the bill color of a juvenile Wood Stork?
839. Which is the most numerous rail of North America?
840. Which nightjar has primaries more than two feet (0.6 m) long?
841. Which colorful, long-tailed Central American bird has a laterally compressed, helmet-like crest that extends forward to cover the base of the bill?
842. Which is the largest Northern Hemisphere duck, and how much does it weigh?
843. How can an observer separate the speck of a soaring California Condor from the speck of a soaring Turkey Vulture?
844. What is a dabbling-duck?
845. Which North American oystercatcher is restricted to the west?
846. Can a dipper walk underwater?
847. Which jaeger is the heaviest, and how much does it weigh?
848. In North America, which is the only gull with a wedge-shaped tail?
849. What color is the Black Guillemot in winter plumage?
850. What is turkey gum?
851. What happens to the plumage of Snowy Owls as the birds age?
852. What color is the mustache of a male Yellow-shafted Flicker?

Questions

853. What is the function of an owl's facial disc?
854. Which North American tern has the whitest plumage?
855. When was the Short-tailed Albatross rediscovered?
856. Which group of birds is the most frustrating for American birders?
857. Are the Old World orioles related to New World orioles?
858. Why do drongos often accompany monkeys?
859. What structure do dippers have that is ten times the size of that of other passerines of comparable size?
860. How do wrens characteristically hold their tails?
861. Which is the only crested corvid found west of the Rocky Mountains?
862. Which is the largest North American titmouse, and how much does it weigh?
863. Where did the early Vikings of Iceland believe Whooper Swans went after the birds migrated?
864. Was the catbird of North America named because of its fondness for feline fur used to line its nest?
865. What is a crake?
866. Which shorebird nests in burrows?
867. What is the culmen?
868. How do Anhingas and darters obtain their food?
869. Which is the largest hummingbird, and what is its length?
870. What is one of the favored nest sites of the neotropical Gartered Trogon (now merged with the Violaceous Trogon)?
871. What is a syndactyl foot, and which birds have feet of this type?
872. Which American birds often glean insects from tree bark by climbing *down* a tree trunk head first?
873. What color is the inner mouth of breeding Black Guillemots and Pigeon Guillemots?
874. Which American bird is called a "wailing bird" or "crying bird," and why?
875. What is a macrosmatic bird?
876. What were Greater Roadrunner eggs formerly sold as to unsuspecting egg collectors?
877. What are wasps in the genus *Pepsis* called?
878. In which waterfowl species do males incubate?
879. Which North American shorebird may carry chicks between its legs and thighs, and possibly even fly carrying young?
880. Where are the Whooping Crane breeding grounds located?
881. What is a bridled murre?
882. How do Coscoroba Swan cygnets differ in appearance from the downy young of other swans?
883. What are frugivorous birds?
884. What do many incubating northern ducks do to their eggs when they leave the nest to feed and drink?
885. What gland do cassowaries, Ostriches and Emus lack?
886. What is a British amateur bird lister called?
887. Which Australian bird is known as a "bushman's clock?"
888. Which South American bird is called a "baker," and why?

Questions

889. Which birds are known as nunbirds or nunlets, and why?
890. Which New World passerines are the weakest fliers?
891. Which is the largest toucan?
892. White tropical forest birds are very unusual, but which especially vocal South American jungle birds are white?
893. Which eagle has been persecuted the most extensively?
894. Which bird is known as the Native Companion Crane?
895. Which corvid has the widest distribution?
896. Which Old World birds are called rainbow birds?
897. What is a mirror?
898. When did the last Dodo disappear?
899. Which is the most numerous bird of England?
900. Flamingos have how many cervical (neck) vertebrae?
901. Which are the most gregarious piciform birds (woodpeckers and their allies)?
902. What do Europeans call the Red Phalarope, and why?
903. Where else besides New Guinea and Australia do birds-of-paradise reside?
904. Which is the only nocturnal-feeding gull?
905. How did the Musk Duck obtain its name?
906. When initially viewed in Australia by Europeans, what structures were believed to be native children playhouses?
907. Which is the heaviest sulid, and how much does it weigh?
908. Which woodpecker is the most terrestrial?
909. John Grisham is well known for his novel *The Firm*, but what other best-seller did he write?
910. In America, which northern tern has the most restricted breeding range?
911. What color are wild Australian Budgerigars?
912. How many of the 15 cranes are endangered or threatened, and which ones are they?
913. Which is the only all white-plumaged gull?
914. How deep can loons dive?
915. What color are the foot webs of a Wilson's Storm-Petrel?
916. How did the nine species of tailorbirds obtain their name?
917. How many species of dippers are recognized?
918. In Wales of a by-gone era, the hooting of an owl foretold what event?
919. Which group of birds includes some of the most highly regarded songsters of all birds?
920. What color are the eggs of the Chilean Tinamou?
921. Which cracid is the largest, and how much does it weigh?
922. Do jacanas occur in the U.S.?
923. What does it mean to pinion a bird?
924. At what height was the highest flying bird recorded, and which bird was it?
925. Which penguin is the most threatened, and what is its population size?
926. What is a female Ruff called?
927. Do loons and grebes construct the same type of nests?
928. What part of a flamingo did hungry ancient Romans especially relish?

Questions

929. Which is the most northerly-breeding passerine?
930. Which is the first bird mentioned in the Bible?
931. Which is the only heron that breeds during its first year?
932. Which is the only native American bird that nests in all 49 continental states?
933. Which alcids are essentially restricted to the Atlantic?
934. Which is the only silky-flycatcher that occurs in the U.S.?
935. How many species of kiwis are described?
936. How many species of sunbirds inhabit Australia?
937. What is a timid, easily frightened person sometimes called?
938. For what was the coal miner's canary used?
939. Does a female Green Kingfisher have a rusty chest-band?
940. What are conspecific birds?
941. Why do only male Emperor Penguins incubate?
942. Do swans mate for life?
943. What is a Bateleur?
944. Is the endangered New Zealand Takahe flightless?
945. Which birds are known as bluebills by American hunters?
946. When was the canary first imported into Europe?
947. Which American cuckoo had a bounty placed on it, and why?
948. In the Bible, which bird was known as an "ossifrage?"
949. Which type of feather was the most highly sought for use as writing pens during American colonial times?
950. Who starred in the TV series *Hogan's Heroes*, and what ultimately happened to him?
951. Which was the last *species* of bird to be declared extinct, and when did this occur?
952. Which is the only shorebird that lays unmarked eggs?
953. Which is the rarest North American seabird?
954. What is the African Stanley Crane also called?
955. To which birds are the northern choughs related?
956. How did the dipper obtain its name?
957. Do hunting hawk-owls depend greatly on their hearing?
958. Since 1600, during what 50-year period of time did the most species of birds become extinct?
959. What are euphonias, and why were the birds so named?
960. Are all cuckoos obligate brood parasitic nesters?
961. What was the size of the Trumpeter Swan population in the contiguous 48 U.S. states at its lowest level in 1933?
962. Did birds evolve from reptiles?
963. What is an eyas?
964. Why has the Northern Fulmar population increased dramatically in the past century?
965. What is a filoplume?
966. Why does the bill of a crossbill cross?
967. What is a Firewood-gatherer?
968. Which large group of flying birds have proportionally smaller breast muscles than any other flying birds?

Questions

969. How did the Northern Wheatear obtain its name?
970. Which flamingo has the most restricted range?
971. Which New Zealand bird is said to *graze like a rabbit?*
972. A number of birds rob other birds of their food, but which ones are considered to be superior at this?
973. Which goose has the shortest incubation and fledging period, how long is it, and why is it so short?
974. Which birds are called "sea clowns" or "sea parrots?"
975. Bat Hawks feed primarily on bats, but on which birds do the specialized raptors prey?
976. What is a florican?
977. What might nestling Hoatzins do if danger threatens?
978. What are "aigrettes?"
979. For what physical feature is the Royal Flycatcher noted?
980. Which bird was formerly called a "solan goose?"
981. What is a *gobbling ground?*
982. What is a gorget?
983. Which birds are known as gouras?
984. Which sea-eagle is the largest, and how much does it weigh?
985. What exactly is the American Great White Heron?
986. What was the fate of most copies of the classic first edition of Bertram E. Smythies' *The Birds of Burma?*
987. Which are the only passerines capable of gliding for extended periods on motionless wings?
988. In South Africa, what is the greatest cause of egg and chick mortality of the Hadada Ibis?
989. Which is the only New World-breeding wagtail?
990. Which is the slowest flying eagle, and why is slow flight advantageous?
991. Which is the only eastern U.S.-breeding swan?
992. At what age are gannet and booby chicks capable of regulating their body temperature?
993. How did wagtails obtain their name?
994. Is a *honker* a rude Latin American driver that continually leans on his horn?
995. Which is the only tern with white mustache-like facial plumes?
996. What is the tomium?
997. Which sulid is endemic to Christmas Island (Indian Ocean)?
998. Which are the most predatory songbirds?
999. How did the European (Common) Starling obtain its name?
1000. Which ibis is on the brink of extinction, and how many of the birds remain?
1001. Which are the only passerines with powder-down feathers?
1002. How does a male Resplendent Quetzal position its 2-foot-long (0.6 m) upper tail covert feathers when incubating in a small cavity?
1003. Which are the only parrots with brush-tongues, and why are their tongues modified?
1004. Which galliform birds have up to four full-sized spurs on each leg?
1005. Is a meadowlark a lark?

Questions

1006. What type of fruit does the shape of a kiwi suggest?
1007. Do oxpeckers drink the blood of their mammalian hosts?
1008. Which ibis has the most widespread distribution?
1009. Which South American galliform bird has a prominent, solid, fig-shaped casque?
1010. What do the Kentish and Snowy Plover have in common?
1011. Which small, colorful African bird was introduced into the West Indies over 300 years ago, and how did the introduction take place?
1012. In England, which bird is known as a "mavis?"
1013. How many species of African forest birds engage in lek displays?
1014. What caused the population crash of the Guam Rail and the Guam Kingfisher?
1015. How many carpal spurs do screamers have on each wing?
1016. Does the Bat Hawk consume prey while on the wing?
1017. Which is the only migratory North American woodpecker where pairs remain together during winter?
1018. How many species of peacocks have been described?
1019. Which birds have the greatest wing area relative to their weight?
1020. What are oropendolas?
1021. Which large mammal is intolerant of clinging oxpeckers?
1022. Which birds are known as "sea-pies?"
1023. Why does the lower mandible of a skimmer have shallow grooves along the side?
1024. What is the most prominent physical character shared by *all* passerines (perching birds)?
1025. Do birds have a knee-cap?
1026. Which two *groups* of pheasants lay a two-egg clutch?
1027. What are the five basic types of feathers?
1028. What is a Pauraque?
1029. What does a dipper do with its eyes when it dips or bobs?
1030. Which bird replaces the Blue-winged Teal in the American west?
1031. Which penguin has the most restricted breeding range?
1032. Which is the only macaw with a completely feathered face?
1033. Which kingfisher preys on barnyard ducklings?
1034. What is neossoptile plumage?
1035. Which birds consistently pluck and swallow their own feathers, and why?
1036. Which South American goose nests in trees?
1037. On what do most Southern Ocean seabirds primarily feed?
1038. Which spectacular raptor was named after a famous German naturalist?
1039. What is the podotheca?
1040. What is the origin of the widespread belief that frightened Ostriches bury their heads in the sand?
1041. Which petrels have a gular pouch?
1042. A bird with an austral range inhabits what area of the globe?
1043. Which Asian bird is called a paddy-bird?
1044. With the exception of male leg spurs, which group of pheasants exhibits very little sexual dimorphism?

Questions

1045. What is a Troupial?
1046. During the height of the DDT controversy in 1970, how many Brown Pelican chicks hatched in western U.S. colonies?
1047. What happened to the Red-shafted Flicker with respect to its status as a species?
1048. Which waterfowl has a large, conspicuous cere?
1049. Which shorebird is the most polymorphic?
1050. Is a Saker an African shorebird?
1051. In slang, how is *to spoil ones chances* often expressed?
1052. Where are propatagium tags or markers attached?
1053. Who was one of Batman's main arch-enemies, and who portrayed him in the popular TV series?
1054. When was the first Migratory Bird Treaty signed between the U.S. and Canada?
1055. How large is the territory of a male Superb Lyrebird?
1056. Did both ratites and penguins evolve from flying birds?
1057. What is a Roulroul?
1058. What large, soaring pelagic seabird has been recorded in Arizona?
1059. How large is the New Zealand Wrybill population?
1060. To which birds are spoonbills most closely related?
1061. What is a spreo?
1062. In what type of habitat would a xerophilous bird most likely be encountered?
1063. The Hudsonian Curlew is better known by what name?
1064. Which three groups of birds have especially large subcutaneous air-sacs?
1065. Are there any flightless passerines?
1066. Which raptor has the longest legs?
1067. Are floating jacana nests anchored?
1068. Are adult male Blue Grosbeaks entirely blue?
1069. Which bird was sacrificed to facilitate man's race to the moon?
1070. What is a clutch?
1071. If a bird is an insular species, what does this imply?
1072. Who was the best known native-born Indian ornithologist?
1073. What is botulism?
1074. What is a precocial bird?
1075. What is a chevron?
1076. What is a bulla?
1077. What location in the lower 48 states is said to contain the fewest birds?
1078. What is a ramphastid?
1079. How much does a drake Magellanic Flightless Steamer-Duck weigh?
1080. What is a commissure?
1081. What is an eider pass?
1082. What is a feather tract?
1083. Which seabird is the most numerous?
1084. Which flightless seabird is the most nocturnal?
1085. How many species of North American fossil birds have been discovered?
1086. Which automobile company manufactures the Thunderbird?

Questions

1087. What usually happens if captive pairs of cassowaries are maintained together?
1088. What is the primary foraging technique utilized by the two turnstones?
1089. Which bird is known as a "vampire finch?"
1090. In what year were the Lesser Snow and Blue Goose recognized as the same species?
1091. What color is favored by a male Satin Bowerbird decorating its bower?
1092. The nest of which North American-breeding bird was the last to be discovered, when did the discovery occur, and what were the circumstances surrounding the discovery?
1093. How many species of native land-birds inhabit Easter Island?
1094. What is the maximum flying speed of a Ruby-throated Hummingbird?
1095. How many of the approximately 142 cuckoos are obligate brood parasitic nesters?
1096. Which *flying* bird has the longest fledging period (hatching to flying stage), and how long is it?
1097. Which Australian birds have become extinct since Europeans settled in 1788?
1098. Which birds have the most secondary flight feathers, and how many are on each wing?
1099. Who was the first person to fly over the South Pole?
1100. When did the Cattle Egret first reach Florida?
1101. Are the vocalizations of adult Brown Pelicans loud?
1102. After whom was Ross' Gull named?
1103. How did the kittiwakes obtain their vernacular name?
1104. At what speed do courting woodcocks fly?
1105. How many Cuban Bee Hummingbirds are required to equal the weight of an adult male Ostrich?
1106. What is a type specimen?
1107. Which birds consistently rank among the top ten *must see* zoo animals?
1108. In former times in China, a pair of which birds was presented as a traditional wedding present, and why?
1109. In U.S. Air Force slang, what does a Big Chicken Dinner (BCD) signify?
1110. Excluding birds, are bats the only group of vertebrates capable of *sustained* powered flight?
1111. According to numerous accounts, what bizarre feeding strategy did the extinct Huia of New Zealand utilize?
1112. What were the harpies of Greek mythology?
1113. Of birds that build tree nests, which large group typically constructs the most flimsy nests?
1114. How were parrots used during World War I?
1115. Very few Southern Hemisphere breeding birds migrate to the Northern Hemisphere, but which group of birds is an exception?
1116. Do birds perspire?
1117. What were the first words spoken from the moon on July 20, 1969?
1118. Compared to other life forms, do birds have superior eyesight?
1119. Are the chicks of any terns polymorphic in color?
1120. Are penguins near-sighted when out of the water?
1121. Which North American bird is known as a "mud bat?"
1122. Why do many gulls have brightly-colored spots on their lower mandible?

Questions

1123. What are bustards also known as in Africa?
1124. Which British birds rob milk from milk bottles on doorsteps?
1125. Why are fresh eggs heavier than incubated eggs, and how much weight is lost during incubation?
1126. Do hummingbirds depend exclusively on nectar?
1127. How many bird families are Madagascar endemics, and which birds are in the families?
1128. Why do insectivorous birds have a more highly developed sense of taste than seed-eating species?
1129. How many orders of birds contain only a single species, and name them?
1130. Do dippers have webbed feet?
1131. How long is the windpipe of an adult Whooping Crane?
1132. Which British breeding bird lays the largest egg?
1133. Do any living birds have teeth?
1134. Which *native* North American land-bird has the longest tail feathers, and how long is its tail?
1135. After whom was Abdim's Stork named?
1136. What is the avian hind toe called?
1137. Where is the southernmost breeding site for a flightless bird located?
1138. When and where did the last Heath Hen disappear?
1139. When did the Elephant-birds of Madagascar disappear?
1140. If stretched out end to end, how many feet of earthworms is an American Robin said to capture daily if feeding a single large chick?
1141. Which is the only bird that has a bladder?
1142. Which is the only goose inhabiting dense tropical forests?
1143. When was the National Audubon Society founded?
1144. Which are the only groups of vertebrates with toothless, horny bills?
1145. What is the most complex skin growth of any animal?
1146. What is the national bird of Rumania?
1147. What is the area between the eye and the bill called?
1148. How did the Bald Eagle obtain its name?
1149. Which is the only spur-winged bird of the U.S.?
1150. What are the remiges?
1151. Why were Lesser Sandhill Cranes formerly called Little Brown Cranes?
1152. What legendary flying horse had feathered wings?
1153. How many flight quills are on each wing of an Ostrich?
1154. What happens to the horn of a Rhinoceros Auklet following the breeding season?
1155. Where and when was John James Audubon born?
1156. Which company manufactures the Firebird sports car?
1157. On what do southern albatrosses primarily feed?
1158. Which penguin is often treated as a white-throated race of the Macaroni Penguin?
1159. Which flamingo is the least numerous, and what is its population size?
1160. Where else besides Hawaii does the Hawaiian (Dark-rumped) Petrel nest?
1161. Which pheasant occurs at the highest altitude?

Questions

1162. Why was the Hawaiian Goose (Nene) given the specific name *sandvicensis*?
1163. Which birds do frigatebirds most often rob of their food?
1164. After whom was the Adelie Penguin named?
1165. In what country is the largest breeding concentration of Common Puffins located, and how many breed there?
1166. In Britain, what is goose grass sometimes called?
1167. Which pelican is the most threatened, and what is its population size?
1168. Which birds are known as horned pheasants?
1169. If only 700-800 pairs of flightless Galapagos Cormorants exist, why is the species not regarded as endangered?
1170. On what does the Great Gray Owl prey almost exclusively?
1171. What happens to avian testes during the breeding season?
1172. Why do the neotropical jacamars have long, slender, pointed bills?
1173. Which shorebird regurgitates food for its chicks?
1174. What is a pelagic bird?
1175. What is the heaviest documented weight of a domestic turkey, and what was the fate of the turkey?
1176. Which alcids dive the deepest, and how deep can they dive?
1177. Pairs of which neotropical gamebirds typically perch so close together that the birds may be in contact with one another?
1178. Which is the largest flying parrot?
1179. Are domestic chickens more numerous than humans?
1180. How did the prions obtain their name?
1181. Do Blue-eyed (Imperial) Shags have blue eyes?
1182. How many of the more than 150 species of waterfowl inhabit Australia?
1183. Which are the most aquatic birds?
1184. A male Australian Jabiru Stork has dark brown eyes, but what color are female's eyes?
1185. Which bird is raised commercially for meat, feathers and skin?
1186. Where do Brown Pelicans fly from coast to coast (from Atlantic to Pacific), and vice versa?
1187. Which is the most numerous penguin of Argentina?
1188. Who starred in the 1941 movie classic *The Maltese Falcon*?
1189. What is a dotterel?
1190. The Ruffed Grouse is the state bird of which state?
1191. Which is the world's rarest parrot, and what is its population size?
1192. With respect to birds, how did Shakespeare describe the city of London?
1193. What does it mean to be *flipped a bird?*
1194. What do painted-snipe often do when threatened or cornered?
1195. Which is the heaviest rail, and how much does it weigh?
1196. Is the Brolga the only Australian crane?
1197. Which are the only birds whose entire population is confined to captivity?
1198. Why are owl ears unique?
1199. Which of the three scoters breeds only in America?

Questions

1200. Which two groups of primarily freshwater diving birds are best adapted to an aquatic lifestyle?
1201. Which endangered predatory bird is restricted to San Clemente Island, off the southern California coast?
1202. What is the difference between a race and a subspecies?
1203. Which bird has the greatest wingspan of any terrestrial bird, and what is its wingspan?
1204. Are there any polyandrous birds in which the female *simultaneously* maintains more than one male?
1205. How do noddies at sea differ behaviorally from terns?
1206. In New Zealand, which bird is called a black teal?
1207. John Cassin (1813-1869), after whom was Cassin's Auklet named, was curator of birds at what institution?
1208. Why are woodpecker skulls strengthened and significantly heavier than other birds of comparable size?
1209. Have eagles been know to carry off children?
1210. Which penguin is the least aggressive, and why?
1211. How far back in time do some of the earliest-known bird cave paintings extend?
1212. What was the name of the last surviving Passenger Pigeon, and after whom was it named?
1213. How did the Killdeer obtain its name?
1214. Why do Alaska-breeding Emperor Geese sometimes straggle to California?
1215. Which North American woodpecker has a curved bill?
1216. What color are freshly-laid Pied-billed Grebe eggs?
1217. How long is the horn of a Horned Screamer?
1218. Are birds color blind?
1219. Where does the Water Thick-knee often lay its eggs?
1220. Does the Corn Crake eat corn?
1221. Which are the only birds in which both sexes of the pair incubate side by side on the nest?
1222. How did the kingbirds obtain their name?
1223. Courting pairs of which American bird race over the water in unison in an upright posture referred to as *rushing?*
1224. What is the columella?
1225. Which shorebirds have eyes set high on their heads, and why?
1226. How many chicken eggs are laid annually worldwide?
1227. What famous ornithologist once said that *it was not really a good day unless he shot 100 birds*?
1228. Are molting cranes flightless?
1229. What type of feathers are known as rectrices?
1230. Which is the largest Australian honeyeater, and what is its length?
1231. Do birds have hair?
1232. What is the state bird of Hawaii?
1233. Which bird is known as Fischer's Eider?
1234. Which American bird is incorrectly called a Wood Ibis?

Questions

1235. Is the Phainopepla truly insectivorous?
1236. Which bird is the most detrimental avian agricultural pest of England?
1237. What are jaegers called in the Old World?
1238. Which two waterfowl breed the farthest north?
1239. How do the foraging techniques of swifts and swallows differ?
1240. What happens to the bill of a breeding American White Pelican?
1241. Which are the only birds that feed their chicks a crop secretion?
1242. What is the population size of the Thick-billed Murre?
1243. In flight, do all birds flap their wings in unison?
1244. How many species of American cuckoos are obligate brood parasitic nesters?
1245. In the Falkland Islands, what is a "Johnny Rook?"
1246. How many species of spoonbills inhabit the New World, and name them?
1247. Which bird is known as a "white nun?"
1248. Which Old World bird carries tortoises aloft and drops them to the rocks below to split them open?
1249. Why do many shorebirds feed extensively after dark?
1250. By what other names is the Kelp Gull known?
1251. Which raptor is held sacred by the Hindus?
1252. How does a feeding Shoe-billed (Shovel-billed) Kingfisher of the mountain forests of New Guinea differ from other foraging kingfishers?
1253. A courting Wandering Albatross occasionally sounds like which large gamebird?
1254. Are any seabirds vegetarian?
1255. What does the ptarmigan generic name *Lagopus* mean?
1256. Does the Laughing Falcon actually laugh?
1257. Do oystercatchers feed their chicks?
1258. How rapidly are a drumming Ruffed Grouse's wings moving?
1259. Is the Brown Pelican endangered?
1260. Which birds are the most efficient plunge-divers, and at what speed do they strike the water?
1261. Which Antarctic-nesting bird is a kleptoparasitic feeder?
1262. Which wading bird may use "bait" to attract fish?
1263. What is the Himalayan Monal commonly called, and after whom was the pheasant named?
1264. Do avocets have bent bills?
1265. What is the fruit of the pokeweed called?
1266. How many of the approximately 650 species of North American-breeding birds are migratory?
1267. Why do birds fly in V-formation?
1268. Was the Eskimo Curlew once known as a *dough bird?*
1269. Why is the Razorbill so named?
1270. Why do some ptarmigans have black in their tails?
1271. Which bird is said to have the foulest smelling nest, and why?
1272. What is the American Marsh Hawk currently called?
1273. Which are the only birds that do not emerge from the egg head first, and why?

Questions

1274. Are all birds warm-blooded?
1275. Which seabird is commonly known as a "paddy?"
1276. Do most of the 95 kingfisher species prey on fish?
1277. After whom was Kirtland's Warbler named?
1278. Do Great Frigatebirds nest in the continental U.S.?
1279. On what do Emperor Penguins primarily feed?
1280. Where are the horns on a Horned Puffin?
1281. To an American birder, what is a LBJ?
1282. How many times a day does a Black Woodpecker pound?
1283. Which is the most numerous goose of New Zealand?
1284. Do Wild Turkeys kill and eat poisonous reptiles?
1285. What was the throne of the King of Persia (Shah of Iran) called?
1286. Which bird is called Steller's Albatross?
1287. What is the greatest *documented* height of a bird flying over North America, and which bird is credited with the feat?
1288. What journal is published by the American Ornithologists' Union?
1289. Which sulids typically nest in trees?
1290. Which North American amphibian is a threat to waterfowl?
1291. What are arena birds?
1292. Which seabird is known as a shoemaker, and why?
1293. How many species were pictured in Audubon's classic work *The Birds of America*?
1294. How deep can the little Cassin's Auklet dive?
1295. Where are dipper nests often located?
1296. What was the middle name of ex-President Richard Nixon's White House chief-of-staff?
1297. What is fratricide?
1298. In Britain, which bird is called a tystie, and why?
1299. Which North American bird has a beard?
1300. What is the origin of the word *bird*?
1301. What is a dacnis?
1302. Which bird was known as a "flying sheep" in some parts of Canada, and why?
1303. Which is the smallest North American falcon, and what does it weigh?
1304. What is a bolus?
1305. Which large American bird is called a "carpenter bird?"
1306. What is an eyespot?
1307. Which American birds are called "black witches" or "death-birds?"
1308. What is urohidrosis, and which birds are known for it?
1309. Do Black Vultures ever *kill* skunks?
1310. Why do penguins molt all their feathers simultaneously?
1311. Which group of birds is known as boatswain or bo'sun-birds, and why?
1312. What does ornithophilous mean?
1313. How many species of lyrebirds are described?
1314. What type of bird is a Veery?
1315. Excluding lapwings, which is the heaviest plover, and how much does it weigh?

Questions

1316. Which diving seabird nests in the middle of Tokyo, Japan?
1317. Do hummingbirds have forked tongues?
1318. The silhouette of which bird was used by ancient Egyptians as the hieroglyphic for the color red?
1319. Which vulture was known as a thunderbird?
1320. What are seedsnipes, and what is the origin of their name?
1321. What is the red line between the upper and lower mandible of a Trumpeter Swan called?
1322. In the Falkland Islands, what is the Gentoo Penguin called, and why?
1323. How did the noddy terns obtain their name?
1324. What is the incubation period of Steller's Eider?
1325. What color are juvenile Little Blue Herons?
1326. Why do birds not prey on adult robber crabs?
1327. In addition to its normal fare of insects, what else might a Chuck-will's-widow eat?
1328. What is a *San Quentin Quail?*
1329. Do skimmers carry fish cross-wise in their bills?
1330. What is the name of the famous tar pits in the city of Los Angeles where many fossil birds have been discovered?
1331. What is the Oldsquaw duck called in Europe?
1332. How much greater is the metabolic rate of a hummingbird than a human's?
1333. Do male woodpeckers incubate?
1334. Which American bird is known as "Whiskey Jack?"
1335. Which bird was pictured on plate Number One in Audubon's *The Birds of America*?
1336. Which birds are the Old World counterparts of the Limpkin, and why?
1337. Which falcon relatives construct extensive nests?
1338. Which birds are said to lay the most variable-colored and patterned eggs?
1339. Which loon has the most northerly distribution?
1340. Which is the only shorebird that lays a single-egg clutch?
1341. Which is the only endangered bird of the Galapagos Islands?
1342. What is adventitious coloring?
1343. What is a Polish Mute Swan?
1344. Which American bird is called a "leatherback," and why?
1345. Which is the largest non-passerine family, and how many species are in it?
1346. Which is the second largest non-passerine family, and how many species are in it?
1347. What is a clay pigeon?
1348. Which large, colorful, aquatic bird is preyed on by African Fish-Eagles?
1349. Which of the three skimmers is the largest?
1350. On what do Antarctic-dwelling Kelp Gulls primarily feed?
1351. What happened in Hawaii around 1900 that ultimately doomed a number of endemic passerines?
1352. What type of bird is a dikkop?
1353. What was the name of Batman's young ward?
1354. How did the Mourning Dove obtain its name?
1355. Can cassowaries swim?

Questions

1356. Which two geese are sexually dimorphic?
1357. Which bird is known as a "king-crab bird," and why?
1358. Which exclusively neotropical terrestrial birds are regarded as superior gamebirds?
1359. What is a *dicky-bird?*
1360. Which bird is a threat to the extremely rare Cahow (Bermuda Petrel), and what does it do that is detrimental?
1361. What is located at the large end of a fertile egg late in incubation?
1362. Aside from being a noted French Antarctic explorer and having the Adelie Penguin named after his wife, for what is Dumont d'Urville known?
1363. Which American vulture is not black, and what color is it?
1364. If gulls are encountered at sea, what does this generally indicate?
1365. When and where did the first successful heavier-than-air, machine-powered flight occur?
1366. Courting males of which small, colorful neotropical passerines produce a series of distinct snaps with their wings, sounding rather like a typewriter?
1367. Which birds are commonly called sawbills, and why?
1368. Which group of colorful tropical fish feeds extensively on coral?
1369. Where was an "artificial" population of Whooping Cranes established, and how was it accomplished?
1370. Which is the most agile penguin over rough, uneven terrain?
1371. The Hermit Ibis is more commonly known by what name?
1372. What does the scientific name of the mockingbird *(Mimus polyglottos*) mean?
1373. What was the name of the German fighting unit that fought during the Spanish Civil War?
1374. What color is the gular pouch of a breeding Brandt's Cormorant?
1375. Which is the largest North American sparrow, and how much does it weigh?
1376. The noted animal behaviorist Konrad Lorenz worked with which large birds?
1377. Which American jay was named for its food preferences?
1378. Which is the heaviest U.S.-breeding cormorant, and how much does it weigh?
1379. How is to talk candidly or bluntly sometimes described?
1380. Who starred in Alfred Hitchcock's 1963 classic thriller *The Birds*?
1381. Males of some wrens have what unusual nesting habit?
1382. Where does the Wompoo Pigeon reside?
1383. Which is the largest domesticated bird?
1384. Are Little Blue (Fairy) Penguins actually blue in color?
1385. Which American raptor digs out nesting storm-petrels and alcids?
1386. What duck is served in Chinese restaurants?
1387. What is the state bird of Minnesota?
1388. How old was *Martha,* the last Passenger Pigeon, when she died in 1914?
1389. Why did American hunters vigorously oppose the conversion from lead shot to steel shot?
1390. Which American sparrow is named after a carnivore, and why?
1391. What are American lawyers commonly called?
1392. Do dolphins prey on penguins?

Questions

1393. How many species of penguins inhabit the South Pole?
1394. What is a group of quail called?
1395. In avicultural terminology, what is a soft-billed bird?
1396. What happens to the corners of the human eye during the aging process?
1397. Which birds are known as colies?
1398. Do Spectacled Owls prey on flying mammals?
1399. Which penguins roost in trees?
1400. Excluding the Ruddy Duck, which only other stiff-tail duck occurs in the U.S.?
1401. Which is the only bird that has chicks with glistening *beads* in their mouths that are said to glow in the dark, and why?
1402. Georg Steller is well known to ornithologists because of Steller's Sea-Eagle, Steller's Eider and Steller's Jay, but which mammals were named after the German naturalist?
1403. Is the Snail Kite restricted to Florida?
1404. Which seabird has been killed as a result of diving into fisherman's boats filled with freshly caught fish?
1405. What is *whitewash?*
1406. Which is the smallest New World bird that feeds extensively on foliage?
1407. Which is the rarest gull, and what is its population size?
1408. What is the cloaca?
1409. What is the more familiar name for the "jack pine warbler?"
1410. What is a *stool pigeon?*
1411. How many species of hummingbirds inhabit Hawaii?
1412. Which crested penguins breed in South America?
1413. Which bird is the cassowary's closest relative?
1414. Are African Bateleur eagles sexually dimorphic?
1415. Who were the two candidates in the 1990 general election for prime minister of Australia, and who ultimately prevailed?
1416. On what does a Gray-headed Albatross feed that differs from the prey of other southern albatrosses?
1417. What type of birds are the African bald-crows?
1418. Which bird was Captain George Vancouver describing in 1791, when he wrote: "A very peculiar one was shot, of a darkish gray plumage, with a bag like that of a lizard hanging under its throat, which smelt so intolerably of musk that it scented nearly the whole ship."
1419. What weapon was commonly used by many North American Indians?
1420. Is the Swallow-tailed Gull endemic to the Galapagos Islands?
1421. Which penguin nests in Africa?
1422. How large is the nail of the outer toe of an Ostrich?
1423. Which is the smallest and darkest of the numerous Canada Goose races?
1424. What is a turkey fish?
1425. Which North American hawk has the widest ecological tolerance?
1426. Which is the slimmest North American owl?
1427. Excluding the Scissor-tailed and Fork-tailed Flycatcher, which North American passerines have the longest tails relative to their size?

Questions

1428. How much did the Great Auk weigh?
1429. Where is the largest colony of Red-footed Boobies located, and how large is it?
1430. Does the Common Gallinule occur in North America?
1431. Good-looking girls are referred to as chicks in the U.S., but what are attractive girls called in England?
1432. Which cranes do not build substantial nests?
1433. What is a gathering of Nightingales called?
1434. Which is the largest North American pigeon, and does it weigh more than one pound (0.45 kg)?
1435. Why are parrot eggs white?
1436. What was the value per ounce of egret plumes collected by plume-hunters at the turn of the century?
1437. Is the Solitary Sandpiper well named?
1438. Do birds eat golf balls?
1439. Which is the most numerous breeding waterfowl of the eastern U.S.?
1440. In Roman mythology, who was known as the winged messenger?
1441. The Mangrove Cuckoo is restricted to which state?
1442. Which is the swiftest terrestrial bird of the New World?
1443. What is the Chinese gooseberry also called?
1444. What nest site is typically selected by the neotropical Orange-fronted Parakeet?
1445. Are Flightless Steamer-Ducks edible?
1446. In folklore, what do owls often symbolize?
1447. Which is the largest North American wren, and what is its length?
1448. Are any of the loons able to take off from land?
1449. Who was the star character in the popular newspaper comic strip *Bloom County*?
1450. When was the American Bald Eagle officially designated as the national bird?
1451. What is the population size of the Spoonbill Sandpiper?
1452. How many species of albatrosses breed in Alaska?
1453. Are King Eiders social nesters?
1454. Which two countries benefit most from the fund-raising activities of the *Ducks Unlimited* organization?
1455. Which bird is most often preyed on by Cooper's Hawk?
1456. What happened to the mythical figure Icarus?
1457. What type of bird is *Ornithorhynchus anatinus*?
1458. Which waterfowl species has adapted to living on lava?
1459. Which booby is the smallest, and how much does it weigh?
1460. What is the only state not inhabited by the bluebird?
1461. What is the name of the original casino and hotel on the Las Vegas Strip, and who built it?
1462. Does the Sanderling breed in the *high* Arctic?
1463. Where do 95 percent of all Heermann's Gulls nest, and how many breed there?
1464. Do Emperor Penguins breed anywhere except in the Antarctic?
1465. What is distinctive about the CV-22 *Osprey* helicopter?
1466. Which is the largest bird of the Galapagos Islands?

Questions

1467. What is an antiphonal song?
1468. Which family of birds has the most distendable gular pouches?
1469. Which birds have the largest flight muscles relative to their size?
1470. Does the Northern Goshawk feed on carrion?
1471. What are Florida seasonal winter tourists called?
1472. How many species of crows breed in the U.S. , and which ones are they?
1473. What is the national bird of Trinidad?
1474. How did the term *booby hatch* originate?
1475. Which is the only gull that lays a single-egg clutch?
1476. Which alcid breeds the farthest south?
1477. Are owls able to move their eyes within their sockets?
1478. Which is the largest heron, and what is its height and weight?
1479. When and where was the fossil proto-bird *Archaeopteryx* first discovered?
1480. Aside from the repugnant odor of their nest cavity, that may function as a predator deterrent, how do Eurasian Hoopoe chicks defend themselves?
1481. Does the Gray-headed Albatross breed annually?
1482. When and where did the first recorded bird strike occur?
1483. How many fish are required to fledge an Atlantic Puffin?
1484. Which are the two largest European raptors?
1485. How does the diet of a Northern Cardinal chick differ from the adult's diet?
1486. Excluding books, how many ornithological papers are annually published worldwide?
1487. Why are ducks forced to waddle?
1488. The theory of evolution was first conceived after the study of which group of birds?
1489. Which North American passerine has the longest fledging period, and how long is it?
1490. Which birds are collectively known as anatids?
1491. Are birds endangered by spider webs?
1492. Which birds were named after Audubon?
1493. What was the incubation period of the extinct Great Auk?
1494. Which birds are the closest waterfowl relatives?
1495. How do mollusks pose a threat to birds?
1496. Which gull breeds the farthest south?
1497. Do both hawks and owls hatch with open eyes?
1498. What measures have been taken to protect nesting Cahows (Bermuda Petrels) from White-tailed Tropicbirds?
1499. Which alcid has an upturned bill?
1500. Do any North American birds prey on toads?
1501. How do jacanas carry their young?
1502. Waterfowl have how many primaries?
1503. Do cormorants transport water for their chicks?
1504. Which vegetable is colloquially known as "sparrow grass?"
1505. What is the name of the main protagonist of the roadrunner of cartoon fame?
1506. Are fledging King Penguins and Emperor Penguins the same size as adults?
1507. Does the Ostrich inhabit Arabia?
1508. What was formerly the favored nesting material of the Chipping Sparrow?

Questions

1509. How did the Downy Woodpecker obtain its name?
1510. Do male avocet bills turn up more than those of females?
1511. How wide is the wing of a Wandering Albatross?
1512. How long does it take for a Cassin's Auklet to dig its nesting burrow?
1513. Which American bird was called a "neighbor's mallard," and why?
1514. Do gannets mate for life?
1515. Which American bird has the highest recorded incidence of albinism?
1516. Which booby is a nocturnal feeder?
1517. The beard of a Wild Turkey consists of what types of feathers?
1518. In what 1975 box office smash movie did Jack Nicholson associate with characters that were as crazy as loons?
1519. Have Wandering Albatrosses ever been recorded *ashore* in North America?
1520. Aside from Ross' and Snow Geese, which is the palest northern goose?
1521. Was Bonaparte's Gull named after Napoleon Bonaparte?
1522. Why is the Mauritius Parakeet (Echo Parrot), which may number no more than 14 birds, so endangered?
1523. What specialized hunting technique has been perfected by the Zone-tailed Hawk?
1524. What type of nest does the Waved Albatross construct?
1525. Which two types of birds have the softest (non-rigid) bills?
1526. The sexes of which American woodpecker are so dissimilar in appearance that they were once considered separate species?
1527. What did the founder of the Tabasco hot sauce company, E. A. McIlhenny, have to do with birds?
1528. Are Black-billed Magpies more social than Yellow-billed Magpies?
1529. Which of the three puffins is the largest, and how much does it weigh?
1530. What was the highest price ever paid for a bird book, and what book commanded the price?
1531. Which state has the most hummingbird species, and how many occur there?
1532. In what year was the North American Migratory Bird Treaty expanded to include Mexico?
1533. Is the American Woodcock also known as a "bleater?"
1534. Which continents do not have a single species of magpie?
1535. Relative to body size, are bird brains large?
1536. Which birds are called "bowhead-birds," and why?
1537. Are birds at risk from *flying* golf balls?
1538. Which American bird is known as a Water Ouzel?
1539. What is a chimango?
1540. Which three North American birds have the greatest wingspans?
1541. Why is the far-carrying voice of a Whooping Crane so loud?
1542. Are all wood-warblers restricted to the New World?
1543. What is *zugunruhe?*
1544. Is the avian thigh always obscured by feathers?
1545. Which group of birds is the most written about, both in the scientific and popular literature?

Questions

1546. When and where were the first House (English) Sparrows introduced into the U.S.?
1547. What percent of its body weight does the heart account for in a hummingbird?
1548. Which country produces the most chicken eggs?
1549. Which species of waterfowl do not copulate in the water?
1550. Have birds ever been killed by lightening?
1551. What substance is mixed with Asian cave swiftlet nests to create bird's-nest soup?
1552. Are birds maintained in apiaries?
1553. What is the avian crop?
1554. Which of the three jaegers is the most numerous?
1555. How many of the 31 orders of birds occur in North America?
1556. Which North American gull has a red bill?
1557. What color is the head of a juvenile Turkey Vulture?
1558. What is the middle name of Captain Robert Scott, the famous British Antarctic explorer?
1559. Is the White-winged Guan, that was rediscovered in Peru in 1979, extremely rare?
1560. Does Ross' Gull breed in North America?
1561. Which abundant American sandpiper has only three toes?
1562. Which western American cormorant has a white flank-patch?
1563. Audubon's *The Birds of America* consisted of how many volumes?
1564. Which is the most graceful American raptor in flight?
1565. At what age are captive-reared waterfowl generally pinioned, and what criteria is used to determine which wing is pinioned?
1566. Do Anhingas prey on crocodilians?
1567. What is a "Wurdemann's Heron?"
1568. Which is the most numerous seabird of the American prairies?
1569. How do breeding Black Guillemots and Pigeon Guillemots differ in appearance?
1570. In terms of overall length, how much larger is the Greater than the Lesser Yellowlegs?
1571. Are most birds anosmatic?
1572. Do cetaceans feed on alcids?
1573. In bowling, what are three consecutive strikes called?
1574. Which American bird is called "sizzle-britches," and why?
1575. Which two exclusively American gamebirds have two distinct color phases?
1576. Which American shorebird during flight produces a whistling sound with its wings and tail called *winnowing*?
1577. Who was the vice-president of the 41st U.S. president?
1578. Since birds first evolved, how many species have existed?
1579. Where do the ranges of Common and Horned Puffins overlap?
1580. How long is the Whooping Crane incubation period?
1581. How long does it take for a pair of Asian cave swiftlets to construct their saliva nest?
1582. Do eagles inhabit Hawaii?
1583. Which is the only western American arboreal cuckoo?
1584. Which bird "saved" the Mormons in Utah?
1585. Which is the least pelagic of the three phalaropes?
1586. Breeding Whiskered Auklets have how many white facial plumes?

Questions

1587. What color are the inflatable air-sacs of cock Lesser Prairie-chickens?
1588. In what classic 1962 black-and-white movie did Gregory Peck portray a distinguished southern lawyer defending an accused rapist in the 1930's in Alabama?
1589. Males of which bird are called stags?
1590. Which bird produces a loud trisyllabic booming roar that has been compared to the roar of a lion?
1591. Which is the heaviest North American kite, and how much does it weigh?
1592. Does the Groove-billed Ani inhabit Florida?
1593. Which sandpiper has a long upcurved bill?
1594. What is an ornithophilatelist?
1595. Which North American raptor has the most varied diet?
1596. Can flying birds carry objects greater than their weight?
1597. What color is the crest of a female Ivory-billed Woodpecker?
1598. What is the origin of the name "Dovekie?"
1599. Which New World bird is featured in an old British family crest, and why?
1600. How do hatching Egyptian Plovers (Crocodile-birds) differ from other shorebirds?
1601. Which raptor frequently nests in prairie dog towns?
1602. What are flufftails?
1603. What was the name of the small cannon used between the 15th and 17th centuries?
1604. Are thrashers related to mockingbirds?
1605. Which North American bird lays the largest clutch , and how large is the clutch?
1606. Where do Kirtland's Warblers winter?
1607. Which is the only North American flycatcher with a black breast?
1608. How did the Savannah Sparrow obtain its name?
1609. Are breeding Waved Albatrosses restricted to the Galapagos Islands?
1610. Who collected the last Labrador Duck, and which sparrow is named after him?
1611. How long is the bill of an American Avocet?
1612. Which American raptor is called a Bay-winged Hawk?
1613. Why are woodpecker nostrils covered with bristle-like feathers?
1614. The Thick-billed Parrot formerly was almost exclusively dependent on which bird?
1615. Why is the bill of a Swallow-tailed Gull white-tipped?
1616. Which country has a bird name?
1617. Which American bird is called a "chalk-line," and why?
1618. Do the neotropical motmots have serrated bills?
1619. What type of fish is favored by foraging boobies?
1620. The newly-hatched chicks of which group of pheasants are the most independent?
1621. Why do ptarmigans and Ruffed Grouse sleep under the snow?
1622. Does the weight of a bird's brain exceed the weight of its eyes?
1623. Which hummingbird is essentially restricted to California as a breeding species?
1624. Which bird has a wattle almost as long as the bird itself?
1625. When were the Canadian nesting grounds of the Whooping Crane discovered?
1626. Which American shorebird is the most abundant?
1627. Which breeding alcid has underparts ranging from almost white to nearly black, and which color is most advantageous?

Questions

1628. Which is the most abundant heron?
1629. Which galliform birds feed their chicks from the bill?
1630. How many species of gannets are recognized, and name them?
1631. How do African Hottentots and the Sudanese use Ostrich eggs?
1632. When and where was the first Bristle-thighed Curlew nest discovered?
1633. Which North American cuckoos share nests?
1634. How many fossil species of birds have been discovered?
1635. What is responsible for the color difference of avian white meat and dark meat?
1636. Which are the only North American owls with long, slender tails?
1637. Was Lincoln's Sparrow named after Abraham Lincoln?
1638. Which is the heaviest endemic African flying bird?
1639. How was it determined that the African Red-chested Pygmy Crake occurred in Togo?
1640. Which four birds are best adapted to Arctic winter conditions?
1641. Which large extinct bird was hunted extensively by the New Zealand Maoris?
1642. Which bird has the longest legs?
1643. How many species of birds are nocturnal or crepuscular?
1644. What is oology?
1645. Is Franklin's Gull the state bird of Utah?
1646. Which raptor is considered the "Cadillac" of raptors by falconers, especially in the Middle East?
1647. How did South Sea natives use live frigatebirds?
1648. What is the fastest documented speed for a racing pigeon?
1649. How long is the tail of an adult male Indian Peacock?
1650. With which group of birds do the flight feathers take up almost the entire wing surface area?
1651. Why are cormorants, darters and Anhingas less waterproof than most diving birds?
1652. Which birds have been recorded at the North Pole?
1653. How fast can a Common (Ring-necked) Pheasant run?
1654. Which North American shorebird nests in old tree nests of other birds, and what type of nests are favored?
1655. Is Cooper's Hawk also called a "grasshopper hawk?"
1656. Are shrikes are the only songbirds to *consistently* prey on vertebrates?
1657. Which are the only birds in the sandpiper family with rounded, rather than pointed, wingtips?
1658. At what age do tree-nesting Marbled Murrelets go to sea?
1659. How many sticks may be used by a pair of African Hamerkops during nest construction?
1660. Which bird lays the largest clutch of any nidicolous species (birds that remain in the nest after hatching), and what is its clutch size?
1661. Do juvenile North American owls resemble adults?
1662. What are raptor claws often called?
1663. Which two gulls nest in the densest colonies?
1664. Which is the only swift that normally occurs east of the Mississippi River?
1665. In the U.S., is the California Thrasher restricted to California?

Questions

1666. Which is the darkest eastern North American shorebird?
1667. Which is the only waterfowl in which the bill is feathered to the nostril?
1668. What expression describes accomplishing two tasks simultaneously?
1669. How much does a set of Audubon's *The Birds of America* weigh?
1670. What is the incubation period of a Eurasian Wryneck?
1671. How many Short-tailed Shearwaters (Muttonbirds) are harvested annually in Australia?
1672. Have humans ever been killed by Ostriches?
1673. If a bird is sedentary, what does this imply?
1674. Which sulid is easily sexed by voice?
1675. Which are the only alcids that do not breed in America?
1676. What is a catastrophic molt?
1677. How much more does the goliath beetle of Equatorial Africa weigh than a male Cuban Bee Hummingbird?
1678. What do Galapagos land iguanas do when approached by a mockingbird or a small ground-finch, and why?
1679. During the frequent demonstrations in the U.S. of the late 1960's, what were the anti-war demonstrators called?
1680. Which hummingbird breeds the farthest north?
1681. In what year were federal duck stamps initially required for American waterfowl hunters?
1682. How did the curassows obtain their name?
1683. Which is the only stork of Australia?
1684. Which bird is known as a "wrinkle-nosed auk?"
1685. Do all woodpeckers have four toes?
1686. Which North American bird feeds its young the most frequently?
1687. How many species of eagles occur in North America?
1688. How did the Australian Apostlebird obtain its name?
1689. Which is the largest North American nightjar, and what is its length?
1690. How many doves do North American hunters shoot annually?
1691. Which is the only polygamous parrot?
1692. Where are the wintering grounds of the Canada-nesting Whooping Cranes?
1693. How many of the 91 larks occur in the New World, and which ones are they?
1694. Who was the wife of the 36th U.S. president?
1695. Do puffins occur in Latin America?
1696. What percent of all bird species do males alone incubate?
1697. How heavy are Wandering Albatross chicks just prior to fledging?
1698. Are grebes related to loons?
1699. What is the national bird of Myanmar (formerly Burma)?
1700. Which alcids have the longest bills, and how long are the bills?
1701. Why are some African turacos called go-away birds?
1702. Which American bird is known as "daddy longlegs?"
1703. Which are the only birds that move their wings solely from their shoulders?
1704. How did the English first receive the news of Napoleon's defeat at Waterloo?

Questions

1705. Why do molting penguins appear so bedraggled?
1706. What percentage of all bird species inhabit tropical regions?
1707. Which rare North American raptor is so dichromatic that the sexes were once considered distinct species?
1708. The strangest corvid nest of record was constructed of what?
1709. How much larger is the eye of an elephant than the eye of an Ostrich?
1710. Which western North American upland gamebird was nearly hunted to extinction?
1711. Which is the only African crane with an all-white neck?
1712. Which is the heaviest North American hummingbird, and what is its weight?
1713. In American slang, what is a *rail-bird?*
1714. How do grebe toenails differ from those of other birds?
1715. Which are the only nocturnal parrots?
1716. Which country produces the most domestic turkeys, and what is the flock size?
1717. What are jesses?
1718. Which birds are capable of reproducing at the earliest age, and at what age does breeding commence?
1719. Did Audubon paint in oils?
1720. How did grosbeaks obtain their vernacular name?
1721. How long was the longest recorded Wild Turkey beard?
1722. How are Bristle-thighed Curlews detrimental to other birds?
1723. Does the African Goliath Heron grab its prey?
1724. Which bird is most likely to be confused with the Downy Woodpecker?
1725. Are the South American seedsnipes good runners?
1726. How were swallows used by the ancient Romans?
1727. What journal was founded by A. O. Hume, the *father of Indian ornithology?*
1728. Why was the goosefish so named?
1729. Which is the only red-billed auklet occurring south of Alaska?
1730. How long does it take a male frigatebird to fully inflate its gular pouch?
1731. Which are the only birds that can fly backwards at will?
1732. Are loon chicks striped?
1733. Which was the first American bird to be banded, who banded it, and when did this take place?
1734. Which is the only all black-plumaged American waterfowl?
1735. Which seabirds lay the largest clutch of eggs?
1736. Is the casque of a cassowary bony or horny?
1737. Why don't South American Black-headed Ducks form pair bonds?
1738. Are all North American tyrant-flycatchers migratory?
1739. What color are the wingtips of a Scarlet Ibis?
1740. In Scotland, which bird is called a "rain-goose?"
1741. Among the Okefenokee swampers, what was the eating of mockingbird eggs said to cure?
1742. How much fish is consumed daily by an adult Brown Pelican?
1743. Is a Golden Eagle heavier than a Bald Eagle?
1744. Are breeding Red-legged Kittiwakes restricted to Alaska?

Questions

1745. What is a downy goose called?
1746. Which albatross may move its egg, and why?
1747. How many acorns may be stored by Acorn Woodpeckers in a single storehouse tree?
1748. Which waterfowl may take over the vacated nest of a Hamerkop stork?
1749. What color is the dorsal skin of a roadrunner, and why is it so colored?
1750. Which is the most numerous eagle, and what is its population size?
1751. Which is the largest North American member of the ibis family?
1752. Which terrestrial Old World cuckoos are sometimes called lark-heeled cuckoos, and why?
1753. Why do roosting ptarmigans often fly directly into snow banks rather than dig out a snow burrow?
1754. Is the bill of a newly-hatched flamingo bent?
1755. The brain of an Ostrich accounts for how much of its overall weight?
1756. Which flightless seabird is sometimes considered a nuisance, and why?
1757. In the legends of King Arthur, which bird was generally perched on Merlin's shoulder?
1758. Which large shorebird nests on roof-tops in central Cairo, Egypt?
1759. Which is the largest falcon, and how much does it weigh?
1760. Which hummingbird weighs more than an ounce (28 gms)?
1761. Which grouse is the smallest, and how much does it weigh?
1762. What was the name of one of the British polar explorers who accompanied Captain Robert Falcon Scott to the South Pole, only to perish on the return in March, 1912?
1763. Which is the smallest North American rail, and how much does it weigh?
1764. Which North American finch is described as a *heavy-billed sparrow dipped in raspberry juice?*
1765. Which passerines are considered the most intelligent?
1766. Do all ducks quack?
1767. How do bitterns differ from herons?
1768. Which birds use geothermal heat for incubation?
1769. Which gull is restricted to Australia?
1770. Which bird did Audubon call the "great-footed hawk?"
1771. Does the Great Gray Owl nest in cavities?
1772. How many species of surviving birds are flightless?
1773. How do nightjars position themselves on a branch?
1774. In Alaska, which bird is called a "soldier duck," and why?
1775. What are groups of downy pelicans in crèches called?
1776. Do bee-eaters ever catch insects on the ground?
1777. When and where did humans first begin to alter habitat on a large scale?
1778. Which incubating bird spends the most time on the nest?
1779. Penguins have how many toes?
1780. Do Pileated Woodpeckers use roosting cavities?
1781. Had Charles Darwin visited Hawaii before the Galapagos Islands, which endemic bird group might have served as the model for the origin of the concept of evolution?
1782. With respect to physical structure, what do skuas and jaegers have in common with birds of prey and parrots?

Questions

1783. The Siberian Red-breasted Goose frequently nests adjacent to which bird, and why?
1784. Are male sandpipers larger than their mates?
1785. Does the Barn Owl inhabit Hawaii?
1786. How did the Evening Grosbeak obtain its name?
1787. How do Egyptian Plovers conceal their chicks?
1788. Which are the most migratory American upland gamebirds?
1789. What group of rays (fish) were named after birds?
1790. How many species of albatrosses breed in the Antarctic?
1791. What character did John Wayne play in the movie *True Grit*, for which he won an Academy Award?
1792. Which are the only seabirds that are generally heavier than the white pelicans?
1793. Which woodpecker is the most social nester?
1794. Do herons and egrets generally defecate in the water?
1795. Which is the most abundant owl of Arizona?
1796. Who initially suggested that birds might hibernate?
1797. Wild Turkeys have how many beards?
1798. In the popular literature, the leopard seal is most often associated with which bird?
1799. On what do Sharp-shinned Hawks primarily feed?
1800. Are all egrets herons?
1801. What unusual Emu-hunting technique was used by Australian aborigines?
1802. Which bird is known as a "cuckoo duck," and why?
1803. In gamebird biologist jargon, what is a *fall shuffle?*
1804. In the military, what does a white feather signify?
1805. Which North American raptor is the most aerial?
1806. Which insular hawk engages in cooperative polyandry?
1807. Which is the only large arboreal bird that feeds its young extensively on foliage?
1808. What is the special tool used to harvest Asian cave swiftlet nests called?
1809. What are the large, slender flies in the genus *Tipula* commonly called?
1810. Why was the huge Amsterdam Albatross overlooked as a species for so long?
1811. What was one of the more distinctive American male hair styles of the 1950's called?
1812. What is bird lime?
1813. Which American sandpiper may dive and remain underwater if attacked by a hawk?
1814. Excluding the American Robin, which common bird has been recorded as having the highest incidence of albinism?
1815. Which bird is known as a pig goose, and why?
1816. Which bird is called a Green Hunting Cissa?
1817. Which is the largest New World kingfisher, and how much does it weigh?
1818. How many frigatebird colonies exceed 10,000 pairs?
1819. Which is the only North American bird known to use a tool?
1820. Which shorebird has the most extensive distribution?
1821. How many consecutive calls of the Whip-poor-will were counted by the famous naturalist John Burroughs?
1822. What is kronism or cronism?
1823. How much water does the pouch of a pelican hold?

Questions

1824. What color are the prominent eyelashes of an Ostrich?
1825. What was "Brewster's Warbler?"
1826. Why do terns seldom swim?
1827. When and where was the first bird banded?
1828. What color are the head feathers of an adult Roseate Spoonbill?
1829. What caused the extirpation of the Masked Bobwhite in the U.S. by the early 1900's?
1830. Is the incubation period of an Emu shorter than that of an Ostrich?
1831. Which American bird is known as a "preacher," and why?
1832. How many bird species were introduced into the continental U.S.?
1833. Do male Ruddy Ducks assist with chick rearing?
1834. The extinct Indian Ocean solitaires were closely related to which bird?
1835. How do egg-eating snakes consume bird eggs?
1836. Which is the smallest North American sandpiper, and how much does it weigh?
1837. Which bird has been used symbolically most often?
1838. Why are geese kept around Scottish whiskey distilleries?
1839. Which is the largest raptor to breed within the city limits of the largest city of the northernmost U.S. state?
1840. Has the Shoebill stork ever been raised in captivity?
1841. Which is the only hummingbird that *regularly* winters in the U.S.?
1842. What type of external nares are gymnorhinal?
1843. How many species of toucans inhabit Chile?
1844. Is the Key West Quail-Dove a common resident of the Florida Keys?
1845. Which are the only two large North American yellow-eyed owls lacking ear tufts?
1846. Are the pelecaniform birds the most arboreal seabirds?
1847. What are Long-billed Corellas, and on what do they primarily feed?
1848. With respect to plumage, what do all juvenile dippers have in common?
1849. Is the Blood Pheasant polygamous?
1850. How many eggs does a domestic chicken lay annually?
1851. Does the Arctic Loon breed in Greenland?
1852. Which small North American land birds migrate the farthest?
1853. Is the California Condor the state bird of California?
1854. Which avian sense is the least developed?
1855. What is a euryphagous bird?
1856. Do *both* male and female Ruffed Grouse have ruffs?
1857. Do antbirds regularly eat army ants?
1858. How much more rapid is the wingbeat of insects than the wingbeat of hummingbirds?
1859. Other than the possibly extinct Ivory-billed Woodpecker, which North American woodpecker has the most specialized habitat requirements?
1860. Which is the largest North American rail, and how much does it weigh?
1861. How many birds are named for U.S. states, and which birds are they?
1862. Which aquatic birds have a claw on the first digit of each wing?
1863. Which birds are called *pedorrera* by Jamaicans, and why?
1864. What is an adherent nest?
1865. Are there any subantarctic snipe?

Questions

1866. How did the name "nightjar" originate?
1867. How far do waterfowl travel to gather nesting material?
1868. Which New World vulture is said to have the most powerful bill?
1869. Why are the toes of a Snowy Egret bright yellow?
1870. Why do nuthatches forage by moving down a tree trunk headfirst, rather than up the tree?
1871. How many House (English) Sparrows inhabit North America?
1872. What is the name of the journal published by the British Ornithologists' Union?
1873. Which gull that nests *exclusively* in North America undertakes the longest migration?
1874. With the exception of owls, which is the only raptor with a reversible outer toe, and why?
1875. What continents are not inhabited by cranes?
1876. Which flamingo is most numerous, and what is its population size?
1877. Which woodpecker lives above the tree line?
1878. Which are the most accomplished avian mimics of Australia?
1879. What is the status of the Short-tailed Albatross, and where do most breed?
1880. How many species of oystercatchers are currently recognized?
1881. Which parrot had the most northerly range in historic times?
1882. Why do birders practice *squeaking*?
1883. Which New World bird has the most limited range?
1884. Which frogmouth is larger than the Australian Tawny Frogmouth?
1885. For what were the feathers of the extinct Great Auk used?
1886. What is to draw back in fear, or to lose heart, courage or spirits sometimes called?
1887. What percentage of all bird species are gregarious?
1888. For what was Dixon Lanier Merrit noted?
1889. Which birds are the closest jacana relatives?
1890. Which large, long-legged birds have a huge, naked, subcutaneous foreneck pouch or air-sac, and a smaller rounded one at the junction of the hindneck and the mantle?
1891. Is the Everglade Snail Kite monophagous?
1892. How many species of birds are obligate brood parasitic nesters?
1893. Do any pelecaniform birds dig their own nesting burrows?
1894. In the 1988 biographical movie *Bird*, directed by Clint Eastwood, who was the subject of the film?
1895. Which is the second largest tern, and what is its wingspan?
1896. How do graminivorous birds and granivorous birds differ?
1897. What is a pinfeather?
1898. Do male hummingbirds ever incubate and assist with chick rearing?
1899. When and where was the Rock (Pigeon) Dove first introduced into North America?
1900. Which familiar, colorful American songbird ceased to exist as a species in 1973?
1901. Do King Penguins breed annually?
1902. In winter plumage, is the Pigeon Guillemot whiter in color than the Black Guillemot?
1903. Does the Demoiselle Crane have a long tail?
1904. Do smaller birds have higher body temperatures than larger birds?
1905. How long is the small intestine of an Ostrich?

Questions

1906. Which birds are pictured on the Japanese 1,000 and 10,000 yen banknotes?
1907. Which North American hawk forms breeding trios?
1908. Which living bird most closely resembles the proto-bird *Archaeopteryx*?
1909. Which North American swift nests under waterfalls?
1910. How much heavier is the heaviest sulid than the lightest?
1911. What is the pygostyle?
1912. Are coots good divers?
1913. Which are the only birds with cat-like, vertical pupils, and why is this advantageous?
1914. Do loons cover their eggs when they are off the nest?
1915. Why was the Woolly-necked Stork formerly known as the Bishop Stork?
1916. When did the last of the moas disappear?
1917. Which is the tallest stork, and how tall is it?
1918. Which passerines have the longest incubation period, and how long is it?
1919. At what age do California Condors assume adult color and plumage?
1920. How many species of birds breed in Great Britain?
1921. What is the incubation period of a Caribbean (Greater) Flamingo?
1922. Oystercatchers have how many toes?
1923. Which tern is most likely to be encountered far from land?
1924. Do fish ever leap out of the water to catch birds?
1925. On which New Zealand coin is the kiwi featured?
1926. What is swan-upping?
1927. Do woodpeckers feed their young by regurgitation?
1928. Although cranes and large herons are similar appearing in flight, how can an observer instantly tell them apart?
1929. What was the first island called the *Island of Penguins*, and why was it so named?
1930. What is a scansorial bird?
1931. Does the American Oystercatcher nest on roof-tops?
1932. Which is the only pelican that has chicks with brownish-black down?
1933. Which is the only fish-duck indigenous to North America?
1934. Do goldfinches feed their chicks insects?
1935. Do pigeons or doves ever lay more than a two-egg clutch?
1936. Who was the first to introduce parrots into Europe from the Far East, and which parrot is named after him?
1937. Which small passerine has the most unusual ornate, paired feathers?
1938. The feathers of some birds have aftershafts attached to the base of the main feather, but which two types of birds have feathers and aftershafts of *equal* length?
1939. Do Anhingas and darters swallow fish underwater?
1940. What is a snood?
1941. Is the ostrich fern an African plant?
1942. Which is the only non-colonial nesting booby?
1943. Are shrikes nocturnal predators?
1944. What is the uncinate process?
1945. Is the bill of a cassowary straight?
1946. What is the *drumstick* of fried chicken fame?

Questions

1947. Which American bird is known as a "bumblebee buzzer," "chuck-duck" and "spatter?"
1948. In Japan, which bird is symbolic of happiness?
1949. Which is the only North American bird that nests twice in two different regions during the same season?
1950. Are cephalopods a threat to birds?
1951. What is the *documented* record number of songs sung by an individual bird in a single day, and which bird accomplished it?
1952. Are female owls always larger than their mates?
1953. Which was the last North American bird discovered, and when was it discoverd?
1954. What are bird-spiders?
1955. How many mud-collecting trips are required by a pair of swallows during nest construction?
1956. Is the Common Murre more numerous than the Thick-billed Murre?
1957. Which European bird is the most gifted singer?
1958. What type of grouse is the European Red Grouse?
1959. Are *breeding* Short-tailed Shearwaters restricted to Australia?
1960. Do drongos occur in the New World?
1961. Who was the only naturalist to see the extinct Spectacled Cormorant alive?
1962. What color is the mouth lining of a nighthawk chick?
1963. Do all birds have crops?
1964. Do swallows often cling to vertical rock walls?
1965. Are hummingbirds and woodcocks stenophagous?
1966. Who played the male lead in the 1962 movie *The Sweet Bird of Youth*, based on the novel by Tennessee Williams?
1967. Which are the only birds that do not propel themselves with their feet when swimming on the surface?
1968. How did Charles Darwin describe the Galapagos Flightless Cormorant?
1969. When catching fish, do Ospreys go completely underwater?
1970. How can Budgerigars be sexed?
1971. How many species of kingfishers inhabit the New World?
1972. Do Bald Eagles defend larger territories than Golden Eagles?
1973. Which bird is called the "Texas Bird-of-paradise?"
1974. Why are there so few captive Imperial Pheasants?
1975. What is a Kaka?
1976. Are cracid chicks brooded on the ground?
1977. Is the South American Painted-Snipe larger and more colorful than the Old World Painted-Snipe?
1978. What is the origin of the name "guineafowl?"
1979. Which bird has the longest laying interval between eggs?
1980. Do tropicbirds carry fish in their bills like terns?
1981. Which is the most social North American blackbird?
1982. Which small sparrow-like birds are agricultural pests in South America?
1983. What is the most accurate way to separate the nearly identical appearing Western and Eastern Meadowlarks?

Questions

1984. Which is the only tool-using shorebird?
1985. In old folklore, which raptors are often regarded as messengers of ill tidings?
1986. Which birds produce Long Island duckling?
1987. Will splitting a crow's tongue make it a better "talker?"
1988. What does the Hopi Indian name *holchko* for the Common Poorwill mean?
1989. Which waterfowl was considered extinct until rediscovered in 1947?
1990. Which northern bird was called a "wamp," and which bird was known as "wamp's cousin?"
1991. Birds in which large family lay the roundest eggs?
1992. Excluding the two condors, which is the largest New World vulture, and how much does it weigh?
1993. What happened to the only specimen of the Cerulean Paradise-Flycatcher known to science?
1994. When a penguin is captured by a leopard seal, does the seal smash it on the surface of the water, literally popping the penguin out of its skin?
1995. Do Northern Fulmars prey on jellyfish?
1996. Which is the largest North American upland gamebird?
1997. Was the extinct Passenger Pigeon the largest North American pigeon?
1998. Which raptor nests in the nest of the African Social-Weaver?
1999. Which is the least-studied North American duck?
2000. On what do Secretary-birds feed extensively?
2001. What does the expression *the goose hangs high* imply?
2002. When and where did the last Carolina Parakeet perish?
2003. How did the Vulturine Guineafowl obtain its name?
2004. Are all 23 species of turacos African endemics?
2005. Are birds more vividly colored than any other class of vertebrates?
2006. Which birds may fly through a wave rather than over it?
2007. In underwater-swimming penguins, which flipper stroke provides forward thrust, the upstroke or the downstroke?
2008. In what 1964 World War II movie did Cary Grant play a grubby beachcomber?
2009. Which group of hummingbirds has the most greatly curved bills?
2010. How does a disturbed incubating Mountain Plover protect its nest, and why did such behavior evolve?
2011. How far away can the drumming of a Ruffed Grouse be heard?
2012. Are any of the trogons migratory?
2013. Is the Whooping Crane the tallest crane?
2014. Which is the most widespread Western Hemisphere grebe?
2015. Which African-breeding shorebird has a disproportionately heavy bill?
2016. Which bird is associated with a clock?
2017. Shoebill storks are most apt to prey on what types of reptiles?
2018. Are most South American wood-warblers sexually dimorphic?
2019. Is the neck of a flying Limpkin retracted?
2020. What is unusual about the British race of the Willow Ptarmigan (Red Grouse)?
2021. What type of bird is a Fieldfare?

Questions

2022. Where is the breeding stronghold of the rare Aleutian Canada Goose located?
2023. Why are many tropical jungle birds brightly colored?
2024. Are nightjars thick skinned?
2025. What is the maximum lifespan of an Ostrich?
2026. What color are adult Egyptian Vultures?
2027. Which bird is featured on the Australian coat-of-arms?
2028. Which North American, non-migratory, chickadee-like birds have the name of a rodent in their collective common name?
2029. Do obligate brood parasitic-nesting cuckoos establish strong pair bonds?
2030. What does the grebe's generic name *Podiceps* mean?
2031. Which endemic Australian long-legged bird has a dewlap?
2032. Is the shell of an Ostrich egg smooth?
2033. Which pheasant is known only from a portion of a single primary feather?
2034. Which American bird is called a "shitepoke," and why?
2035. What was unusual about the flight ability of the extinct Spectacled Cormorant?
2036. Which is the largest, most powerful eagle, and how much does it weigh?
2037. Which birds are collectively called ruffed pheasants?
2038. Which bowerbird is a tool-user, and what type of tool does it use?
2039. Which wading birds hold food in their foot like parrots?
2040. Can domestic ducks lay as many eggs annually as chickens?
2041. Which was the first bird featured on a U.S. postage stamp?
2042. Where on a bird are the smallest contour feathers located?
2043. What is the first book devoted to birds written in English, and when was it published?
2044. What is the difference between a mannikin and a manakin?
2045. How much heavier are female kiwis than their mates?
2046. What is a Nukupuu?
2047. How many species of mammals lay eggs, and which ones are they?
2048. Do woodswallows inhabit Africa?
2049. What is the only bird endemic to the island of Corsica?
2050. Which live bird commanded the highest price, and for much was it sold?
2051. Which North American heron is the weakest flier?
2052. How many fish can a Common Puffin carry crosswise in its bill at a time?
2053. Which tropicbirds nest in trees, and where does tree-nesting take place?
2054. Is the Central American Ocellated Turkey gregarious?
2055. Do Black Vultures have longer wings than Turkey Vultures?
2056. Do shrikes kill more than they can consume?
2057. What are coverts?
2058. Which is the only black sandpiper?
2059. Which Antarctic-breeding bird is the best adapted to a terrestrial lifestyle?
2060. Which distinctive New Zealand bird had penny-sized, waxy, orange wattles?
2061. Which part of an Ivory-billed Woodpecker was a valued item of American Indian trade?
2062. Are grouse nostrils visible?
2063. Which neotropical river-dwelling waterfowl has carpal spurs?

Questions

2064. How much longer is the upper mandible of a newly-hatched woodpecker chick than its lower mandible?
2065. Why do foraging skimmers often backtrack?
2066. In slang, what is a person called who has been swindled, fleeced or cheated?
2067. Do Manx Shearwaters breed in the U.S.?
2068. What was Rosalinda more affectionately called in Johan Strauss' light opera *Die Fledermaus*?
2069. How many species of grebes are flightless, and which ones are they?
2070. How do the feet of the Puna (James') and Andean Flamingo differ from the feet of other flamingos?
2071. How much more vividly colored are male Australian Pink-eared Ducks than females?
2072. Is a palearctic species restricted to the temperate and Arctic regions of the Old World?
2073. Is the Hamerkop stork nocturnal?
2074. What are *Buffalo wings,* and how did the name originate?
2075. Which is the most widely distributed northern gull?
2076. Which bird was responsible for the saga that ultimately became known as *the worst journey in the world*?
2077. Which American hawk undertakes the longest migration?
2078. What is the vent?
2079. What is a casting?
2080. What famous avian personality hangs out on Sesame Street?
2081. Which is the smallest bird of New Zealand?
2082. Which long-legged, carnivorous African bird has benefitted to some extent by humans?
2083. Which passerine puffs up its naked red throat to grotesque proportions like a displaying male frigatebird?
2084. Do most raptors kill prey with their feet, rather than their bills?
2085. Aside from some albatrosses and petrels, do any other birds require more than seven years prior to successful breeding?
2086. Why is the Australian crane called a Brolga?
2087. When and where was the last U.S. Eskimo Curlew shot?
2088. What is the typical clutch size of a duck-billed platypus?
2089. On what does the Crab Plover primarily feed?
2090. Which northern birds are the ecological counterparts of the South American steamer-ducks?
2091. Which federally confiscated birds were released in Arizona in 1986, in an attempt to re-establish the species in the U.S.?
2092. Does the American Woodcock range as far south as Mexico?
2093. How many pairs of air-sacs are connected to the lungs of a bird?
2094. Which seabird is the spiritual symbol for a clan of southeast Alaskan Tlingit Indians ?
2095. Approximately how many species of birds are migratory?
2096. Which are the only stilts that are not black-and-white plumaged?
2097. On which island do the most species of albatrosses breed, and which species breed there?

Questions

2098. In the Bahamas, which color morph of the Reddish Egret is the most numerous?
2099. What type of nest is often utilized by breeding African Red-and-yellow Barbets?
2100. How much do the nests of the Asian Edible-nest Swiftlet, of bird's-nest soup fame, sell for per pound?
2101. Is the Prairie Warbler a prairie species?
2102. With respect to its corvid population, how does the West Indies differ from Central and South America?
2103. Which aquatic birds have the shortest tails?
2104. Did the King Vulture occur in Florida in historical times?
2105. How do birds navigate?
2106. How many species of flying birds have less than nine primaries?
2107. What is a heterodactyl foot, and which are the only birds with a foot of this type?
2108. How many species of extinct penguins are smaller than the Little Blue (Fairy) Penguin of Australia and New Zealand?
2109. How long can the egg of some petrels, such as the Manx Shearwater, be left unattended and still hatch once incubation resumes?
2110. What is definitive plumage?
2111. What percent of an American Woodcock's diet consists of earthworms?
2112. Which gull undertakes the longest migration?
2113. What does the dorsal down of a recently-hatched Ostrich resemble?
2114. What are mews?
2115. Why does the endangered Red-cockaded Woodpecker pock the bark around the entrance of its nesting or roosting hole?
2116. What is the condition of a bird's bill when it requires coping?
2117. Are all adult New World vultures mute?
2118. Do all fish-eating birds use their beaks to capture prey?
2119. Where does the Gray Gull of the Humboldt Current sector of South America nest?
2120. Which bird was the ancestor of domestic canaries?
2121. Which fish found in California is called a starling?
2122. How often does an American Redstart turn its eggs?
2123. Which parrot formerly inhabited the subantarctic island of Macquarie?
2124. What does the Gentoo Penguin specific name *papua* imply?
2125. How did the existence of the enormous eggs of the Madagascar Elephant-bird first become known to the outside world?
2126. What is the name of the largest-known butterfly, and how much more does it weigh than the smallest bird?
2127. In which country does the Northern Bald Ibis (Hermit Ibis or Waldrapp) breed?
2128. What type of sounds are produced by vocalizing curassows?
2129. In the Bible, which bird was referred to as a "lapwing?"
2130. What group of islands accommodates the most breeding species of penguins, and which penguins breed there?
2131. On what does the Australian Pink-eared Duck feed?
2132. Which portions of a bird are called the soft parts?
2133. Which group of North American birds is best known for a sinusoidal flight style?

Questions

2134. What did Audubon call the Tricolored (Louisiana) Heron?
2135. In what 1985 movie about a medieval knight and his lady did Ruger Hauer and Michelle Pfeiffer star?
2136. How many species of birds have five toes?
2137. Which is the heaviest pelican, and how much does it weigh?
2138. Which birds do pipits most closely resemble?
2139. Which is the only planktivorous alcid of the Atlantic?
2140. The chicks of which type of birds lack an egg tooth?
2141. Does the Osprey breed in South America?
2142. Which is the only waterfowl indigenous to New Guinea?
2143. Do the Sooty Terns of Ascension Island breed annually?
2144. What color is the bare facial skin of an adult Hoatzin?
2145. Which is the only heron that is well established in the high Andes?
2146. How many species of birds are considered at risk?
2147. What did the legendary Roc of Madagascar feed its young?
2148. What is the incubation period of the Ivory-billed Woodpecker?
2149. Which American aquatic bird is known as a "hairyhead?"
2150. Which exclusively aquatic birds have a hunchbacked profile in flight?
2151. Do Spectacled Eider ducklings have prominent spectacles?
2152. With respect to its diet, what type of bird is the Marabou Stork?
2153. Which birds are known as brain-fever birds, and why?
2154. Are injured giant-petrels attacked and consumed by their own kind?
2155. How long can European Swift chicks go without food?
2156. Which African eagle has the longest incubation period, and how long is it?
2157. Are Painted Petrels (Cape Pigeons) cavity nesters?
2158. How did the ground-nesting Samoan Tooth-billed Pigeon survive the introduction of terrestrial predators, especially pigs?
2159. Which of the six flamingos is the reddest in color?
2160. When do male Ostriches incubate?
2161. Breeding Limpkins are restricted to which U.S. states?
2162. Do Hyacinth Macaws forage on the ground?
2163. Are birds more sexually dimorphic than any other class of vertebrates?
2164. Why were shearwaters given the generic name *Puffinus*?
2165. Do birds have more contour feathers during the winter than in the summer?
2166. Which birds feed on the highly venomous Portuguese man o' war jellyfish?
2167. Do the wingtips of foraging skimmers touch the water?
2168. Which African bird is called a "monkey bird," and why?
2169. What do recently-hatched kiwis resemble?
2170. After whom was the Australian Gouldian Finch named?
2171. How were Gila Woodpeckers beneficial to the Apache Indians?
2172. Which is the only jacana that has a non-breeding plumage?
2173. Do crabs prey on birds?
2174. Do Jackdaws nest in cavities?
2175. Do oystercatchers swallow mollusks whole?

Questions

2176. Which hummingbird occurs the farthest south?
2177. What type of birds are avadavats, and what is the origin of their name?
2178. What is Howard Hughes's famous huge plywood seaplane popularly called?
2179. Which corvid is the smallest?
2180. How do birds chew their food?
2181. How many species of parrots inhabit Australia?
2182. Do the recently-hatched downy chicks of any terns resemble adults in color and pattern?
2183. Was the European Robin ever introduced into America?
2184. Do Hamerkops construct a single huge nest annually?
2185. Do Brown Pelicans fish cooperatively?
2186. Which is the largest African eagle, and how much does it weigh?
2187. Which South American bird is called a "snake crane?"
2188. What type of bird is a Malleefowl?
2189. What is the crane symbolic of in China?
2190. What does the name "fulmar" mean?
2191. What is the typical hummingbird clutch size?
2192. Why were guineafowl formerly called turkeys?
2193. How did the Short-tailed Albatross manage to survive despite the butchering of every breeding bird?
2194. Is the Lappet-faced Vulture the largest Old World Vulture?
2195. According to legend, how did pelicans feed their young?
2196. Which Jamaican bird is known as a doctor bird?
2197. Do piculets have woodpecker-like tails?
2198. What color is the skin of a newly-hatched King Penguin?
2199. Does the Rufous Hornero reuse its oven-shaped mud nest year after year?
2200. Who starred in the 1940 movie *The Bluebird?*
2201. What is the primary physical difference between the Northern Jacana and the closely-related Wattled Jacana?
2202. Which is the only parrot that uses a lek display ground?
2203. Is the smallest bird smaller than the smallest mammal?
2204. What happened to the population of Red-breasted Geese, when the Peregrine Falcon population crashed as a result of pesticide poisoning in late l960's and early 1970's?
2205. What is the normal curassow clutch size?
2206. What are the red or yellow wattles at the base of the bill of the African Saddle-billed Stork sometimes called?
2207. What does the 12th of August signify for British aristocracy and the rich?
2208. What was the Blackburnian Warbler formerly called?
2209. Are loon bones hollow?
2210. What is the average nesting density for Common Murres on broad, flat ledges?
2211. Is the Roseate Tern actually rosy colored?
2212. Which sulid has two distinct color morphs?
2213. Which group of aquatic birds has laterally-compressed, blade-like legs, and why is this advantageous?

Questions

2214. Which is the rarest *breeding* raptor of England?
2215. Which group of birds is said to have the most unusual courtship?
2216. At what time of day does the Egyptian Plover (Crocodile-bird) incubate?
2217. Can hummingbirds perch on twigs that are angled upward more than 45 degrees?
2218. Where else besides Christmas Island does the Christmas Island Frigatebird nest?
2219. What is cited as one of the primary reasons for the eastward range extension of breeding Horned Larks?
2220. How do flying loons position their feet?
2221. Which American shorebird is known as a "squatter," and why?
2222. What is the most distinct physical feature of the Blue-faced Honeyeater of Australia and New Guinea?
2223. Do tropicbirds have serrated bills.
2224. Do male curassows brood chicks under their wings?
2225. Does the Black Skimmer have a forked tail?
2226. If arranged in a single column, how many feet of feces are excreted daily by a capercaillie?
2227. Do both species of Three-toed Woodpeckers occur in Alaska?
2228. Are snails cracked open in the gap of the bill of the openbill storks?
2229. Following a plunge-dive by a gannet or booby, are fish ever taken from beneath as the birds rise to the surface?
2230. When was the Cooper Ornithological Society founded, and after whom was it named?
2231. How much larger is the largest known dinosaur egg than the egg of the extinct Elephant-bird?
2232. Do Wrentits mate for life?
2233. Why are Tawny Frogmouths often run over by cars?
2234. Which bird may migrate south with Arctic Terns, and why?
2235. Which is the smallest African waterfowl?
2236. As of 1992, how many species of African birds were considered to be endangered or threatened?
2237. How is speaking bitterly or reproachfully, or to complain vociferously, often described?
2238. How many birds would theoretically result from a single pair of American Robins after ten years if the sex ratio of the resultant young were equal, all young survived and bred, and pairs raised two broods of four chicks annually?
2239. Do tropicbirds always prey on fish in the water?
2240. How much larger is the brain of a small perching bird than the brain of a lizard of comparable size?
2241. Which North American woodpecker often forages on the wing for insects like a flycatcher?
2242. To what island are breeding Ipswich Sparrows restricted?
2243. What color are the ankle joints of the African Saddle-billed Stork?
2244. What are male Wild Turkeys called?
2245. When *exactly* do Short-tailed Shearwaters lay their eggs?
2246. Are flamingos good swimmers?

2247. What is the population size of the endangered Philippine Monkey-eating Eagle?
2248. How is to gain the upper hand sometimes described?
2249. Is the second deepest lake in the world, Lake Tanganyika, remarkable for the bird life it supports?
2250. Does Abbott's Booby breed annually?
2251. Which parrot nests the farthest south?
2252. What is a sclerotic ring?
2253. What was the falconer's name for a male Gyrfalcon?
2254. How many species of South American ratites are recognized?
2255. Do walrus prey on birds?
2256. Which three swans were combined and are currently simply known as the Tundra Swan?
2257. Where is the largest collection of bird study skins in the New World housed, and how large is the collection?
2258. The males of which two groups of birds have an external penis during copulation?
2259. Male Andean Condors can be sexed because of the prominent fleshy caruncle on the head that females lack, but how can California Condors be visually sexed?
2260. In American slang, what is a coward sometimes called?
2261. Does the Eurasian Capercaillie currently breed in Great Britain?
2262. Does a holarctic bird have a circumpolar distribution?
2263. When fishermen say that a bird is *drying its sails,* to which bird are they generally referring?
2264. Do Long-billed Dowitchers have longer wings than Short-billed Dowitchers?
2265. Which of the 17 penguins is the most numerous?
2266. What type of nest site is typically selected by the Asian Rufous Woodpecker?
2267. Which bird is represented on the Lufthansa German Airlines logo?
2268. According to the Roman scholar Pliny, what fate befell the Greek poet Aeschylus?
2269. Which is the only shorebird with nidicolous young?
2270. Which are the only two endemic, insular subantarctic ducks?
2271. What was the oddest location for a Carolina Wren's nest?
2272. Why are the birds-of-paradise not superior singers?
2273. Which group of Arctic birds is the most herbivorous?
2274. Which bird has an alarm call only uttered when an obligate brood parasitic-nesting Brown-headed Cowbird appears near its nest?
2275. What do carpenters call a small headless nail?
2276. After fledging, are swallow young fed by their parents?
2277. How much larger are male skuas than females?
2278. What is the spread made from the livers of specially fattened geese called?
2279. On what do Torrent Ducks feed almost exclusively?
2280. Is the Giant Coot of the South American highlands a powerful flier?
2281. Are the finch-shaped bills of the three South American plantcutters serrated?
2282. How do grebes alter their specific gravity?
2283. Which European sandpiper makes use of old tree nests?
2284. Which is the largest waterfowl of South America?

2285. How many species of storks are currently recognized?
2286. How high up in the Andes do hummingbirds occur?
2287. Do cranes nest in trees?
2288. Are molting grebes flightless?
2289. How large are the wattles of the Little Wattlebird of Australia?
2290. When a Razorbill is sitting upright, does the auk brace itself with its tail?
2291. When and where did the most disastrous North American oil spill occur, and how much oil was spilled?
2292. Do all 24 drongos have forked tails?
2293. How large were the wings of the extinct moas?
2294. Why were Carolina Parakeets hunted so extensively?
2295. What is the population size of the Magellanic Plover?
2296. Is a Royal Albatross more likely to follow a ship than a Wandering Albatross?
2297. Including seabirds, how many species of birds have been recorded for Kenya?
2298. Do the four diving-petrels spend less time brooding their chicks than most other small petrels?
2299. Which American bird is called a "topsy-turvy bird?"
2300. Which types of birds are feathered uniformly throughout?
2301. Are the shells of duck eggs greasy?
2302. How do the roosting habits of gannets and boobies differ?
2303. What was the *plumed serpent?*
2304. Which group of large neotropical gamebirds engages in courtship feeding?
2305. Which northern ducks are the most carnivorous?
2306. Which birds are the South American counterparts of the Old World pittas?
2307. Have rheas ever been raised commercially?
2308. Which are the only piciform birds that do not hatch naked?
2309. For what is Phillip Island, near Melbourne, Australia, best known?
2310. Do ratites nest in underground burrows?
2311. Are female frigatebirds larger than their mates?
2312. How many of the 177 owls are completely nocturnal?
2313. In terms of weight, how much fish is consumed annually by the Peruvian guano-producing seabirds?
2314. Which ptarmigan nests south of Canada?
2315. Do fish-owls plunge into the water like Ospreys?
2316. What is the more accepted name for the Black Chachalaca?
2317. What is the origin of the name "Gentoo" Penguin?
2318. How many storks have decurved bills, and which ones are they?
2319. What is the Cain and Able syndrome (battle)?
2320. Which is the tallest flying bird?
2321. Which is the only large Arctic-breeding petrel?
2322. Which is the rarest duck of Europe, and what is its status?
2323. Where is the largest collection of bird skins housed, and how many skins are maintained in the collection?
2324. In sports jargon, what is a goose egg?

Questions

2325. What color is the African Verreaux's Eagle?
2326. Which bird adorns the NBC TV logo?
2327. What is a Jacky-winter?
2328. In American folklore, what happens if a crow flies over a house and croaks three times?
2329. What is a cryptic bird?
2330. Do Common Murres use the same *nest* site year after year?
2331. Which bird was responsible for the name "finch?"
2332. Do penguin chicks hatch with their eyes open?
2333. What is the more familiar name for the Buff-backed Heron?
2334. What is *fleyg* netting?
2335. What is another name for breeding plumage?
2336. What type of nest does the Bushtit construct?
2337. For what is Richard Bach best known?
2338. How do American fishermen sometimes refer to a backlash?
2339. How far south does the neotropical Oilbird range?
2340. Which raptor has the longest tail, and how long is the tail?
2341. What is the size of the Glaucous Macaw population?
2342. Which rail preys on nesting petrels?
2343. Excluding the Harpy and Philippine Monkey-eating Eagles, which eagle is the largest?
2344. Which group of pheasants has beautiful, brilliantly-colored, bib-like throat wattles or lappets?
2345. Which country has the greatest number of threatened or endangered birds?
2346. Do bustards ever roost in trees?
2347. To which birds are finfoots most closely related?
2348. Which parrot puffs itself up like a football and booms?
2349. Do loons generally winter at sea?
2350. What is the Red-crowned Crane also called?
2351. Do fledging Emperor Penguins go to sea while still partially covered with down?
2352. Which American raptor has a heart-shaped face?
2353. Which waterfowl are called loggerheads, and why?
2354. How many specimens made up the personal mounted hummingbird collection of the ornithologist and painter John Gould, and what ultimately happened to the collection?
2355. Is the avian respiratory system the most efficient of all vertebrates?
2356. Do all three skimmers have black-tipped bills?
2357. Does the Osprey have a larger preen gland than most other birds of prey?
2358. What is the obsolete name for the handle of a plow?
2359. Which African bird looks almost exactly like an American meadowlark?
2360. Do Lammergeiers ever kill their prey?
2361. How do African and Asian finfoot chicks differ from the chicks of the Sungrebe?
2362. Are bird eggs always laid pointed-end first?
2363. Does the Worm-eating Warbler feed on earthworms?
2364. What might an incubating male Pheasant-tailed Jacana do if disturbed, or its nest is in danger of flooding?

Questions

2365. What type of nest site is favored by the Common Miner of South America?
2366. Which Asian waterfowl is well established as a feral species in Great Britain, and what is its status?
2367. When beating out a tattoo on a tree, how many blows a second have been recorded from the Great Spotted Woodpecker?
2368. Do the young of some birds weigh *twice* as much as their parents?
2369. What is a pamprodactyl foot, which birds have them, and why are they advantageous?
2370. Which colorful Siberian thrush is established in the U.S.?
2371. How are Ostrich eggs used by the Ethiopian Orthodox Church?
2372. Do Andean Condors occur as far south as Cape Horn?
2373. Do loons spear fish underwater?
2374. During the 4-month breeding season, how many lemmings may be taken by a pair of Snowy Owls rearing nine owlets?
2375. Which are the only pelecaniform birds with eggs lacking the chalky covering typical of other members of the order?
2376. Were homing pigeons flown by the U.S. Army during WW II?
2377. Which upland gamebird introduced into North America is called a "Hun?"
2378. Do cormorants soar?
2379. Does the American Avocet have the shortest legs of the four avocets?
2380. How is the Hamerkop regarded by African natives?
2381. Where is the alcid center of distribution located?
2382. Hummingbirds have how many neck (cervical) vertebrae?
2383. Which bird has a bill in which the upper mandible is very long and decurved but the lower mandible is short and wedge-shaped, and what is the purpose of such a design?
2384. In terms of percent of body weight, how much fish is required daily in fish-eating birds?
2385. Which is the only waterfowl that engages in ritualized courtship feeding?
2386. Is the Hyacinth Macaw a Bolivian resident?
2387. If the largest living bird, the Ostrich, weighs nearly 100,000 times as much as the smallest bird, the Cuban Bee Hummingbird, how much more does the largest mammal, the blue whale, weigh than the smallest mammal, Kitti's hog-nosed bat?
2388. Which is the largest raptor of Australia?
2389. Are the South American screamers noted for soaring?
2390. Excluding the adult Egyptian Vulture, which is the only predominately white Old World vulture?
2391. What was peculiar about the first California record of a White-tailed Tropicbird?
2392. Including the air-sacs, how much of the total volume of a bird is taken up by the respiratory system?
2393. Do falconets occur in the Western Hemisphere?
2394. Which is the only tern with fleshy wattles around the gape?
2395. What is the national bird of South Africa?
2396. How often are powder-down feathers molted?
2397. How many species of storks inhabit the Western Hemisphere, and which ones are they?

Questions

2398. After whom was Clark's Grebe named?
2399. Which American waterfowl is known as a "poacher," and why?
2400. Is the shimmering display of colors known as iridescence due to feather pigmentation?
2401. Why do most passerines that rear more than a single brood a year build a new nest with each new brood?
2402. Do screech-owls hatch both red and gray-color morph chicks in the same brood?
2403. What color is the skin of a White (Fairy) Tern?
2404. In American folklore, what does the bluebird symbolize?
2405. Which birds are capable of moving the tip of the upper mandible, and why is this advantageous?
2406. In what television series did Susan Dey have a feature part prior to her role on the hit TV series *L. A. Law*?
2407. How large are the scapulas (shoulder blades) of an Ostrich?
2408. Which large soaring bird is Georg W. Steller credited with discovering?
2409. What is an ornithichnite?
2410. How long does it take a shrike to digest a mouse?
2411. What type of bird is the Australian Whistling-eagle?
2412. To which birds are cranes most closely related?
2413. Is the Arctic Tern the only tern to breed in the Arctic?
2414. What is a pantropical bird?
2415. On what does the Red-breasted Pygmy-Parrot primarily feed?
2416. With respect to aviculture, what do the Zebra Finch and Java Sparrow have in common?
2417. In America, which bird is called a "Black-heart Plover?"
2418. How can some birds, such as the Asian Rufous Woodpecker, nest in active termitaries without being attacked by defending ants?
2419. Do birds have a diaphragm?
2420. Are birds more resistant to cold than mammals?
2421. Is the endemic Kagu of New Caledonia nocturnal?
2422. Which northern passerine buries itself in the snow?
2423. If a bird is an indeterminate layer, what does this mean?
2424. Which birds are called jewel thrushes?
2425. At what age do male Satin Bowerbirds attain sexual maturity?
2426. Buttonquail (hemipodes) have how many toes?
2427. Why is the incubation period of an Ostrich so much shorter than that of the substantially smaller Emu?
2428. Why is the bill of the Broad-billed Hummingbird broader than most other hummingbirds?
2429. Which frigatebird is at risk, and why?
2430. What unique hunting technique is the African Hawk-Eagle said to utilize?
2431. What theory has been proposed regarding the possible origin of the obligate brood parasitic-nesting habits of the American cowbirds?
2432. How many species of New Zealand birds have become extinct since the arrival of the Polynesians, around 950 A.D.?

Questions

2433. Do tropical birds generally lay larger clutches than birds breeding in more temperate regions?
2434. How many museum specimens of the U.S. race of the Ivory-billed Woodpecker exist?
2435. Which relatively abundant Indian Ocean bird was not described until 1955, and where are its breeding grounds?
2436. What is the name of the publication of the Royal Navy Bird-watching Society?
2437. Which bird is regarded as symbolic of the Holy Spirit?
2438. Which swan has the longest legs?
2439. Are wild Ostriches wary of humans?
2440. Which is the only swallow that *regularly* consumes a substantial amount of plant food?
2441. Are the magnificent head plumes of the male King of Saxony Bird-of-paradise erectile?
2442. Do most pheasants roost in trees at night?
2443. Have Black-footed Albatrosses ever been recorded in the Southern Hemisphere?
2444. Why does the Australian Gray Goshawk have an all-white color phase?
2445. Which white, long-tailed seabirds have exceptional flight endurance in still air?
2446. Do Common Ravens occur in Costa Rica?
2447. What is peculiar about the habitat preference of the two African rockfowl?
2448. Which bird has been recorded flying at the greatest altitude over Great Britain?
2449. Which European raptor is noted for extreme plumage variation?
2450. How are frigatebirds a threat to sea turtles?
2451. Is the Turkey Vulture the most numerous New World vulture?
2452. What is the status of the introduced flightless Weka rail on subantarctic Macquarie Island?
2453. What are Emu-wrens?
2454. Is a Loggerhead Shrike larger than a Northern Shrike?
2455. Which country is best known for eider duck farms?
2456. At what age do Craveri's Murrelet chicks go to sea?
2457. When over-heated, do storks resort to gular fluttering?
2458. What is the typical clutch size of a Brown Kiwi?
2459. How does Blakiston's Fish-Owl of northeast Asia differ from other fish-owls?
2460. Is the Razorbill a cavity nester?
2461. Birds in which family do the thick-knees (stone curlews) superficially resemble?
2462. How many species of birds are polyandrous?
2463. Which extinct bird was once known as a "hooded swan" or a "bastard ostrich?"
2464. How many male Sage Grouse may gather at a single lek?
2465. How are penguins able to survive in tropical areas such as the Galapagos Islands, South Africa and South America?
2466. On what event was Ken Follet's nonfiction best seller *On Wings of Eagles* based?
2467. Immediately after alighting, what does a Northern Jacana generally do with its wings?
2468. Which ptarmigan has no black in its tail?
2469. How many species of Old World vultures are recognized?
2470. In America, what is the Arctic Skua called?

Questions

2471. How do the feeding habits of owls and diurnal raptors differ?
2472. In the huge order Passeriformes, what is the difference between the oscines and the suboscines?
2473. The eggs of which gamebird are sold in Asian markets?
2474. Do Common Starlings nest on the ground?
2475. Which terrestrial neotropical gamebirds have especially rich vocalizations?
2476. Which is the most conspicuous Falkland Islands passerine?
2477. Is the American Kestrel (Sparrowhawk) related to the Old World sparrowhawks?
2478. Which Central American animal is known as the *chicken of the trees,* and why?
2479. What is peculiar about the incubation behavior of the European Red-legged Partridge?
2480. What is the main function of a Hoatzin's long, broad tail?
2481. Do kiwis call most frequently on moonlit nights?
2482. Why do New Guinea natives keep cassowaries?
2483. What is the Shy Mollymauk also called?
2484. What is a gathering of goldfinches called?
2485. Is the Congo Bay-Owl of Zaïre an abundant species?
2486. Why do the long legs of crane and stork chicks, both of which are capable of flight at about 70 days of age, not lengthen at the same rate?
2487. What is the wingbeat per second of a Common Puffin?
2488. How does a hummingbird feed its chicks?
2489. In New Zealand, which introduced parrot is regarded as an agricultural pest?
2490. Do insects catch and kill birds?
2491. What type of instrument is known as a crane's-bill?
2492. How did the African Bateleur eagle obtain its name?
2493. Cassowary footprints have been compared to the footprints of which animal?
2494. On what do Lesser Flamingos primarily feed?
2495. Why do Australian tiger snakes enter occupied Short-tailed Shearwater burrows?
2496. Which is the only eider with a speculum?
2497. Which of the three gannets is most likely to nest on precipitous cliff faces, rather than on relatively flat terrain?
2498. The Horned Coot of the Bolivian and northern Chilean highlands has what strange nesting habit?
2499. Which is the rarest vulture of Europe?
2500. Why are the largest-known dinosaur eggs smaller than the largest-known bird eggs?
2501. Which grouse has a distinctive curled, lyre-shaped tail?
2502. Which are the only North American thrushes that regularly nest in cavities?
2503. Are penguin flippers *completely* covered with feathers?
2504. Are rails especially vocal?
2505. Who starred in the 1975 movie *Three Days of the Condor*?
2506. What is the favored prey of the Gyrfalcon?
2507. Which is the only storm-petrel that breeds in both the North Pacific and the North Atlantic?
2508. What is the plant *Heliconia* commonly called?
2509. Do bustards commonly rest standing on one leg?

Questions

2510. Do filter-feeding flamingos pump with their tongues?
2511. When nesting in Chinstrap Penguin colonies, what do Snowy Sheathbills typically use to line their untidy nests?
2512. Are all northern cranes migratory, wintering well south of their breeding grounds?
2513. Which New World alcid is the least known?
2514. Which African bird has the slowest reproductive potential?
2515. How much does the second largest penguin weigh?
2516. What do the swamp-dwelling people of Georgia call the Hooded Merganser?
2517. What color is the gular pouch of a California Brown Pelican *early* in the breeding season?
2518. Which birds can move their eyes within the socket?
2519. How many species of birds display in leks or arenas?
2520. Do skimmers swim well?
2521. What is a *cloacal kiss?*
2522. Do foraging Magellanic Plovers dig into the sand?
2523. If its bizarre-shaped bill is discounted, which bird does the tropical American Boat-billed Heron most closely resemble?
2524. In Jamaica, which large raptor is preyed on by the introduced Indian mongoose?
2525. What type of birds are coucals?
2526. Do owls have ceres?
2527. Does the Hadada Ibis nest in especially large colonies?
2528. Once pipping commences, how long does it take a Wandering Albatross chick to emerge from the egg?
2529. Are thick-knees (stone curlews) nocturnal?
2530. Which gull is a southern African endemic?
2531. Which pheasant is restricted to the submontane forests of Borneo?
2532. Which is the commonest breeding stork of northern tropical Africa?
2533. Why do many shorebirds lay four similar-sized, pyriform-shaped eggs?
2534. Which American bird is called "John Crow?"
2535. What state has the most breeding bird species, and how many breed there?
2536. Are all true nuthatches restricted to the Northern Hemisphere?
2537. Is the tip of the lower mandible of a skimmer rigid?
2538. How many species of kookaburras are described?
2539. As of 1990, how many Kirtland's Warblers existed?
2540. Which two birds have the most fearsome weapons?
2541. How long is the small intestine of a Ruby-throated Hummingbird?
2542. Who was Leda of Greek mythology?
2543. Are most seabirds efficient gliders?
2544. Which of the approximately 75 species of true wrens is not restricted to the New World?
2545. Which of the four banded or jackass penguins have only one chest-band, rather than two?
2546. How many species of birds-of-paradise inhabit Australia, and which ones are they?
2547. Is Bernier's Teal a Seychelles Islands endemic?

Questions

2548. What is a Scottish measure of fresh herring, equal to 45 gallons (170 liters), called?
2549. How did the name "skua" originate?
2550. Which three American raptors were the most negatively impacted by DDT and other chlorinated hydrocarbon pesticides?
2551. Which is the only storm-petrel that nests on the mainland?
2552. How did the Boobook (Morepork) Owl of Australia and New Zealand obtain its name?
2553. How long is the tail of an Asian Pheasant-tailed Jacana?
2554. Do female frigatebirds select the nest site?
2555. Do Spoonbill Sandpiper chicks have the distinctive spoon-shaped bill of their parents?
2556. Which is the only flightless bird with naked, blind and helpless young?
2557. Does a Trumpeter Swan have yellow on its bill?
2558. What color are the legs of an Andean Flamingo?
2559. Are Hoatzins powerful fliers?
2560. How many species of plovers occur in North America?
2561. What is the ideal incubation temperature for most penguins?
2562. How many rock hyraxes (dassies) are taken in a single year by a pair of African Black (Verreaux's) Eagles rearing a chick?
2563. Which goose is the most brightly colored?
2564. What is a currawong?
2565. What is the national bird of Finland?
2566. Which North American bird has a breast sponge, and what is its function?
2567. What are turacos generally called in Africa?
2568. Does the Belted Kingfisher prey on birds?
2569. Which precocial birds have the shortest incubation period, and how long is it?
2570. What is the emblem of the British Ornithologists' Union?
2571. What is harrying?
2572. How did the name "penguin" originate?
2573. What is the Vulturine Fish-Eagle more commonly called?
2574. Do terns catch flying insects in the air?
2575. Which is the rarest falcon, and what is its status?
2576. Which bird did the artist Vincent van Gogh paint a few hours before he shot himself?
2577. How many types of birds are mentioned in the Bible?
2578. What is the female Black Grouse commonly called?
2579. Which waterfowl have the most restricted ranges?
2580. Are kiwi vocalizations loud?
2581. How did Funk Island in the North Atlantic obtain its distinctive name?
2582. Are vulture nests generally foul smelling?
2583. How did the Cattle Egret reach Hawaii?
2584. Which American bird is known as a "sea-dog," and why?
2585. Which swan is the tallest and most statuesque?
2586. Do loons copulate in the water?
2587. In most birds, what percent of the total number of eggs laid will result in breeding adults?

Questions

2588. Where do hummingbirds occur west of mainland South America?
2589. Are woodpeckers good songsters?
2590. What type of nesting site is preferred by all perching-ducks?
2591. How many species of penguins typically breed on the two main islands of New Zealand, and which ones are they?
2592. Which is the largest finch of Japan?
2593. What is the state bird of New Mexico?
2594. Do grebes attempt to retrieve eggs displaced a short distance from their nest?
2595. How many species of moas existed?
2596. Which is the only albatross *regularly* observed off the North American coast?
2597. Which sex of painted-snipe constructs the nest?
2598. Do sandgrouse have ceres?
2599. Do jacanas frequent saltwater habitats?
2600. Which are the only birds that feed in the same manner as flamingos?
2601. What color is the head of the Australian Gouldian Finch?
2602. Where in South America does the Gray-headed Albatross nest?
2603. What is the wingspan of the Great Black-backed Gull?
2604. Which is the only North American hummingbird with a distinct decurved bill?
2605. How many species of dippers inhabit Africa?
2606. What specialized technique do Sharp-shinned and Cooper's Hawks sometimes utilize to kill their prey?
2607. How did the term *hoodwinking* originate?
2608. Are cassowaries wary?
2609. Which is the only stork that lays a single-egg clutch?
2610. Which type of birds do jacamars most closely resemble?
2611. Which is the largest stiff-tail duck, and how much do drakes weigh?
2612. What type of bird is a Twite?
2613. Do Tawny Eagles prey on *flying* flamingos?
2614. What is jizz?
2615. What is the length of the stride of a running Ostrich?
2616. Are humans responsible for the scarcity of the Galapagos Lava Gull?
2617. In old falconry terminology, what were hawks trained to hunt herons called?
2618. How did lovebirds obtain their name?
2619. Do *all* petrels and albatrosses lay a single-egg clutch?
2620. After whom was Brandt's Cormorant named?
2621. Which two birds adorn the coat-of-arms of Trinidad?
2622. What was Clark's Nutcracker formerly called?
2623. Why do tropical penguins have bare facial skin?
2624. How did the babblers obtain their name?
2625. What was the former name of the National Audubon Society publication *American Birds?*
2526. Are all egrets white?
2627. What is a tippet?
2628. Which sex of Snowy Owl is more heavily barred?

Questions

2629. Is the Crab Plover exclusively coastal?
2630. Which seabirds are called "vultures of the sea?
2631. How territorial are breeding male honeyguides?
2632. What is one of the favored foods of the Red-throated Caracara?
2633. Do juvenile tropicbirds have long central tail feathers?
2634. Rothschild's Myna is the only endemic on what island?
2635. Which is the largest North American spotted thrush?
2636. Rough-winged Swallows may utilize the nest of which type of bird?
2637. Are most hummingbirds monogamous?
2638. Are Sooty Terns at sea generally encountered in fairly large groups?
2639. Must a cassowary kneel to drink?
2640. Are toucans related to kingfishers?
2641. In England, why is the Red-legged Partridge called a "Frenchmen" or "French partridge?"
2642. Do Tundra (Whistling) Swans whistle?
2643. Which birds feed their young the greatest length of time after fledging, and how long are juveniles fed?
2644. How much of the overall weight of a kiwi is accounted for by its powerful, muscular legs?
2645. What is unusual about the eye color of the Philippine Monkey-eating Eagle?
2646. Do threatened vultures spew up stomach contents?
2647. Why is the Common Nighthawk known as a "bull bat?"
2648. How many species of vultures inhabit Australia?
2649. After whom was Lady Ross' Turaco of Africa named?
2650. The Redhead duck frequently lays its eggs in the nests of other ducks, but which host species is favored?
2651. By merely examining the base of a tree containing breeding hornbills, how can Asian natives determine the date of incubation of the hornbills?
2652. Why is the Ruddy Shelduck regarded as sacred in India?
2653. Why do Greater Roadrunners occasionally eat their young?
2654. The Carrion Crow is considered a subspecies of which bird?
2655. What is the average distance covered by a Ruffed Grouse during a single flight?
2656. In terms of weight, what was the per capita consumption of turkey in the U.S. in 1993?
2657. What kinds of birds are coquettes, mangos, thornbills, sylphs and Incas?
2658. What color is the prominent fleshy knob on the bill of a breeding King Eider drake?
2659. In medieval Europe, which corvid was considered a bird of evil omen?
2660. If a bird is a lutino, what color is it?
2661. Which tropicbird is most widespread and numerous?
2662. How do Great Skuas obtain kittiwake chicks?
2663. Bennett's (Dwarf) Cassowary has how many wattles?
2664. Sage Grouse are definitely polygamous, but do males ever breed with more than a single female on the same day?
2665. Which are the only birds that engage in creophagy?

Questions

2666. How many species of fish-owls inhabit the New World?
2667. Is the cutting edge of the horn-colored upper mandible of a Wandering Albatross lined with black?
2668. Which North American bird gathers in groups and spends the night in a protected depression, with the birds in a circle with their heads pointing outwards, and why?
2669. What was the population of the Florida Everglade (Snail) Kite in 1965, when the birds were nearly extinct?
2670. Do any jays occur in Africa south of the equator?
2671. What does the Laughing Falcon do with a venomous snake prior to taking it to the nest?
2672. Do all hornbills have short legs?
2673. Which bird does the juvenile Masked Booby resemble?
2674. The Indian Peacock is symbolic of what?
2675. Is lanneret a falconer's name for a female Lanner Falcon?
2676. Was *Archaeopteryx* the size of a pheasant?
2677. How many holes do industrious sapsuckers drill daily?
2678. Which is the only North American bird ever to be domesticated?
2679. What popular dance step of the late 1930's had the name of a seabird?
2680. Are grebes solitary breeders?
2681. Do the males of most raptors supply all the food required for their incubating mates?
2682. Are all waterfowl monogamous?
2683. What percent of island rails, living or recently extinct, are, or were, flightless?
2684. How soon after hatching can curassow chicks fly?
2685. What are the names of the four American flyways used by migrating birds?
2686. Was the extinct Elephant-bird carnivorous?
2687. With what do flamingos line their nests?
2688. Which aberrant passerine, currently in a monotypic family (Dulidae), is restricted to the Caribbean island of Hispaniola and adjacent Gonâve and Saona Islands?
2689. Is the Carpentarian Grasswren a well-known Australian bird?
2690. Which bird is the European counterpart of the American Redhead duck?
2691. Which seabird has a bill resembling the bill shape of the Boat-billed Heron?
2692. What color is the bare orbital skin of a Greater Roadrunner?
2693. What color is the down of a Turkey Vulture chick?
2694. Do most hummingbirds have black or dusky-colored bills?
2695. Do any birds use both their wings and feet for underwater propulsion?
2696. What is the more familiar name for the Blue-faced Booby?
2697. Which two waterfowl have greatly reduced foot webs, almost to the point of being nearly non-web-footed.
2698. Why do male Kakapos (Owl Parrot) greatly outnumber females?
2699. What is the oldest *international* conservation organization, and when was it founded?
2700. Which North American diurnal raptor is most skilled at pursuing prey through the forest?
2701. Do Limpkins generally nest on the ground?
2702. How does a Calliope Hummingbird camouflage its tiny nest?

Questions

2703. Are the cutting edges of a gannet's bill serrated?
2704. What form of nest deception does the Penduline-Tit use?
2705. Does heavy rain inhibit birds from singing?
2706. Which is the only carnivorous stiff-tail duck?
2707. Is the head of a juvenile ibis more feathered than that of an adult?
2708. Which is the only small North American owl with dark eyes?
2709. Is the Swallow-tailed Gull of the Galapagos Islands a seasonal breeder?
2710. Can Hawfinches crack open cherry stones?
2711. Excluding the American west, where else does the Scrub Jay occur in North America?
2712. How did the King Vulture obtain its name?
2713. How does the Blue Petrel differ in appearance from the superficially similar prions?
2714. What is the Black-necked Stork called in Australia?
2715. How many species of sheathbills are currently recognized, and name them?
2716. How much longer is the bill of a female kiwi than the bill of a male?
2717. Which waterfowl has been observed flying over the highest peaks of the Himalayas?
2718. Does the Bateleur eagle feed on carrion?
2719. What is the advantage of a deeply-forked tail, such as that of a frigatebird?
2720. Do hummingbirds display in leks?
2721. Do frigatebirds prey on chicks of their own kind?
2722. Why were some of the birds pictured in Audubon's classic work *The Birds of America* in rather contorted poses?
2723. Are both sexes of Blue-eyed (Imperial) Shags vocal?
2724. What type of bird is a baza?
2725. How many species of birds breed in Madagascar?
2726. The *horn* of a Horned Screamer consists of what?
2727. Do grebes feed feathers to their chicks?
2728. What is the function of the small pellets of bright red waxy material that tip the secondaries and, to a much lesser degree, the tail of the waxwings?
2729. What unusual habit does Eleonora's Falcon have with respect to captured prey?
2730. What is a Chiffchaff, and why was it so named?
2731. After what was the Manx Shearwater named?
2732. What position do sleeping Ostriches and rheas assume?
2733. Which was the last crane species discovered and described, and when did this occur?
2734. Which petrels have exceptionally long legs?
2735. Which of the two American dowitchers is most likely to frequent freshwater habitat?
2736. What journal is published by the American Ornithologists' Union?
2737. Does Sabine's Gull migrate close to shore?
2738. Does a female American Kestrel have a barred tail?
2739. Which animal is the national emblem of Australia?
2740. What is unusual about the extinct gallinule of Lord Howe Island (Tasman Sea)?
2741. What is the siphon or crooked pipe used for drawing liquor out of a cast called?
2742. Are the nests of the Old World broadbills inconspicuous?
2743. Which bird-of-paradise is the smallest, and how much does it weigh?
2744. Do many species of corvids store or hide food?

Questions

2745. Which mammalian predator preys on Magpie Geese?
2746. Does a feeding spoonbill chick thrust its bill down the throat of a parent?
2747. Why do raptors migrate during the day?
2748. On what does the Kelp Goose primarily feed?
2749. Do skuas prey on Wandering Albatrosses?
2750. Do jacana eggs have non-shiny shells?
2751. How many species of thick-knees inhabit New Zealand?
2752. How many species of sandgrouse are African endemics?
2753. What do large thrushes often do with prey prior to swallowing it?
2754. Are foraging Australian Gang-gang Cockatoos wary?
2755. How long does a female cassowary care for its chicks?
2756. Which pelican has the blackest wings?
2757. Do Little Auks (Dovekies) breed in Iceland?
2758. Are both Black Vultures and Turkey Vultures often killed on highways?
2759. In falconry, what is a haggard?
2760. What color is the cere of a Black Scoter drake?
2761. How much does a Greater Flamingo weigh?
2762. Was Wilson's Storm-Petrel named for the Antarctic explorer Dr. Edward Wilson?
2763. What states have selected the Western Meadowlark as their state bird?
2764. Why were tityras and becards, long considered with the cotingas, transferred into the tyrant-flycatcher family?
2765. At what time of day or night are Hermit Thrushes most vocal?
2766. Which New World vulture has the widest range?
2767. Does the speculum of a Gadwall have a metallic sheen?
2768. Which tern is the most abundant?
2769. When large areas of Africa were sprayed with organochlorine pesticides such as DDT, what was the primary target species?
2770. How many species of New Guinea birds are endangered?
2771. Which group of large digging pheasants has iridescent, brilliantly-colored metallic plumage?
2772. What is the loud noise produced by feeding flamingos and waterfowl called?
2773. Are swans ever swept over Niagara Falls to their death?
2774. Are the West Indian todies related to motmots?
2775. Why do altricial birds generally nest in trees?
2776. When nesting skuas dive on intruders, how fast are the birds flying?
2777. What did Audubon call the Ivory-billed Woodpecker?
2778. How many of the New World quail are migratory?
2779. Which long-legged wading birds were formerly sought as prey by European falconers?
2780. What happens to the color of an Emu egg with the passage of time?
2781. Is the long, thin tongue of a toucan serrated?
2782. Which alcid is the heaviest, and how much does it weigh?
2783. In Europe, what is the white-phase Gyrfalcon sometimes called?
2784. Do foraging avocets up-end in deep water?

Questions

2785. Which close relative of the Canada Goose is not migratory?
2786. Do Great Egrets occur on all continents except Antarctica?
2787. Are there many breeding species of gulls in the central Pacific?
2788. Do hummingbirds have terete bills?
2789. Are guineafowl eggs thin-shelled?
2790. Why do most small subantarctic petrels, such as prions, generally approach and depart their breeding islands at night?
2791. Are the South American seedsnipe sexually dimorphic?
2792. Do owls lift food to their beaks with one foot?
2793. Which colorful neotropical birds often perch with their long tails cocked off to one side?
2794. Are there more species of birds than any other group of vertebrates?
2795. How many bird families have representatives that are obligate brood parasitic nesters, and what type of birds are in the families?
2796. Which bird is sometimes known as a "falcon-gentle?"
2797. What is the incubation period for the egg-laying duck-billed platypus?
2798. How did the neotropical Sunbittern obtain its name?
2799. How many species of woodpeckers inhabit Australia?
2800. Which is the only flightless bird that uses a lek?
2801. How did the Blacksmith Plover obtain its name?
2802. Which is the only bird known to be poisonous?
2803. Which Alaska seabird has the most restricted range?
2804. Which Arctic-nesting petrel has two color morphs?
2805. Did the extinct Dodo have a long straight tail?
2806. Are birds in the only vertebrate class in which no species give birth to live young?
2807. What is the primary avian prey of the Long-tailed Jaeger?
2808. Which American upland gamebird was successfully introduced into New Zealand?
2809. Do bee-eaters have decurved bills?
2810. Commencing about 1880, until the trade was prohibited, how many bird-of-paradise skins were annually exported from New Guinea for the European feather trade?
2811. Will Turkey Vultures and Black Vultures re-nest if a nest is lost?
2812. Which swan has the most yellow on its bill?
2813. What type of bird does a flying pratincole resemble?
2814. Do Three-banded Plovers nest in the Falkland Islands?
2815. On what do Roulrouls (Crested Partridges) primarily feed?
2816. According to European folklore, how did the Old World Robin obtain its bright red breast?
2817. Are all the neotropical cracids monogamous?
2818. Does the Cinereous Vulture prey on live animals?
2819. How many species of spoonbills are described?
2820. Do birds have eardrums?
2821. Who was the famous soprano of the 1800's popularly known as *The Swedish Nightingale?*
2822. Which group of parrots were formerly called bat parrots?

Questions

2823. Was the Muscovy Duck first domesticated in Mexico?
2824. Are molting South American screamers flightless?
2825. What color is the bare facial skin of a breeding American White Ibis?
2826. Do sheathbills nest in cavities or crevices?
2827. What type of musical instruments were fashioned from vulture bones?
2828. Is the Redhead the largest pochard?
2829. Which hummingbird primary feather is the longest?
2830. How far out of the water can an Adelie Penguin leap to reach an overhanging ledge?
2831. How does a Surfbird protect its nest?
2832. Does the Laughing Kookaburra occur in Tasmania?
2833. Is Gurney's Pitta extinct?
2834. Why were Passenger Pigeons live-trapped?
2835. Do females of most northern ducks outnumber males?
2836. What did the U.S. Air Force do to reduce air strikes between albatrosses and aircraft at Midway Island?
2837. What caused the decline of the African Black-footed (Jackass) Penguin?
2838. How did Trinidad plume-hunters attract hummingbirds?
2839. What are caiques?
2840. With what color is the white plumage of some tropicbirds tinged?
2841. Which New World wren is the most widespread?
2842. Was Cooper's Hawk named for James Graham Cooper, after whom the Cooper Ornithological Society was named?
2843. Which sulid has a black-tipped bill?
2844. Which European bird do African mousebirds superficially resemble?
2845. What is a Franklin's Grouse?
2846. Is the Lesser Yellow-headed Vulture restricted to South America?
2847. Which passerines have a very distinctive delta-wing flight style?
2848. Where is the largest North American colony of Common Murres located, and how large is it?
2849. Do adjutant storks breed in Africa?
2850. Are Cape Pigeons (Painted Petrels) carrion feeders?
2851. Do all ibises have bare faces?
2852. Does the Red Phalarope breed in Iceland?
2853. In terms of volume, when and where did the worst oil spill occur?
2854. Do all waterfowl lay unmarked eggs?
2855. What is the basic coloration of the Painted Stork?
2856. What percent of all species of sandpipers, snipes, turnstones and allies breed in the Northern Hemisphere?
2857. Where is the only "mainland" albatross colony located, and how many albatross breed there?
2858. Are Anhingas or darters likely to be found on salt water?
2859. How long can dippers remain underwater?
2860. Which is the only woodpecker that both regurgitates food for its chicks and carries food in its bill?

Questions

2861. How fast is the American Wild Turkey able to sprint?
2862. Do the newly-hatched chicks of most birds have dark-colored eyes?
2863. What does the expression *as the crow flies* imply?
2864. Which is the only parrot not known to eat seeds or nuts?
2865. How many endemic *species* of waterfowl inhabit New Zealand, and which ones are they?
2866. In birds, what is the dominant sense?
2867. Why do Rooks have bare faces?
2868. Do the neotropical tinamous have long tails?
2869. Which is the rarest shorebird, and what is its status?
2870. Which small North American finch is essentially all blue?
2871. How did the Malleefowl obtain its name?
2872. Was Stephanie of Princess Stephanie's Bird-of-paradise fame from Holland?
2873. What is the national bird of Jamaica?
2874. Excluding their bills, what other body part do woodcocks and snipe use when extracting prey from soft substrates?
2875. How often is the domestic chicken or cock mentioned in the Old Testament of the Bible?
2876. Do the four southern hemisphere diving-petrels migrate?
2877. Which raptor is possibly a link between the fish-eagles and the Old World vultures?
2878. Which North American wren is least likely to cock its tail up in typical wren-like fashion?
2879. Why is the bill of an Australian Shrike-tit so large and powerful?
2880. How high can a cassowary leap from a standing position?
2881. How many species of birds occur in Australia?
2882. The rare Tule Goose is a subspecies of which bird?
2883. Which alcid is named after a large endangered mammal?
2884. Do swifts fly at speeds exceeding 100 mph (161 km/hr)?
2885. Which South American birds are the ecological counterparts of the African Secretary-bird?
2886. What is the clutch size of the New Zealand Wrybill?
2887. Do Cattle Egrets actually ride on the backs of cattle?
2888. What is the primary means used by birds to increase their heat production while at rest?
2889. How do storks communicate?
2890. Do most birds have monocular vision?
2891. Which common American shorebird is said to have a division of labor where the female incubates, and the male cares for chicks?
2892. Which of the two giant-petrels is most likely to follow ships?
2893. Who first exported live Budgerigars from Australia, and when did this occur?
2894. With respect to its eyes, how does a kiwi differ from most other nocturnal birds?
2895. How did the name "cotinga" originate?
2896. Which conspicuous bird is known as a "grand gosier" in Louisiana and the West Indies?

Questions

2897. Do shrikes have a toothed upper mandible?
2898. Is a Sungrebe or finfoot capable of springing directly from the water into the air?
2899. Do lyrebirds inhabit Tasmania?
2900. The population of which formerly scarce North American raptor increased significantly during the 1960's and 1970's, only to decline again?
2901. Do birds outnumber individuals in all other vertebrate classes?
2902. What is the bracket on the side of a ship, such as used for supporting spars, called?
2903. What does the bower of the Fire-maned Bowerbird resemble?
2904. What is the size of the endangered Waldrapp (Hermit or Northern Bald Ibis) population?
2905. Are puffins more closely related to skuas than to petrels?
2906. Who is the most famous personality in the fast-food business?
2907. Is the rhea didactyl?
2908. Australian currawongs have what shrike-like habit?
2909. What were "Hollywood finches?"
2910. What was the original name of the magazine published by the National Audubon Society?
2911. In death, what do birds do that is not typical of other animals?
2912. How many species of nightjars inhabit New Zealand?
2913. What are bezoars, and for what were they used?
2914. Which New World vulture has white eyes?
2915. Which albatross has the subtle coloring of a Siamese cat?
2916. Which raptor is regarded as holy by Tibetan Buddhists?
2917. Which sandpiper spends the most time running?
2918. How many species of petrels were believed extinct, only to be rediscovered later?
2919. With which birds do foraging Galapagos (White-cheeked) Pintail often associate, and why?
2920. Why is the bill of the Malachite Sunbird serrated?
2921. Which two types of seabirds swim so low in the water that their backs are awash?
2922. Do juvenile parrots remain with their parents until the next breeding season?
2923. Do screamers inhabit Trinidad?
2924. Which arboreal hornbill has the longest tail, and how long is its tail?
2925. Why are sheathbills so named?
2926. Can American Robins hear earthworms underground?
2927. What percent of the number of all species of vertebrates do birds represent?
2928. What is a skein?
2929. What is unusual about the tongues of the 11 woodswallows?
2930. In what 1962 World War II movie did Charton Heston star?
2931. Most hummingbirds have how many secondary feathers?
2932. What is the diameter of the base of a Trumpeter Swan nest?
2933. What are euryoecious birds?
2934. Which American vulture is the most gregarious?
2935. Are cuckoo-shrikes related to either cuckoos or shrikes?
2936. What is the status of the Madagascar Pochard?

Questions

2937. Why, after feeding, do Ospreys often flop along the sea surface dragging their feet in the water?
2938. Which group of waterfowl is most likely to establish a huge territory and defend it?
2939. Was the Chukar named for its calls?
2940. At what age do kiwis attain adult size?
2941. Which pigeon or dove is the rarest?
2942. How large is the Abbott's Booby population?
2943. What are the greenish-colored bulbuls often called?
2944. Which waterfowl may feed on the remains of human bodies in the Ganges River (India)?
2945. When did the first commercial Ostrich farm commence operation?
2946. How many of the more than 1,800 species of fleas are known to parasitize birds?
2947. Why do tropical penguins nest in burrows?
2948. What are the beaters called that flush Red Grouse during the big Scottish grouse shoots?
2949. Are Barn Swallows currently more numerous in North America than before the arrival of Europeans?
2950. How do sleeping Ruddy Ducks maintain a stationary position when in the water?
2951. Why do Crab Plovers nest in burrows?
2952. Which is the largest petrel that breeds in Chile?
2953. Can the pendant-like pouch of a drake Musk Duck be inflated with air?
2954. How many times an hour does an American Song Sparrow sing?
2955. What is a Burdekin Duck?
2956. Do oystercatchers actually feed on oysters?
2957. Do pelicans float high on the water?
2958. Which aquatic passerines cock their tails up like a wren?
2959. Do all waterfowl fly with extended necks and trailing legs?
2960. Are Guanay Cormorants tree nesters?
2961. Do Spruce Grouse have engorged combs over their eyes?
2962. Which animal is referred to as the *chicken of the sea?*
2963. How many species of birds occur in New Guinea and its surrounding satellite islands?
2964. Are rollers predominantly green?
2965. Which extant parrot has the most northerly distribution, and in what country is its northernmost distribution reached?
2966. Most birds have bills of what color?
2967. How many of the 31 pittas inhabit Australia, and which ones are they?
2968. Which is the only surviving Indian Ocean flightless rail, and what island does it inhabit?
2969. Does the African Palm-nut Vulture have a naked neck?
2970. Which is the largest Caribbean parrot, and what is its status?
2971. The feathers of which bird denoted rank among Chinese mandarins?
2972. Do Boat-billed Herons have long crests?
2973. Which American bird is known as the "old man of the sea?"
2974. What color is the underwing of a New Zealand Kea?

Questions

2975. The head profile of a Canvasback most closely resembles the head profile of which other duck?
2976. What was the highest price ever paid for a mounted bird, and which bird commanded the price?
2977. Which birds do the Asian snowcocks most closely resemble?
2978. Natives of the Sudan tie the stuffed head of which animal on top of their heads to enable them to more easily stalk game?
2979. How do flying passerines hold their toes?
2980. How many species of birds occur in California, and how many breed there?
2981. Why does the Northern Pygmy-Owl have two black spots on the back of its head?
2982. Which of the 20 bowerbirds are monogamous?
2983. Do all grouse nest on the ground?
2984. How did the Passenger Pigeon obtain its name?
2985. Which large Antarctic petrel has an erratic bat-like flight?
2986. In what year was the first edition of Roger Tory Peterson's *Field Guide to the Birds* published?
2987. Are the South American screamers exclusively vegetarian?
2988. What type of bird is the Bird of Providence, and how did it obtain its distinctive name?
2989. In what 1940 classic black-and-white movie did W. C. Fields and Mae West star?
2990. Which is the most numerous tropical albatross, and how large is its population?
2991. How many of the 11 species of coots inhabit South America?
2992. What part of a captured kittiwake does a Great Skua generally consume first?
2993. How many New World vultures construct nests?
2994. Which New Zealand passerines routinely consume feathers?
2995. What color feathers surround the eyes of most toucans?
2996. Which neotropical birds do the two Australian lyrebirds most closely resemble?
2997. Do male rheas gather harems?
2998. When Europeans first viewed huge megapode nests in northern Australia, what did they believe the nests represented?
2999. In Europe, what is the Bank Swallow commonly called?
3000. Do tropicbirds follow ships?
3001. Which herons are the least gregarious?
3002. Do sandgrouse inhabit Hawaii?
3003. Which animal was for many years considered to be the source of the peculiar calls of the nocturnal Common Potoo?
3004. Where in the West Indies are woodpeckers best represented?
3005. In America, what is the European Black-necked Grebe called?
3006. What happens to the horn-colored bills of Wandering and Royal albatrosses when the birds are cold?
3007. Do newly-hatched crossbills have crossed mandibles?
3008. In nautical slang, what is a blue pigeon?
3009. How many breeds of domestic chickens are known?
3010. How many species of Galapagos mockingbirds are recognized?

Questions

3011. How often are Asian cave swiftlet nests collected annually for the bird's-nest soup trade, and has nest harvesting negatively impacted the birds?
3012. Do thrushes occur in New Zealand?
3013. Are painted-snipe nocturnal?
3014. What is a hummer moth?
3015. Do Eclectus Parrots roost colonially?
3016. What type of nests do the two African rockfowl construct?
3017. What are Emden, Pilgrim and Toulouse geese?
3018. Do puffins occur in Africa?
3019. How many specimens of the extinct Spectacled Cormorant exist?
3020. Which toucan is known from only a single specimen?
3021. Are South American bellbirds related to Australian bellbirds?
3022. Do most Clapper Rails dwell in freshwater marshes?
3023. Do *both* sexes of pigeons and doves produce *pigeon milk?*
3024. What color are the wings of the Eurasian Wallcreeper?
3025. On what does the Mississippi Kite primarily feed?
3026. Where is the entrance of a Hamerkop nest located?
3027. What striking Asian feline is primarily a bird-eater?
3028. Where does the New Zealand Long-tailed (Cuckoo) Koel winter?
3029. Do any terns have bare-colored skin around the eyes like many gulls?
3030. In reality, what are the whiskers of a bird?
3031. Which is the only cavity-nesting tanager?
3032. Which seabird lays the smallest egg relative to its weight?
3033. Which is the only endemic African tern, and what is its status?
3034. Which is the only penguin that regularly breeds on the North Island of New Zealand?
3035. Do insular avian endemics tend to be brighter in color than their mainland relatives?
3036. In India, which bird is called a "King Vulture?"
3037. Which New Guinea parrot has a vulture-like head?
3038. Do crossbills have short fledgling periods?
3039. What is the bill shape of an Australian Pink-eared Duck?
3040. Why do flying juncos flick their white outer tail feathers?
3041. What color does the bill of a Scarlet Ibis change to during the breeding season?
3042. What is the population size of Steller's Sea-Eagle?
3043. Which is the only surviving cormorant with the name of an individual in its common name?
3044. Do both sexes of most cavity-nesting birds incubate?
3045. What is remarkable about the eggs of most brood obligate parasitic-nesting cuckoos?
3046. Do molting birds have higher temperatures than non-molting birds?
3047. What color is the tuft of plush feathers on the back of the head of a breeding-plumaged drake Steller's Eider?
3048. How were hummingbirds collected by plume hunters?
3049. What cheap wine is often consumed by American winos?
3050. Which sex of Andean Condor has a white neck-ruff?
3051. Which American bird is known as a "butterball?"

3052. Which northern gull is named after a country?
3053. Do Oilbirds establish permanent pair bonds?
3054. Do the closely related Wood Ducks and Mandarin Ducks hybridize in captivity?
3055. Which northern predatory birds are called "whip-tails?"
3056. What is the display of a male Wild Turkey called?
3057. Penguins have how many tail feathers?
3058. Does the Great Horned Owl have corniplume feathers?
3059. Which American bird endures the coldest temperature?
3060. Are the feet of the South American screamers webbed?
3061. Are the voices of most hummingbirds high-pitched?
3062. In Florida, with which foraging birds do Belted Kingfishers sometimes associate, and why?
3063. In slang, what is a *kite flier?*
3064. The adult males of which frigatebirds are not totally black-plumaged?
3065. Which are the only birds with skins of commercial value?
3066. Do flamingos fly with extended necks and retracted legs?
3067. Is the Willet a tattler?
3068. Do noddies carry fish crosswise in their bills?
3069. Was the Dodo considered good eating?
3070. Do swifts often perch on telephone wires?
3071. In Norse mythology, what were the names of the two ravens that perched on the shoulders of the god Odin?
3072. Which swimming shorebirds bob their heads backwards and forwards?
3073. Which pelicaniform birds may be flightless during the molt?
3074. Which raptor rides on the back of cattle, and why?
3075. Do birds have many skin muscles?
3076. Do Fork-tailed Flycatchers roost colonially?
3077. Which American seabird is called a "little striker," and why?
3078. Is a female Wandering Albatross whiter than a male?
3079. Which northern gliding mammal may prey on nestling birds?
3080. Which dove occurring in the U.S. has the name of a reptile in its common name?
3081. Which whistling-duck is at risk?
3082. Are both gnatcatchers and gnatwrens restricted to the New World?
3083. Do Brown Pelicans prey on sting-rays?
3084. If a penguin is brownish in color, what does this imply?
3085. On what island do pigeons and doves attain their greatest diversity with respect to size, form and ecology, and how many species occur there?
3086. Which European bird was regarded as the *king of the birds* in medieval times?
3087. Which birds have the most taste buds?
3088. Which bird is known by its Swaheli name of *kanga kanga?*
3089. Why do the flightless rheas have long wings?
3090. Which is (or was) the largest Hawaiian honeycreeper?
3091. Which raptor feeds extensively on mistletoe fruit?
3092. Are most larks boldly patterned?

Questions

3093. How many species of birds occur in Japan?
3094. Do Eurasian Capercaillie and Black Grouse readily hybridize?
3095. Which is the only bee-eater that is not completely carnivorous?
3096. Which shorebird nests near breeding pairs of Golden Eagles in eastern Siberia, and why?
3097. Which nocturnal shorebird was believed extinct, but was rediscovered in India on January 12, 1986, after not being seen since 1900?
3098. Which group of parrots does not engage in mutual preening or partner feeding?
3099. Does a juvenile Royal Albatross have dark plumage?
3100. Which bird is called "mooruk" by New Guinea natives?
3101. Who initially introduced the Indian Peacock to Europeans?
3102. Is the European Bullfinch an agricultural pest?
3103. Besides Australia and New Zealand, where are petrel chicks harvested for human consumption?
3104. In India, which bird is called "nukta?"
3105. Which is the only cormorant that breeds in Costa Rica?
3106. The female of which waterfowl is considered to be more beautiful than the drake?
3107. How did the name "Lammergeier" originate?
3108. In bird lingo, specifically that of raptors, how might avoiding *burning both ends of the candle* be expressed?
3109. What are the dimensions of a duck-billed platypus egg?
3110. What color is the bill of a breeding Whiskered Auklet?
3111. In what 1933 World War I movie did Cary Grant, Fredric March and Carole Lombard star?
3112. What type of bird is an Amazon?
3113. Is the neotropical Boat-billed Heron sedentary?
3114. What is the Maned Duck also called?
3115. Which crane has ornamental white ear-tufts?
3116. Do fulmars have a juvenile plumage stage?
3117. Do Groove-billed Anis associate with cattle?
3118. How long should a 16-pound (7.3 kg) stuffed turkey be roasted at 350°F (177°C)?
3119. What type of bird is a Rosy Pastor?
3120. Which aquatic birds have very dense plumage with a satiny sheen?
3121. Which bird is credited with being the most fluent talking bird on record, and how extensive was its vocabulary?
3122. In what year was the last U.S. Whooping Crane nest seen?
3123. Which are the only temperate zone passerines that *consistently* frequent fast-flowing streams?
3124. Which Old World bird is called an erne?
3125. What was the instrument formerly used in dental surgery for extracting teeth called?
3126. Which group of American birds were formerly known as "greenlets?"
3127. What is the population size of the Snares Islands Crested Penguin?
3128. Is the House Sparrow well established in Greenland?
3129. How do hawk and owl ovaries differ from the ovaries of most other birds?

3130. Which Australian goose perches in trees?
3131. Is hummingbird skin thin and delicate?
3132. Why does a male Indian Peacock quiver its massive train?
3133. What color is the seriously endangered Crested Ibis?
3134. Do *all* petrels feed their chicks by regurgitation?
3135. Which insects are the swiftest fliers?
3136. What is the fledging period of Ross' Gull?
3137. Are Australian lyrebirds powerful fliers?
3138. Which American ornithologist referred to the hummingbird as *a glittering fragment of the rainbow?*
3139. How many species of geese are named after a country?
3140. Do most passerines breed in their first year?
3141. What typically surrounds an Oilbird nest?
3142. Where in Africa do Eurasian Great Bustards breed?
3143. What large, diversified Australasian passerine family includes birds resembling warblers, thrushes, sunbirds, hummingbirds, bee-eaters, corvids and orioles?
3144. Which waterfowl nests adjacent to breeding Gyrfalcons?
3145. Do tropicbirds generally fly low over the sea?
3146. How much does the nest of an Old World Rock Nuthatch weigh?
3147. What position do sleeping toucans typically assume?
3148. Where do most Australian Gannets breed?
3149. Which North American bird was once believed to be a carrier of the disease anthrax?
3150. Which tanager is the most migratory?
3151. Do heron chicks hatch asynchronously?
3152. Do female Wild Turkeys have beards?
3153. Which temperate birds plaster the inside of their nesting holes with mud to reduce the entrance size?
3154. Why are the bones of Pacific Eiders tinted pink or purple?
3155. What type of nest does the Blue-faced Honeyeater of Australia and New Guinea construct?
3156. What is unusual about the greater wing coverts of an Australian Black Swan?
3157. What are rhea flight feathers used for in South America?
3158. Are African turaco nests substantial?
3159. Which is the only pheasant that is not Asian in origin?
3160. Do the young of Royal Albatross and Wandering Albatross require the same length of time to fledge?
3161. Does the nocturnal Kakapo (Owl Parrot) have large eyes?
3162. What is an individual called who tends swans?
3163. Which corvid is the most terrestrial?
3164. Which is the only storm-petrel that is not nocturnal when ashore?
3165. Who was the first elected president of the American Ornithologists' Union?
3166. What is the Hedge-sparrow commonly called?
3167. Does a skimmer chick hatch with a lower mandible that is longer than the upper?
3168. Is the voice of a Greater Flamingo higher-pitched than that of a Lesser Flamingo?

Questions

3169. Which tanager exhibits the most extreme seasonal plumage variation?
3170. Which waterfowl has especially *deep* furrows formed by its neck feathers?
3171. Is a Laysan Albatross more inclined to follow ships than a Black-footed Albatross?
3172. Were the three South American trumpeters named for their vocalizations?
3173. How many species of birds are not known to breed in any other U.S. state except California, and which ones are they?
3174. In terms of the individuals, how many birds live in African deserts?
3175. Which eider is the least numerous, and what is its population size?
3176. What type of bird is the Australasian Magpie?
3177. Which endangered bird is adversely impacted by the Pearly-eyed Thrasher, and why?
3178. Which bird is called a "toad duck," and why?
3179. What color is a male African Amethyst Starling?
3180. When foraging, which bird has been described as . . . *moving slowly, tapping its bill on the ground like a blind person feeling ahead with a stick?*
3181. Which New Guinea parrot is the most threatened?
3182. Are all kingfishers cavity nesters?
3183. Can an albatross fly on windless days?
3184. Which swan has proportionally the longest neck?
3185. Do foraging sandpipers rely chiefly on visual, rather than tactile, clues?
3186. Do all species of cormorants have strongly hooked bills?
3187. Are many mammals named after birds?
3188. Was Franklin's Gull named after Ben Franklin?
3189. Which pheasants display in flight?
3190. Which is the most numerous non-endemic New Zealand land bird?
3191. In Trinidad, which bird is known as a "gray ibis" or a "black ibis?"
3192. What is the incubation period of the neotropical King Vulture?
3193. Which president created the Hawaiian Islands National Wildlife Refuge, and why?
3194. In which group of parrots are females larger than males?
3195. Where does the West Indian Palmchat build its bulky nest?
3196. What is albatross cloth?
3197. How has the shape of the eggs of the extinct Pink-headed Duck of India been described?
3198. Who starred in the hit movie *Good Morning Vietnam*?
3199. Are megapodes powerful fliers?
3200. What surrounds the eye of an adult White (Fairy) Tern?
3201. What does falcated mean?
3202. Do Sooty Tern chicks fledge rapidly?
3203. What color stomach-oil is ejected by Cape Pigeon and Snow Petrel chicks?
3204. The hatching of European Blue Tits is timed to coincide with what annual event?
3205. Are Lesser Flamingos often bred in captivity?
3206. Which group of birds is regarded as the most colorful of tropical American birds?
3207. How often is the International Ornithological Congress held?
3208. At what age do Abbott's Boobies commence breeding?
3209. What type of bird is a Pukeko?

3210. What color are the legs and feet of breeding American White Pelicans?
3211. Are vultures clean birds?
3212. Which is the least numerous Australian waterfowl, and what is its population size?
3213. What organization publishes *The Condor*, and when was it first published?
3214. Do birds have long large intestines?
3215. In Germany, which birds were known as *rackel fowl?*
3216. Do Anna's Hummingbirds display on cloudy or sunless days?
3217. Which New Zealand reptile preys on petrels?
3218. The music from what Simon and Garfunkle hit song was taken from the classic South American composition *El Condor Pasa*?
3219. Which cervid might eat Snow Goose eggs?
3220. What is the average Wild Turkey clutch size?
3221. How many National Football League teams are named after raptors, and which ones are they?
3222. Which is the smallest American east coast alcid?
3223. Which shorebirds can be considered seabirds during the 9-month-long non-breeding season?
3224. When both Black Vultures and Turkey Vultures are feeding on a carcass, which species is dominant?
3225. Which stork might feed at night?
3226. Which bird is featured in the Japan Airlines logo?
3227. Which is the most common raptor of Chile?
3228. Which neotropical bird has strange forward-curling head feathers resembling shavings of glossy black horn?
3229. Is the bill of a Red-legged Kittiwake longer than the bill of a Black-legged Kittiwake?
3230. Which parrot has distinct rictal bristles?
3231. With the exception of humans, which are the only animals that produce rhythmic instrumental sounds, using something *other* than a part of the animal itself?
3232. Does the Limpkin inhabit the West Indies?
3233. Are coucals obligate brood parasitic nesters?
3234. What is a Taiko?
3235. Are most buttonquail (bustard quail) solitary?
3236. How did the western subspecies of the Carolina Parakeet differ from the nominate race?
3237. Which is the only flamingo named after an individual?
3238. Can a Great Black-backed Gull swallow a puffin whole?
3239. What is the favored food of the New Caledonian Kagu?
3240. What color are the rumps of all hanging-parrots?
3241. Do the folded wings of a Northern Goshawk extend beyond its tail?
3242. Which are the only petrels that feed ashore?
3243. What is the contradictory scientific name of the Japanese race of the Great Tit?
3244. Which bird is called "big cranky" in the Carolinas and Florida?
3245. Is the Australian Plains-wanderer especially elusive?
3246. Do owls have crops?

Questions

3247. How many species of brown pelicans are recognized?
3248. Why were lyrebirds formerly hunted?
3249. The American Ring-necked Duck would be more accurately named if it were called what?
3250. Most petrels have what type of odor?
3251. Which is the rarest Alaskan alcid, and what is its population size?
3252. Which bird is the Eurasian counterpart of the Ruffed Grouse?
3253. Which South American swan is especially wary?
3254. In Japan, which bird is known as a "star crow?"
3255. Do Cliff Swallows forage in flocks?
3256. Does a Secretary-bird have long toes?
3257. Which is the commonest sulid of the West Indies?
3258. What is the forward dive called in which the legs are held straight and together, the back is curved, and the arms are stretched out to the sides?
3259. Which bird is called a Thick-billed Penguin?
3260. Do all cranes have short tails?
3261. In India, which birds are called "likhs" and "houbaras?"
3262. Are dark-color phase Ferruginous Hawks more numerous than pale-color phase birds?
3263. Why do carpenter bees sometimes pursue small birds?
3264. Why is the Great Blue Heron sometimes called "Long John?"
3265. When do male Mourning Doves generally incubate?
3266. Which western American gamebird was introduced into Chile?
3267. What is peculiar about the horn of the Horned Coot?
3268. The Stanley (Blue) Crane is most closely related to which bird?
3269. Which is the brightest-colored North American kingbird?
3270. Are the South American seedsnipe arboreal nesters?
3271. Which bird is sometimes called an Antarctic pigeon?
3272. What color are the two long, central tail feathers of the male Ribbon-tailed Astrapia?
3273. Does the Mountain Duck (Chestnut-breasted Shelduck) inhabit the mountains?
3274. Do anis generally forage on the ground?
3275. How many species of Old World orioles breed in Europe?
3276. Which gull may *kill* new-born lambs in Tierra del Fuego?
3277. What color is the nape of a Jackdaw?
3278. What is the current name of the White-bellied Stork of Africa?
3279. Which is the only U.S. state in which the White-faced Glossy Ibis and the Glossy Ibis *regularly* occur together?
3280. Is the nest of a Giant Hummingbird proportionately larger than the nests of other hummingbirds?
3281. In terms of weight, how much material does a pair of Australian Malleefowl shift about annually when adding or removing forest litter to control the temperature in their nesting mound?
3282. Are fish held crosswise in the bill of a flying kingfisher?
3283. Are rayaditos South American flycatchers?
3284. Is the voice of a female Snowy Owl higher-pitched than that of a male?

Questions

3285. Which group of North American gamebirds carries tularemia (rabbit fever)?
3286. In West Africa, which bird is known as a "gon-gon?"
3287. What color is the sparse down of most parrot chicks?
3288. Where does the Great Skua nest in North America?
3289. What did Aristotle believe happened to wintering Redstarts?
3290. Which tanager ranges the farthest north?
3291. During the feather-trade era, how many Snowy Egrets were sacrificed for every ounce (28 gms) of plumes collected?
3292. Is the Greater Scaup confined to the New World?
3293. In South Korea, where is the last major sanctuary for cranes located?
3294. How does the avian aortic arch differ from the aortic arch of mammals?
3295. Do sunbirds construct enclosed nests?
3296. In Florida, which bird is called a "Spanish buzzard," and why?
3297. Which birds may take over the domed nest of an occupied African Hamerkop, ejecting the rightful owner?
3298. Which bird was known as a "pied duck" or "skunk duck?"
3299. How much did the heaviest egg laid by a goose weigh?
3300. Is the Spectacled Owl of South America diurnal?
3301. Do Brahminy Kites snatch fish from the water's surface?
3302. Is the entire head of *both* sexes of adult Red-headed Woodpeckers bright red?
3303. What do Old World kestrels do with slugs prior to consumption?
3304. Which is the only sulid named after an individual?
3305. On what does Sanford's Sea-Eagle of the Solomon Islands primarily feed?
3306. Which American rail migrates the farthest, and how far does it travel one way?
3307. Which southern goose was named after a famous explorer?
3308. Which two starlings are named after a large mammal?
3309. Which tapaculo has the most northerly range, and how far north does it occur?
3310. Do female Golden Pheasants have longer tails than female Ring-necked Pheasants?
3311. Are Australian wattlebirds considered gamebirds?
3312. Is the North American Dipper polygamous?
3313. Which is the largest endemic Hawaiian passerine?
3314. Which is the only pelican with a black tail?
3315. What color are the outer tail feathers of a meadowlark?
3316. Do kingfishers have especially functional tongues?
3317. Which American bird is called a "Labrador twister?"
3318. Do Indian Peacocks feed on snakes?
3319. Which bird and its eggs were specifically protected by King Henry VIII?
3320. Which American passerine is called a "skunk bird?"
3321. On Midway Island, which passerine evicts Black Noddies and White (Fairy) Terns from their nests, consuming their eggs?
3322. When was the first Marbled Murrelet nest discovered in Canada?
3323. Are mousebirds noted for mutual preening?
3324. Do sandgrouse have large supra-orbital salt glands?
3325. Which bird is pictured on the national flag of Zambia?

Questions

3326. Which seabird was named after the largest inland body of water?
3327. Which South American birds are called anvil-birds?
3328. Are the voices of male and female Emus similar?
3329. Which bird do American hunters call a greenhead?
3330. Is the neotropical Jabiru stork a colonial nester?
3331. An American Robin has how many feathers?
3332. What is a bird sonogram?
3333. How is to upset or irritate someone often described?
3334. Do hummingbirds have hollow tongues?
3335. Do pittas locate prey by smell?
3336. What color is the cere of an adult Bateleur eagle?
3337. Why are seriemas kept by South Americans?
3338. Is the Yellow-crowned Night-Heron more nocturnal than the Black-crowned Night-Heron?
3339. Do Sooty Terns feeding chicks carry fish cross-wise in their bills?
3340. Do ibises fly in V-formation?
3341. Which are the only sexually dimorphic tanagers?
3342. What is the small English geranium with a flower shaped like a bird's foot commonly called?
3343. Which is the only one of the four Australian friarbirds without a knob on its bill?
3344. Which bird is known as a "howling jackass?"
3345. Is a Kalij an African partridge?
3346. How do woodswallows often roost at night?
3347. Do turnstones feed on carrion?
3348. Which bird was known by the archaic name *laverock?*
3349. Is the Limpkin a good swimmer?
3350. The slaughter of southern baleen whales was so extensive that some naturalists suggest the removal of so many major krill-consumers has made available enough krill to support how many additional penguins?
3351. Are the rudimentary wings of a Galapagos Flightless Cormorant longer than its tail?
3352. What is the function of a toucan's long, thin, bristly tongue?
3353. Which distinctive family of primarily nocturnal carnivorous birds is named after an amphibian, and why?
3354. Which flightless waterfowl is restricted to the Auckland Islands?
3355. Do Golden Eagles prey on frogs?
3356. How did the neotropical Squirrel Cuckoo obtain its name?
3357. Do all hornbills seal up their nest-cavity entrances?
3358. Do carnivorous birds have longer small intestines than seed-eating birds?
3359. How many swans are named for their calls, or lack of them, and which ones are they?
3360. What is the eye color of the two African oxpeckers?
3361. Which seabird families have flightless representatives?
3362. Do the African turacos feed on plantains or bananas?
3363. Which is the smallest ratite, and how much does it weigh?
3364. Are most terns diurnal feeders?

Questions

3365. How does the Australian Brush-turkey discourage food-seeking competitors, such as monitor lizards?
3366. As of 1990, how many acres of U.S. wetlands have been lost since the American Revolution?
3367. In Britain, what is the Lapland Longspur called?
3368. Excluding humans, what is the only predator capable of reaching the nest of an Asian cave swiftlet?
3369. How much of the planet's land surface does the House Sparrow inhabit?
3370. Are extinct penguins generally larger than surviving penguins?
3371. How many species of storm-petrels breed in California, and which ones are they?
3372. Do American White Pelicans ever roost in trees?
3373. Do Dunlins winter in North America?
3374. Are wrens powerful fliers?
3375. Which North American hummingbird nests earliest in the year?
3376. Do oxpeckers feed on swine ectoparasites?
3377. Are sandgrouse nomadic?
3378. Which northern goose is the smallest, and how much does it weigh?
3379. Why do fledging Common Puffins generally take their first flight at night?
3380. Which country has the most forms of surviving flightless birds, and how many occur there?
3381. In the New World, what is the Slavonian Grebe called?
3382. How much richer in fat is pigeon milk than cow's milk?
3383. What color is the bill of an adult Steller's Sea-Eagle?
3384. Is the Yellowhammer an Old World tit?
3385. Which piciform birds are somewhat ventriloquial?
3386. Which large parrot has a distinct facial disc?
3387. Which caprimulgids are the weakest fliers?
3388. Does Hall's (Northern) Giant-Petrel have two color morphs?
3389. Do all adult waxwings have yellow-tipped tails?
3390. Which passerines have webbed toes?
3391. Does the Japanese Night-Heron have an unusually long bill?
3392. What do flushed incubating eiders often do to discourage egg predators?
3393. Which duck was the first to be banded, and when and where did this occur?
3394. How long do cowbirds brood their nestlings?
3395. Which albatross preys on birds?
3396. Which group of bark-foraging birds is most specialized?
3397. Do White Storks breed in Japan?
3398. In which group of petrels are females larger than males?
3399. How many North American diurnal raptors have the name of an individual in their common name, and which ones are they?
3400. Are pelican chicks vocal?
3401. How rapidly can a flock of feeding vultures strip all of the soft tissues from a large African antelope?
3402. What posture is assumed by a vocalizing kiwi?

Questions

3403. What is the average weight of a Wild Turkey gobbler?
3404. Which sulid lays as many as four eggs, often successfully rearing all the young?
3405. What is the opening or space in a hoisting block or pulley through which a rope passes called?
3406. Which nocturnal, island-nesting birds have often been mistaken for evil spirits, and why?
3407. Is the Harlequin Duck a common Japanese bird?
3408. In Britain, which bird is known as a "peewit?"
3409. Do the neotropical trumpeters have long tails?
3410. To early Christians, the Indian Peacock was symbolic of what?
3411. Is the Capuchinbird a bird-of-paradise?
3412. How much more rapidly does a Rock Dove in flight breathe than when at rest?
3413. Why does the Kakapo not breed annually?
3414. Which type of pheasants are tree nesters?
3415. Which large group of birds has feathers that detach easily, and why is this advantageous?
3416. Prior to the prolonged drought of late 1980's and early 1990's, about 100 million American waterfowl migrated south in the fall, but how many of the migrants made it back to the breeding grounds in the spring?
3417. During the non-breeding season, do Wild Turkeys segregate into flocks of single sex birds?
3418. Which is the only cormorant that constructs a stone nest, like some penguins?
3419. Do Turkey Vultures commonly prey on Great Blue Herons?
3420. Was the extinct Rodrigues Solitaire a long-necked bird?
3421. Do all snake-eagles swallow snakes head first?
3422. Why do ptarmigan follow caribou during the winter?
3423. After whom was Lucy's Warbler named?
3424. An adult Killdeer has how many chest-bands?
3425. Which American long-legged aquatic bird is named after a state?
3426. Which bird was listed first in the U.S. Endangered Species Act of 1969?
3427. What color is the South American Plumbeous Ibis?
3428. Do all weaver birds construct roofed nests?
3429. Which aberrant African corvid is a cooperative breeder?
3430. Do pratincoles hawk insects on the wing?
3431. Which bird is most often represented on national flags?
3432. *Archaeopteryx* had how many primary feathers?
3433. Do antbirds have hooked bills?
3434. Which is the most distinctive and handsome endemic New Caledonian columbid?
3435. Which group of birds was at one time considered so different that they were placed in a separate superorder (Impennes)?
3436. Which large gamebird is considered sacred in many parts of India?
3437. What is *rookooing?*
3438. Is the second egg of the clutch of a crested penguin smaller than the first egg?
3439. Is the Greater Yellow-headed Vulture blacker than the Turkey Vulture?

Questions

3440. Does the Horned Screamer use its horn when fighting?
3441. What was the primary cause of the decline of the Aleutian Canada Goose?
3442. What color is the endangered Rothschild's (Bali) Myna?
3443. Do Arctic Loons occur in the eastern U.S.?
3444. Which heron snaps its mandibles together producing a distinct popping sound?
3445. Which are the only non-passerines with nine primaries?
3446. Does the Wild Turkey occur in California?
3447. When and where were the breeding grounds of Hutton's Shearwater discovered?
3448. Which type of feathers were most valued by the American Plains Indians?
3449. Is the American Woodcock larger than a Eurasian Woodcock?
3450. How does the juvenile plumage of an Abbott's Booby differ from that of an adult?
3451. What is the origin of the name "tanager?"
3452. Is a Cornish game hen a large partridge?
3453. Do foraging cormorants venture far from land?
3454. In American slang, what is it called when one prods suddenly and playfully the backside of an individual, startling them?
3455. Following each breath, how much residual air remains in the avian lungs?
3456. Do toucans prey on nestling birds?
3457. How do walking bustards generally hold their bills?
3458. What is the state bird of South Carolina?
3459. Which bird is on the national flag of Papua New Guinea?
3460. Do the Old World rollers have long necks?
3461. How many species of tits are not cavity nesters?
3462. Why was the Ruddy Duck sometimes called a "dumb-duck?"
3463. What was the press run of the initial 1934 printing of Roger Tory Peterson's *Field Guide to the Birds,* and how long did it take for the first printing to sell out?
3464. How do combative penguins fight?
3465. Which group of three distinctive seabirds can barely walk?
3466. Which seabirds circle high in thermals like raptors?
3467. Do birds yawn?
3468. Which is the only breeding bird extirpated from California since European settlement?
3469. Do Razorbills carry fish head first like murres?
3470. What body part does a megapode use to test the temperature of its incubation mound?
3471. Are all barbets in the same family?
3472. Which bird was formerly called a "morillon?"
3473. What is a swan hook?
3474. Most hummingbirds have how many tail feathers?
3475. Are all nesting albatrosses gregarious?
3476. What is peculiar about the posture of all three South American trumpeters?
3477. Which is the only phalarope that regularly feeds by wading, rather than swimming?
3478. Do female Coscoroba Swans have higher-pitched calls than their mates?
3479. What is the state bird of South Dakota?
3480. What is the difference between a pinnated grouse and a prairie-chicken?
3481. Which American raptor is named after a gamebird?

Questions

3482. Which bird is called "tammie norry" in Iceland and St. Kilda?
3483. Do most buteo hawks hunt from perches?
3484. Which is the only aquatic bird to become extinct in North America in historical times?
3485. Is a tyrannulet a small African flycatcher?
3486. How did sailors get live Great Auks aboard their ships?
3487. Why do griffon vultures have such long necks?
3488. Which tiny Jamaican bird is sometimes called a robin?"
3489. Do woodswallows have swallow-like bills?
3490. What is the head color of a breeding Black-headed Gull?
3491. How many species of birds have been recorded in Panama?
3492. Why does the Common Merganser have white down, whereas the Red-breasted Merganser has dark brown down?
3493. Does a Wallcreeper have strong feet?
3494. Which bird may force an incubating Ostrich to desert its nest?
3495. Which American meadowlark has the superior song?
3496. Which are the only primarily frugivorous birds of the north temperate zone?
3497. What does the old English expression *A goose-quill is more dangerous than a lion's claw* mean?
3498. How did the Cape Sable Sparrow almost make ornithological history?
3499. Do Gyrfalcons and Peregrine Falcons ever nest on the same cliff face?
3500. What color is the beak of an adult Barn Owl?
3501. How many American diurnal raptors are named after states, and which ones are they?
3502. In Great Britain, which shorebird nests at the greatest altitude, and how high does it breed?
3503. How long does it take an Oilbird chick to fledge?
3504. What type of bird was the extinct pterodactyl?
3505. Which are the largest birds that sustain flight wholly through active flapping?
3506. Is a group of titmouses more accurately known as titmice?
3507. Which cuckoo-shrike is named after an insect?
3508. Are guillemots bottom feeders?
3509. What is an owl chick commonly called?
3510. Which noddy is the least pelagic?
3511. How many seabirds perished during the *Torrey Canyon* oil spill at Land's End, England, on March 18, 1967?
3512. Which Eurasian bird is called a "yaffle?"
3513. Why do Osprey carry fish with the head directed forward?
3514. To maintain a constant body temperature, are penguins more dependent on their plumage than their blubber?
3515. How is counting on something that may not materialize sometimes expressed?
3516. When was the Cattle Egret first recorded in California?
3517. Do Turkey Vultures make good pets?
3518. Which American sparrow is called "weary willie?"
3519. Which are the only two land birds of Costa Rica in which males assume a female-type plumage following the breeding season?

Questions

3520. Does the Greater Snow Goose have a blue-color morph?
3521. Which is the only northern albatross with color on its head and neck, and what is the color?
3522. Can small downy crane chicks swim?
3523. How many waterfowl are annually banded in America?
3524. What percent of a griffon vulture's weight can be stored in its capacious crop?
3525. Which is the only bird parasitized by the Screaming Cowbird?
3526. Why do swifts lack the rictal bristles typical of many birds that hawk insects on the wing?
3527. What classic line was penned by Charles Warner, a noted writer of the late 1800's, regarding the abuses of the feather trade?
3528. Is the tip of a spoonbill's bill equipped with sensitive nerve endings?
3529. What is the normal swimming speed of most waterfowl?
3530. Which North American flycatcher preys on birds?
3531. Was the Eurasian Collared-Dove introduced into Europe by humans, or did it become established on its own?
3532. How do perched jacamars hold their bills?
3533. What is the gonys?
3534. Was the extinct Labrador Duck shy and difficult to approach?
3535. Do nuthatches use their tails as a brace when foraging on vertical tree trunks?
3536. Who starred in the 1933 movie classic *Duck Soup*?
3537. Is a bird's nuchal collar located on the neck?
3538. How did the name "auk" originate?
3539. What is the unguis?
3540. Which flat fish is named after a colorful bird?
3541. What do hornbills and toucans commonly do with their food prior to swallowing it?
3542. How many California Condors have been killed by hailstones?
3543. How often do incubating northern hummingbirds leave the nest each day to feed?
3544. What is the wingbeat speed per second of a Ring-necked (Common) Pheasant?
3545. Do harriers prey on bird eggs?
3546. Which American songbirds regurgitate pellets of fur, feathers and other indigestible parts?
3547. What color are the bills of adult oystercatchers?
3548. What is a penguin duck?
3549. Which is the largest North American member of the mockingbird family?
3550. Which North American hawk has feathered legs to the toes?
3551. What is a *yardbird?*
3552. How did the Conservation Department of the New Zealand Wildlife Service save the rapidly vanishing Chatham Islands Black Robin?
3553. Do trogons have "toothed" lower mandibles?
3554. Do swimming cormorants ever cock their tails up?
3555. What color are the wingtips of an Australian Black Swan?
3556. Where is the largest colony of Emperor Penguins located, and how many birds occupy the rookery?

Questions

3557. What do most African antelope do if they see an Ostrich running?
3558. How wary are sheathbills?
3559. Which is the rarest British breeding seabird?
3560. Who was the first biologist to publicly raise the alarm about DDT and related pesticides and the risk to birds?
3561. Are megapode eggs harvested for human consumption in Africa?
3562. Which bird preys extensively on the Galapagos Storm-Petrel?
3563. What is a vocal mobile peddler called?
3564. What was the clutch size of the extinct Passenger Pigeon?
3565. Which common African bird is known as a Dioch?
3566. In America, which pair of birds is known as the "lord-and-lady?"
3567. Which is the only Australasian honeyeater that nests in tree holes?
3568. Which was the first Australian parrot to be illustrated in color, and when did this occur?
3569. Which neotropical raptors follow vultures to carrion?
3570. When did management of the Hialeah racetrack in Miami, Florida, first import flamingos, and from where did the founding stock originate?
3571. Which type of bird does the flight style of the African Hamerkop stork resemble?
3572. Are the heads and necks of Emperor Geese orange?
3573. Do female Keas have longer upper mandibles than males?
3574. How far south does the Mourning Dove range?
3575. Do both red and white-color morph Reddish Egret chicks hatch in the same clutch?
3576. What is the fastest reported flying speed for a waterfowl, and what species was it?
3577. Do Brown Pelicans submerge completely when plunge-diving?
3578. Are tail feathers considered flight feathers?
3579. Can a hummingbird attain top speed as soon as it lifts off a perch?
3580. What color are the feet of a Pink-footed Shearwater?
3581. Which bird is represented on the flag of Uganda?
3582. According to American folklore, what was the cure for a backache?
3583. Is a Turkey Vulture more stable in flight than a Black Vulture?
3584. Flamingo vocalizations sound rather like the calls of which type of birds?
3585. Where in the Old World do Sandhill Cranes nest?
3586. Is the Sage Thrasher really associated with sage brush?
3587. The chicks of which ibises have vestigial claws on their wings?
3588. Are all male hummingbirds brightly colored?
3589. What is the foresail or mainsail of a schooner called?
3590. Do motmots typically perch on exposed branches?
3591. Do rollers consume insects while in flight?
3592. What generally clutters the unlined nesting chamber of a jacamar?
3593. Does the South American Horned Screamer readily perch in trees?
3594. After whom did Joseph Sabine name Sabine's Gull?
3595. Is the Ring Ouzel a dipper?
3596. How many states have selected the Eastern Meadowlark as the state bird?
3597. Why is the White-winged Scoter called a "half-moon eye?"

Questions

3598. Does the Lapland Longspur occur in Lapland?
3599. Do Whooping Cranes commonly rear two chicks?
3600. Do hummingbirds suck nectar using their tongues like straws?
3601. Can Golden Eagles capture birds as large as geese and cranes in *flight*?
3602. Which is the least territorial swan?
3603. Which North American owl has serrations on the claw of the middle toe?
3604. Are dippers odoriferous?
3605. How fast can Wild Turkeys fly?
3606. What was the bacterial disease chlamydiosis formerly called?
3607. What are birders called in Australia?
3608. During the Middle Ages, which bird was often served at formal banquets?
3609. Why did the Guadalupe Storm-Petrel disappear?
3610. In which non-passerine order were the Australian lyrebirds at one time included?
3611. Do birds require more oxygen than all other life forms?
3612. Do trogons have strong feet?
3613. Which cotingas are the largest?
3614. How many alcid species breed in California, and which ones are they?
3615. Which pelican has the longest fledging period.
3616. What is a triller?
3617. How far back in time does the practice of collecting cave swiftlet nests for bird's-nest soup extend?
3618. Do all birds have a single brood patch?
3619. Which is the only stiff-tail duck with a speculum?
3620. The plumage of which group of birds does the plumage pattern of a Eurasian Wryneck resemble?
3621. How did the name "loon" originate?
3622. Is the Least Bittern relatively non-vocal?
3623. What types of sounds do alarmed kiwis produce?
3624. Prior to the conversion to steel shot, how much lead shot fell into U.S. wetlands each fall?
3625. Is the Crab Plover a swift flier?
3626. How do the distensible throat pouches of gannets differ from those of the closely-related boobies?
3627. Does the Black Vulture have a flat flight profile?
3628. Was Carl Linnaeus, the father of the modern biological classification system, an ornithologist?
3629. Which is the only *eastern* North American woodpecker with black-and-white, ladder-like barring across its back and wings?
3630. How many quail species have become extinct in historical times?
3631. How many birds were named after the noted American ornithologist Alexander Wilson, and which birds are they?
3632. Do African Fish-Eagles ever nest on the ground?
3633. Do the neotropical tinamous prey on rodents?
3634. Which type of bird does a Chinese Pond-Heron at rest resemble?

Questions

3635. Which is the only non-gregarious New World vulture?
3636. In Spain, what unusual nest site might be selected by a pair of Marbled Teal?
3637. Which passerines have the most developed sense of smell?
3638. What was the name of a popular early 20th-century ballroom ragtime dance?
3639. Which birds are commonly called *living jewels?*
3640. Why are the three tiny pygmy-geese, which are really perching-ducks, called geese?
3641. Where on a bird is the malar stripe located?
3642. The name "mollymauk" is applied to many smaller southern albatrosses, but which bird was initially called a mollymauk or mollymoke?
3643. Do all neotropical puffbirds have hooked bills?
3644. How do adult male manucodes (birds-of-paradise) differ from all other passerines?
3645. Are Eastern Meadowlarks paler colored than Western Meadowlarks?
3646. Which neotropical raptor is named after a long-legged bird?
3647. Do any mammals have gestation periods as short as the shortest incubation period of birds?
3648. What color is the bill of a breeding Tufted Puffin?
3649. What are the stones on which thrushes pound snails called?
3650. Why do trogons have relatively large eyes?
3651. Which was the first songbird to be honored by a monument, where is it, and when did the dedication take place?
3652. In former times, what did the fishermen of Brittany believe would happen to skippers who treated their crews badly?
3653. How did the Rock Dove (Pigeon) obtain its name?
3654. What is the "official" bird of Washington, D.C.?
3655. How many species of birds did the American ornithologist Frank M. Chapman record on hats while walking about on a single day in New York City, in 1896?
3656. Which is the largest songbird parasitized by cowbirds?
3657. Do storks migrate in flocks?
3658. With respect to its size, which native British bird has the longest tail?
3659. Which country has the most species of hummingbirds, and how many occur there?
3660. Do *birds of a feather* generally flock together?
3661. What color is the crest of a male Macaroni Penguin?
3662. What color are the head feathers of the two African rockfowl?
3663. Do defensive albatrosses spit oil at one another?
3664. Which raptor preys on pelican and flamingo eggs?
3665. How many of the six New World kingfishers occur in Costa Rica?
3666. Do swifts fly during heavy rainstorms?
3667. Which large North American finch has a pair of food-carrying pouches in its mouth enabling it to carry extra food?
3668. Is a cassowary depicted on the official *crest* of Papua New Guinea?
3669. How did the Surfbird obtain its name?
3670. Which bird did American Indians encourage to nest by putting up hollow gourds.
3671. Which are the only toucans known to lodge socially in tree holes throughout the year?
3672. How did the Clapper Rail obtain its name?

Questions

3673. Which American hawk often nests in tall saguaro cactus?
3674. Where in the U.S. do the ranges of Western and Eastern Bluebirds overlap?
3675. What color are Ross' Goose goslings?
3676. What type of birds do Sunbitterns superficially resemble?
3677. Does a fusiform-shaped egg taper at only one end?
3678. What is the upward stroke of a flying bird's wing called?
3679. How does the foraging behavior of treeswifts differ from true swifts?
3680. Are the black primaries of a resting Siberian White Crane visible?
3681. Does the African Martial Eagle have bare legs?
3682. Which bird is most closely related to the Mute Swan?
3683. In Europe, which bird is known as "cuckoo's leader" or "cuckoo's mate," and why?
3684. Was the extinct Pink-headed Duck a prime table bird?
3685. Why are pigeons able to drink without lifting their heads?
3686. Which is the only loon capable of standing upright?
3687. Do most birds have poorly developed salivary glands?
3688. Which is the only heron that occurs north of the Rio Grande River that does not breed in the West Indies.
3689. Do all birds have external nares or nostrils?
3690. Which is the only heron *species* with completely black plumage?
3691. Does the Saw-whet Owl have a distinct juvenile plumage?
3692. Is the Mallard the most numerous wild waterfowl?
3693. Why does the Crane Hawk have such long legs?
3694. The King Penguin occasionally breeds in the colonies of which other penguin?
3695. What is *roding?*
3696. Do African Shoebill storks soar?
3697. Are kiwis swift runners?
3698. Does the Bar-tailed Godwit have a straight bill?
3699. What do skimmer chicks often do to make themselves less conspicuous?
3700. What took place in Transvaal that enabled Martial Eagles and Lanner Falcons to nest where neither raptor previously bred?
3701. What color is the tip of a Keel-billed Toucan's bill?
3702. How did the African helmet-shrikes obtain their name?
3703. Who named Bewick's Wren in honor of Thomas Bewick, the famous English artist and wood engraver?
3704. How can African ground-hornbills be sexed?
3705. What is the flexible fixture for supporting a desk lamp called?
3706. Why do many birds flock?
3707. How many North American rails were named after a color, and which ones are they?
3708. Did *Archaeopteryx* have a short tail?
3709. Which bird is known only to breed on Cabbage Tree Island, Australia?
3710. Which South American parrot has an erectile fan of feathers at the back of its head?
3711. Aside from respiration, what is the main function of avian air-sacs?
3712. How many species of kites inhabit North America, and which ones are they?
3713. How much heavier is an Ivory-billed Woodpecker than a Pileated Woodpecker?

Questions

3714. What happened to the last stuffed Dodo?
3715. Which three groups of forest birds use leks?
3716. How can petrels be attracted at sea?
3717. Which bird has been described as *the goose that nests beyond the north wind?*
3718. Are crossbill chicks fed only by the female?
3719. How does a newly-hatched Northern Fulmar chick treat its parents?
3720. What are the two main threats to the endangered Philippine Monkey-eating Eagle?
3721. Who founded the Wildfowl Trust at Slimbridge, England?
3722. What bird product was formerly carved into ornate snuff bottles?
3723. How long does it take for a Bewick's Swan to fledge?
3724. Is the Great Blue Heron the second largest heron?
3725. How far north does the South American Torrent Duck range?
3726. Is a Yellow-billed Magpie larger than a Black-billed Magpie?
3727. If a bird is said to be apteroid, what does this imply?
3728. Which birds have heads and necks of nearly the same circumference?
3729. Are all pochard ducks accomplished divers?
3730. Which African lake supports the greatest number of birds, and how many birds utilize the lake?
3731. Is the White-breasted Guineafowl an abundant bird?
3732. In which group of northern passerines does the tongue play an important part in extracting seeds from fir cones?
3733. Which is the only heron with a relatively short neck and a wide bill?
3734. Which is the only *resident* crane of India?
3735. In what type of mammal burrow do Hume's (Tibetan) Ground-Jays nest?
3736. After whom was Cook's (Blue-footed) Petrel named?
3737. What do American hunters often call the Scaled Quail?
3738. Which seabird was named after an Indian empire?
3739. Which crane is the most aquatic?
3740. Why was the Western Meadowlark given the specific name *neglecta*?
3741. Is the African Double-toothed Barbet especially gregarious?
3742. Do kingfishers feed on snakes?
3743. Which four birds-of-paradise were named by the Germans after royal personages?
3744. Where did the name "Sanderling" originate?
3745. What happens to quail if the birds consume large quantities of poisonous hemlock seeds?
3746. What color is favored by most foraging hummingbirds?
3747. How many exotic birds were introduced into South America, and how many became established?
3748. Which was the last bird species to be discovered in Hawaii, and what is its current status?
3749. What is peculiar about the plumage of a Sebastopol Goose?
3750. Which rare bird is called Toki by the Japanese?
3751. Who illustrated the birds in the field guide *Birds of North America* by Robbins, Bruun and Zim, first published in 1966.

3752. Do more than 40 species of parrots inhabit Peru?
3753. Are loon chicks commonly lost as a result of sibling rivalry?
3754. Which is the only shearwater that breeds in Panama?
3755. Which South American bird has such a musky odor that natives call it the "stinking bird?"
3756. Which thrasher has a short, straight, slender bill?
3757. Do shrikes ever seize birds in mid-air with their feet?
3758. Do Boreal Owls occur in New Mexico?
3759. What do migrating geese, swans and cranes have in common?
3760. Is a "field martin" a swallow?
3761. What is the popular name for the Sikorsky UH-60A helicopter?
3762. Some tropical seabirds are greatly dependent on what type of predators?
3763. Which drongo is named after a corvid?
3764. What should one do if he or she hears *fore* shouted out at a golf course?
3765. What is a saltator?
3766. Can the huge screamers walk over floating vegetation?
3767. Are Western Hemisphere warblers better songsters than Old World warblers?
3768. How does the white on the wing of an adult Andean Condor differ from the white on the wing of an adult California Condor?
3769. Does a feeding kiwi probe into soft earth with the *full* length of its long bill?
3770. Which caprimulgid, known only from two 19th-century museum specimens, was rediscovered in Brazil in August, 1988?
3771. In what country is the ornithological journal *Notornis* published?
3772. Do incubating hummingbirds become torpid at night?
3773. In the U.S., which venomous reptile is most often preyed on by the Glossy Ibis?
3774. What color is the crown of a Lesser Yellow-headed Vulture?
3775. During May, which is the most abundant bird of California?
3776. Do any cranes have the name of an individual in their common name?
3777. If the egg of a White Tern is balanced on a limb, how are nest exchanges accomplished?
3778. Which aquatic passerines eject pellets of indigestible material?
3779. Was the Lucifer Hummingbird named after the devil?
3780. During spring aerial waterfowl surveys in Alaska, how can a Trumpeter Swan nest be differentiated from a superficially similar Tundra (Whistling) Swan nest?
3781. What color is the breast of a male Coppery-tailed (Elegant) Trogon?
3782. Why was the Australian Letter-winged Kite so named, and what unusual foraging technique does it employ?
3783. With which other albatross does the Black-browed Albatross frequently nest in mixed colonies?
3784. What does the scientific name of the Magpie Goose (*Anseranas semipalmata*) mean?
3785. Are woodswallows timid birds?
3786. Which macaws have become extinct, and when did the parrots disappear?
3787. Is the Ovenbird of North America related to the South American ovenbirds?
3788. Do all trogons have square-tipped tails?

Questions

3789. Where is the southernmost range limit for breeding passerines located?
3790. What color is the bare face of an adult Egyptian Vulture?
3791. Do obligate brood parasitic-nesting cuckoos generally parasitize songbirds?
3792. On what Old World island do Snow Geese breed?
3793. What was the cause of death of Charles Leslie McKay, after whom McKay's Bunting is named?
3794. How did the barbets obtain their collective name?
3795. Which Old World bird is known as a reedling?
3796. For what type of marching were the Nazis known?
3797. What is the national bird of Nigeria?
3798. What type of nest do South American cocks-of-the-rock construct?
3799. The European Goldfinch commonly breeds in which southern country?
3800. Do dippers have large feet?
3801. Do hole-nesting birds tend to have young that develop more slowly than birds with exposed nests?
3802. How much larger is a Hoatzin's crop than its stomach?
3803. Is the Bobolink a warbler?
3804. Which are the only birds obligated to feed with their heads held upside down?
3805. Which is the largest native bird of the West Indies?
3806. How many of the true vireos occur in the Old World?
3807. In dense colonies, how many Lesser Snow Goose nests may be concentrated within a single square mile (2.6 sq km)?
3808. Are barbets generally brightly colored?
3809. Do night-herons have a distinct juvenile plumage?
3810. Which endangered eagle is named after Spain, and what is its population size?
3811. Which is the only North American swallow with a deeply-forked tail?
3812. The Magellanic Plover is probably most closely related to which birds?
3813. How many species of owls inhabit Australia?
3814. Was the Bobwhite named for its call?
3815. Do pittas flick their tails like rails?
3816. Do most waterfowl migrate at night?
3817. What is egg neglect?
3818. Which seabird can fly while still largely covered with down?
3819. Which birds have the largest and fleshiest tongues?
3820. Which bird is said to be able to spit oil through the egg prior to hatching?
3821. How was the alarm sounded when Rome was attacked by the Gauls in 390 B.C.?
3822. Do archer fish "shoot" down small birds?
3823. Which is the only stork that roosts on the ground at night?
3824. Which tit is the smallest, and what is its length?
3825. What physical feature do curassows share with raptors and parrots?
3826. How many skins of the extinct Pink-headed Duck survive?
3827. What avian bone is known as the plowshare bone?
3828. Which ground-nesting owl may commence egg-laying while snow is still on the ground?

Questions

3829. Do any alcids produce more than one brood a season?
3830. Is the endangered Chatham Islands Black Robin of New Zealand long-lived?
3831. Does a feeding oxpecker use its tail as a brace?
3832. When one has eyes expressive of tenderness, how is this often expressed?
3833. Are female woodcocks larger than males?
3834. As of 1990, how many copies of Roger Tory Peterson's bird field guides had been sold?
3835. What is the feeding action of skimmers commonly called in the Old World?
3836. Is avian blood richer in hemoglobin than mammalian blood?
3837. Which two American birds are called "mosquito hawks?"
3838. When was the indiscriminate use of DDT banned in the U.S.?
3839. Who was probably the first individual to captively rear Greater Prairie-chickens?
3840. Which ratites prey on birds?
3841. Why is the American Goldfinch called a "thistle-bird?"
3842. How did the viral disease puffinosis get its name?
3843. On what does the Thick-billed Parrot primarily feed?
3844. To which type of birds are cotingas most closely related?
3845. Was William the Conqueror active in aviculture?
3846. Which crane is most likely to be found near salt water?
3847. Which gull is named after a valued substance?
3848. How many species of birds occur in China?
3849. Which American bird is called "silver tongue?"
3850. Why is the Red-throated Loon able to breed on smaller ponds than those favored by the Arctic Loon?
3851. Which is the most common raptor of Borneo?
3852. Do bustards have an oil (uropygial) gland?
3853. At what time of day or night do kiwis generally vocalize?
3854. The young of which northern goose gather in crèches?
3855. Do Wild Turkeys roost in trees?
3856. Do most species of snipe have aerial displays?
3857. Which American bird is known as a "sky-gazer," and why?
3858. Are Pyrrhuloxias good songsters?
3859. Is the incubation period of most non-parasitic-nesting cuckoos long?
3860. What is the state bird of Alaska?
3861. Do cormorants swallow prey underwater?
3862. Are each of a storm-petrel's external nostrils completely encircled within a separate tube?
3863. Which ratites are vulnerable to forest fires?
3864. Have waterfowl been known to kill humans?
3865. What does the specific name *novaehollandiae* for the Cape Barren Goose indicate?
3866. Do flying peacocks commonly glide?
3867. Does Kittlitz's Murrelet have a longer bill than a Marbled Murrelet?
3868. The Red Crossbill has what type of flight style?
3869. Does the mouth of a swift extend back under its eyes?

Questions

3870. How many species of anis are recognized?
3871. Do Red-headed Woodpeckers store acorns for future use?
3872. What type of person is described as a ravener?
3873. Which group of distinctive New Guinea parrots is named after a carnivore?
3874. Are breeding male Willow Ptarmigan especially combative?
3875. How gregarious are the neotropical potoos?
3876. Do tree-nesting White (Fairy) Terns always select low branches?
3877. Which Alaskan bird is called a "beach goose," and why?
3878. In the famous old English nursery rhyme, how many blackbirds were baked in the pie, and what type of birds are the "blackbirds?"
3879. How are nesting cormorants detrimental to trees?
3880. Which piciform birds have shimmering metallic-green plumage?
3881. On what do African Hamerkop storks primarily feed?
3882. What did eggers do at colonies to ensure that all seabird eggs they collected were fresh?
3883. What is a group of starlings called?
3884. When was the Mute Swan domesticated?
3885. Which penguin nests in rainforests?
3886. Do most birds molt annually?
3887. In what year was the California Condor depicted on a U.S. postage stamp?
3888. When Bahama Woodstars (Caribbean hummingbirds) appear in Florida, are most of the visitors males?
3889. Did the Carolina Parakeet breed south of the U.S.?
3890. Where do Willets nest in South America?
3891. Are northern woodpeckers often observed in flocks?
3892. Which Alaskan auklet is the most numerous?
3893. Is the bill of a South American seedsnipe conical-shaped, like that of a seed-eating finch?
3894. Which group of birds banded by biologists traditionally yields the highest number of band returns, and why?
3895. Is a Sanderling egg larger relative to the size of the female than the eggs of most other shorebirds?
3896. In what year was the Royal Australian Ornithologists' Union (RAOU) founded?
3897. For what was oil from the extinct Great Auk used?
3898. Do northern geese molt twice a year?
3899. Which of the three jaegers is least inclined to harry birds for their food?
3900. Are northern grebe pair bonds formed during the winter?
3901. How do Wood Storks generally forage?
3902. Are hummingbirds known as *butterflies of the bird world?*
3903. Do birds prey on pufferfish?
3904. Are cassowary chicks spotted?
3905. Which exclusively North American-breeding bird has the name of two shorebirds in its common name?
3906. What type of bird is the rare New Zealand Saddleback?

Questions

3907. What was a *goose quill gentleman*?
3908. Are cuckoo-falcons well named?
3909. Do swallows call in flight?
3910. Which myna is best known for its talking ability?
3911. How many skins of the extinct Stephen Island Wren exist?
3912. Does the Oilbird have a strongly hooked bill?
3913. What is the eye color of most adult antbirds?
3914. Which is the most numerous breeding bird of the American prairie?
3915. Do leeches kill Trumpeter Swans?
3916. The Verdin is most closely related to which birds?
3917. Do screech-owls cache food in winter-roosting cavities?
3918. Was Bell's Vireo named after Alexander Graham Bell?
3919. Why are many rails called hens, such as waterhens and swamphens?
3920. What do agitated kiwis sometimes do with their feet?
3921. Which large northern bird was known as a "wobble?"
3922. Relative to its size, which bird has the longest tail, and how long is it?
3923. Which large group of predatory birds has eyes shaded by a bony shield?
3924. Which is the only brown-eyed North American grebe?
3925. What type of nest does a dipper construct?
3926. Do tropicbirds hover momentarily like terns prior to diving on fish?
3927. In military slang, what is an amphibious motor vehicle called?
3928. When was the Wild Bird Society of Japan (WBSJ) founded?
3929. Which motmots do not have racket tails?
3930. If captured by hand, what might a Turkey Vulture do?
3931. Is the Snow Bunting polygamous?
3932. Does the Black Guillemot breed south of the Bering Strait?
3933. What color are male cocks-of-the-rock?
3934. Which North American grebe is the largest, and how much does it weigh?
3935. Which is the only crane that maintains a winter feeding territory, and why?
3936. Which plovers breed in Costa Rica?
3937. When a bird is stricken with a severe case of botulism, what term has been coined to describe the paralytic symptoms?
3938. Does a breeding Gull-billed Tern have a white cap?
3939. Is the Lark Bunting a *native* North American songbird?
3940. What promising, popular young Hollywood personality died of a drug overdose at 1:51 a.m., November 1, 1993?
3941. What bird was named in honor of the late neotropical ornithologist Ted Parker?
3942. What may form on the wings of flying birds and force them down?
3943. Which birds construct the largest tree nests relative to their size?
3944. Are curlews noted for a seasonal plumage change?
3945. Is a Greater Sandhill Crane heavier than a Whooping Crane?
3946. According to Iroquois Indian legend, wild moccasin flowers were considered to be what?
3947. Does the Boat-billed Heron use its bill as a scoop?

Questions

3948. What was the name of the original publication of the Wilson Ornithological Society?
3949. What color is the knob on the bill of a Mute Swan?
3950. In what type of tree does a Kea generally nest?
3951. Can hummingbirds fly upside down?
3952. Is the Yellow-billed Magpie less timid than the Black-billed Magpie?
3953. Which North American gull often nests in trees?
3954. How deep can a Common (Atlantic) Puffin dive?
3955. Why do many North American woodpeckers drill their nest holes facing south or east?
3956. Which tanager resembles a small corvid?
3957. What is the length of the Asian Ibisbill?
3958. Which is the largest wading bird of the Galapagos Islands?
3959. When did the domestic chicken reach Egypt?
3960. Why are stone-curlews called thick-knees?
3961. After whom was Xantus' Murrelet named?
3962. Why do the parents of most altricial chicks remove egg shells from the nest?
3963. In what year was the first American Ornithologists' Union *Checklist of North American Birds* published?
3964. What is the estimated weight of the biomass of all birds occurring south of the Antarctic Convergence?
3965. Does a Ruffed Grouse usually drum at the large end of the drumming log?
3966. Of all vultures, both Old and New World, which is the only true forest resident?
3967. What is a place of execution, such as the gallows, sometimes called?
3968. Do the larger neotropical parrots, such as the Amazons and macaws, pair for life?
3969. Why are all the flight feathers of the larger alcids, such as murres and the Razorbill, molted simultaneously?
3970. How did the Lark Sparrow obtain its name?
3971. Which is the only swan with an entirely feathered face?
3972. How many North American quail have the name of an individual in their common name, and which ones are they?
3973. What physical structure do honeyguide chicks have that nestlings of other birds lack?
3974. What is peculiar about a dipper's nictitating membrane?
3975. What is a cacique?
3976. How long do Sooty Terns live?
3977. Which large, long-legged African bird nests alongside nesting crocodiles?
3978. Do flying dippers ever dive directly into a stream?
3979. On what type of reptiles do Ospreys prey?
3980. How much larger is a male coucal than its mate?
3981. Is the Bronzed Cowbird a common Canadian bird?
3982. What was Strickland's Woodpecker formerly called?
3983. By what other names is Major Mitchell's Cockatoo known?
3984. How many of the 2,600 species of lice parasitize birds?
3985. How many species of pelicans inhabit Africa, and which ones are they?
3986. Which type of bird was raised in what was formerly the USSR to fight like fowl of cockfighting fame?

Questions

3987. Is the Great Argus Pheasant elusive?
3988. What color are *all* the flight feathers of a neotropical King Vulture?
3989. Which is the only freshwater, marsh-nesting Clapper Rail?
3990. Which goose is said to be the most vociferous?
3991. How is the Egyptian Plover misnamed?
3992. Who wrote the World War II novel *Where Eagles Dare*?
3993. How often does an incubating male Emu eat or drink?
3994. Do the Old World white-eyes (silver-eyes) have white eyes?
3995. Which passerines lay the largest eggs, and how much do their eggs weigh?
3996. Why is the Easter-egg chicken so named?
3997. Which galliform birds may share nests?
3998. How did the Shy Albatross obtain its name?
3999. Which raptors are the most vigorous flappers in flight?
4000. What famous frontiersman befriended John James Audubon, who reportedly gave him valuable tips on shooting?
4001. Birds have how many pairs of cranial nerves?
4002. Do plovers probe for food like many sandpipers?
4003. Of the 19 extant grebes, are more than half equatorial or southern hemispheric in distribution?
4004. Which small, greatly threatened Southern California passerine is dependent on riparian habitat?
4005. Is a bird represented on the state flag of Wyoming?
4006. Do any bitterns frequent saltwater habitat?
4007. Which pigeon has a bill superficially resembling the bill of the extinct Dodo?
4008. Which are the only birds that lack an uncinate process?
4009. Where is the Erect-crested Penguin's breeding stronghold?
4010. Is the Mountain Plover a mountain inhabitant?
4011. Did ancient Egyptian hieroglyphic writings contain many bird characters?
4012. Is the echo-locating ability of birds, such that of the Oilbird or cave swiftlets, equal to that of bats?
4013. Do swallows feed at night?
4014. Who directed the 1952 movie *The Big Sky*, starring Kirk Douglas, and the 1951 thriller, *The Thing*, with James Arness?
4015. Do Barn Owls ever plunge into the water after fish like Ospreys?
4016. Chicks of which large, extinct Indian Ocean bird gathered in crèches?
4017. How did the Bufflehead duck obtain its name?
4018. Which heron is a *consistent* canopy-feeder, and what is the purpose of such a feeding technique?
4019. Where are most of the Asian cave swiftlet nests collected for bird's-nest soup sold, and how many pounds of nests are sold annually?
4020. What is the apparatus called that is attached to the front of a hay-mowing machine that has saved the lives of thousands of birds in North America?
4021. Which raptor primaries are the longest?
4022. Which bird is called a "skunk-head coot?"

Questions

4023. Which sandpipers breed in the Southern Hemisphere?
4024. Which Australian bird is known as a Galah?
4025. What is wiffling?
4026. Which are the only North American-nesting gulls with relatively short, thin, unmarked yellow bills?
4027. Which Old World passerines sometimes feed on bats?
4028. How does the crest of a male Javan Peacock differ from that of a male Indian Peacock?
4029. Which gamebird is named after a body fluid?
4030. Australian bird-watchers refer to which bird simply as a BFCS or *"bifcus?"*
4031. Which birds are collectively called hookbills?
4032. What is a bird blind called in the Old World?
4033. What is brailing?
4034. What is a raft?
4035. During the heat of the day, do Ostriches seek shade?
4036. How did the Old World parrotbills obtain their name?
4037. Do cranes fly in V-formation?
4038. Do Carolina Wrens eat snakes?
4039. Which is the smallest North American curlew, and how much does it weigh?
4040. Does the Western Grebe spear its prey underwater?
4041. Which New Zealand birds have shaggy, hair-like plumage?
4042. Do Northern Goshawks reuse the same nest season after season?
4043. Are bitterns nocturnal feeders?
4044. Which shorebird has a large crop that is distended during courtship displays?
4045. Which long-legged birds have a brush-like mass of lamellae inside their bills?
4046. Are the horns of a Horned Lark always conspicuous?
4047. Which penguin nests in such a manner that breeding pairs are not visible to one another?
4048. Which woodpeckers are the most efficient climbers?
4049. When male frigatebirds display with inflated pouches, what is this sometimes called?
4050. Is the sheath of a Snowy Sheathbill more prominent than that of the Black-faced Sheathbill?
4051. Is the Wrybill wary?
4052. How often is the call of a Whimbrel generally repeated?
4053. Are any of the herons sexually dimorphic?
4054. Are Australian butcherbirds related to shrikes?
4055. The song of which tiny western songbird is heard on the sound-tracts of many movies filmed in southern California?
4056. What type of ectoparasites might be found in the throat (gular) pouches of pelicans and cormorants?
4057. Do Chimney Swifts actually roost in chimneys?
4058. What was the name of the awkward and credulous schoolmaster in Washington Irving's *Legend of Sleepy Hollow?*
4059. Are all the neotropical puffbirds cavity nesters?

Questions

4060. What is peculiar about a kiwi's droppings?
4061. Are the Old World rollers accomplished walkers?
4062. Do *both* sexes of the neotropical White-headed (Blue-throated) Piping-Guan have a white crest?
4063. Why is the southern population of the Marbled Murrelt considered engangered?
4064. Which moth has the greatest wingspan?
4065. Which North American cuckoo produces mechanical sounds by snapping its mandibles together?
4066. Which is the most common vulture of eastern Africa?
4067. How rapidly is a hummingbird's tongue moving when the bird is extracting nectar from a flower?
4068. Which three American waterfowl undertake the longest migrations?
4069. Do barbets cling, woodpecker-fashion, to tree trunks?
4070. What world-famous impressionist painter was sometimes inspired by birds?
4071. In 14th-century England, which raptor was flown only by falconers of royalty?
4072. What term describes a set of eggs?
4073. Which are the only two hummingbirds with upturned bills?
4074. What is a group of teal called?
4075. How did the name "titmouse" originate?
4076. Do treeswifts perch in trees?
4077. As befits their larger size, do martins have deeper voices than swallows?
4078. Do dark-color phase parasitic jaegers tend to migrate farther south than pale-color phase birds?
4079. According to American folklore, what malady could be cured if Turkey Vulture skull bones were strung around the neck?
4080. How far do the calls of an American Bittern carry?
4081. Which American bird is called a "French mockingbird?"
4082. What is the oldest avicultural organization, and when was it founded?
4083. How many ibises have the name of an individual in their common name?
4084. Why was the Greater Prairie-chicken given the specific name *cupido?*
4085. Why is the African Harrier-Hawk "double-jointed?"
4086. How many species of pratincoles or coursers inhabit the Western Hemisphere?
4087. Who wrote the earliest known systematic accounts of bird behavior?
4088. Is the Yellow-crowned Night-Heron primarily a bird of marine habitats?
4089. Which two endangered pheasants are restricted to Taiwan?
4090. Do Sulphur-crested Cockatoos congregate in large flocks?
4091. What does the old saying *A fox should not be on the jury when a goose is tried* imply?
4092. In breeding plumage, which is the whitest North American songbird?
4093. What do the flags of Austria and Bolivia have in common?
4094. After whom was Scott's Oriole named?
4095. Do Northern Harriers roost in solitude?
4096. With what is the bulky, domed nest of a lyrebird lined?
4097. Which are the only countries that have penguins breeding along the entire coastline?
4098. What is peculiar about the egg tooth of an American Woodcock chick?

Questions

4099. Do Arctic-nesting shorebirds migrate farther than waterfowl?
4100. Which raptor may rob a Brown Pelican of its catch?
4101. When and where were wild Carolina Parakeets last observed?
4102. What is a *puddle duck?*
4103. Which type of herons may attempt to stab one another during aerial combat?
4104. What is the incubation period of an American Robin?
4105. Which white-eye of Borneo has a contradictory common name?
4106. In poultry jargon, what does booted imply?
4107. How many species of lapwings are crested, and which ones are they?
4108. When and where was the Eurasian Tree Sparrow introduced into North America?
4109. To which birds are trogons most closely related?
4110. Which booby is most inclined to plunge-dive into extremely shallow inshore waters?
4111. The chicks of which tern have particularly well-developed claws, and why?
4112. What is unusual about the bill of a Kagu?
4113. Do jaegers dig lemmings out of the ground with their feet?
4114. Courting males of some of the larger species of which bird family appear to turn themselves inside out?
4115. Including seabirds and the birds of Cocos Island, how many species of birds have been recorded for Costa Rica, and how many are permanent residents?
4116. What is a training airplane with small wings and a small motor, but cannot actually be flown, called?
4117. Chicks of which puffin have dark ventral down?
4118. Which is the smallest wood-warbler, and how much does it weigh?
4119. How do lovebirds transport nesting material?
4120. Where were Wild Turkeys first domesticated?
4121. With which mammals is the Trumpeter Swan often associated?
4122. What is a chlorophonia?
4123. Should the feathers of both wings be cut when clipping a bird to render it temporarily flightless?
4124. On what do wood-warblers primarily feed?
4125. Which type of birds are most closely related to the African turacos?
4126. Which American bird is called a "wart-nosed wavy?"
4127. How long are the spurs of a Wild Turkey?
4128. Which is the largest American goose, and how much does it weigh?
4129. According to superstition, what would happen to a farmer's cow if he destroyed a swallow's nest?
4130. In grebes, what is vigorous preening sometimes called?
4131. What famous American conservationist wrote *Everyone knows that the autumn landscape in the north woods is the land, plus a red maple, plus a Ruffed Grouse. In terms of conventional physics the grouse represents only a millionth of either the mass or the energy of an acre. Yet, subtract the grouse and the whole thing is dead.*
4132. Why were Gray Herons protected during the Middle Ages?
4133. If a Brewer's Blackbird has a yellow eye, what sex is it?
4134. Do the large, flightless Weka rails of New Zealand have degenerate wings?

Questions

4135. Is the American Yellow Rail larger than a sparrow?
4136. Which American gull may construct a floating nest?
4137. Do most falcons have dark-colored eyes?
4138. Which crane is the least aquatic?
4139. Do gnatcatchers often flick their tails up and down?
4140. What do Michigan, Wisconsin and Connecticut have in common?
4141. What color is the crest of the Crested Partridge (Roulroul)?
4142. Concentrating falcons have what owl-like habit?
4143. What is the function of the large, naked, grotesque throat sac of the Marabou Stork?
4144. Who was the most famous medical personality of the Crimean War (1854-1856)?
4145. Which member of the gull family requires the greatest amount of time to attain breeding age?
4146. Why do the bills of geese have distinct serrations on the upper mandible?
4147. Does the Turkey Vulture have a longer, slimmer tail than the Black Vulture?
4148. With the exception of humans, which other animal consumes the saliva nest of the Asian cave swiftlets?
4149. Which American-breeding shorebird is named after a metal?
4150. What is the wingspan of a Ruby-throated Hummingbird?
4151. What is the name of the Indian journal that is the most important local medium for the dissemination of knowledge of Indian birds?
4152. The black on the yellow breast of a meadowlark is in the shape of what letter?
4153. Which American heron is most likely to run about in shallow water chasing prey?
4154. Does a Razorbill ever carry two or more fish crosswise in its bill?
4155. Why are Anhingas and darters able to strike prey so rapidly?
4156. Do Crested Caracaras nest at high altitudes?
4157. When and where was the first International Ornithological Congress held?
4158. What type of nests do frogmouths construct?
4159. What is counter singing?
4160. Which bird was formerly known as Richardson's Skua?
4161. How did the name "dowitcher" originate?
4162. What is candidiasis, a mycotic infection of the digestive tract of birds, also called?
4163. Are there any documented records of a California Condor attacking a live animal?
4164. Do Bronzed Cowbirds frequently associate with cattle?
4165. How much longer is a male Riflebird's bill than that of a female?
4166. Were live Dodos ever sent to Europe?
4167. What do calling Crested Caracaras do with their heads?
4168. Are loon chicks more agile ashore than adults?
4169. Can swifts take off from level ground?
4170. Which wren commonly nests in cholla cactus?
4171. Do any wild geese have a dewlap or feathered wattle?
4172. Which New Guinea bird has shiny, black, coarse, bristle-like, drooping plumage?
4173. Which is the heaviest bird-of-paradise, and how much does it weigh?
4174. If a person walks with his or her feet turned in, what is this condition called?
4175. Are all 24 drongos glossy black?

Questions

4176. Which heron is the most widespread?
4177. Why are North American hummingbirds attracted to red flowers?
4178. Do shrikes impale their prey with the heads directed up?
4179. How does a flying pelican position its bill, and why?
4180. In the U.S., what is the Mexican Grebe called?
4181. How many species of woodcocks have been described?
4182. Does the nocturnal feeding Oilbird forage in groups?
4183. Are lyrebird chicks raised in the domed nest for the entire fledgling period?
4184. Do any alcids have brightly-colored plumage?
4185. To many Americans, the first whistle of which upland gamebird is the harbinger of spring?
4186. Why are ticks generally restricted to a bird's head and neck?
4187. Do the salivary glands of Asian cave swiftlets serve a digestive function?
4188. What is the function of the "beard" of downward-pointing bristles on both sides of the beak of a Lammergeier?
4189. How many chicks does the Hooded Grebe of Patagonia rear per nesting attempt?
4190. What is eider yarn?
4191. Which heron has a distendable gular pouch?
4192. Which crane has the longest bill relative to its size?
4193. What was the first California Condor sanctuary called, and when was it established?
4194. Do wrynecks climb vertical tree trunks?
4195. Dippers have what type of plumage?
4196. Where was the breeding range of the Labrador Duck?
4197. Which is the largest owl of Japan?
4198. In ancient Greek legends, what were sirens?
4199. Which heron is most apt to fly in V-formation?
4200. Which is the only crested shorebird of South America?
4201. Do owls have long necks?
4202. Do adult pratincoles and coursers feed their chicks?
4203. Do male geese have higher-pitched calls than females?
4204. Which diurnal African raptors were named after canines?
4205. How long does it take for male birds-of-paradise to assume full adult plumage?
4206. Which American cuckoo has a wide vocal repertoire?
4207. How were flying birds beneficial to Columbus?
4208. What color is the speculum of a Bufflehead duck?
4209. What do disturbed or alarmed nightjars do with their eggs or chicks?
4210. Are the wing bones of a swallow more massive than those of a swift?
4211. Which African dove has the same name as a musical instrument?
4212. Which American-breeding tern has still not fully recovered from the turn-of-the-century slaughter by plume-hunters?
4213. Is the Asian Ibisbill a lowland resident?
4214. Which are the only three herons that *frequently* forage away from water?
4215. Which abundant British bird is known as a "cushat?"
4216. Is the Great Gray Owl among the most nocturnal of North American owls?

Questions

4217. Where do Siberia-nesting Sandhill Cranes winter?
4218. Relative to the size of the egg, which two groups of birds lay eggs with the largest yolks, and what percent of the egg consists of yolk?
4219. Of the 42 states where Chukar partridge were introduced, in how many did the birds become established?
4220. Are shearwaters noted as ship followers?
4221. What color is the face of a juvenile Ivory Gull?
4222. Are caracaras migratory?
4223. The polymorphic goslings of Ross' Goose are yellow or gray, but what color predominates?
4224. Is the Roseate Spoonbill the only spoonbill in which the plumage is not white?
4225. Which of the three trumpeters is restricted to Brazil?
4226. Why is the Glaucous Gull called an "owl gull?"
4227. Is the bill of a toucan hollow?
4228. Who "discovered" the Tundra (Whistling) Swan?
4229. How long after copulation are turkey hens still capable of laying fertile eggs?
4230. At what time of year does the Australian Pelican breed?
4231. Which is the most numerous wintering goose of Great Britain?
4232. How many species of birds occur in Puerto Rico, and how many are regular breeders?
4233. Do Darwin's finches construct open nests?
4234. What is the origin of the name "Budgerigar?"
4235. Which North American swallow is most sexually dimorphic?
4236. Which North American goose nests in trees?
4237. Where in the U.S. does the White-crowned Pigeon occur?
4238. What is the origin of the name of the Nankeen Night-Heron and Nankeen Kestrel?
4239. Which is the smallest British seabird?
4240. Which insect-eating swifts might forage at night?
4241. If a bird is pied, what does this imply?
4242. Wintering Golden-crowned Sparrows frequently associate with which other sparrow?
4243. When emerging from an underwater dive, can a murre or puffin literally explode out of the water and become airborne?
4244. Is the Hawaiian Crow holding its own?
4245. When was cockfighting banned in England?
4246. Are there many saltatorial birds?
4247. When was the Whooping Crane given federal protection?
4248. What position do sleeping African mousebirds often assume?
4249. After whom was Grace's Warbler named?
4250. With the ancient Mayans and Aztecs, what was the penalty for killing a quetzal?
4251. Kublai Khan had how many Gyrfalcons?
4252. Which jaeger breeds the farthest north?
4253. Which are the two smallest British birds , and which one constructs the smallest nest?
4254. How many species of bee-eaters nest in tree cavities?
4255. How many species of birds have been introduced into Australia, and how many became established?

Questions

4256. Which bird of the Bible was called a "glede?"
4257. Do any African fish-owls have ear tufts?
4258. How many North American-breeding shorebirds have the name of an individual in their common name, and which ones are they?
4259. Is the Madagascar Serpent-Eagle extinct?
4260. How do overheated nightjars cool themselves?
4261. Why do Harlequin Ducks and harlequin beetles have the same name?
4262. What color are the eyes of an Oilbird?
4263. What is the throat color of a male Costa's Hummingbird?
4264. Are chickadees tits?
4265. Are frigatebird nests substantial?
4266. Do Ostriches inhabit Australia?
4267. Which North American oriole is the largest, and what is its length?
4268. In New Zealand, what do small triangular-shaped holes in the ground indicate?
4269. Are spadebills Old World flycatchers?
4270. If someone is run out of town on a *rail*, with what is he covered?
4271. Does the Shoebill generally fledge more than one chick?
4272. What is the rhamphotheca?
4273. Has more than a single Bachman's Warbler ever been sighted at the same time during the past quarter-century?
4274. How many of the approximately 64 species of Greenland birds depart the island entirely during the winter?
4275. After whom did Audubon name the Common Poor-will?
4276. Which North American woodpecker may dump its eggs into the nests of other birds?
4277. Is the Least Bittern restricted to the Western Hemisphere?
4278. In the North American bird-watchers world, what is the 600 Club?
4279. Do caracaras soar?
4280. Are night-calling birds prone to be more vocal when the moon is full?
4281. Where is the most famous Northern Gannet rookery located?
4282. Which of Australia's two lyrebirds is the superior mimic?
4283. What causes the loud claps when a pigeon takes off?
4284. What does the gull generic name *Larus* mean?
4285. Do birds ever get drunk?
4286. Which raptor is the most numerous and obvious in the warmer parts of the Old World?
4287. What are House Finches commonly called?
4288. Do chicks call from within the egg prior to pipping?
4289. Do hummingbirds have long, sharp, hooked nails?
4290. Are wintering Black Turnstones restricted to the west coast of North America?
4291. What is a francolin?
4292. Is the Jamaican Streamertail a declining species?
4293. Does a breeding Royal Tern have rosy underparts?
4294. What are parrotbills also called?
4295. Do hole-nesting birds lay larger clutches of eggs than birds with exposed nests?
4296. Which bird attacks adult Emus?

4297. Do any birds lay more than one egg within a 24-hour period?
4298. On which reptiles do some Galapagos Islands birds perch?
4299. Which kingfisher is named after a long-legged bird?
4300. Are Black Phoebes generally encountered in dry habitat?
4301. Males of some species of which group of birds pass food to their mates in the air?
4302. Do lapwings have longer legs than plovers?
4303. Can dippers swim on the surface of the water?
4304. What type of nest site is utilized by a Laughing Falcon?
4305. Does a frogmouth have a large tongue?
4306. Who played the leading man in the 1957 movie *The Wings of Eagles,* directed by John Ford?
4307. Was falconry widely practiced in Great Britain *prior* to the Norman conquest?
4308. Which waterfowl were named after states?
4309. How did the two Asian adjutant storks obtain their name?
4310. Which is the smallest passerine?
4311. Do sandgrouse commonly perch in trees?
4312. Do all North American quail attain sexual maturity in their first year?
4313. Which American wood-warbler has the most specialized feeding technique?
4314. Which eagle-owl is primarily insectivorous?
4315. Who initiated the Christmas bird counts, and in what year were the counts started?
4316. How does the chick behavior of obligate brood parasitic-nesting African whydahs differ from chick behavior of obligate brood parasitic-nesting cuckoos?
4317. Which adult boobies do not have white bellies?
4318. Do juvenile birds wander more extensively than adults?
4319. Are the tropical American solitaires flycatchers?
4320. How did bird's-eye maple get its name?
4321. Which bird was known as an "Ember Goose" or "Immer Goose?"
4322. In the Orient, how are Tree Sparrows generally cooked?
4323. What was the American folklore cure for heart disease?
4324. Which is the only caprimulgid that hatches naked?
4325. Which American bird is called a "robin dipper?"
4326. Why do sleeping birds tuck their bills into their feathers?
4327. How did the Cattle Egret obtain its specific name *ibis?*
4328. Why does the Gray Gull of South America fly at night?
4329. At what angle is the body of a hovering hummingbird held?
4330. Was the Zitting Cisticola named after an Australian river?
4331. Is a dipper capable of rising straight out of the water after a dive and flying off?
4332. How much of the weight of a newly-hatched Ostrich chick consists of the unabsorbed internal yolk sac?
4333. How many species of dowitchers are recognized?
4334. Which bird is the eastern U.S. counterpart of the Black-headed Grosbeak?
4335. What is the size of the American Gyrfalcon population?
4336. Did *Archaeopteryx* have teeth in both jaws?
4337. Why are young boys hired by golfers in India?

4338. Is a kiwi egg thick-shelled?
4339. What is the synsacrum?
4340. When and where was the "extinct" Campbell Island Flightless Brown Teal rediscovered, and what is its status?
4341. Do foraging plovers rely chiefly on tactile clues?
4342. Kittlitz's Murrelet chicks often reach the sea by what different means than other alcids?
4343. Do falcons ever overtake a bird from below, flipping upside down to snatch it out of the air with their feet?
4344. Which is the smallest North American grebe, and how much does it weigh?
4345. How do gauchos (South American cowboys) catch rheas?
4346. Why are many shorebirds swift runners?
4347. How does a jacana keep its eggs dry in the damp, floating nest?
4348. Do loons swimming on the surface bring both feet forward simultaneously?
4349. Does the African Hamerkop have a hooked bill?
4350. Are all swallows diurnal?
4351. Is the bill of a plover swollen?
4352. What is the normal flying speed of most passerines?
4353. Which is the only corvid with well-developed salivary glands, and why?
4354. Where did the ancient Greeks believe kingfishers nested?
4355. Which gull was named after a fish?
4356. In Latin America, which birds are often called banana-birds, and why?
4357. Do the Australasian woodswallows build substantial nests?
4358. What percent of a Sharp-shinned Hawk's diet consists of birds?
4359. Which American shorebird has the name of a cervid in its common name?
4360. Why was there a bounty on New Zealand Harriers?
4361. Which bird is known as a "speckle-belly" by American hunters?
4362. Do Turkey Vultures have a syrinx?
4363. Does the kiwi have a high body temperature?
4364. What were the circumstances surrounding the discovery of the Spectacled Cormorant by Georg W. Steller in 1741?
4365. Do hanging-parrots sleep upside down?
4366. Do Franklin's Gulls commonly follow the plow?
4367. In the 18th century, what did many people believe swallows did during the winter?
4368. After whom was the Argus Pheasant named?
4369. Which North American seedeater is best adapted to desert conditions?
4370. Is the Wrentit migratory?
4371. What does a Shovel-billed Kingfisher (Kookaburra) nest resemble?
4372. What is a munia?
4373. Which is the only North American bird that is completely red ventrally with a blue head?
4374. With respect to anatomy, what do white-eyes and honeyeaters have in common?
4375. What type of behavior is called *tidbitting?*
4376. Which American falcon is notorious for frequency of coition?

Questions

4377. Which loon can be decoyed by waving a cloth and shouting?
4378. Do frigatebirds prey on jellyfish?
4379. Do American Robins ever nest on the ground?
4380. Which bird is said to drive large African vultures off carcasses?
4381. How many of the 42 species of jays inhabit the New World?
4382. Are grouse commonly maintained in captivity?
4383. Are the African rockfowl regarded as gamebirds?
4384. Which birds are known as thermometer birds, and why?
4385. How does the African Bateleur eagle catch birds?
4386. The down from how many Common Eider nests is required to equal one pound (0.45 kg) of down?
4387. What initially limited the American Tree Swallow population?
4388. Do oxpeckers hang upside down on African ungulates?
4389. When was the Cooper Ornithological Society founded?
4390. Which loon is most gregarious during the non-breeding season?
4391. How many species of kingbirds occur in North America, and which ones are they?
4392. Which birds are mentioned most often in the Bible?
4393. What is the feathery or hairy crown of some seeds, such as the down of a thistle, sometimes called?
4394. Which Old World shorebird continues caring for its young *after* fledging?
4395. Does the Palm Warbler generally nest in palm trees?
4396. Are the Australian lyrebirds frugivorous?
4397. How do overheated hummingbirds cool off?
4398. Why do many herons have *kinked* necks?
4399. Do all larks nest on the ground?
4400. What body part do ptarmigan shed during the winter?
4401. Do Limpkins fly often?
4402. Which waterfowl are most inclined to mutually preen?
4403. What color are juvenile King Vultures?
4404. Is the huge nest of the Hamerkop plastered with mud?
4405. What color are male Northern Harriers?
4406. Are *all* Hawaiian honeycreepers endemics?
4407. Why do rollers congregate at African grass fires?
4408. Do all mousebirds have pronounced crests?
4409. Do most passerines have dark brown eyes?
4410. Do caracaras have long thin wings?
4411. Which woodpeckers deliver the weakest blows when pounding?
4412. Most falcons lay reddish-colored eggs, but what color are those laid by the Pygmy Falcon and falconets?
4413. What is the most southerly record for a shorebird, and which bird was it?
4414. What is the status of the San Clemente Island (California) Sage Sparrow?
4415. Why did the New Brunswick game department close the hunting season on American Woodcocks in 1970?

Questions

4416. Does the white morph of the Lesser Snow Goose tend to nest earlier than the Blue Goose?
4417. Are Great Crested Flycatchers cavity nesters?
4418. Why does the Wrybill plover's bill curve to the right, and not to the left?
4419. Which spectacular birds nested on the Sun Life Building in Montreal, Canada, between 1936 and 1952?
4420. Are sandgrouse legs feathered to the feet?
4421. Do juvenile Sunbitterns resemble adults?
4422. Pellets from which birds are especially resistant to decay?
4423. How many species of Asian sandpipers and plovers have been recorded in the Aleutian Islands, Alaska?
4424. How tall is an African Secretary-bird?
4425. To which birds are the two neotropical peppershrikes most closely related?
4426. Which group of small, non-passerine, African birds is considered an agricultural pest?
4427. Do swifts lay round eggs?
4428. Do the crests of titmice remain erect at all times?
4429. Which stiff-tail duck has been introduced into England?
4430. In Wetmore's more traditional taxonomic arrangement, how many non-passerine families contain only a single species?
4431. Do grebes yodel?
4432. Which American bird produces a sound called *thundering?*
4433. Which neotropical bird is known as a "curassow hawk?"
4434. How fast can Gambel's Quail run?
4435. Since the inception of the federal duck stamp program, which three waterfowl have been represented most often on a stamp, and how many times has each been pictured?
4436. Which bird is featured on the national flag of Fiji?
4437. What color is the yolk of a flamingo egg?
4438. Which North American gull has a *slightly* forked tail as a juvenile?
4439. What is the internal temperature of a megapode nest mound?
4440. Do Australian Black-breasted Buzzards drop stones from a height on Emu eggs to break them open?
4441. Were chickens introduced into the New World by Spanish explorers?
4442. What does raptor mean in Latin?
4443. Does Australia have more species of parrots than Brazil?
4444. When was the Aleutian Tern first recorded in the Aleutian Islands?
4445. What do Lewis' Woodpeckers do with acorns prior to stuffing them into tree-bark crevices?
4446. Is the Long-tailed Tit a tame and confiding bird?
4447. Which group of alcids lays the heaviest eggs relative to their weight?
4448. Are sunbirds polygamous like most hummingbirds?
4449. Is there an audible snap of a flycatcher's bill when it snaps up a flying insect?
4450. Was the extinct Labrador Duck considered good eating?
4451. Do all three phalaropes have lobed toes?
4452. Are goldfinches noted for clean nests?

Questions

4453. According to American folklore, what was the ground-up bill of a bittern said to induce?
4454. Bonaparte's Gull is named after what city?
4455. What is peacock coal?
4456. What is the spinning feeding style of phalaropes called?
4457. The calls of an Elf Owl have been compared to the vocalization of which animal?
4458. What is Cuvier's beaked-whale also called?
4459. Which large African wading bird utters sounds described as resembling the bellowing of a calf or the barking of a baboon?
4460. A plump refers to a group of which type of birds?
4461. Which shorebird is named after a group of native Americans?
4462. Which birds are in the last non-passerine family?
4463. Which mammals may be preyed on by potoos?
4464. Which three penguins are known as brush-tailed or long-tailed penguins?
4465. Which American bird is called a "tarweed canary?"
4466. What hunting technique does the African Pale Chanting-Goshawk use?
4467. How does a flushed American Bittern hold its legs?
4468. Do tits often hang upside down?
4469. Are most starlings cavity nesters?
4470. All sulids have what shaped tail?
4471. Why does the Wrybill have a laterally deflected bill?
4472. Are the neotropical toucans relatively non-vocal?
4473. Are neotropical tapaculo nests conspicuous?
4474. Does Darwin's Woodpecker Finch have a long, woodpecker-like bill?
4475. What man-made object does the communal nest of the Social Weaver resemble?
4476. At what age do murre chicks go to sea?
4477. Are nightjars soft-plumaged?
4478. What is the incubation period of the Limpkin?
4479. Are Greater Yellowlegs more gregarious than Lesser Yellowlegs?
4480. How did the name "nuthatch" originate?
4481. Does the Long-tailed Jaeger have two color phases?
4482. Does a migrating hummingbird lay on a heavy layer of subcutaneous fat?
4483. In bygone days, how were American winter roosts of crows often destroyed?
4484. Which American-breeding bird is called a "quawk," and why?
4485. Do American Ruddy Ducks ever nest in close association with breeding mammals?
4486. The primarily vegetarian African mousebirds are especially fond of which insects?
4487. Other than the Inca Tern, which tern may nest under cover?
4488. How long does it take a male Superb Lyrebird to develop its distinctive tail plumes?
4489. On what do wrynecks mainly feed?
4490. On which American gamebirds do Northern Goshawks prey extensively?
4491. Which North American hummingbird may reuse its nest of previous seasons?
4492. Are plovers less dependent on water than lapwings?
4493. When nightjars move their eggs or chicks, how is this accomplished?
4494. Does a penguin chick have a special name, such as duckling, gosling, cygnet or owlet?

Questions

4495. Which are the only passerines lacking fused clavicles, thus lacking wishbones?
4496. Do hornbills catch insects in flight?
4497. Are swallows among the earliest fall migrants in the temperate zone?
4498. Which continent is the barbet stronghold?
4499. What do singing grassquits frequently do?
4500. Do nighthawks nest on roof tops?
4501. How large is the hind toe of a thick-knee (stone-curlew)?
4502. Do treeswifts construct large nests?
4503. Is the Marabou Stork a colonial nester?
4504. What mammal dwelling is used by nesting Common Shelducks?
4505. Do Merlins nest on the ground?
4506. What is the bitter liquid called that remains after the crystallization of salt from brine?
4507. What is the origin of the name "becard" for some of the small cotingas?
4508. Who named Queen Victoria's Riflebird, one of the birds-of-paradise?
4509. Which penguin has the longest bill, and how long is it?
4510. Why do pygmy-parrots have stiff tails?
4511. Is the New Zealand South Island (Rock) Wren a powerful flier?
4512. Which heron is the *most* terrestrial?
4513. How was the Northern Goshawk re-established in Britain?
4514. How did the Magnolia Warbler obtain its name?
4515. Are kingfisher eggs oblong?
4516. Which American bird is called a "feathered wildcat?"
4517. Is the left lobe of the bi-lobed avian liver larger than the right lobe?
4518. Do the Old World drongos have exposed nostrils?
4519. Do pratincoles have relatively short legs and bills?
4520. What percent of Palearctic migrants to Africa remain north of the equator?
4521. When and where was the last Passenger Pigeon nest seen?
4522. Are Old World orioles insectivorous?
4523. In what 1939 movie about heroism and cowardice among British officers serving in the Sudan did Ralph Richardson star?
4524. How high have flying Secretary-birds been recorded?
4525. Are swifts known for the age-old reputation as harbingers of spring?
4526. Did *Archaeopteryx* have hollow bones?
4527. Are puffbirds active birds?
4528. Which vireo is the most widely distributed?
4529. Are Northern Harriers polygamous?
4530. How are Trumpeter Swan parents responsible for cygnet mortality?
4531. Who made the observation *one swallow does not make a spring?*
4532. Are grebes especially vocal?
4533. Which North American woodpecker has the name of a man-made object in its common name?
4534. Which loon is the most widely distributed?
4535. Do tinamous have decurved bills?
4536. Are the neotropical woodcreepers cavity nesters?

Questions

4537. Are the South American seriemas powerful fliers?
4538. How does the Common Grackle open acorns?
4539. Has any single zoo or institution bred all 15 species of cranes?
4540. Is the courtship display of a male Scissor-tailed Flycatcher primarily aerial?
4541. In American slang, what is Thanksgiving day often called?
4542. Which is the most numerous wintering gull of southern U.S. coasts?
4543. What is unusual about the clutch size of a lyrebird?
4544. Does the female Anna's Hummingbird have any red on its throat?
4545. What does the expression *eagle-eyed* imply?
4546. How much does a large Osprey nest weigh?
4547. How often does an incubating potoo turn its egg?
4548. Do female kiwis ever incubate?
4549. Can the South American trumpeters swim?
4550. How many species of birds breed in the Arctic?
4551. Do Ocellated Turkeys have beards?
4552. What color are the wattles of the rare New Zealand Kokako?
4553. What is a hedge called with a ditch on one side that is too high for a horse and rider to jump?
4554. What does the name of the Belgium journal *Le Gerfaut* mean?
4555. How did the Antarctic explorer, Sir Ernest Shackleton, know he was nearing the subantarctic island of South Georgia in 1916, during his famous open-boat journey from Elephant Island?
4556. How many species of neotropical sharpbills are described?
4557. Which parrot is regarded as one of the most musical birds of the New Zealand forest?
4558. How did the harriers obtain their name?
4559. Skins of Black-necked Swans were formerly used for what?
4560. Which is the least numerous American alcid, and what is its population size?
4561. When standing, which storks hold their heads hunched down between their shoulders?
4562. Do mousebirds have an erratic, weaving flight style?
4563. Which Australian bird is called a "gray jumper," and why?
4564. Which large group of aquatic birds has tails that are loose and hair-like?
4565. How does the African Black-chested Snake-Eagle hunt?
4566. Are all the tropical cotingas large birds?
4567. Which birds have the largest hearts relative to their size?
4568. Do wrynecks dig out their own tree-nesting holes?
4569. Do bustards fly high?
4570. How many macaws have the name of an individual in their common name, and which ones are they?
4571. Which bird was sacred to Pallas Athene, the Greek goddess of wisdom?
4572. Do all eggs in a grebe's clutch hatch within a few hours of one another?
4573. Why have guineafowl not been introduced more often as gamebirds?
4574. Are all pipits almost entirely insectivorous?
4575. Which northern goose has the most luxurious down?
4576. Do Ospreys nest on the ground?

Questions

4577. How many species of shrikes are native to South America?
4578. Do cranes frequently glide?
4579. Which long-legged aquatic North American bird was nearly hunted to extinction?
4580. Is the Common Raven most abundant south of the tree line?
4581. Do foraging Wrybills always circle in a counterclockwise direction?
4582. Do sheathbill chicks remain in their nesting crevice until fledged?
4583. Which huge seabirds nest on neatly cut lawns and golf courses at Midway Island?
4584. Which is the heaviest swift, and how much does it weigh?
4585. What color is the tail of an African Gray Parrot?
4586. The chicks of which North American-nesting gulls remain in the nest until fledged?
4587. Do the Australian lyrebirds fly often?
4588. How long do dipper chicks remain in the nest?
4589. What color is the down of a newly-hatched mousebird?
4590. What is a group of curlews called?
4591. Which icterid has the most northerly distribution?
4592. Is the long hind claw of a Horned Lark curved?
4593. Why are American Wigeons and Gadwalls less affected by lead poisoning than many waterfowl?
4594. Where did the word "lek" originate?
4595. Do female Harpy Eagles weigh twice as much as their mates?
4596. What was reportedly one of the favored foods of the Tudor king, Henry VIII?
4597. Are Barn Owl legs completely feathered to the feet?
4598. Which Old World birds regarded as hummingbird counterparts are named after an arachnid?
4599. What type of insects does the generic name of the Ivory-billed Woodpecker suggest that the bird favors?
4600. Which bird in flight has been described as a flying cross?
4601. Do most hummingbirds have dark-colored eyes?
4602. Are all fringillid finches monogamous?
4603. Do pittas always nest on the ground?
4604. Did the Carolina Parakeet have a long pointed tail?
4605. What organization publishes *The Elepaio*?
4606. How many *species* of albatrosses have the name of an individual in their common name, and which ones are they?
4607. Are barbets powerful fliers?
4608. How did the European Hawfinch obtain its name?
4609. Which American grebe was adversely impacted by pesticide pollution?
4610. Do American Coots have twice as many feathers as Bald Eagles?
4611. In Japanese folklore, the turtle is said to live ten thousand years, but how long is the crane supposed to live?
4612. Why do bee-keepers dislike kingbirds?
4613. Do avocets build more extensive nests than stilts?
4614. Which American birds are known as bay-ducks?
4615. Do all passerines hatch naked and helpless chicks?

Questions

4616. Which North American tanager tears open paper-wasp nests to feed on larvae, pupae and adult wasps?
4617. How did the cocker spaniel obtain its name?
4618. Does the tiny Elf Owl have an exceptionally loud voice?
4619. Most passerines have how many cervical vertebrae?
4620. Do bustards have long tails?
4621. Which type of Old World shorebird commonly stands on tip-toes to obtain a better view?
4622. Do Scissor-tailed Flycatchers migrate in flocks?
4623. Which merganser has the shortest bill?
4624. In flying hummingbirds, which wing stroke provides the power?
4625. Can sandgrouse swim?
4626. Why does a shrike capture mice with its bill rather than its feet?
4627. Who first released goats in Hawaii that ultimately destroyed critical avian habitat?
4628. What color are the legs and feet of an adult Ross' Gull?
4629. How many endemic European birds have become extinct in historical times?
4630. After whom was Albert's Lyrebird named?
4631. Are Sunbitterns solitary?
4632. Does a Gyrfalcon have rapid wingbeats?
4633. Are the smallest of the parrots Asian birds?
4634. Does the Rough-legged Hawk feed on carrion?
4635. Do Ostriches bathe regularly?
4636. Which is the only honeycreeper that occurs in the U.S.?
4637. Has the largest American upland gamebird been introduced into Hawaii?
4638. When approached in their burrows, why do some penguins, such as Magellanic and Humboldt, twist their heads from side to side?
4639. Are waxwings especially gregarious?
4640. Are chickens considered to be pheasants?
4641. Are fruit-pigeons arboreal?
4642. Do screech-owls actually screech?
4643. Why might some people think the Bearded Reedling is related to the parrots?
4644. Which American bird is called a "little meadowlark?"
4645. Are bustards wary birds?
4646. Which eagle is most commonly used in falconry?
4647. Do potoos occur in the West Indies?
4648. What physical feature is shared by larks, pipits and wagtails?
4649. What color do hummingbirds appear to be in poor light?
4650. Do Red-legged Kittiwakes follow ships?
4651. How many North American woodpeckers are named for their head color, and which ones are they?
4652. Are newly-hatched sandgrouse chicks down covered?
4653. Why should more than one hummingbird feeder be provided in a backyard?
4654. If a person has an aquiline nose, what does this imply?
4655. Do Dovekies remain near the ice-pack during the winter?

Questions

4656. Was Forster's Tern named after a global explorer?
4657. When large numbers of live cocks-of-the-rock were shipped from South America during the mid-1960's, why were so few females exported?
4658. Do owls often feed on carrion?
4659. How did the Macaroni Penguin obtain its name?
4660. How many species of birds occur in Guatemala?
4661. Which corvids are *most* apt to store and hoard food?
4662. How many Ring-necked (Common) Pheasants do U.S. hunters shoot annually?
4663. Which are the only two nocturnal, inland-breeding alcids?
4664. Do American Kestrels prey on flying mammals?
4665. Does the Pied-billed Grebe tolerate American Coots and Common Moorhens in its breeding territory?
4666. What color are the hindnecks of all three gannets?
4667. How long is the lyrebird fledgling period?
4668. In which group of parrots do both sexes incubate?
4669. How much longer is the middle toe of an Osprey than its other toes?
4670. Why do American Goldfinches nest in the late summer?
4671. By what other name is the Lesser Rhea known?
4672. Do ticks plague Antarctic penguins?
4673. What is the prize called, usually a ridiculous one, that is awarded to whoever had done worst in a game, race, or contest?
4674. Do Ospreys prey on shorebirds?
4675. When was the first bird-of-paradise skin taken to Europe?
4676. Which is the largest tern of the central Pacific?
4677. Which two bird families are restricted to the West Indies?
4678. How do the winter movements of Blue Grouse differ from the movements of most other highland birds?
4679. How did the ancient Romans receive word of Caesar's conquests of Gaul?
4680. How did the Swallow-Tanager obtain its name?
4681. Why does the Verdin make no attempt to conceal its nest?
4682. In feudal Japan, what did royal purple jesses on a Northern Goshawk signify?
4683. How many species of nuthatches occur in the Western Hemisphere, and which ones are they?
4684. What color are nightjar eyes when struck by a beam of light at night?
4685. Which cranes lay unmarked eggs?
4686. Which are the only three U.S. states in which the Crested Caracara is found?
4687. Which of the pochard ducks is the most marine?
4688. How large are the eyeballs of a Snowy Owl?
4689. Do bustards build substantial nests?
4690. After whom was the Calliope Hummingbird named?
4691. In the most recent taxonomic revision, the loons are placed between which two families?
4692. Which birds have the most pairs of powder-down feathers?
4693. Do pygmy-owls catch small birds in flight?

Questions

4694. Which birds copulate on the backs of moving mammals?
4695. Which North American tanager retains its color year-round?
4696. Do hornbills prey on eggs and young birds?
4697. Why has Bachman's Warbler all but disappeared?
4698. Do Shoebill storks feed on carrion?
4699. What is the caeca, and what is its function?
4700. Are Craveri's Murrelet chicks at sea attended by *both* parents?
4701. Which African raptor is sometimes regarded as an aberrant snake-eagle?
4702. Do Ospreys ever feed on dead fish?
4703. Which birds were called Numidian birds by the ancient Greeks and Romans?
4704. What unusual bird did Audubon keep as a pet?
4705. Which American woodpecker is named after a nut?
4706. The two tiny eggs of a Vervain Hummingbird constitute what percent of the female's weight?
4707. Do the monotremes (duck-billed platypus and echidnas) incubate their eggs in a pouch?
4708. Are Bananaquits fond of bananas?
4709. When a Northern Goshawk strikes a bird in the air, does the raptor generally hang on and ride it to the ground?
4710. Which extinct bird could have been the first victim of environmental pollution?
4711. Was Virginia's Warbler named after the state of Virginia?
4712. Which American raptor is most feared by aviculturists?
4713. Which is the most commonly hunted grouse?
4714. Does the Northern Harrier have a gray rump?
4715. Which tribe of South American Indians is responsible for many bird names that have common usage in English?
4716. Do incubating Yellow-billed Loons face the water?
4717. Is the endangered New Caledonian Kagu a lowland bird?
4718. Are African Red-billed and Yellow-billed Hornbills timid?
4719. Are griffon vultures colonial nesters?
4720. Do oystercatchers have especially loud calls?
4721. What does the cuckoo-shrike family name Campephagidae mean?
4722. Which cetacean is known as a sea canary, and why?
4723. Why is the shell of a kiwi egg less porous than the egg shells of most other birds?
4724. Do Greenland and Siberian-nesting Baird's Sandpipers migrate through the U.S.?
4725. Do skuas ever settle on the water?
4726. Besides having only three toes, what other feature separates the Black-backed and Three-toed Woodpeckers from all other North American woodpeckers?
4727. Do *both* sexes of courting cranes dance?
4728. Are there any brightly-colored nocturnal birds?
4729. Do mousebird chicks remain in the nest until fledged?
4730. For what did ancient Roman physicians prescribe Ostrich gizzard stones?
4731. What color is the head of an adult Australian White-headed Egret (Pied Heron)?
4732. Which North American dabbling duck is most prone to dive?

Questions

4733. Are sandgrouse considered good gamebirds?
4734. On which hand do western falconers wear the leather gauntlet used for carrying raptors?
4735. Do most crakes have short bills?
4736. Does North America have the worst record of any comparable land mass for exterminating its bird life?
4737. Is the bill of a Great Egret black?
4738. Do pigeons have bills that are thickened toward the tip?
4739. Why do kingfisher chicks approaching fledging have a hedgehog-like appearance?
4740. Who described and gave scientific names to the Turkey Vulture, Bald Eagle and Pileated Woodpecker?
4741. How do jacanas threaten one another?
4742. Why were shrikes so named?
4743. Which is the most numerous crane, and what is its population size?
4744. What physical feature of the Aleutian Canada Goose distinguishes it from the other races of Canada Geese?
4745. Which of the two yellowlegs is the most approachable?
4746. Why were hummingbirds heavily exploited during the 19th-century feather trade?
4747. How is to grow rich by taking advantage of the circumstances sometimes described?
4748. Was the Stonechat named because of its habitat preference?
4749. Do Peregrine Falcons ever catch fish?
4750. Which are the only animals above reptiles on the evolutionary scale that do not depend on their own body heat to incubate their eggs or brood young?
4751. What was the primary reason for the decline of the Everglade (Snail) Kite?
4752. How many species of quail inhabit southern South America?
4753. Did Peruvian guano island wardens regard the sighting of an Andean Condor as a good omen?
4754. Are nighthawks active during the day?
4755. Do loons call while in flight?
4756. Which American bird is called a "tadpole duck?"
4757. At sea, what does the smell of ammonia indicate?
4758. What was the ultimate fate of Meriwether Lewis, of the Lewis and Clark expedition and Lewis' Woodpecker fame?
4759. How many species of parrots occur in China?
4760. Do Barn Owls scream?
4761. How many of the 1,850 species of African birds are resident south of the Sahara?
4762. Which North American nightjars were named for their calls?
4763. What is the major predator of unattended Xantus' Murrelet eggs and chicks on Santa Barbara Island, California?
4764. Does the Yellow-billed Magpie generally live near water?
4765. What is unusual about the head of an Australian Partridge Pigeon?
4766. How heavy a fish can an Osprey carry?
4767. Are the tropical Old World orioles good mimics?
4768. How rapidly is a feeding Greater Flamingo's tongue pumped?

Questions

4769. How do superstitious sailors regard the albatross?
4770. How does the breeding behavior of Spotted Sandpipers differ from most sandpipers?
4771. What do most pigeons do with their necks when vocalizing?
4772. Do Le Conte's Thrashers generally feed in low vegetation?
4773. Are the Eurasian choughs noted for aerial acrobatics?
4774. How many species of galliform birds can swim?
4775. Do Western Grebes ever carry their chicks overland?
4776. Do sandgrouse have feathered toes?
4777. What is typically located just outside the nest of an Australian Noisy Pitta?
4778. The members of which distinct group of South American birds were formerly believed to be sparrows?
4779. What is the extinct Spectacled Cormorant also called?
4780. What position do courting cock Ostriches assume?
4781. When did the Glossy Ibis commence breeding in the U.S.?
4782. Does the Boat-billed Heron prey on small mammals?
4783. How high into the air do dancing Sandhill Cranes leap?
4784. How did the tragopans obtain their name?
4785. Were Great Auk chicks taken to sea by their parents?
4786. Who initially introduced the Common (Ring-necked) Pheasant into England?
4787. After her death, what happened to the carcass of *Martha*, the last Passenger Pigeon?
4788. After whom was Brunnich's Guillemot (Thick-billed Murre) named?
4789. How far south do Black Skimmers winter?
4790. Do skuas rear two young?
4791. How did the name "siskin" originate?
4792. Despite control measures claiming up to a billion African Red-billed Queleas a year, how does the population maintain such incredible numbers?
4793. Males of which ptarmigan have a black eye stripe during the winter?
4794. How many of the 19 storks are migratory, and which ones are they?
4795. How many American grouse were named for their color, and which ones are they?
4796. Are most parrots powerful fliers?
4797. Do Green Sandpipers nest in association with mammals?
4798. Are many of the bulbuls crested?
4799. Are African mousebirds relatively silent birds?
4800. What color are the wing flight feathers of a cock Ostrich?
4801. Is the Ancient Murrelet one of the most abundant offshore murrelets of the U.S. Pacific coast?
4802. What do aviculturists call the Mexican Elegant Quail?
4803. What is the striped wrasse (a fish) also called?
4804. Are Ospreys ever found far from water?
4805. Do the South American seriemas nest on the ground?
4806. Are jaeger bills particularly hard near the base?
4807. Which is the only U.S. state where the Colima Warbler occurs?
4808. Do hanging-parrots build nests?
4809. Do foraging Brown Pelicans generally dive downwind?

Questions

4810. Why are Ostriches not regarded as good draft animals?
4811. Which are the only gregarious New World cuckoos?
4812. In what year was the Bald Eagle given federal protection in the lower 48 states?
4813. After hatching in its subterranean nesting mound, how much sand must a Malleefowl chick dig through to reach the surface?
4814. Is the New Zealand Brown Teal nocturnal?
4815. Does a neotropical King Vulture fly with its wings held at a dihedral?
4816. Is the Jamaican Owl restricted to Jamaica?
4817. Are Northern Jacanas polymorphic?
4818. How can Common Nighthawks be sexed?
4819. In terms of percent, how long is the bill of a female Stewart Island Brown Kiwi relative to the size of the bird?
4820. Does the South American Hoatzin have a crest?
4821. What is the wingspan of an Elf Owl?
4822. What do panicked tinamous sometimes do that can be fatal?
4823. Have flamingos been introduced into Hawaii?
4824. What is the name of the mechanical device used for lifting and moving heavy objects with a movable projecting arm?
4825. When Lesser Snow Geese and Blue Geese form mixed pairs, which color morph predominates in the resultant goslings?
4826. Which is the only non-crested member of the waxwing and silky-flycatcher family?
4827. Are loon wings positioned well forward on their body?
4828. Does the Bobwhite quail range as far north as Canada?
4829. Which large group of familiar American birds are among the few well-known birds whose scientific name has become widely accepted as their common name?
4830. Which new bird was discovered wintering in central Thailand in 1968?
4831. Which birds are also called "titlarks" or "fieldlarks?"
4832. Is a Black Oystercatcher heavier than an American Oystercatcher?
4833. What is called *ho-ting* by the Chinese?
4834. Do Secretary-birds consume bird eggs?
4835. What mammal dwelling do Great Horned Owls often utilize?
4836. Is the Pine Warbler associated with pine trees?
4837. In the feather-trade, what were "Osprey plumes?"
4838. What is the shape of a Cliff Swallow's nest?
4839. Do western North American quail roost on the ground?
4840. How does the guano of the Lammergeier differ from the guano of most raptors?
4841. How many species of fringillid finches nest colonially?
4842. Who first imported Indian Peacocks into the Holy Land, and when did this occur?
4843. Do hummingbirds push off a perch with their feet when taking flight?
4844. Which predatory North American songbird mimics the calls of other birds?
4845. What does the specific name of the Laysan Albatross imply?
4846. Does the head of a breeding drake Lesser Scaup have a greenish sheen?
4847. How many species of birds occur in what was formerly the USSR?
4848. How does a flamingo brood its chick?

Questions

4849. Did Cattle Egrets reach Australia on their own?
4850. Are West Indian todies seen more often than heard?
4851. Which birds were formerly called shell storks, and why?
4852. Do Secretary-birds use the same nest year after year?
4853. What is parrot coal?
4854. Which are the only two North American grebes with unstriped chicks?
4855. What happens to the bill of an excited New Zealand Blue Duck?
4856. How many North American owls have an individual's name in their common name?
4857. Why did the Spectacled Cormorant vanish?
4858. Do all potoos have dark eyes?
4859. Do Shoebill storks clap their bills like other storks?
4860. When and where was the last Bachman's Warbler nest seen?
4861. Are skimmers colonial nesters?
4862. Which is the only reddish-colored North American oriole?
4863. Which New Zealand bird is considered to have the most beautiful and musical of songs?
4864. Which New World warbler is the largest, and what does it weigh?
4865. How large are the leg spurs of a guineafowl?
4866. Which cormorant has the most extensive range?
4867. Are loon eggs spotted?
4868. Which towhee is the smallest?
4869. Cassowaries have how many brood-patches?
4870. In Australia, which bird is known as a "policemanbird?"
4871. Why was the Acorn Woodpecker given the specific name *formicivorus*?
4872. Are kiskadees highly vocal?
4873. Which is the smallest Old World vulture?
4874. Are Torrent Ducks restricted to the Andean highlands?
4875. Which is the most abundant owl?
4876. What is the name of the bay at Martha's Vineyard, Masachusetts, where the last Heath Hen disappeared in 1932?
4877. In terms of overall number, how many birds occur in Great Britain?
4878. What do flying Anhingas and darters appear to lack?
4879. Do Great Horned Owls ever attack porcupines?
4880. Do toucans fence with one another with their long bills?
4881. Why were kiwi skins sought by European immigrants?
4882. What is the diameter of the entrance hole of a West Indian tody nesting burrow?
4883. What color is the bare orbital skin of a Hyacinth Macaw?
4884. Are flamingos vocal in flight?
4885. In American slang, if something is *peachy* or *fine,* how is this sometimes expressed?
4886. Which bird was symbolic of ancient Rome's power?
4887. Do drongos often perch on cattle?
4888. What must a pelican do after catching a fish, prior to swallowing its catch?
4889. Where are the nesting grounds of the Eskimo Curlew?
4890. Why does the male Tooth-billed Catbird have a toothed bill?

Questions

4891. Which order of birds exhibits the greatest contrast in terms of the size of the birds in the order?
4892. Can a Brown Skua fly carrying a King Penguin egg?
4893. At what time of day do murre chicks flutter down from their nesting ledges to the sea, and why?
4894. Do mousebirds use their bills like parrots when scurrying through the trees?
4895. What color is the pouch of a non-breeding male frigatebird?
4896. Do parrots have large eyes?
4897. Do Burrowing Owls breed in Tierra del Fuego (Chile and Argentina)?
4898. During the time of the Roman Empire, which very large bird was considered a fitting main course for the Emperor's feasts?
4899. Do any mammals possess a cloaca like birds and reptiles?
4900. What is a heavy, swelling wave that breaks on the shore called?
4901. Which neotropical gamebirds are considered to be amongl the most primitive of birds?
4902. When a grebe leaves its nest before the entire clutch is laid, are the eggs covered?
4903. Do swimming Limpkins float low in the water?
4904. What is the eye color of *both* sexes of Satin Bowerbirds?
4905. Which Mexican corvid lives in social groups?
4906. Which bird is responsible for the word pedigree?
4907. Which cuckoos construct domed nests?
4908. How did the tropical American solitaires obtain their name?
4909. Which is the most widely distributed *native* North American galliform bird?
4910. Which sex of Burrowing Owl incubates?
4911. Which family of land birds is best represented on Southwest Pacific islands?
4912. Which small North American finch has a pair of food-carrying pouches in its mouth enabling the bird to carry extra food to its chicks?
4913. Which North American cuckoo is named for its habitat?
4914. In New Zealand, which bird is called a "pied piper?"
4915. Which passerines are a nuisance around fish hatcheries?
4916. How many North American birds have common names that honor people, such as Kirtland's Warbler or Ross' Goose?
4917. What does the Bald Eagle on the great seal of the United States hold in its left foot?
4918. How many male Cuban Bee Hummingbirds are required to equal one ounce (28 gms) in weight?
4919. Are chicks more susceptible to roundworms than adults?
4920. Which is the most widespread American wood-warbler?
4921. In what 1963 movie musical did Ann Margaret and Dick Van Dyke star?
4922. What habitat is favored by a breeding Ibisbill?
4923. Do the African woodhoopoes have crests?
4924. Why did the Spanish call the pelican *alcatraz?*
4925. Did English explorers discover the Australian Black Swan, and when was the bird discovered?
4926. What coin is called the American double-eagle?
4927. Why is it believed that the Shoebill formerly inhabited Egypt?

Questions

4928. Are any pigeons or doves flightless?
4929. How did egrets obtain their name?
4930. How far south do tits and titmice range in the New World?
4931. Which is the only raptor with nostrils that can be closed at will?
4932. What prompted the formation of the Royal Society for the Protection of Birds in 1889 in England?
4933. In what 1949 movie did Joan Crawford play a woman who dominated the weak men in her life?
4934. Are manakins insectivorous?
4935. Do Barn Owls have large eyes?
4936. Are the necks of all jabiru storks naked?
4937. Do cranes feed extensively on fish?
4938. Were the ancient Greeks and Romans aware of the talking ability of the African Gray Parrot?
4939. Are all partridges native to the Eastern Hemisphere?
4940. Do Common Nighthawks have rictal bristles?
4941. Does the African Spur-winged Goose have goose-like vocalizations?
4942. What common American expression describes something that is extremely rare or nonexistent?
4943. What color is the substrate of many Chinstrap and Adelie Penguin colonies, and what causes the color?
4944. Do North Island Brown Kiwis dig deep nesting burrows?
4945. Was the *father of American ornithology,* Alexander Wilson, an accomplished singer?
4946. What animal does a threatened wryneck mimic?
4947. What is the name of the game that is rather like bowling, but played with smaller pins and balls?
4948. In America, what is a fence lizard often called?
4949. Do Hamerkops always nest in trees?
4950. Why does a serpent-eagle have proportionally short toes?
4951. Was Meller's Duck of Madagascar named after a military officer?
4952. Does the Cahow (Bermuda Petrel) line its nest?
4953. Are buttonquail powerful fliers?
4954. Which neotropical bird is known as a "peacock bittern?"
4955. What is the state bird of Colorado?
4956. Do any pheasants have inflatable throat sacs like grouse?
4957. What color are the eyes of a Jackdaw?
4958. Do newly-hatched King Vulture chicks have downy heads?
4959. With respect to conservation, what is the greatest threat facing the planet?
4960. Prior to its reintroduction, when did the White-tailed Sea-Eagle last breed in Great Britain?
4961. Do stilts ever lie on their bellies with their long legs stretched out behind?
4962. Are Trumpeter Swan cygnets carried on their parent's backs?
4963. What is a group of wigeons sometimes called?
4964. Do landing loons use their feet?

Questions

4965. What is the most effective way to conduct a rail census?
4966. What color is the wattle of an Australian Eastern Brush-turkey?
4967. Are Chukar partridge sexually dimorphic?
4968. Which are the only two northern ducks that invariably have a black inverted V on their throat?
4969. Which parrot inhabits the mountainous regions of Tibet?
4970. Do all species of shearwaters have long, slender bills?
4971. Do Ruffs generally display on flat terrain?
4972. Which birds did the New Zealand Maoris seek for their feather cloaks?
4973. On what reptile do albatrosses prey?
4974. How much more brightly colored were male Dodos than their mates?
4975. Which New Zealand river duck is especially territorial?
4976. How did the South American miners (ovenbirds) obtain their name?
4977. How far south in the Americas do Ospreys breed?
4978. Are Antarctic Blue-eyed (Imperial) Shags migratory?
4979. Do finfoots and Sungrebes inhabit fast-moving streams?
4980. Do hummingbirds have bladders?
4981. During the winter, which loon is most likely to gather in large, compact flocks?
4982. Entire undamaged skulls of small rodents are commonly found in the regurgitated pellets of which type of birds?
4983. Do most petrel chicks have two different coats of down?
4984. Are tinamous rather slim birds?
4985. After whom was Sprague's Pipit named?
4986. Which Seychelles Islands passerine preys on White (Fairy) Tern eggs?
4987. How does the foraging flight of skimmers differ from the flight of other seabirds?
4988. What is located in the center of the breast of most adult motmots?
4989. Which is the only sexually dimorphic North American owl?
4990. During storms and high winds, where do storm-petrels generally fly?
4991. Which American raptor is called a "little blue corporal?"
4992. How many species of guineafowl range north of the Sahara Desert?
4993. Why are some tropicbirds tinged pink or golden?
4994. Do most passerines build relatively simple, cup-shaped nests?
4995. Which is the only stork that regularly nests on the ground?
4996. Are nightjars noted for polymorphism?
4997. Why do toucan chicks have prominent, thickened heel pads?
4998. Which land mass is called the *bird continent*?
4999. What is the difference between a roller pigeon and a tumbler pigeon?
5000. Which tern preys extensively on frogs and other amphibians?
5001. How many penguins have the name of an individual in their common name, and which ones are they?
5002. Is the Barn Swallow a common British bird?
5003. What is the American paddlefish sometimes called?
5004. Which is the only North American cormorant with a partially yellow bill?
5005. Do owlets depart the nest before they are capable of flight?

Questions

5006. Do chicks of brood parasitic-nesting cuckoos mimic the calls of the young of their host?
5007. At what age do cassowaries assume adult plumage coloration?
5008. Does a Secretary-bird lay large eggs for its size?
5009. In Borneo, what was the significance of a Helmeted Hornbill ear-plug?
5010. Is the African Lizard Buzzard well named?
5011. When first observed by outsiders in Australia, Channel-billed Cuckoos were believed to be what type of birds?
5012. What type of bird is a laughingthrush?
5013. Do most quail have short, rounded tails?
5014. Which sulid is most likely to pirate fish from other birds?
5015. Are both Ruddy and Black Turnstones coastal birds?
5016. Are Palm Cockatoos especially gregarious?
5017. Along the east coast of the U.S., which bird is most likely to steal food from Brown Pelicans?
5018. Do foraging oystercatchers ever probe into the turf?
5019. What sex was the last Heath Hen that disappeared at Martha's Vineyard in March, 1932?
5020. Female Broad-tailed Hummingbirds are almost identical to the female of which other North American bird?
5021. Does the upper mandible of all pheasants overhang the lower?
5022. On what do many rural Puerto Ricans incorrectly believe screech-owls feed extensively?
5023. How does the nest site of a Laughing Kookaburra differ from the nest of most kingfishers?
5024. Is the Common Snipe regarded as a gamebird in North America?
5025. Which bird has the longest Latin name, and what is it?
5026. Why are guillemots more effective at preying on bottom-dwelling fish than other alcids?
5027. How far from shore do Galapagos Flightless Cormorants venture?
5028. When do female Ruffed Grouse display like males, with spread tail, extended neck-ruff, and drooped wings?
5029. Does an Osprey have a shield of bone over its eyes?
5030. Which penguin has the longest toenails?
5031. Besides diving continuously on an intruder, what else might defensive terns do to discourage a potential predator?
5032. On which raptor does the Great Horned Owl prey?
5033. With respect to plumage, how do the Rose-ringed Parakeet and Crimson Rosella differ from most parrots?
5034. Which striking Central American bird was formerly known as a "trainbearer?"
5035. What is unusual about the feeding behavior of recently-fledged White-fronted Nunbirds?
5036. Were cranes domesticated by the ancient Greeks?

Questions

5037. Is the New Zealand Kea generally silent in flight?
5038. Prior to recent controls, how extensive was the incidental drift-net kill of seabirds?
5039. Which Guatemalan bird was locally known as *Poc*?
5040. Do Satin Bowerbirds use the same bower site year after year?
5041. Do Brown Pelicans breed in the Galapagos Islands?
5042. What color is the mask of a male Masked Bobwhite quail?
5043. What disease is most often associated with pigeons?
5044. Are tropical and temperate penguins sedentary?
5045. What part of a drake King Eider is considered a delicacy by some Eskimos?
5046. Do the neotropical tinamous gather in flocks?
5047. Does a Wandering Albatross have a black eye-ring?
5048. How can a West Indian tody be attracted?
5049. Which eider, when swimming, often cocks up its tail?
5050. What disease adversely impacted skua colonies in Antarctica?
5051. Why do female phalaropes depart their breeding grounds before the peak date of hatching?
5052. What type of canine is known as a barbet?
5053. On what is Buddha sometimes depicted riding?
5054. Do mousebirds have especially water-repellent plumage?
5055. Which land bird undertakes the longest migration?
5056. In American slang, what is a person in an exposed and dangerous situation called?
5057. Which North American mockingbird relative is named after a feline?
5058. In terms of weight, how much food does a pair of Golden Eagles rearing chicks require a year?
5059. Do Caspian Terns soar?
5060. How many birders participate in the Christmas bird counts?
5061. Do nuthatches maintain and defend winter territories?
5062. Which are the only two African lovebirds not named for plumage color?
5063. What is the highest Boy Scout rank?
5064. What is the clutch size of the New Zealand Takahe?
5065. Courting male eiders sound like which other birds?
5066. Do *adult* Ostriches lie down with their legs stretched out behind them?
5067. Did dinosaurs have hollow bones, like modern birds?
5068. In old poetry, was the Nightingale considered feminine?
5069. Do Rhinoceros Auklets winter as far south as Central America?
5070. Is the name "potoo" Portuguese in origin?
5071. What color is the neck-ring of a drake Ring-necked Duck?
5072. Which is the largest Philippine parrot?
5073. Do Egyptian Vultures ever nest in trees?
5074. What is a group of hawks sometimes called?
5075. Is a Pseudonestor a gamebird biologist term for a bird pretending to be nesting?
5076. Who was the most dominant player on the Boston Celtics basketball team of the 1980's?
5077. Which penguins incubate in a standing position?
5078. Who wrote the foreword for the multi-volume set *The Birds of Africa,* the first volume of which was published in 1982?

Questions

5079. Which is the only endemic Japanese corvid?
5080. At the turn of the century, which birds were collectively known as the Pygopodes?
5081. Which is the most widespread Western Hemisphere owl?
5082. What is the origin of the name "manakin?"
5083. Why did some American Indians keep caged or tethered eagles?
5084. Do turnstones ever eat eggs of their own kind?
5085. When and where was the first captive breeding flock of Whooping Cranes established?
5086. Which American edentates prey on bird eggs?
5087. Why are there so few captive Oldsquaw ducks?
5088. Which frigatebird is most inclined to fly with its gular pouch fully inflated?
5089. What is the largest mammalian prey killed by Secretary-birds?
5090. What state first opened a hunting season on the introduced Chukar, and when did this occur?
5091. What is the roughened condition of the human skin called that results when one is cold, apprehensive, or frightened?
5092. Do bustards have long, sharp toenails?
5093. In the West Indies, which bird is known as the "jumbie bird?"
5094. What unusual shorebird-nesting strategy do Subantarctic Snipe sometimes utilize?
5095. What is the national bird of Great Britain?
5096. If an Abbott's Booby chick falls from its tree nest, will the parents feed it on the ground?
5097. How many of the 30 *species* of Amazon parrots were named after localities or countries, and which ones are they?
5098. Do swallow chicks nearing fledging weigh more than adults?
5099. Was Reeves' Pheasant named after a famous British diplomat?
5100. What is the throat color of a male Williamson's Sapsucker?
5101. Which American bird is known as a "Nonpareil Bunting?
5102. In American slang, what type of individual is called a woodcock, and why?
5103. Do *all* Western Hemisphere quail have more or less serrated bills?
5104. Which American cuckoo has the greatest vocal repertoire?
5105. How are electric power lines detrimental to African Cape Vultures?
5106. Was the Hispaniolan Palmchat so named because of its fondness for palm trees?
5107. Do shorebirds fly in V-formation?
5108. Is a sheathbill able to break open a King Penguin egg?
5109. Do captive birds have maximum lifespans that are well over twice as long as those of captive mammals of comparable size?
5110. Do skimmers typically feed in deep water?
5111. Is the King Bird-of-paradise fairly common in *lowland* New Guinea forests?
5112. How many pounds of hamburger must a 170 pound (77 kg) human consume daily to equal the daily consumption of a hummingbird?
5113. In by-gone days, how did many people theorize that small birds migrated?
5114. The name "skua" originated in what country?
5115. What was the first name of John James Audubon's wife?
5116. After which bird was the New Zealand kiwi named?
5117. Do Gyrfalcons roost in sea caves?

Questions

5118. Do pittas roost in trees?
5119. What color is the cap of an adult Chipping Sparrow?
5120. Why don't foraging plovers concentrate in large numbers, even in intertidal areas?
5121. Which bird in America is known as "hunkie?"
5122. Where and when was the first Blue Goose nest discovered?
5123. In South Africa, which bird is called a "turkey buzzard?"
5124. Do nuthatches have relatively short bills?
5125. What caused the near extinction of the Cahow?
5126. Which birds did Japanese Shoguns hold sacred?
5127. Do male Rock Ptarmigans assist with chick rearing?
5128. Which alcid is named after a man-made object?
5129. What is one of the most prevalent childhood diseases?
5130. In what two states do hunters account for the greatest Sandhill Crane kill?
5131. During the 35-hour, non-stop flight of the golden-plover from the Aleutian Islands, Alaska, to its Hawaiian wintering grounds, how many wingbeats are required per bird?
5132. Do Least Bitterns breed in California?
5133. Do the large, flightless New Zealand Weka rails breed in their first year?
5134. On what does the smallest diurnal raptor primarily feed?
5135. What drastic means are used in Africa to control the Red-billed Quelea population?
5136. Do Surf Scoters actually forage in breaking surf?
5137. What sound is produced by a Mourning Dove taking flight?
5138. Do Old World trogons tend to be less insectivorous than trogons of the Neotropics?
5139. Does the obligate brood parasitic-nesting Common Cuckoo remove the eggs of its host prior to laying its own egg?
5140. In what year was the Aransas Wildlife Refuge, the Whooping Crane wintering grounds in Texas, established?
5141. Why do nightjars swallow stones?
5142. Is the Pitt Island Shag restricted to Pitt Island, and what is its population size?
5143. Does a chicken have lips?
5144. Were hummingbirds named because of their humming songs?
5145. Was the Chestnut-sided Warbler well known to Audubon?
5146. Does a Golden Eagle have a wedge-shaped tail?
5147. How much did the extinct Spectacled Cormorant weigh?
5148. When at a lek, what do female Ruffs do that indicates to a male he has been selected?
5149. When feeding young, how many insects can a Eurasian Alpine Swift carry in a compact ball in its throat?
5150. Which five gulls occurring in North America have no black in their wingtips as adults?
5151. What is a railway train called that runs late at night after regular hours of service have been discontinued?
5152. Which two cormorants are known for two distinct color phases?
5153. How far from the breeding grounds do American White Pelicans fly one way for food?
5154. Which long-legged African bird poses a threat to adult flamingos?
5155. Does the Short-eared Owl inhabit Australia?
5156. How many North American hummingbirds engage in lek displays, and name them.

Questions

5157. What is a gaggle?
5158. For what were Leach's Storm-Petrels formerly used?
5159. Do the cavity-nesting sheathbills lay white, unmarked eggs?
5160. Do all alcids have short tails?
5161. Which falcon relative patrols highways seeking carrion?
5162. Are the two African rockfowl (*Picathartes* or bald-crows) colonial nesters?
5163. How long does it take an African Bateleur to assume full adult plumage?
5164. Was the extinct Auckland Islands Merganser flightless?
5165. Do the West Indian todies have round bills?
5166. What is the average number of feathers per square inch on an Adelie Penguin?
5167. Does the Pine Grosbeak breed in Europe?
5168. Which is the smallest Australian bird, and how much does it weigh?
5169. Which endangered seabird may nest in parking lots?
5170. Why was the Reed Cormorant of Africa so named?
5171. Which is the largest Mexican toucan?
5172. What is stretching one's neck sometimes called?
5173. What color is the bill of an African Saddle-billed Stork?
5174. Is the Least Flycatcher the smallest North American flycatcher?
5175. What is the incubation period of a domestic chicken?
5176. Is the Australian Noisy Scrub-bird a common species?
5177. The crop is an enlargement of what structure?
5178. For what was William Elford Leach, after whom Leach's Storm-Petrel is named, best known?
5179. What percentage of all bird species depend on insects to some degree as a food source?
5180. Why is the incubation period of the Superb Lyrebird so long?
5181. How many American grouse were named for a plant, and which ones are they?
5182. What is the value of the annual retail parrot business in the U.S. alone?
5183. How did the French suggest eagles could be used during the First World War?
5184. Which American bird is known as a "meadow chicken?"
5185. On which day are the Cliff Swallows at the mission at San Juan Capistrano supposed to commence their southbound migration?
5186. Which American waterfowl is most apt to prey on crayfish?
5187. Do the Palmchats of Hispaniola nest anywhere but in palm trees?
5188. Many people are aware that the feathers of trogon study-skins fade, but which colors are most susceptible to fading?
5189. On the guano-rich Peruvian islands of Las Chinchas, how many tons of guano are deposited per acre (0.4 hectare) annually?
5190. Is the tip of an American Wild Turkey's tail white?
5191. Which bird may take over an occupied Bald Eagle nest, evicting the eagles?
5192. Which is the most abundant breeding seabird of Great Britain, and what is its population?
5193. What is an eaglestone?
5194. With respect to their names, what do the Brent and Barnacle Goose have in common?
5195. Is the North American Spotted Owl wary of humans?
5196. At what time of day does a neotropical Laughing Falcon commence calling?

Questions

5197. Is a woodcock egg more elongate than most shorebird eggs?
5198. Is the Northern Ovenbird an arboreal nester?
5199. Which state was formerly known as the eagle state, and why?
5200. What is a person called who follows cockfighting as a business or sport?
5201. Is the Dunlin one of the more southerly-nesting northern sandpipers?
5202. When the German naturalist/explorer Alexander von Humboldt made his historic canoe voyage up the Orinoco River, what pet bird accompanied him?
5203. Does a kiwi's upper mandible overlap the lower mandible?
5204. How did the coot obtain its name?
5205. Do Sungrebes and finfoots have long, stiff tails?
5206. Why do obligate brood parasitic-nesting cuckoos lay smaller eggs relative to their size than non-parasitic cuckoos?
5207. Are Snowy Owls capable of catching ducks on the *wing?*
5208. Do House Finches (Linnets) cause crop damage?
5209. How does the head of a juvenile Sandhill Crane differ from the head of an adult?
5210. Sheathbills have what booby-like habit?
5211. Was William Swainson, after whom Swainson's Hawk was named, a medical doctor?
5212. When is the horn on top of the upper mandible of breeding American White Pelicans shed?
5213. Which falcons nest in North American cities?
5214. By looking at the bill of an Adelie Penguin, how can one deduce it is from the Antarctic?
5215. Which African bird in flight resembles a large bird followed by two smaller birds, and why?
5216. Do male Brown Kiwis incubate continuously?
5217. During the late 1980's and early 1990's, for which National Football League team did the scrambling quarterback, Randall Cunningham, play?
5218. How do the Indians of the Andes capture adult flamingos?
5219. The males of which sunbirds incubate?
5220. Which is the heaviest bird in the largest bird of prey genus *(Accipiter)*?
5221. Do avian bones have marrow?
5222. Do downy Sunbittern chicks resemble bittern young?
5223. Do swallows generally have longer tails than martins?
5224. What is the name of the cloth that is rather like canvas, but finer and lighter in weight?
5225. What ultimately happened to William Gambel, after whom Gambel's Quail is named?
5226. Do Cockatiels feed extensively on the ground?
5227. Which North American grosbeak sings while in flight?
5228. Relative to its size, which North American-breeding grebe has the largest head?
5229. Which is the only American bird whose breeding range is restricted to the Edwards Plateau region of central Texas?
5230. Which are the only tiger-herons that inhabit coastal habitat?
5231. Do Ruffed Grouse have a crest?
5232. What color is the base of a White (Fairy) Tern's black bill?
5233. What were early snipe hunters called?
5234. Are Black Guillemots heavier than Pigeon Guillemots?

5235. According to ancient legend, what did sentinel cranes do at night to remain awake?
5236. Are the tail feathers of an Ostrich mostly white?
5237. Which American-breeding gull relative has a common name of something that is eaten?
5238. What color is the eye-ring of an Iceland Gull?
5239. Which American waterfowl often nests in flicker holes?
5240. Which large South American nightjar relative clings to rock walls like a swift?
5241. What is unusual about the nostrils of the South American tapaculos?
5242. Do most birds-of-paradise lay a two-egg clutch?
5243. How do male Indian Yellow-rumped Honeyguides guarantee themselves a mate?
5244. What color is the lower border of the red shoulder of a male Tricolored Blackbird?
5245. What is the English stone fly sometimes called?
5246. Do Cooper's and Sharp-shinned Hawks often inhabit the same woods?
5247. How did the name "gannet" originate?
5248. What is the range of a hummingbird's body temperature?
5249. Does a group or gathering of penguins have a special name, such as a gaggle of geese or a covey of quail?
5250. How many specimens exist of the extinct Labrador Duck?
5251. Are tody nesting burrows invariably straight?
5252. Is the Crested Auklet one of the least vocal auklets?
5253. Do spoonbills have large, functional tongues?
5254. Do the feet of a flying Black-crowned Night-Heron extend beyond its tail?
5255. What is a group of crows called?
5256. Are African broadbills more colorful than those of Asia?
5257. In India, which raptor is a serious threat to nesting darters?
5258. Which American birds are known as "butter-billed coots?"
5259. What is a bal chatri?
5260. Why do both sexes of King Penguins incubate, whereas in the closely related Emperor Penguin, only males incubate?
5261. At what age do Bald Eagles assume white heads and tails?
5262. Does the endemic South Georgia Pintail feed on carrion?
5263. Which North American gamebird constructs a domed nest?
5264. Which *flying* bird of the Antarctic steals nesting material from adjacent nests?
5265. Are honeyguides brightly colored?
5266. Which pittas are endemic to Borneo?
5267. Do Ospreys occur in Hawaii?
5268. What color is the cap of a male Wilson's Warbler?
5269. Why are whydahs (weaver-birds) called widowbirds?
5270. Do juvenile Northern Harriers resemble females?
5271. Have Bachman's Warblers ever been sighted in Cuba?
5272. Was Montezuma, after whom the Montezuma Quail was named, the *last* Aztec emperor of Mexico?
5273. Which Antarctic-nesting bird has carpal spurs?
5274. On what crustacean do Barred Owls prey?
5275. Do Sooty Terns feed at night?

Questions

5276. Jaguars are said to be able to imitate the call of which bird?
5277. Rails lead the list of recently extinct birds with 11 species, but which two families have suffered the next highest losses?
5278. What percent of birds species are colonial nesters?
5279. Which are the only coraciiform birds (kingfishers and allies) with down-covered chicks?
5280. Are sibias Old World warblers?
5281. When and where was the first colony of the Puna Flamingo discovered?
5282. In which U.S. state besides Florida might the Short-tailed Hawk be encountered?
5283. Is the New Guinea Shovel-billed Kingfisher one of the smaller kingfishers?
5284. How do Secretary-birds carry nesting material?
5285. After whom was Say's Phoebe named?
5286. Is the California Gull the only gull named for a state?
5287. Is the bill of a male Western Grebe significantly longer than the bill of a female?
5288. Who was the director of the New York Aquarium between 1902 and 1937, and which bird is named after him?
5289. Are waxwings nocturnal migrants?
5290. Is the Horned Grebe more gregarious than the Eared Grebe?
5291. How do Asian falconers flying Golden Eagles travel when hunting?
5292. When Europeans first viewed bird-of-paradise skins, what ludicrous assumption was made regarding the breeding habits of the birds?
5293. Is the Zigzag Heron one of the better-known herons?
5294. What is a *rat run?*
5295. Is the Vasa Parrot restricted to Madagascar?
5296. In Africa, which bird was called a "scissor-billed tern?"
5297. On what does the Red-shouldered Hawk feed almost exclusively in Florida?
5298. Do Elf Owls and pygmy-owls both have short tails?
5299. Which is largest member of the sandpiper family, and how much does it weigh?
5300. Which is the least vocal heron of North America?
5301. What is cuckoo spit?
5302. Do cock Ostriches often fight?
5303. Does the flightless, nocturnal New Zealand Owl Parrot (Kakapo) feed on live prey?
5304. What is the incubation period of the Marbled Murrelet?
5305. How much more does Anna's Hummingbird weigh than a Calliope Hummingbird?
5306. Do skimmers skim sand?
5307. Which is the smallest European upland gamebird?
5308. Which European raptor nests in rabbit burrows?
5309. Are the crests of *both* sexes of the South American cocks-of-the-rock permanently erect?
5310. Which is the only western North American cormorant that flies with its neck kinked?
5311. Are bee-eaters wary?
5312. What type of bird is officially named Bluebonnet?
5313. Which animal was introduced into Hawaii to control introduced rats, but instead became an efficient bird predator?
5314. Is Kirtland's Warbler one of the smaller warblers?

5315. Of all migratory shorebirds wintering in Southwest Pacific islands, which is the only one not from the north?
5316. Females of how many birds-of-paradise have ornate plumes?
5317. What do agitated owls often do to express anger?
5318. In Alaska, on what does the Willow Ptarmigan feed extensively?
5319. Do Cave Swallows actually nest in caves?
5320. What is a group of peacocks called?
5321. Which two birds are regarded as the most remarkable ornithological discoveries in Africa this century?
5322. Which New World fish-duck is among the earliest spring migrants?
5323. Most passerines have how many tail feathers?
5324. Why don't condors commence flying at dawn?
5325. Is the Helmeted Guineafowl well named?
5326. Which American city is named after a mythical bird?
5327. Do curassows breed in their first year?
5328. Was James' (Puna) Flamingo named after a distinguished lawyer?
5329. Which bird was known as a "bastard Baltimore oriole?"
5330. With which bird does the Glaucous-winged Gull frequently hybridize?
5331. What is the purpose of the loud roar or whirring of the wings of a quail startled into flight?
5332. Does the neotropical King Vulture kill live prey?
5333. Which African gamebirds are often used as watchbirds?
5334. Are the black-tipped outer primaries of a White Ibis only visible when its wings are extended?
5335. Early naturalists believed the upturned bill of the Parakeet Auklet was used for what purpose?
5336. Which insect accounts for Bobwhite egg and chick mortality?
5337. With what animal product do Burrowing Owls sometimes line their nesting tunnels?
5338. How many North American flycatchers have the name of a plant in their common name, and which ones are they?
5339. How long is the tail of a Swallow-tailed Kite?
5340. Which is the only North American warbler with a chestnut-colored rump?
5341. What is cuttlebone, and what is its commercial use?
5342. Why do the paired nostrils of diving-petrels open upward?
5343. Do juvenile Secretary-birds have long tails?
3344. Which American bird is known as a "voodoo bird?"
5345. Why do lyrebirds build mounds?
5346. How do Andean Indians make use of downy flamingos?
5347. Is the voice of a male crane higher than that of a female?
5348. Do female Bobwhites have white throats?
5349. Is the Red-throated Caracara a solitary bird?
5350. Do smaller hummingbirds have proportionally larger hearts than larger hummingbirds?
5351. Is the Galapagos Lava Gull a colonial nester?
5352. Which bird was introduced to the subantarctic island of South Georgia, and why?
5353. Which gamebird is entirely dependent on sagebrush?

Questions

5354. Do both species of northern wigeons often feed at night?
5355. How did the Beardless-Tyrannulet obtain its name?
5356. How many pounds of fish are required to fledge a Brown Pelican chick?
5357. The chicks of which Galapagos bird gather in crèches?
5358. In Australian slang, what is a *cockatoo squatter?*
5359. Which is the only North American hummingbird with red iridescence on both its throat and crown?
5360. Which is the only North American blackbird with a brown head?
5361. Does the African Shoebill stork ever nest in trees?
5362. What is the Gray Plover more commonly called in America?
5363. Which seabird is known as a "sea cat?"
5364. What is a Koloa?
5365. Which African bird was described as having a smell like putrid carrion with flesh that tastes like it smells?
5366. Which island accommodates the most breeding species of alcids, how many breed there, and which ones are they?
5367. What was the Gray Jay formerly called?
5368. Which large, long-legged bird was often kept as a pet by the natives of Borneo?
5369. Do the ranges of Eastern and Western Kingbirds overlap?
5370. What do foraging Marabou Storks do at Red-billed Quelea colonies?
5371. Is the Limpkin an elusive species?
5372. What does a Yellow Warbler often do if its nest is parasitized by a Brown-headed Cowbird?
5373. Do female Surf Scoters have swollen bills?
5374. How do juvenile Adelie Penguins differ in appearance from adults?
5375. Which country has the highest percent of its native bird species considered endangered, and what is the percentage?
5376. Why are many Red Crossbills killed by traffic?
5377. Which African cuckoo is named after a jewel?
5378. What is the shape of a screamer's bill?
5379. Does the Tennessee Warbler breed in Tennessee?
5380. How far into the ground do Crab Plover burrows extend?
5381. What color are female Summer Tanagers?
5382. Are African Gray Parrots responsible for crop damage?
5383. Which is the only pelican having its eyes totally encircled by feathers, rather than bare lores?
5384. Why are mousebirds often mobbed by small birds?
5385. Where was Alexander Wilson, the *father of American ornithology,* born?
5386. Why do honeyguides have thick skins?
5387. How does an Anhinga or darter announce its intention to fly?
5388. In America, which birds are known as "sea snipe?"
5389. Do underwater-swimming cormorants or shags paddle both feet alternatively?
5390. Do petrels kill mammals?
5391. What is often located in front of a Rock Wren's nest?

5392. Do Black Vultures kill livestock?
5393. Do all hornbills have casques?
5394. Which seabird is known as a "bone-shaker?"
5395. Do any Australian or New Guinea catbirds construct bowers?
5396. What ultimately was responsible for the death of John Kirk Townsend (1809-1851), after whom Townsend's Solitaire and Townsend's Warbler are named?
5397. Which two cotingas are the most terrestrial?
5398. Are Sharp-tailed Grouse well named?
5399. Are drongos prone to gather in flocks?
5400. What do thirsty pelicans sometimes do in the rain?
5401. If an individual is in very good humor, health or form, how is this often expressed?
5402. Do threatened Montezuma Quail fly, or are the birds more inclined to run?
5403. Why was the name of the Ring-necked Pheasant changed to Common Pheasant?
5404. When was the Auckland Islands Merganser last observed?
5405. What might a Common Snipe do to escape attacking raptors?
5406. Why do Monk Parakeets construct large communal nests?
5407. Are the bowerbirds known as catbirds named for their plumage pattern?
5408. How long do skuas live?
5409. Do flying oystercatchers vocalize?
5410. Do Gentoo Penguins always breed adjacent to the coast?
5411. How much do Secretary-birds weigh?
5412. Does the Western Sandpiper breed in Siberia?
5413. What direction is the eagle on the great seal of the United States facing?
5414. Are waxwings especially nomadic?
5415. Is the Mississippi Kite most numerous in Mississippi?
5416. How does the casque of the Abyssinian Ground-Hornbill differ from all other hornbill casques?
5417. Does the New Zealand Wrybill breed along the coast?
5418. The female Blue-winged Teal is nearly identical to the female of which bird?
5419. The Coppery-tailed Trogon is a subspecies of which bird?
5420. How did the Sandwich Tern obtain its name?
5421. Do flushed snipe have a straight flight?
5422. Which is the largest state bird?
5423. Do Rhinoceros Auklets land in trees?
5424. What is the function of the crop of a female painted-snipe?
5425. Which is the only eastern U.S. state with Burrowing Owls?
5426. Why do King Vultures typically soar higher than Black or Turkey Vultures?
5427. What do male juvenile hummingbirds resemble?
5428. Which North American Indian tribe was named after a bird?
5429. How many of the 23 species of African turacos lack crests?
5430. The range of which northern quail barely dips into Mexico?
5431. Do Dunlins wet their food before swallowing it?
5432. Nesting Kirtland's Warblers abandon dense stands of young jack pine trees when the trees reach what height?

Questions

5433. Which passerine group is second only to the New Guinea birds-of-paradise with respect to ornamentation?
5434. When killed by traffic in Utah, on what are migrant Rough-legged Hawks generally feeding?
5435. What are lady bugs also called?
5436. When was the kiwi officially protected in New Zealand?
5437. Are bluebirds nocturnal migrants?
5438. What color are freshly-laid hummingbird eggs?
5439. In what year did John James Audubon die, and how old was he?
5440. When taking flight, does a crane require a running start?
5441. Which Old World vulture is the most migratory?
5442. Which American waterfowl is known as a "redleg?"
5443. What color is the bill of a breeding Tricolored (Louisiana) Heron?
5444. Are jacanas good swimmers?
5445. Why do African Fork-tailed Drongos mob flying Hamerkops?
5446. How did the Dickcissel obtain its name?
5447. Do spoonbills forage with their heads completely submerged?
5448. Who wrote the famous fairy tale *The Ugly Duckling*?
5449. Is the Pink-backed Pelican the only pink pelican?
5450. Which bird is known as *grootflamink* by South African Afrikaaners?
5451. Which cuckoos are polyandrous?
5452. Which is the most widespread gamebird?
5453. Which American passerine was known as a "ricebird?"
5454. Which is the only thrasher named for a plant?
5455. Which North American swift nests in sea caves?
5456. Excluding Christmas Island, where did Abbott's Booby formerly breed?
5457. Which is the smallest North American dove, and what is its length?
5458. Are any of Darwin's finches colorful?
5459. Are rheas restricted to the lowlands?
5460. When passerines sing, does the tongue play a part?
5461. What are several of the lightweight divisions in the "sport" of boxing called?
5462. Which are the only two herons that lay a single-egg clutch?
5463. Which American bird has become the center of controversy between environmentalists and Pacific northwest loggers, and why?
5464. How do African snake-eagles feed their young?
5465. Why should honey never be used in hummingbird feeders?
5466. Do Antarctic Terns migrate north of the equator?
5467. When a male Scissor-tailed Flycatcher is perched, are its *scissors* held together?
5468. When and where was the last European Great Auk captured, and what ultimately became of it?
5469. When leaping from an elevated nest cavity, do goldeneye ducklings use their wings?
5470. What is the more familiar name for a "gray ternlet?"
5471. Is the Yellow-billed Loon more numerous in Eurasia than in North America?
5472. Is a Pyrrhuloxia more inclined to feed on the ground than the related Northern Cardinal?

Questions

5473. How do the calls of the Pileated and Ivory-billed Woodpecker differ?
5474. On what type of reptile does the White-bellied Sea-Eagle prey?
5475. Which is the smallest *dabbling-duck* of North America?
5476. Do Barn Owls commence incubating with the first-laid egg?
5477. Which western American woodpecker preys on lizards and bird eggs?
5478. Does the Great Kiskadee occur in North America?
5479. Which bird does a walking Northern Waterthrush recall?
5480. What is a group of pheasants called?
5481. Who was the first person to take New World parrots to Europe?
5482. Do juvenile Sabine's Gulls have forked tails?
5483. What color are the inflatable throat sacs of a male Sharp-tailed Grouse?
5484. Which is the heaviest Alaskan auklet, and how much does it weigh?
5485. Which birds were formerly called *oakum boys,* and why?
5486. Do wood-warblers warble?
5487. Do juvenile Limpkins remain with their parents throughout the winter?
5488. What is unusual about the paired wattles of an African Wattled Crane?
5489. Which New Zealand bird is known as an "apteryx?"
5490. How do the nests of euphonias differ from the nests of most other tanagers?
5491. Do rollers prey on mammals?
5492. Does the Red-billed Hornbill feed primarily in trees?
5493. What was John L. Le Conte's profession, and which western U.S. bird is named after him?
5494. Is the Indian Skimmer a coastal resident?
5495. Do Cattle Egrets perch on African elephants?
5496. Which region has been described as *the land of sheep, geese and high winds*?
5497. What color is the hind-neck of a Green Jay?
5498. Are Bald Eagles restricted to North America?
5499. Does the endangered, flightless New Zealand Takahe grip its food in one foot like a parrot?
5500. Do owls range in size from large sparrows to eagles?
5501. Which gamebirds were introduced into Hawaii from Panama?
5502. What expression describes an accomplishment which gives distinction?
5503. Which is the only seabird with a long red tail?
5504. Are Vermilion Flycatchers generally associated with water?
5505. Do the Old World sunbirds typically hover at flowers like hummingbirds?
5506. What is the favored prey of the Golden Eagle?
5507. What is one of the major causes of Red-crowned Crane mortality?
5508. In many birds-of-paradise, why is the most colorful part of the plumage on the under surface of the bird, whereas in the closely related bowerbirds, colors and ornamentation, such as crests and capes, are on the dorsal surfaces?
5509. Which vulture has a very long, diamond-shaped tail?
5510. Which colorful thrush is a common cagebird in Asia?
5511. Where does the Eurasian Dotterel breed in North America?
5512. Why do most California Brown Pelicans nest on the ground, rather than in trees?
5513. Where else besides South America does the White-faced Whistling-Duck occur?

Questions

5514. Which North American woodpecker is the most gregarious year-round?
5515. How far south does the Northern Goshawk migrate?
5516. In what 1983 TV mini-series did Richard Chamberlain play an outback Australian priest?
5517. Do Bald Eagles take prey as small as a Whiskered Auklet?
5518. Which New World kingfisher has a long, heron-like bill?
5519. Do Cooper's Hawks take songbird chicks from nests?
5520. Why are pinioned cranes not completely flightless?
5521. In winter plumage, which murre has black below its eyes?
5522. How deep can Gentoo Penguins dive?
5523. Are sheathbills good runners?
5524. Which is the most social of the Australian grass finches?
5525. Which North American-breeding kite is sexually dimorphic?
5526. When was the bird-of-paradise feather-trade prohibited?
5527. Most raptors squirt their guano away at an angle, but which *group* of raptors does not?
5528. Do California Quail and Gambel's Quail commonly hybridize?
5529. Do both sexes of African Saddle-billed Storks have dark eyes?
5530. Are Golden-crowned Kinglets entirely insectivorous?
5531. What was the name of the character played by Alan Alda in the popular television series *M*A*S*H*?
5532. Why are megapode eggs apparently always buried in the mound with the large end up?
5533. Which passerine feeds its chick the greatest length of time after fledging?
5534. Why are flickers also known as "yuckers?"
5535. Which rail relative is a significant Western Grebe egg predator?
5536. What color are juvenile Tundra (Whistling) Swan bills?
5537. Does the Great Horned Owl nest in winter?
5538. How was the extinct Spectacled Cormorant cooked by the natives of Kamchatka?
5539. When something is viewed from high above, how is this often expressed?
5540. How do the external nostril tubes of a Wandering Albatross differ from those of a Royal Albatross?
5541. What do alarmed little Green Herons do with their tails?
5542. Which is the only American nuthatch not named for its color?
5543. Why do American golfers dislike coots?
5544. In the game of chess, what piece has a bird name?
5545. What are halcyon days?
5546. At what age can most North American quail young fly?
5547. Has the Eskimo Curlew ever been recorded in Russia?
5548. After what animals were the Canary Islands named?
5549. How did the Australian Wonga Pigeon obtain its name?
5550. Did Passenger Pigeons occur in the West Indies?
5551. How do breeding European Green-winged Teal drakes differ in appearance from American Green-winged Teal drakes?
5552. Are the folded wings of a resting grebe discernible?
5553. Do grouse have relatively long legs?
5554. Do Caspian Terns pirate fish from other birds?

5555. What is the Olivaceous Cormorant also called?
5556. What is the wingspan of a female Harpy Eagle?
5557. Which American arboreal bird is known as a "logcock?"
5558. Do Lesser Yellowlegs winter in Australia?
5559. Which tropicbird has a distinctive black collar?
5560. Do cracids build large nests relative to their size?
5561. In slang, what is a slacker in time of war called?
5562. Which of the three jaegers is the most pelagic?
5563. Do recently-fledged wrens return to the nest to sleep?
5564. Does a hummingbird have a large gape?
5565. On what does the Common Black-Hawk feed almost exclusively?
5566. Who designated the Mute Swan as royal game in old England, and when did this occur?
5567. What two technological events of the late 1800's ultimately doomed the Passenger Pigeon?
5568. What is a snipe hunt?
5569. Do pelicans prey on birds?
5570. Until recently, which grebe was regarded as a pale-color phase of the Western Grebe?
5571. Which birds did the 1989 *Bahia Paraiso* oil spill in the Antarctic most adversely impact?
5572. According to Chinese legend, how did the souls of the dead reach heaven?
5573. How far south does the Black Guillemot migrate?
5574. Have nesting Thick-billed Parrots ever been recorded in the U.S.?
5575. Are juvenile Gray Jays paler than adults?
5576. Which scoter is the largest, and how much does it weigh?
5577. How many passerine species are known only from museum specimens?
5578. In Shakespeare's time, what did a birder do?
5579. How many bird species occupy the Indian subcontinent (India, Pakistan, Bangladesh, Sri Lanka and Nepal)?
5580. Do landing grebes often plane across the surface of the water for some distance with their wings raised?
5581. Is a Hairy Woodpecker less shy than a Downy Woodpecker?
5582. Males and females of the highly sexually dimorphic Eclectus Parrot were for years regarded as separate species, but how was it determined that both were actually the same species?
5583. What type of duck is the Old World Ferruginous Duck?
5584. Do all courting eagles engage in aerial displays?
5585. Which is the only American warbler with a white rump?
5586. How have small passerines been known to start fires?
5587. Which Arctic gull dives, and may even swim underwater, to capture prey?
5588. What color are juvenile Whiskered Auklets?
5589. To reach the ground with its bill, is it necessary for a stilt to bend its legs?
5590. What is a group of coots called?
5591. What habitat is favored by Bachman's Warbler?
5592. Which is the smallest North American titmouse, and what is its length?
5593. Which are the two largest North American herons?

5594. What is a group of owls called?
5595. Which North American-breeding shorebird is flightless for about two weeks during the molt?
5596. Is a Great Gray Owl prone to attack a human if its nest is approached?
5597. In what African country is the Lammergeier most numerous?
5598. Which northern gamebird was introduced into the high Sierra?
5599. Which bird is almost identical in appearance to the Short-tailed (Slender-billed) Shearwater?
5600. What popular Australian beers are named after birds?
5601. Which American alcid is commonly known as the "fog lark," and why?
5602. Where was the type specimen of the Gull-billed Tern collected?
5603. Why are the songs of birds living in dense growth generally more complex than those of birds living in the open?
5604. Is the nocturnal, neotropical Oilbird black in color?
5605. What was primarily responsible for the decline of the Hawaiian Goose (Nene)?
5606. Do hummingbirds have long secondaries?
5607. Which was the last bird-of-paradise discovered, and when did the discovery occur?
5608. What is unusual about the sternum of the Galapagos Flightless Cormorant?
5609. In Britain, what is a dog called used to hunt hares and rabbits?
5610. Which is the only Western Hemisphere grosbeak not named for its color?
5611. Into which European country has the Wild Turkey been relatively successfully introduced?
5612. What color is the crest of a breeding-plumaged American White Pelican?
5613. What is unusual about the hunting technique of Eleonora's Falcon?
5614. During what month is duck hunting legal in New Zealand?
5615. Which is the only *dabbling* duck with non-patterned ducklings?
5616. What were Horned Screamers formerly called in Trinidad?
5617. Do kiwis occur as high up as the snow line?
5618. What is the head color of adult Brown Pelicans during the winter?
5619. With respect to its scientific name, the Rough-legged Hawk is named after what animal?
5620. How much do Wood Storks weigh?
5621. On what do Red Crossbills feed exclusively?
5622. Was John James Audubon's mother from France?
5623. Which bird is called the *great white ghost of the north*?
5624. If two species have allopatric ranges, do their geographic ranges overlap?
5625. Do waxwings hawk insects in flight?
5626. Do digging Himalayan Monal pheasants use their feet?
5627. Why is the North American Black Duck decreasing in number?
5628. Are titmice nostrils completely exposed?
5629. Do frigatebirds migrate great distances?
5630. Is the tail of a Blue-winged Kookaburra blue?
5631. At what time of day do Vesper Sparrows generally sing?
5632. In the far north, what is the major Snow Goose predator?
5633. Which is the only sexually dimorphic North American-nesting swallow?

Questions

5634. How do Greenland natives prepare Dovekies for consumption?
5635. Why do the Old World fantails fan their tails?
5636. Do foraging tinamous use their feet to scrape the ground in quest of food?
5637. Do Rusty Blackbirds associate with Red-winged Blackbirds?
5638. Is Wilson's Plover a flocking species?
5639. Are hummingbirds wary of bees and wasps?
5640. Are Thick-billed Murres "blacker" than Common Murres?
5641. Which family of passerines exhibits the greatest size extremity in terms of weight, and how much heavier is the largest species than the smallest?
5642. Which rodent is a Sage Grouse nest predator?
5643. Which Old World bird is known as a "brahminy duck?"
5644. Do stilts build floating nests?
5645. Are birds-of-paradise still hunted?
5646. How long was the strongly-hooked bill of a Dodo?
5647. Do juvenile hawks have darker eyes than adults?
5648. Which American bird is called a "torch bird" or "fire-brand?"
5649. How many pounds of food is required to fledge a single Adelie Penguin chick?
5650. The bill of which dove or pigeon curves downward to the greatest extent?
5651. What color is the distal end of a breeding Horned Puffin's bill?
5652. How deep can cormorants dive?
5653. Are corvids well represented in southern South America?
5654. What is American slang, for a person who acts as a watchbird while his confederate steals?
5655. The female of which waterfowl has a more developed crest than that of the male?
5656. Do parrots occur in the Falkland Islands?
5657. Which African pheasant roosts in trees?
5658. Are Red Junglefowl still numerous in the wild?
5659. Can hummingbirds scratch themselves while in flight?
5660. Does a breeding drake Barrow's Goldeneye have a greenish glossed head?
5661. What is the origin of the name "ptarmigan?"
5662. Does the Song Thrush inhabit Australia?
5663. Which North American warbler is named after a recluse?
5664. What is unusual about the plumage of the Red Avadavat during the non-breeding season?
5665. Which raptor readily takes to artificial nesting platforms?
5666. Which state has the largest population of breeding Piping Plovers?
5667. Is the Woolly-necked Stork restricted to tropical Africa?
5668. In America, which bird is known as a "jiddy-hawk?"
5669. Do incubating murres generally face the cliff, or do the birds face the ocean?
5670. Do female and juvenile Red-faced Warblers resemble males?
5671. Which starlings inhabit Vancouver, Canada?
5672. How did the Peterson bird field guide aid the war effort during World War II?
5673. What do all four Papua-Australian birds that were formerly in the family Grallinidae (Magpie-lark or Mudlark, Torrentlark, White-winged Chough and Apostlebird) have in common?

Questions

5674. Is the white on the wing of a White-winged Dove evident both in flight and at rest?
5675. If someone is referred to as a rook, what does this imply?
5676. In what type of trees do Secretary-birds generally nest?
5677. Are most parrots highly gregarious?
5678. Do Greater Roadrunners generally run with their crests elevated?
5679. What is the origin of the name "whydah?"
5680. What does an attacking crane generally do as it approaches its opponent?
5681. Which is the most aggressive and powerful American owl?
5682. Which bird is the Eurasian counterpart of the Blue-winged Teal?
5683. Which shorebird has variable-colored legs and bills?
5684. Which bird has been described as *the voice of Africa*?
5685. Which Australian bird is called a "zebra duck?"
5686. What is the basic color of a cock Japanese Pheasant?
5687. Which is the only European country where the Hoopoe might be encountered year-round?
5688. Does the Whooping Crane account for substantial Red-throated Loon egg mortality?
5689. How long does it take a male frigatebird to acquire adult plumage?
5690. Which is the smallest and most abundant Australian owl?
5691. Which is the largest North American swift, and how much does it weigh?
5692. Which flamingo has the most yellow on its bill?
5693. Aside from humans, what is the largest New Guinea animal?
5694. Do forest-dwelling owls have proportionally shorter wings than owls inhabiting more open areas?
5695. Who starred in the 1925 movie *The Eagle?*
5696. Which of the large falcons is best adapted to a desert lifestyle?
5697. Are trogons restless birds?
5698. Which northern bird is called a "King Kongfisher?"
5699. Why are the Trumpeter Swans of Red Rock Lakes Wildlife Refuge, Montana, non-migratory?
5700. Which northern woodpecker has a very direct flight, tending not to undulate in typical woodpecker fashion?
5701. What are the only two annual events obligating penguins to come ashore or up onto the ice?
5702. Which group of flying seabirds fights viciously over nest sites?
5703. How do cuckoos handle noxious-tasting hairy caterpillars?
5704. Excluding Alaska, where else do Emperor Geese breed?
5705. In 1990, how many parrots were smuggled into the U.S.?
5706. Which African passerines are called "feathered locusts?"
5707. Is a Shikra a falcon?
5708. How does a nesting Congo Peacock differ from nesting Indian and Java Peacocks?
5709. What is an irregular flaw in a precious stone called?
5710. Prior to the recent reintroduction attempts, when was the Thick-billed Parrot last reported in southwestern U.S.?
5711. Why is the scientific name of the Great Frigatebird misleading, and why was it so named?
5712. What happened to Guy Bradley, one of the first National Audubon Society wardens?

Questions

5713. What color objects do catbirds (bowerbirds) favor for their bowers or stages?
5714. The eggs of which bird are often taken by Sabine's Gulls?
5715. When and where were kiwis first bred in captivity?
5716. After whom was Cory's Shearwater named?
5717. Which bird has negatively impacted the Tahitian Lory?
5718. Does the dark-color morph of the Reddish Egret predominate in Texas?
5719. Do pygmy-geese graze?
5720. Which American bird did Thoreau believe had the most beautiful of avian voices?
5721. Do juvenile White Storks have red bills?
5722. When Common (Atlantic) Puffins were reintroduced into Maine, where was the founding stock collected?
5723. Do Red-tailed Hawks inhabit the tundra?
5724. Which fish-duck is most apt to frequent salt water?
5725. Which gamebird is commonly known as the Valley Quail?
5726. Is the Mourning Dove especially resistant to high temperatures?
5727. Is the Horned Grebe more inclined to feed on fish than the Eared Grebe?
5728. Which are the largest South American passerines?
5729. How did the name "goshawk" originate?
5730. Is Forster's Tern an exclusive North American nester?
5731. Which American bird was known as a "dusky duck?"
5732. Do penguins occur in Norway?
5733. Which new bird species was described from a mist-netted bird in Somalia, and released without a specimen being collected?
5734. When moving in a group, which bird was once described as *a bunch of hunch-backed old men with their feet tied*?
5735. Which bird adorns the logo of the International Council for Bird Preservation (ICBP)?
5736. With what country was the ornithologist Walter L. Buller associated, and which birds were named after him?
5737. Which American shorebird was known as a "sicklebill?"
5738. What color was the fabled mythical Phoenix?
5739. Which Australian bird is known as a "needle-beak shag?"
5740. Do toucans preen one another?
5741. What term is used to describe the calls of a courting Blue Grouse?
5742. Which northern auklet does not breed in large colonies?
5743. Which North American flycatcher has the name of a non-metallic chemical element in its common name?
5744. What is the Short-billed Marsh Wren commonly called?
5745. Do guillemots engage in collective mutual displays?
5746. Which is the largest North American finch, and what is its length?
5747. How did the name "cormorant" originate?
5748. The introduction of which animal adversely impacted the American Redhead duck?
5749. At what age do Rhinoceros Auklet chicks fledge?
5750. What do some female hummingbirds do to their nests as their chicks increase in size?

Questions

5751. Do Burrowing Owls ever dig their own burrows?
5752. Loons have how many brood-patches?
5753. In what kind of tree does Kirtland's Warbler nest?
5754. What is a bar sharpened at one end and often fitted with a claw at the other end, and is used as a lever for prying, called?
5755. Which neotropical waterfowl may nest in kingfisher burrows?
5756. Was Henslow's Sparrow named after an American historian?
5757. What color is the tip of an adult Ivory Gull's bill?
5758. Is a White Stork more apt to bill-clatter than a Black Stork?
5759. Do Snowy Plovers nest in Chile?
5760. How did the Northern Cardinal obtain its name?
5761. Do winter-plumaged Ross' Gulls have a neck-band?
5762. Which large wading bird preys on penguins?
5763. Are Plain Chachalacas wary?
5764. If handled, what might an Elf Owl do?
5765. When Passenger Pigeons were abundant, what was burned beneath a nesting tree to suffocate the birds?
5766. Do Anhingas occur in Oklahoma?
5767. Which African geese perch in trees?
5768. Are the neotropical ovenbirds monogamous?
5769. To which group of birds are the two Australian figbirds closely related?
5770. Do Black Vultures ever nest on the ground?
5771. What color is the center of the *eyes* of the long trailing train of a male Indian Peacock?
5772. Which white Antarctic bird is prone to fly with its legs dangling?
5773. What is unusual about the long, central tail feathers of an adult Pomarine Jaeger?
5774. In the U.S., do the ranges of the Smooth-billed and Groove-billed Ani overlap?
5775. Why did some American Indians not eat Wild Turkeys?
5776. Does the African Wattled Crane wade in deep water?
5777. Which is the most abundant eastern North American woodpecker?
5778. How long is the tail of a breeding male South American Lyre-tailed Nightjar?
5779. After what were many months in the Persian calendar named?
5780. Which bird was called a "pencilled cormorant," and why?
5781. Were Sacred Ibis reared in captivity in ancient Egypt?
5782. What do incubating Least Flycatchers do if approached by a human?
5783. In North America, which of the three scoters is most abundant and widely distributed?
5784. In the Antarctic, with which bird is a flying Snowy Sheathbill most likely to be confused?
5785. Do the tiny euphonias and chlorophonias have shorter incubation and rearing periods than the larger tanagers?
5786. How do roosting Long-eared Owls conceal themselves?
5787. Do sub-Antarctic and Antarctic-dwelling skuas prey on mammals?
5788. Where in the continental U.S. do Black Francolins occur?
5789. Why, at the age of 18, was Audubon sent by his father from France to Pennsylvania?

5790. With respect to plumage, how are juveniles of all North American thrushes similar?
5791. Which distinctive Australian pigeon often hunches down on its perch, assuming an almost hawk-like appearance?
5792. Why are Red Crossbills able to nest so early in the year?
5793. Does the Caspian Tern prey on tern eggs and chicks?
5794. Which American bird has been described as a giant chickadee, and why?
5795. Why do Eleonora's Falcons breed late in the season?
5796. Do female Phainopeplas have white wing-patches?
5797. During World War II, what were ladies serving in the Women's Royal Naval Service called?
5798. Which upland gamebird is confined to the British Isles?
5799. Which of the three guillemots is the largest?
5800. Do Black-billed Magpies prey on mammals?
5801. Which long-legged Florida bird can extract snails without breaking open the shells?
5802. How many Bald Eagles inhabit Alaska?
5803. How many species of nightjars breed in North America?
5804. What might a screech-owl do if approached by a flock of sparrows?
5805. What is the population size of the Humboldt Penguin?
5806. Do migrating Common Cuckoos of Eurasia fly continuously non-stop?
5807. How many species of pittas breed in Japan, and name them?
5808. Which country is the Booted Eagle's stronghold?
5809. What color are juvenile flamingos?
5810. Which American bird is known as a "gray cardinal?"
5811. Thayer's Gull was formerly regarded as a subspecies of which bird?
5812. Which American bird is called a "zebraback?"
5813. Which is the only North American blackbird not named for its color?
5814. What is McCormick's Skua more commonly called?
5815. Were females of the extinct moas larger than males?
5816. Do the larger species of hornbills lay a smaller clutch of eggs than smaller hornbills?
5817. Which is the rarest European-breeding gull, and what is its population size?
5818. In Latin America, which birds are called *picaflores?*
5819. Which northern bird is known as an "ice partridge?"
5820. In what 1990 movie did Goldie Hawn star as a hotshot corporate laywer, and Mel Gibson as a mystery man from her past?
5821. Which petrel nests on buildings?
5822. Which New Zealand honeyeater is the most accomplished mimic?
5823. Where do most McKay's Buntings breed?
5824. Do Rockhopper Penguins usually dive into the water head first?
5825. How did the Calfbird (Capuchinbird) gets its peculiar name?
5826. Is the bill of a pigeon soft at the tip?
5827. Do Andean Condors breed at sea level?
5828. Which cassowary is most likely to occur at high altitudes?
5829. Which finch is *essentially* restricted to California as a breeding species?
5830. What color is the bald head skin of the African Gray-necked Rockfowl?

Questions

5831. Do long-distance migrating grouse flock?
5832. Up to 1981, when rediscovered, which bowerbird had never been seen alive, and was known only from three, possibly four, trade-skins, all of which were males?
5833. Is a Laysan Albatross larger than a Black-footed Albatross?
5834. What type of bird is the Northern Waterthrush?
5835. Do voles prey on alcids?
5836. How were humans responsible for the extinction of the Auckland Islands Merganser?
5837. Do Evening Grosbeaks travel in flocks?
5838. What is the length and weight of the largest turaco?
5839. Do both South American cocks-of-the-rock have especially short incubation periods?
5840. Which African bird is so similar in appearance to the Rattling Cisticola that it can only be separated by voice in the field?
5841. In which herons does the male alone construct the nest?
5842. Which Australian marsupial has the name of a familiar aquatic bird in its name?
5843. How does a female Violet-crowned Hummingbird differ in appearance from a male?
5844. What is the Salmon-crested Cockatoo also called?
5845. What is a group of woodcock called?
5846. Do flying skuas and jaegers have rapid wingbeats?
5847. Do male oropendolas generally outnumber females?
5848. In Europe, what is the Arctic Loon called?
5849. In the Aleutian Islands (Alaska), with which two petrels does the Ancient Murrelet often nest?
5850. Are the toes of a Barn Owl feathered?
5851. What is the range of the Scaly-sided Merganser?
5852. Do both sexes of Secretary-birds incubate?
5853. Who starred in the 1970 movie *The Owl and the Pussycat*?
5854. Do Steller's Sea-Eagles generally nest on cliffs?
5855. In Australia, at what time of year are muttonbird (Short-tailed Shearwater) chicks harvested?
5856. Do swimming drake Oldsquaw ducks generally hold their tails flush with the water?
5857. Which birds are called "tree-nighthawks?"
5858. How long do Brown Noddy chicks remain in the nest after hatching?
5859. What color is the down of a Waved Albatross chick?
5860. In Australia, what type of bird is the Hardhead?
5861. How did the Wood Stork obtain its name?
5862. Which are the only ground-nesting North American icterids?
5863. When nesting, which of the three jaegers is least likely to attack an intruder?
5864. A foraging Black-capped Vireo has what titmouse-like habit?
5865. Where in Russia do Aleutian Terns breed?
5866. Why are kiwi burrow entrances surprisingly small?
5867. Which alcid often nests far inland?
5868. With which bird is the Ruddy Duck most often associated?
5869. When and where was Hawk Mountain Sanctuary established?
5870. Are swifts monogamous?

5871. What were Willow and Alder Flycatchers formerly called?
5872. Do migrating hummingbirds travel in groups?
5873. Is the Eurasian Wryneck a large bird?
5874. Do most swallows have flattened bills?
5875. What color is the powerful bill of the probably extinct Mexican Imperial Woodpecker?
5876. Which of the four most familiar North American tanagers is not named for its color?
5877. Which white-fronted goose has the most conspicuous eye-ring?
5878. Which are the only birds that normally swim with only their heads and necks above the water?
5879. How many species of grebes are endemic to Africa?
5880. Why is a winter-plumaged Black-throated Blue Warbler one of the easier warblers to identify?
5881. Which American bird was known as a "devil downhead?"
5882. What color is the gular pouch of a Brown Pelican during the early part of the breeding season?
5883. What color are the feet of an African Black Heron?
5884. How many species of waterfowl occur naturally in Britain?
5885. Which American bird is called a "chow-chow bird," and why?
5886. How far away from their nesting caves do Oilbirds forage?
5887. What color is the bill of an adult Inca Tern?
5888. How do wide rivers pose a risk to South American trumpeters?
5889. Is the Lesser Prairie-chicken more numerous than the Greater Prairie-chicken?
5890. In bygone years, why was crane fat placed in the ears?
5891. How long is the fledging period of a Common Loon?
5892. With which plant is the Lucifer Hummingbird generally associated?
5893. Are there any endemic waterfowl species in the Galapagos Islands?
5894. Where was the founding stock of Indian mongooses obtained that was foolishly introduced into Hawaii in 1883?
5895. In the South American Andes, which birds are sometimes known as "mountain hummingbirds," and why?
5896. What is a long tuft of kinky hair, which grows on a horse and cannot be made to lie smooth, commonly called?
5897. When tropicbird young take their first flight, do they immediately depart the nesting island?
5898. How do adult Australian Freckled Ducks supposedly differ physically from all other ducks?
5899. Are Sanderlings heavier than Dunlins?
5900. Do any typical nightjars construct nests?
5901. Are bee-eaters accomplished singers?
5902. What color do the adults of all three jaegers have on their heads?
5903. In the book *Jurassic Park* by Michael Crichton, which dinosaur was credited with being the most intelligent?
5904. What do fasting male Emperor Penguins that have incubated continuously throughout the Antarctic winter do if their chicks hatch before the females return?

Questions

5905. After whom was the endangered Rothschild's Myna named, and what institution did he found?
5906. In Australia, which bird is called a "tippet goose?"
5907. In what state is a Bald Eagle most likely to feed on road-kills?
5908. Do megapodes have large, powerful claws?
5909. How many feathers make up the crest of a Mountain Quail?
5910. The two goldeneye ducks are most closely related to which bird?
5911. Does a Razorbill in winter plumage retain the distinctive white vertical stripe on its bill?
5912. In South America, which bird is known as a "chaja," and why?
5913. On what do the large West Indian cuckoos in the genus *Saurothera* feed primarily?
5914. Which two families of gallinaceous birds are considered to be the most primitive?
5915. What is the popular name of the AIM-7 radar-guided air-to-air missile?
5916. Is the Ostrich the most fleet-of-foot of all two-legged animals?
5917. Does the specific name of the Black-necked Stilt imply that the first described specimen originated from Mexico?
5918. According to folklore, how did the Least Bittern produce its peculiar calls?
5919. What are flocks of Snow Buntings sometimes called?
5920. Semipalmated in the name of the Semipalmated Sandpiper refers to what?
5921. Mountain-dwelling birds occupy habitat equivalent to that found in more northern latitudes, but how many feet of altitude equals a degree of latitude?
5922. What is Scoresby's Gull of southern South America more commonly called?
5923. What is the population size of Spix's Macaw?
5924. Is the Purple-throated Fruitcrow highly social?
5925. Which two British birds fledge in the shortest period of time, and how much time is required?
5926. The alarm call of the Arctic Loon rather sounds like the call of which bird?
5927. Do nest-building hummingbirds use glue-like saliva?
5928. How does the breeding behavior of the Monetzuma Quail differ from most northern quail?
5929. Do adult crested penguins have dark brown eyes?
5930. Have Torrent Ducks ever been bred in captivity?
5931. Which American bird is known as a "desert sparrow?"
5932. Why do many tropical antbirds have bare blue skin around their eyes?
5933. Did the extinct Great Auk breed in Greenland?
5934. Which of the three jaegers is the most versatile feeder?
5935. In the U.S., do the breeding ranges of the Great-tailed and Boat-tailed Grackle overlap?
5936. How much more energy does a trotting rhea expend than a rhea at rest?
5937. Do Xantus' Murrelets have darker underwing linings than the nearly identical, but slightly smaller, Craveri's Murrelet?
5938. What color is the bill of a juvenile Yellow-billed Magpie?
5939. How long do Limpkin chicks remain in the nest?
5940. Do Pied-billed Grebes in winter plumage have a dark, vertical band on their bills?
5941. Do roosting nightjars keep their eyes open during the day?
5942. What is the most effective way to conduct a Kirtland's Warbler census?
5943. Is the Double-crested Cormorant really double-crested?

Questions

5944. Do Trumpeter Swans hold their necks in a graceful curve?
5945. Do all four avocets have seasonal plumage changes?
5946. How much does the nest of a Superb Lyrebird weigh?
5947. Are the Antarctic-nesting Gentoo Penguins migratory?
5948. Are Black Vultures abundant in Baja California?
5949. Which raptors prey on Dovekies?
5950. Do all North American upland gamebirds have broad, cupped wings?
5951. Is the nest of an African mousebird conspicuous?
5952. Which avian order is best represented in New Zealand, and how many species have been recorded there?
5953. Is the Northern Gannet an abundant breeding bird of Iceland?
5954. Do Burrowing Owls nest colonially?
5955. How many geese are *legally* shot by U.S. hunters?
5956. Which is the largest pigeon of Europe?
5957. What is the observation deck on the superstructure (island) of an American aircraft carrier commonly called?
5958. Which bird is credited with being the *dumbest*?
5959. How many North American corvids have the name of an individual in their common name, and which ones are they?
5960. How long is the fledging period of the Edible-nest Swiftlet?
5961. Why do newly-hatched tragopan and Congo Peacock chicks have well-developed primaries?
5962. How does the chick-feeding behavior of tropicbirds differ from all other pelecaniform birds?
5963. On what mammal dwelling might a Black Tern nest?
5964. When approached by a potential predator, does a nesting Ruddy Turnstone utilize a distraction display?
5965. At the height of the plume-hunter era in the late 1800's, how many people were employed in the feather-trade in Paris and environs alone?
5966. Does the Swallow-tailed Kite migrate in large flocks?
5967. Which bird is called an "oatmeal duck" or "monkey duck?"
5968. Which two states have the greatest numbers of breeding Common Loons south of Canada?
5969. Do swimming phalaropes float high on the water?
5970. Which three wood-warblers are known as "swamp warblers?"
5971. Which is the least social Atlantic alcid?
5972. Where the two species overlap, is the South Polar Skua more inclined to feed on penguin chicks than the Brown Skua?
5973. At the Cincinnati Zoo, which other birds shared the aviary with *Martha,* the last of the Passenger Pigeons?
5974. How do western Rufous-sided Towhees differ in appearance from the eastern races?
5975. What does the collective term pinniped for seals and sea lions mean?
5976. Which hosts are favored by the obligate brood parasitic-nesting African whydahs?
5977. Are kingfishers related to hornbills?
5978. When was the first Emperor Penguin egg collected?

Questions

5979. Compared to their size, do newly-hatched hummingbirds have large bills?
5980. Which is the only North American-nesting cormorant with a white throat-patch?
5981. Do the ground-nesting meadowlarks construct open nests?
5982. How many species of cockatoos inhabit Australia?
5983. When displaying Greater Prairie-chicken cocks stamp their feet, how rapidly are their feet moving?
5984. Is the pigment responsible for the green color of African turacos the only true green pigment known in birds?
5985. Which of the three long-tailed penguins is said to be the most aggressive?
5986. What is the high lookout on a ship called?
5987. Which bird is known as *beskrypaya gargarka* in Russia?
5988. Is the long bill of a Limpkin decurved?
5989. Do Eurasian Capercaillie hybridize with Ring-necked (Common) Pheasants?
5990. Why were New Zealand Weka rails formerly hunted?
5991. What is the major predator of the Auckland Islands Flightless Brown Teal?
5992. The Zebra Finch is native to which country?
5993. When the Kagu of New Caledonia was discovered around 1860, it was believed to be what type of bird?
5994. How many Dodo eggs are deposited in museum collections?
5995. Which is the most recently described grebe?
5996. Where do breeding pairs of terns generally copulate?
5997. Why was the Ancient Murrelet so named?
5998. Is fowl pox the same disease as avian diphtheria?
5999. Is the African Crowned Hawk-Eagle noted for noisy flight?
6000. Is the Sarus Crane thriving in India?
6001. What is a group of murres called?
6002. Which is the largest Mexican parrot?
6003. In obsolete usage, how was boasting described?
6004. Is the Long-eared Owl also known as a "bog owl?"
6005. Which American bird was called "Wilson's Thrush?"
6006. Which type of bird does the African Long-tailed Hawk superficially resemble?
6007. Are Leach's Storm-Petrels inclined to follow ships?
6008. At its lowest level in the late 1940's, what was the population of Hawaiian Geese (Nene)?
6009. Are any Australian raptors migratory?
6010. Do mousebirds have long, thin bills?
6011. Do flightless steamerducks nest in cavities?
6012. Which southern seabird is called a "nelly?"
6013. Who was the kicker for the Los Angeles Raiders NFL in the late 1980's and early 1990's?
6014. Do most nightjars inhabit the tropics?
6015. Which North American grackle is most apt to frequent saltwater habitat?
6016. Which penguin has the longest tail?
6017. Is the voice of a Great Gray Owl powerful?
6018. Which hornbills prey on tortoises?
6019. Are hummingbird eggs large relative to the female's size?

6020. What was the origin of the name "pheasant?"
6021. Why were House (English) Sparrows introduced into Brazil?
6022. What color is the breast of a male Resplendent Quetzal?
6023. In winter plumage, did the Great Auk retain the conspicuous white patch in front of its eye?
6024. Which American bird is called a "mud hen?"
6025. What is a building or group of buildings sometimes called that are old and dilapidated and house many people, such as a tenement house or tenement district?
6026. Other than the two northern magpies, which is the only black jay?
6027. Do Anhingas and darters have serrated bills?
6028. In old American slang, what was a discarded, partly-smoked cigarette or cigar called?
6029. Why is the Tahitian Lory called a "coconut-bird?"
6030. Based on its scientific name, was the first described Ruddy Duck collected in Mexico?
6031. How many species of cranes occur in the New World?
6032. Do all species of frigatebirds indulge in bill-clattering?
6033. Do Snowy Owls occur in Florida?
6034. Do foraging Brown Creepers only creep *up* a tree trunk?
6035. Do tobogganing penguins use their flippers, or do they use their feet?
6036. Are Yellow Rails *particularly* secretive?
6037. Is the Barred Owl abundant in Florida?
6038. What was the former name of the Solitary Vireo?
6039. Which colorful cockatoo is often killed by traffic on Australian highways?
6040. In Europe, do Lammergeiers feed at rubbish dumps?
6041. Do Wilson's Warblers hawk insects in flight?
6042. Which neotropical cotinga has the most elaborately modified feathers?
6043. Which bird is known as an "eves swallow?"
6044. What is the weight of a dried Brown Pelican skeleton?
6045. What is the cranberry named after?
6046. Is a Snow Bunting whiter in color during the winter than during the summer?
6047. Do all tinamous have only three toes?
6048. Do blinking birds use their nictitating membranes?
6049. How many Galapagos Hawks exist?
6050. Which is the most abundant North American cormorant?
6051. How much more efficient is a foraging Cattle Egret when associated with cattle than when foraging alone?
6052. How many species of grouse inhabit the New World?
6053. What color is the eye of a drake Canvasback?
6054. In Massachusetts, which bird is known as a "shad bird" or "shad spirit," and why?
6055. Do Common Ravens mate for life?
6056. Which icterid has the name of a man-made object in its name?
6057. Who recommended Charles Darwin for the job of naturalist on the HMS *Beagle*?
6058. What is a politician called during his or her last term in office?
6059. Are all storks exclusively carnivorous?
6060. Do any bowerbirds inhabit the arid interior of Australia?

6061. What is a puffin colony sometimes called?
6062. How do oystercatchers kill crabs?
6063. Do any hummingbirds have pleasant voices?
6064. In the Bible, what was the Egyptian Vulture called?
6065. Why were many Peregrine Falcons shot during World War II?
6066. Are fruitcrows corvids?
6067. Is the African Goliath Heron a colonial breeder?
6068. What is the only cockatoo with a red head and wispy crest?
6069. Which phalarope has the shortest and stoutest bill?
6070. Which albatross is sometimes a kleptoparasitic feeder?
6071. Which type of bird do Xantus' and Craveri's Murrelets superficially resemble?
6072. Do kingfishers prey on African giant earthworms?
6073. Do *both* sexes of the South American Horned Screamer have the distinctive long horn?
6074. Which petrel may cause significant Common (Atlantic) Puffin egg losses, and why?
6075. Does a Pine Grosbeak have a longer tail than an Evening Grosbeak?
6076. Do New Zealand kiwis avoid steep terrain?
6077. Do Crested Caracaras reuse the same nest year after year?
6078. What are rosellas?
6079. Which bird preys on Darwin's finch eggs?
6080. Which alcids whistle?
6081. Why do Ostriches become sodden in the rain?
6082. In Colombia, which bird is symbolic of liberation from Spanish rule?
6083. Why do most birds lay their eggs early in the morning?
6084. Do breeding drake Common Goldeneye ducks have a conspicuous crescent-shaped white spot in front of the eye?
6085. How quickly can an American Robin digest a cherry?
6086. Is the Orchard Oriole especially shy?
6087. In the New World, who were the first Europeans to exploit the Great Auk?
6088. Were pewees named for their small size?
6089. How many species of American quail have spurs?
6090. With which bird does the Hooded Crane hybridize?
6091. What is the usual clutch size of the Royal Tern?
6092. Foraging Limpkins have what rail-like habit?
6093. What is the basic color of the male Java Peacock?
6094. Are molting petrels flightless?
6095. Which is the only Arctic fish-eating waterfowl?
6096. On what type of insects does Wahlberg's Eagle feed?
6097. Which jaeger has the widest distribution?
6098. Which bird nests at the lowest altitude?
6099. Are birds-of-paradise powerful fliers?
6100. Does the Green Jay occur in Arizona?
6101. Why is a flying tinamou especially inefficient at maneuvering?
6102. After which bird was the Mexican resort island of Cozumel named?
6103. Does the Montezuma (Mearns') Quail lay an unusually small clutch of eggs?

6104. Which shorebirds prey on lizards, and even small mammals?
6105. Do foraging cranes scratch the ground with their feet?
6106. Did John James Audubon co-author a book on reptiles?
6107. Is the European Tawny Owl tolerant of humans?
6108. In obsolete military jargon, what was a large artillery bombardment used in sieges and defensive works called?
6109. In the folklore of some European countries, what did the appearance of a hoopoe signify?
6110. Which owl breeds in the Falkland Islands?
6111. Are newly-hatched flamingo chicks naked?
6112. In ancient Egypt, what did an Ostrich feather symbolize?
6113. Which corvid pirates food from puffins?
6114. What color are the eyes of a Black-necked Stilt?
6115. Which South American bird is called a mossy-throated bellbird, and why?
6116. Which Australian bird is known as a dragoon bird?
6117. How many species of birds inhabit Afghanistan?
6118. Which of the three long-tailed penguins (Adelie, Chinstrap and Gentoo) is most apt to forage nearest its colony?
6119. How many of the approximately 45 birds-of-paradise inhabit the lowlands?
6120. Are trogons weak fliers?
6121. Which is the most widespread New Zealand rail?
6122. Which seabird does a small downy Shoebill stork chick superficially resemble?
6123. What nest sites are favored by the tiny falconets?
6124. Is the Pied-billed Grebe often encountered on salt water?
6125. Is the Screaming Piha a highly vocal hummingbird?
6126. The Roman emperor Augustus had what type of pet bird?
6127. Are Sunbitterns ground nesters?
6128. What color is the eye of a White (Fairy) Tern?
6129. Where in the tropics have Emperor Geese been recorded?
6130. Has any state selected a hummingbird as the state bird?
6131. In America, which bird is known as an "ermine owl?"
6132. Why was the Myrtle Warbler so named?
6133. Which southern albatross is the most abundant, and what is its population size?
6134. Do penguins sleep in a standing position?
6135. Does the Least Storm-Petrel have a forked tail?
6136. Which American waterfowl nests in house chimneys?
6137. How long is the nestling period of the two African ground-hornbills?
6138. How did the New Zealand Parsonbird obtain its name?
6139. What is the size of the U.S. breeding population of Anhingas?
6140. What color is the bill of a Sora rail?
6141. What are downy puffin chicks called?
6142. Does the Australian Brush-turkey have a bare head and neck?
6143. Do Red-faced Warblers invariably nest on the ground?
6144. Which American bird is called a "crippled bird," and why?
6145. Which is the only true Arctic-breeding crane?

Questions

6146. Do leopard seals prey on the heaviest seabird?
6147. Which four exotic waterfowl are established in Britain?
6148. How did the curlews obtain their collective name?
6149. Do all guineafowl have unfeathered heads and necks?
6150. How large is a male Tricolored Blackbird's territory?
6151. In the 18th century, which bird was called a lesser pelican?
6152. Which is the only towhee named after an individual?
6153. Which two North American-breeding hummingbirds occur east of the Great Plains?
6154. Does the journal published by the Cooper Ornithological Society come out six times a year?
6155. What was the record high count of the Everglades (Snail) Kite in recent years?
6156. The Ipswich Sparrow is a race of which bird?
6157. How many birds are shot annually in Italy in the fall?
6158. Were the kites flown by children named after the birds, or were the raptors named for the kites?
6159. Which colorful bird has been described as the most spectacular bird of the New World?
6160. What caused the death of Adolphus Lewis Heermann, after whom Heermann's Gull is named?
6161. Do Great Gray Owls have short, strongly-curved claws?
6162. Which American bird was formerly called a "hair-bird" or "hair sparrow," and why?
6163. Which petrel breeds at the highest elevation?
6164. Which bird exhibited the greatest sexual dimorphism with respect to bill size and shape?
6165. Which small birds feed on the sugary sap secreted around Yellow-bellied Sapsucker holes?
6166. What are ossuaries?
6167. What was the Great Egret formerly called?
6168. Which birds might be the Old World counterparts of the neotropical jacamars?
6169. In Roman legend, who was *Rhea* Silvia?
6170. Why were Eskimo Curlews called "prairie pigeons?"
6171. Which is the most widespread European woodpecker, and why is it so widespread?
6172. Are penguin hybrids common?
6173. Which American bird is called a "bastard broadbill?"
6174. How do immature male bowerbirds learn the basics of bower building?
6175. In New England, which birds are known as "mackerel gulls," and why?
6176. Which Alaskan auklet breeds the farthest north?
6177. Are whistling-ducks sexually dimorphic?
6178. What is unusual about the female African Rueppell's Parrot?
6179. Which jaeger has the most distinct chest-band?
6180. Why do vultures have such large crops?
6181. Which grebe is the most numerous worldwide?
6182. Do frigatebirds catch airborne flying fish?
6183. Which passerine lays the greatest number of eggs each breeding season?
6184. Do Short-eared Owls hoot?
6185. How many species of raptors breed in New Zealand, and which ones are they?

Questions

6186. Which is the smallest albatross?
6187. Do bee-eaters actually favor bees over other types of prey?
6188. What is the more familiar name for the Revillagigedo Shearwater?
6189. Is the Coral-billed Nuthatch a common Asian bird?
6190. Which is the southernmost-nesting raptor, and how far south does it breed?
6191. What color is the breast of a female Mountain Bluebird?
6192. Which American thrush has a specific name suggesting it was named after a mustelid?
6193. Do rollers often capture prey on the ground?
6194. If penguin guano is greenish, what does this imply?
6195. Do female North American pigeons and doves tend to be duller in color than their mates?
6196. Who is credited with coining the phrase *You don't have to be a hen to recognize a bad egg*?
6197. Do fighting coots use their feet?
6198. Do skuas have long, sharp claws?
6199. What percent of threatened bird species occur in the tropics or oceanic islands?
6200. What was the major reason for the decline of the Laysan Duck during the early 20th century?
6201. In Southern California, which bird is known as a "palm-leaf oriole?"
6202. Are Least Grebes noisy during the breeding season?
6203. Do King and Common Eiders hybridize?
6204. Which is the only South American inland-nesting cormorant?
6205. Why is the deep-sea fish *Eurypharynx pelecanoides* called a pelican fish?
6206. Which raptor do frigatebirds harass for food?
6207. Between 1890 and 1899 alone, how many tinamous were sold in the markets of Buenos Aires?
6208. Do Painted Buntings ever rear six broods a year?
6209. Which type of non-related wading bird does a foraging Spotted Sandpiper resemble?
6210. Juvenile Scarlet Ibises resemble the young of which bird?
6211. After whom is Attwater's Prairie-chicken named?
6212. Do American Kestrels (Sparrowhawks) feed extensively on sparrows?
6213. To which group of birds are the birds-of-paradise probably most closely related?
6214. How much more do kingfisher chicks approaching fledging weigh than adults?
6215. What color is the crest of an Erect-crested Penguin?
6216. What was the U.S. ten-dollar gold coin called?
6217. Why is the Palm-nut Vulture believed to be related to fish-eagles?
6218. Which alcids have the longest legs relative to their size?
6219. Which birds are sometimes known as "mimic-thrushes?"
6220. Which raptor has a conspicuous orange caruncle on top of its bill?
6221. Do female Willow Ptarmigan engage in courtship flights?
6222. Are insects longer-lived than birds?
6223. Why does a foraging African Shoebill prefer water with a low oxygen content?
6224. Are tinamous arboreal?
6225. Do kingfishers regurgitate pellets of indigestible parts of their prey?
6226. How does the winter plumage of a Western Grebe differ from its summer plumage?

Questions

6227. How many bowerbirds have head wattles?
6228. Do Pomarine Jaegers breed in colonies?
6229. Does the black on the wings of an American White Pelican extend onto the secondaries?
6230. Why do trumpeters gather under trees frequented by howler monkeys, parrots and toucans?
6231. Which is the smallest *species* of North American goose?
6232. What is the Orange-chinned Parakeet also called?
6233. Is the Chihuahuan Raven more inclined to gather in flocks than the Common Raven?
6234. How many U.S. colleges and universities have adopted an eagle as a mascot?
6235. Which bird has the largest toenail?
6236. What is commonly known as *Jewish penicillin?*
6237. What is the average heartbeat rate of a chicken?
6238. Which eider has the most extensive circumpolar range?
6239. How many species of gulls breed on tropical oceanic islands, and which ones are they?
6240. At what age do cranes commence breeding?
6241. What country constitutes the largest market for exotic, wild-caught birds?
6242. The name of which North American flycatcher contains the name of a man-made object?
6243. How does the behavior of the Forest Wagtail differ from the other wagtails?
6244. Are any partridges arboreal?
6245. Does the Merlin soar?
6246. Which is the largest heron that occurs in Alaska?
6247. According to theoretical predictions, how much larger is a kiwi egg than it should be?
6248. Is the American Painted Redstart a lowland bird?
6249. Do Hamerkops gather in groups?
6250. Which is the smallest nocturnal bird of Australia?
6251. What is a sea gooseberry?
6252. In summer plumage, the male of which ptarmigan is the most reddish in color?
6253. What is a person called that repeats or imitates another, often without understanding the meaning?
6254. Which southern U.S. raptor is known for laying a large percentage of eggs that fail to hatch?
6255. Where is the southernmost Gentoo Penguin rookery located?
6256. Is the Sage Grouse a swift runner?
6257. Do Cape Pigeons (Painted Petrels) dive underwater after prey?
6258. After which bird is the Royal Air Force jump-jet named?
6259. Where is the largest concentration of nesting Black-browed Albatross located, and how many breed there?
6260. Do both sexes of the endangered, flightless New Zealand Takahe incubate, and how long is its incubation period?
6261. Do adult Red Phalaropes have black bills?
6262. Which hummingbird has the shortest bill, and how long is it?
6263. Which South American seabird is known as a piquero?
6264. Do potoos have greatly curved claws?
6265. Are owls inhabiting thick woodlands more vocal than open country owls?

6266. Which of the three jaegers is most likely to hover?
6267. Which alcid is most apt to stray *well* inland?
6268. Are openbill storks primarily diurnal feeders?
6269. Are toucan eggs relatively large compared to the bird's size?
6270. What is the national bird of Colombia?
6271. Which was the only bird new to science discovered on Audubon's 1883 trip to Labrador?
6272. Which North American owl projects a falcon-like appearance when flying?
6273. Did *Archaeopteryx* have fused vertebrae like modern birds?
6274. What significant World War II sea and air battle in the Pacific was dubbed the *Marianas Turkey Shoot?*
6275. Which is the smallest North American swift, and how much does it weigh?
6276. Do hawks and eagles consume the red meat of their prey prior to the abdominal organs?
6277. Which American waterfowl is most dependent on fruits and nuts?
6278. Which European explorer is credited with discovering the Oilbird, and in which country was the unique bird discovered?
6279. If a species is polytypic, what does this imply?
6280. What was one of the objectives of the *Tanager* expedition to Laysan Island in the early 1920's?
6281. Do *both* sexes of rollers engage in spectacular courtship flights?
6282. How long is the nestling period of a Black-footed Albatross?
6283. Is the Marbled Murrelet about the same size as an American Robin?
6284. How do Crested Caracaras know exactly where to dig up turtle nests?
6285. Which tern carries a number of small fish crosswise in its bill like a puffin?
6286. Is the Northern Saw-whet Owl less numerous than commonly believed?
6287. Which goose has the smallest range?
6288. How many species of nightjars occur north of 60°?
6289. Is the flightless Kakapo (Owl Parrot) the only New Zealand bird that gathers in a lek to court?
6290. Which bird adorns the logo of the Jersey Wildlife Preservation Trust (Great Britain)?
6291. What part of an Ostrich did ancient Egyptians consider a gastronomic delight?
6292. How many Wild Turkeys are legally shot annually?
6293. Do trumpeting (calling) Emperor Penguins point skyward with their bills?
6294. Which shorebird has a heavy, black, laterally-compressed bill?
6295. Most motmots have what color ear-patches?
6296. What is the primary mammalian predator in Australia of the Double-wattled Cassowary?
6297. Do Komodo dragons prey on birds?
6298. What color is the large, bare ring of orbital skin of a male Frilled Monarch of northern Australia and New Guinea?
6299. According to Greek mythology, who was the daughter of Uranus and Gaea who, because she was the mother of Zeus, Hestira, Demeter, Juno, Pluto and Poseidon, was known as the *mother of the gods*?
6300. Do grouse commonly rear more than one brood a year?
6301. What colors does a male Regent Bowerbird adorning its bower favor?
6302. What was the fate of the last Laysan Honeycreeper?

6303. Which are the only storks that retract their necks in flight?
6304. According to Alabama slave folklore, what would happen if a crane circled a house three times?
6305. Which is the only New World nightjar to become extinct in historical times, and was its demise the result of depredation by introduced predators?
6306. Which petrels do not have slender bills?
6307. Which bird do the Zulus of Africa call a "warrior bird?"
6308. Which institution owns all but three of the priceless life-sized water-color paintings from which the plates of the double elephant folio were engraved for Audubon's *The Birds of America*?
6309. Do kiwi pairs sleep together in the same burrow?
6310. Which birds have the greatest wingspan relative to their weight?
6311. Which cormorant breeds the farthest north, and where is the northernmost colony located?
6312. Why is it advantageous for hanging-parrots to sleep upside-down?
6313. Do Common and Thick-billed Murres often hybridize?
6314. What is the origin of the booby generic name *Sula*?
6315. The Red-throated Caracara is most likely to be confused with which type of bird?
6316. Does an ibis often circle and soar?
6317. Which icterid undertakes the longest migration?
6318. Is there only a single species of roadrunner?
6319. Which pochard ducks have crests?
6320. Do swifts have flat skulls?
6321. Are male Palmchats more colorful than females?
6322. Why has the South American Hoatzin seldom been captively maintained?
6323. What is the more familiar name for the Massena Trogon?
6324. Does the Egyptian Vulture have a square tail?
6325. Do American Arctic Loons winter primarily along the Atlantic coast?
6326. When Ospreys are fishing, are most strikes successful?
6327. How many species of parrots inhabit Bolivia?
6328. How many birds are so specialized that they feed solely on flowers?
6329. Are wrens gifted songsters?
6330. Of the ten Mexican cotingas, which is the only brilliantly-colored one?
6331. What color is the neck-ruff of a neotropical King Vulture?
6332. Do Keas dig into the ground in quest of insects?
6333. When was the endangered Cahow initially protected?
6334. What is unusual about the crest shape of a female Mexican Imperial Woodpecker?
6335. On what do wintering Blue Grouse mainly feed?
6336. How were Imperial (Blue-eyed Shags) of South Georgia formerly captured at sea by whalers?
6337. Which is the most familiar Old World weaver?
6338. Which icterid is the largest, and how long is it?
6339. Are Snowy Owl nests well lined?
6340. According to U.S. National Research Council estimates, how many metric tons of oil enter the sea annually as a result of human activities?

6341. Which American bird was called a "corn thief?"
6342. Do foraging bulbuls ever descend to the ground?
6343. How did the name "puffin" originate?
6344. Do Bald Eagles nest along the Texas gulf coast?
6345. Which Old World warbler breeds in the U.S.?
6346. Are breeding Red-winged Blackbirds almost entirely insectivorous?
6347. Do male wrens incubate?
6348. Do sandgrouse have long tails?
6349. How did the Cuban Grassquit reach the Bahamas?
6350. What journal is published by the Royal Australian Ornithologists' Union?
6351. Which two American grebes are the most abundant?
6352. What are young lovers often called?
6353. Is the endemic duck of Laysan Island flightless?
6354. Which raptor occasionally lands on ships far out to sea?
6355. Do bee-eaters have a swallow-like flight style?
6356. At what age are young Rock Ptarmigan capable of flight?
6357. Which North American grebe may rear two broods annually?
6358. Why are many male hawks smaller than their mates?
6359. How long are the tail quills of a cassowary?
6360. Are Common Loons common?
6361. Which ratites are preyed on by felines?
6362. What color are both wing and tail flight feathers of a male Andean Cock-of-the-rock?
6363. Which raptor steals food from Marabou Storks?
6364. Do male hummingbirds feed their chicks after fledging?
6365. Is the Osprey noted for rapid wingbeats?
6366. When did the Welcome Swallow reach New Zealand from Australia?
6367. When Milky or Lesser Adjutant storks thrust their entire bill down a hole in a mudflat, what prey are they generally seeking?
6368. Are Egyptian Plovers (Crocodile-birds) wary of humans?
6369. What is the population size of the Fiordland Crested Penguin of New Zealand?
6370. Which are the only charadriform birds that do not feed primarily on animal life?
6371. Which bird is featured on the coat-of-arms of Colombia?
6372. Which are the only oceanic shorebirds?
6373. Which bird has been described as the most gaudily-colored North American bird?
6374. How often are attacking Peregrine Falcons successful in capturing prey?
6375. Is the Long-eared Owl one of the least nocturnal owls?
6376. Which state constitutes the heart of the current range of the Greater Prairie-chicken?
6377. What was the Bronzed Cowbird formerly called?
6378. Do Oilbirds have long toes and sharp nails?
6379. Do White (Fairy) Terns generally forage far from shore?
6380. Are West Indian todies usually encountered in pairs?
6381. Which American bird is called a "moose bird," and why?
6382. How many species of birds have been recorded in Cyprus?
6383. What is the average weight of a wild Muscovy Duck drake?

Questions

6384. What type of birds often flock with chickadees and titmice?
6385. Which is the only shorebird confined to riverbeds *throughout* the year?
6386. What is the current American name of the bird formerly known as Richardson's Owl?
6387. Do newly-hatched Australian Pink-eared Ducklings have spatulate bills?
6388. Is a carrier pigeon also known as a homing pigeon?
6389. How many species of longspurs inhabit North America, and which ones are they?
6390. Does the Greater Prairie-chicken prefer drier habitat than the Lesser Prairie-chicken?
6391. How much did the extinct Carolina Parakeet weigh?
6392. The Eurasian Wallcreeper is most closely related to which birds?
6393. The vocalizations of the South American Sharpbill are said to resemble the call of which arboreal animal?
6394. What color are the bills of both sexes of *all* races of the South American Torrent Duck?
6395. Which northern bird reportedly has nested on large ice-floes?
6396. Do titmice construct small nests?
6397. Do most corvids have feathered nostrils?
6398. What famous naturalist originated the phrase *gloried reptiles*, and what does it mean?
6399. Which New Zealand bird is called "old swampie" or "pook?"
6400. Which bird is a major predator of Flightless Cormorant chicks?
6401. What is peculiar about the eyes of a sandgrouse?
6402. Bald Eagles prey on which mustelid?
6403. What is the spoutshell, a marine mollusk with an outer lip elongated like a siphon, also called?
6404. Do cormorants ever engage in cooperative feeding?
6405. What unusual nesting habit do Snail Kites have that is not known for any other bird?
6406. Which bird hybridized with captive Passenger Pigeons?
6407. Excluding owls, which five raptors have the most northerly breeding range?
6408. Do flying hornbills flap continuously?
6409. Why do Australian lyrebirds dig extensively?
6410. With what do oxpeckers line their nests?
6411. How long would a hummingbird's wing be if the bird was as large and as heavy as a swan?
6412. During 19th-century plume-hunter times, what were the white downy breast feathers of the Great Crested Grebe called?
6413. Which birds did Thoreau describe as *true spirits of the snowstorm*?
6414. Which American bird was called a "black-capped thrush?"
6415. Do Aplomado Falcons hunt alone?
6416. Why did the Norsemen, even prior to 880 A.D, take live birds with them on their long ocean voyages?
6417. How many fossilized skeletal remains of *Archaeopteryx* have been discovered?
6418. With respect to the number of species, which group of Arctic-nesting birds outnumbers all other birds breeding in the far north?
6419. Does the Emperor Penguin have a distinct juvenile plumage?
6420. Who starred in the 1926 silent black-and-white movie *Sparrows?*
6421. Why is a kiwi egg elongate in shape, rather than round?
6422. When were pigeons first used to carry messages?

Questions

6423. What must backyard bird-feeders be stocked with to attract Pileated Woodpeckers?
6424. Do Lammergeirs (Bearded Vultures) fly at high altitudes?
6425. How much does an Anhinga weigh?
6426. Where do Ospreys breed south of the equator?
6427. Do titmice pluck hair from mammals?
6428. Which is the most numerous *resident* Australian shorebird?
6429. Do Northern Harriers often forage at dusk?
6430. What does Toroshima, the name of the island stronghold of the Short-tailed Albatross, mean?
6431. Does the Great Horned Owl inhabit the Amazon rainforest?
6432. Which is the third largest penguin, and how much does it weigh?
6433. Which group of birds has the most highly-developed brain?
6434. How have recreational fishermen negatively impacted Mute Swans in Great Britain?
6435. Which are the only birds that do not brood their young?
6436. Does the Palm Warbler forage on the ground?
6437. Are female neotropical bellbirds vocal?
6438. How many mice does an American Kestrel consume annually?
6439. How did the name "thrasher" originate?
6440. What is the number of Tricolored Blackbird nests per acre (0.4 hectare) in dense colonies?
6441. Which bird is listed first in the sixth edition of the A.O.U. *Check-list of North American Birds*?
6442. Why are South American gauchos often accompanied by dogs?
6443. Are seabirds more colonial than terrestrial birds?
6444. Why are many seabirds white?
6445. Do puffins copulate at the nest site?
6446. Why do vultures have less powerful feet than other raptors?
6447. Which is the largest heron of Japan?
6448. Do Swamp Sparrows often gather in flocks?
6449. Where do Arctic-breeding Common Ravens generally nest?
6450. Did the parrot name *lory* originate with Australian aborigines?
6451. The sexes of which high-altitude gamebird were at one time considered different species?
6452. Why is fishing often better downstream from a heron colony than upstream?
6453. Do the tropical toucans mob potential predators?
6454. Is the heart-rate of a turkey more rapid than the heart-rate of a pelican?
6455. Which group of birds is most likely to consume hairy caterpillars?
6456. Is the bill of a New Zealand kiwi rigid?
6457. In Australia, shearwater (muttonbird) oil is used in what cosmetic product?
6458. Do aracaries have proportionally shorter tails than the larger toucans?
6459. Which group of long-legged wading birds is named after a large mammalian carnivore?
6460. In Australia, which large bird is sometimes called a "mountain pheasant?"
6461. Do female hummingbirds have shorter bills than males?
6462. What did the ancient Greeks believe owls could do?
6463. How did the Cheer Pheasant obtain its name?
6464. Which continent has the most species of eagles, and how many occur there?

Questions

6465. Does the large, fleshy caruncle of a male Andean Condor extend onto its bill?
6466. What do Ancient Murrelet parents feed their chicks?
6467. Which bird does the Sharp-shinned Hawk most closely resemble?
6468. Which American bird is known as a "cutwater?"
6469. In Texas, is a Bronzed Cowbird more likely to parasitize a Northern Oriole nest than a Brown-headed Cowbird?
6470. Which sea-duck is most capable of taking off from the water without pattering over the surface?
6471. What is the clutch size of the Great Argus pheasant?
6472. What type of bird is a Green Jery?
6473. Which American duck often grazes on western golf courses?
6474. What is a group of sparrows called?
6475. According to Indian legend, why is the breast of the male Resplendent Quetzal bright red?
6476. Does the Masked Booby breed in Mexico?
6477. Is the Hyacinth Macaw wary around humans?
6478. The call of which New Zealand bird might be mistaken for the vocalizations of a kiwi?
6479. In what European country was the southernmost Great Gray Owl nest discovered?
6480. How did the shearwaters known as muttonbirds by Australians and New Zealanders get the distinctive name?
6481. Which African eagle is the most powerful?
6482. What professional U.S. hockey team is named after a seabird?
6483. What is the Australian Short-billed Corella also called?
6484. Are Red Crossbills vocal in flight?
6485. Do pigeons and doves hatch with opened eyes?
6486. Does the Lesser Noddy nest in trees?
6487. Is a Wedge-tailed Eagle more apt to spook Brolga cranes into flight than a White-bellied Sea-Eagle?
6488. Why were Mexican Thick-billed Parrots never in high demand in the pet-trade?
6489. Which northern gull spins around on the water like a feeding phalarope?
6490. Are ptarmigans the only *group* of birds that turn completely white during the winter?
6491. Which large New Zealand bird is one of the boldest and most inquisitive of all birds?
6492. Do the breeding ranges of the Black-billed and Yellow-billed Magpie overlap?
6493. What color are the foot webs of a White (Fairy) Tern?
6494. Along the American west coast, are most Ruffed Grouse gray-color phase birds?
6495. What does the female Superb Lyrebird do with its chick's droppings?
6496. Do diurnal migrants flock more than nocturnal migrants?
6497. Does the duck-billed platypus swim faster than the normal swimming speed of most ducks and geese?
6498. Which two groups of closely related birds comprise the family Motacillidae?
6499. Which American bird is locally known as a "garoo?"
6500. Do American Ruddy Ducks rear two broods a season?
6501. Is the beard of a Lammergeier (Bearded Vulture) evident from a distance?
6502. Which is the only North American land bird that is dark ventrally and pale-colored dorsally?

Questions

6503. How long are the leg spurs of a Sage Grouse?
6504. Why was the Slender-billed Prion considered vermin in the Falkland Islands?
6505. What is a dog called with a long, slender, pointed nose?
6506. Where is the densest concentration of nesting raptors located, and which raptors are the most numerous?
6507. Which gamebirds undoubtedly inspired some of the dances of the American Plains Indians?
6508. Do nest-building flamingos use their feet?
6509. Do most tree-nesting birds construct nests within 6-8 feet (1.8-2.4 m) of the ground?
6510. Do Secretary-birds have bare, bright-red faces?
6511. Feathers consist primarily of what substance?
6512. Which is the only Hawaiian-breeding seabird that does not use a nest?
6513. What is the tarsometatarsophangeal joint?
6514. What do Australian Banded Stilt chicks have in common with the young of flamingos, some penguins and Common Eiders?
6515. Which is the largest bird of the subarctic?
6516. Where does an incubating murre position its egg?
6517. Which birds come closest to actually laying eggs in the water?
6518. Is the abdomen of a Spotted Owl spotted?
6519. Which is the only entirely white Asian crane, except for its black primaries?
6520. Do ibises have a distinct breeding plumage?
6521. On what does the Madagascar Serpent-Eagle primarily feed?
6522. Are waxwings wary birds?
6523. What does the nightjar family name Caprimulgidae mean?
6524. Which is the most intensively-studied American songbird?
6525. Which sex of Common (Atlantic) Puffin does the most digging when nest burrows are dug?
6526. In Newfoundland, which gamebird is known as a "rocker?"
6527. Which Australian waterfowl has a prominent cranial knob?
6528. Are birds the most admired and the most studied of all animals?
6529. A two-egg clutch accounts for what percent of a Golden Eagle's weight?
6530. What does the name *Archaeopteryx* mean?
6531. Are defensive nesting Peregrine Falcons vocal?
6532. What is the rictus?
6533. Will Little Blue (Fairy) Penguins incubate eggs other than their own?
6534. Why is the Florida Scrub Jay at risk, and how many birds remain?
6535. Why do larks and pipits have such a long hind claw?
6536. Which of the two American meadowlarks is most apt to nest on damp ground?
6537. In terms of impact on the overall population, which bird was most severely impacted by the devastating 1989 *Exxon Valdez* oil spill in Alaska?
6538. Do precocial birds lay larger eggs than altricial birds?
6539. Which American raptor was named after a passerine?
6540. Which endangered Puerto Rican bird was negatively impacted by *Hurricane Hugo* in 1989?

Questions

6541. Which northern waterfowl is said to walk along stream bottoms against the current?
6542. Do fruit-picking South American Oilbirds hover?
6543. How does the shape of Snowy Owl eggs differ from the egg shape of most owls?
6544. Is a carrier pigeon a variety of domestic pigeon?
6545. Which is the largest North Atlantic shearwater?
6546. If a human expended energy at the same rate as a hovering hummingbird, how many pounds of perspiration would have to be evaporated an hour to keep the skin temperature below the boiling point of water?
6547. Does the Florida Scrub Jay mate for life?
6548. Do Common Nightingales sing at night?
6549. Who wrote the popular novel *The Red Badge of Courage*?
6550. Are neotropical cotinga chicks naked at hatching?
6551. What color are the toes of an American Wood Stork?
6552. What country has the largest Peregrine Falcon population?
6553. Are Ancient Murrelets colonial nesters?
6554. Which bird symbolizes the British Royal Society for the Protection of Birds (RSPB)?
6555. What is the name of the fifth satellite of Saturn?
6556. Where do most North American Eared Grebes winter?
6557. What is the national bird of the Bahamas?
6558. When White-tailed Sea-Eagles were reintroduced into Scotland, from what country did the stock originate?
6559. Are incubating Common Ravens fed by their mates?
6560. Do egg-laying reptiles have longer incubation periods than birds?
6561. Do herons have exceptionally large oil (uropygial) glands?
6562. In Great Britain, what is a truck called?
6563. Does the Bufflehead duck only breed in North America?
6564. How much pressure per square inch can the Hawfinch exert with its enormous bill?
6565. Does the Short-eared Owl breed in the Arctic?
6566. Which sex of racing homing pigeon is the swiftest?
6567. During the days when New Caledonian Kagus were exploited, which feathers were the plume-hunters specifically seeking?
6568. What event is described as the greatest adventure in a bird's life?
6569. In ancient mythology, which African bird was believed to represent the original Phoenix?
6570. King Solomon, the wisest of men, was reputed to be able to talk to the birds and beasts, but how was this accomplished?
6571. When did American plume-hunters commence killing egrets?
6572. Which of the six white pelicans are least inclined to feed cooperatively?
6573. In America, which bird is called a "spectral owl," and why?
6574. Do Bald Eagles migrate entirely within the northern hemisphere?
6575. Why is the Firewood-gatherer troublesome to Argentine power companies?
6576. Do Northern Goshawks consume their prey in trees?
6577. What is the dominate cock at a Sage Grouse lek called?
6578. Why must swans run over the surface of the water prior to becoming airborne?
6579. How much area is required to support a single pair of Ivory-billed Woodpeckers?

6580. Which large, captively-reared bird has been reintroduced into Colombia?
6581. Are the nests of most small birds constructed within a week or less?
6582. During the course of evolution, did insects achieve flight prior to birds?
6583. Do Ross' Gulls nest on small offshore Arctic islands?
6584. What is the national bird of Argentina?
6585. Are toucan eggs spherical in shape?
6586. Are any southern hemisphere mergansers sexually dimorphic?
6587. How much thicker is the shell of an Ostrich egg than the eggshell of a hummingbird?
6588. What is the major predator of fledging albatross of the northwestern Hawaiian Islands?
6589. The milk of which mammal is said to be chemically similar to pigeon milk?
6590. According to Exodus in the Bible, which bird saved the children of Israel from starvation?
6591. Which American bird is known as a "smoking duck?"
6592. What is a group of herons or bitterns called?
6593. Which of the two murres is superior at terrestrial locomotion?
6594. Why do Great Horned Owls seldom prey on drumming Ruffed Grouse?
6595. Do Brown-headed Cowbirds have longer bills than Bronzed Cowbirds?
6596. What was the primary cause for the drastic decline of Attwater's Prairie-chicken, and how many remain?
6597. Do tyrant-flycatchers eat fruit?
6598. Which endemic parrot did early European settlers of New Zealand consider a delicacy?
6599. Why was General Winfield Scott, after whom Scott's Oriole is named, called *Fuss and Feathers*?
6600. How many of the approximately 91 species of larks occur in Africa?
6601. In American slang, if someone has been gulled, what has happened to him?
6602. Which is the smallest Australian parrot?
6603. Are the South American plantcutters especially wary?
6604. Are female Rock (Pigeons) Doves longer-lived than males?
6605. Do Snowy Egrets stray as far north as Alaska?
6606. Do the insectivorous African woodhoopoes prey on nestlings?
6607. The Humboldt Penguin is restricted to which two countries?
6608. Are loons fed by their parents after fledging?
6609. Do Barn Owls consume prey on the ground?
6610. Is the Sharp-tailed Grouse a forest bird?
6611. Early European settlers believed the Australian Cape Barren Goose was which type of bird?
6612. What is the principal predator of birds in rural America?
6613. Which is the only vulture from whom the Marabou Stork does not steal scraps?
6614. Do hornbills establish strong pair bonds?
6615. What is a group of chickens sometimes called?
6616. How many species of barbets breed in Costa Rica, and which ones are they?
6617. What unusual male hair style of the 1970's was named after an American Indian tribe?
6618. In dense colonies, how many nesting Lesser Flamingo pairs are concentrated in a square meter?
6619. Are the most modestly-colored birds among the most accomplished singers?

6620. Which is the most numerous Australian parrot, and what is its population size?
6621. Do mousebirds carry their crests erect?
6622. Is the Wrentit a chaparral resident?
6623. Do the New Zealand kiwis have tiny ear openings?
6624. Do the neotropical woodcreepers use their tails as a brace?
6625. What is the eye color of an adult Whooping Crane?
6626. Do female tragopans have lappets?
6627. Which insect is often called the state bird of Minnesota?
6628. Breeding Bewick's Swans are restricted to what country?
6629. Are meadowlark nests conspicuous?
6630. Is the Cedar Waxwing a breeding bird of Europe?
6631. Which songbird is a threat to the endangered California Least Bell's Vireo?
6632. In North America, which goose breeds the farthest north?
6633. Why did Eskimos believe the Boreal Owl was easy to capture?
6634. Do any thrashers occur in South America?
6635. Do swift wings curve backwards?
6636. What is the normal clutch size of most antbirds?
6637. Why are the nests of many cotingas reduced in size compared to the size of the bird?
6638. What do falconers call the disease trichomoniasis?
6639. Are the shells of gannet eggs exceptionally thick?
6640. Does Virginia's Warbler nest high up in trees?
6641. Do woodpeckers bathe frequently?
6642. Do *all* North American quail gather in coveys during the non-breeding season?
6643. Which Hawaiian-nesting seabirds typically have the lowest breeding success?
6644. When and where has the International Ornithological Congress been held in the Southern Hemisphere?
6645. Do Dovekies generally copulate in the water?
6646. What is a tern colony often called?
6647. Why do phalaropes float higher on the water than other swimming shorebirds?
6648. Do the Palmchats of Hispaniola roost in their communal apartment nests?
6649. During the winter, which is the most numerous of the spot-breasted thrushes in the U.S.?
6650. Which small birds was Homer, the ancient Greek poet, particularly fond of eating?
6651. How many Keas inhabit New Zealand?
6652. In Florida, which icterid feeds extensively on snails?
6653. What happens to a Blue Jay's plumage in heavy rainstorms?
6654. Is Pallas' Fish-Eagle a coastal bird?
6655. The Hooded Crane winters primarily in what country?
6656. Does the Iceland Gull breed in Iceland?
6657. In the southern U.S., where the Barred Owl inhabits swampy woodlands, what are its primary food items?
6658. Which American bird is known as a "horse-duck?"
6659. What is the origin of the Dalmatian Pelican's name?
6660. What is the origin of the crane's generic name *Grus*?
6661. Do juvenile West Indian todies have the characteristic red throat of adults?

6662. Do neotropical Jabiru storks have black on their wings?
6663. The fat of a rhea was formerly used for what?
6664. How many bird species occur in the Netherlands Antilles?
6665. Do White-tailed Sea-Eagles generally nest in trees?
6666. Which birds are the major predators of Great Cormorant chicks in Africa?
6667. Do loons fly at higher altitudes than most waterfowl?
6668. Which ratites are often associated with sheep and cattle?
6669. Where do flamingos breed in Mexico?
6670. Do male swallows always help feed their chicks?
6671. The feeding habits of which African mammals most closely parallel those of the Lammergeier?
6672. Is the Hawaiian Crow black?
6673. Adolphus L. Heermann, of Heermann's Gull fame, had what debilitating disease?
6674. Was the Australian Sandstone Shrike-Thrush named for its habitat preference?
6675. When courting Galapagos Flightless Cormorants swim around one another with their necks twisted, what is this behavior called?
6676. Which large South American bird was introduced into the Ukraine prior to World War II?
6677. Aside from the Wandering, Royal and Amsterdam Albatrosses, which albatross is the largest?
6678. What nest site is favored by the Australian Golden-shouldered Parrot?
6679. Do Everglade (Snail) Kites nesting in cattails construct substantial nests?
6680. Did the name "goose" formerly only apply to the gander?
6681. The Australian Regent Bowerbird was formerly believed to be which type of bird?
6682. During the winter, what percent of bird species found in Panama consists of northern migrants?
6683. Do newly-hatched loon chicks have variable-colored bare head spots?
6684. In the most recent taxonomic revision, the Hoatzin is considered to be related to which birds?
6685. Do Steller's Sea-Eagles prey on albatrosses?
6686. When a female hornbill emerges from its sealed nest hole, leaving the nestlings behind, does she re-seal the hole?
6687. Does the Razorbill incubate in an upright position like a murre?
6688. Which North American hummingbirds do not have black bills?
6689. Which familiar western American desert bird does the Long-tailed Ground-Roller resemble?
6690. Do Brent geese fly in V-formation?
6691. What is one of the favored foods of the Australian Willie-wagtail?
6692. What percent of buteo hawks fledged survive to breeding age?
6693. The vocalization of which bird is said to be similar to the loud ringing call of an Impeyan Pheasant (Himalayan Monal)?
6694. Just after the turn of the century, did the Great Egret feathers sold to London milliners alone exceed a ton in weight?
6695. Why is the scientific name of the Laughing Kookaburra misleading?
6696. Do penguins have distinctive bill plates like petrels?

Questions

6697. When pursued by an aerial predator such as a Northern Goshawk, how do Ruffed Grouse use snow banks?
6698. Which large African bird is known as a "malagash?"
6699. When Scarlet Ibis were introduced into Florida in 1960, how was this accomplished?
6700. Which New World tropical animals forage in the manner of a fish-owl?
6701. What was the mythical griffon?
6702. What color is the inner eggshell of New World vultures?
6703. What color is the inner shell of a falcon's egg?
6704. What is a Mute Swan colony called?
6705. In what 1981 movie did Sylvester Stallone play an undercover policeman in pursuit of an international terrorist?
6706. What is a gathering of Rooks called?
6707. Are territorial Red-winged Blackbirds especially aggressive?
6708. Which of the two murres is better adapted to feeding on bottom-living fish and invertebrates?
6709. Do grebes generally dive deeper than loons?
6710. Do flushed high-altitude gamebirds, such as pheasants and snowcocks, generally fly downhill?
6711. Which bird is the Eurasian counterpart of the Great Blue Heron?
6712. Which booby is most gregarious at sea?
6713. Do the neotropical jacamars have especially melodious songs?
6714. Which bird was formerly known as "Cabot's Tern?"
6715. Is the Blackpoll Warbler one of the most abundant North American warblers?
6716. Do *both* sexes of African Saddle-billed Storks have a pair of yellowish or reddish pendent wattles at the base of the bill?
6717. Which North American owl juvenile has conspicuous white triangular patches on its forehead?
6718. Do *all* coots have chicks with black down?
6719. Which African waterfowl was prominently depicted in ancient Egyptian art?
6720. The Eurasian Hoopoe has what type of flight style?
6721. Do kittiwakes have a larger rear toe than other gulls?
6722. Do the second largest birds in the world have colorful naked heads?
6723. Why do falcons lay their eggs in a shallow hollow?
6724. Do South American trumpeters bathe often?
6725. With respect to head plumage, what do the juveniles of all frigatebirds have in common?
6726. Which birds may roost overnight on the backs of mammals?
6727. How does the chick-rearing behavior of thick-knees differ from most shorebirds?
6728. What is the favored prey of the Australian Barking Owl?
6729. When competing for nesting sites, which species is dominant, the Wood Duck or Black-bellied Whistling-Duck?
6730. Is the under surface of an Ostrich's foot hard?
6731. What portion of the brain is less developed in flightless birds than in flying birds?
6732. When a male Tooth-billed Catbird positions a leaf at its bower, is the shiny side exposed?
6733. Are first-year Whooping Cranes white?

Questions

6734. How large was the largest Passenger Pigeon colony, where was it, and how many birds occupied the colony?
6735. What is the clutch size of the neotropical King Vulture?
6736. Do crossbills build intricate nests?
6737. Does South America have more motmot species than Central America?
6738. Are *female* Spruce Grouse territorial?
6739. What percent of the planet's approximately 9,700 species of birds occur in Africa?
6740. Where does the New Zealand Rifleman often place its nest?
6741. Which American bird is called a "flycatching thrush?"
6742. Do nest-constructing Steller's Jays use mud?
6743. Do harriers lay small clutches for diurnal raptors?
6744. What is unusual about the plumage structure of the Australian Rufous Bristlebird?
6745. What color are juvenile Andean Condors?
6746. Which bird of Florida is known as a "Cuban parrot?"
6747. Which European owl is described as the most musical?
6748. Do Poorwills forage relatively close to the ground?
6749. Were cranes eaten by the ancient Romans?
6750. How do Greater Roadrunners catch swifts?
6751. How many countries have breeding populations of penguins, and which countries are they?
6752. Which of the seven fish-eagles is the smallest?
6753. Is the White-throated Swift a colonial nester?
6754. Which American passerine often snatches crayfish from the bill of a Glossy Ibis?
6755. Why is the flight of a drongo noisy?
6756. Which bird do the Alaskan Aleuts call "toporkie?"
6757. What color are the eyes of a New Zealand kiwi?
6758. Are any of the neotropical ovenbirds crested?
6759. What large, endangered, endemic New Caledonian bird lays a single-egg clutch?
6760. After striking the water when diving for fish, how do Ospreys rid their plumage of water?
6761. Are the eyes of a yawning bird partially or wholly closed?
6762. Why was there a bounty on Alaskan Bald Eagles?
6763. How many species of birds occur in Brazil?
6764. The Lapland Longspur is a common tundra species, but what habitat is preferred by the other three American longspurs?
6765. What is the favored prey of the Madagascar Cuckoo-Roller?
6766. Two sons of what famous American naturalist married two daughters of John Bachman, after whom Bachman's Warbler is named?
6767. Are antbirds solitary birds?
6768. Is the Prairie Falcon more likely to take prey on the ground as opposed to in the air?
6769. Anna's Hummingbird may rear how many broods a year?
6770. Which American organization publishes *The Migrant*?
6771. In Europe, which bird is simply known as the teal?
6772. Which North American grebe is said to have the most penetrating voice?
6773. In what year was the Mississippi Sandhill Crane designated a distinct subspecies?

Questions

6774. Band-tailed Pigeons feed extensively on what in the fall?
6775. How did the coursers obtain their collective name?
6776. Why do Pied-billed Grebes have such large, stout bills?
6777. How many species of seabirds breed in Hawaii?
6778. Do Purple Martins consume large numbers of mosquitos?
6779. Which American bird is called a "prairie dove?"
6780. Do hornbills have unusually smooth eggs?
6781. Why was the Parasitic Jaeger so named?
6782. Which is the only alcid that can be fairly reliably sexed in the field on the basis of external morphology?
6783. How do male harriers feed their young?
6784. For what did fishermen formerly use Razorbills?
6785. Which extinct North American bird roosted collectively in tree hollows?
6786. Why was the House Wren so named?
6787. What color are the wattles of a cock Ring-necked (Common) Pheasant?
6788. How accurate is a Lammergeier's aim when aiming at a specific spot to drop a bone?
6789. In what 1966 movie comedy did Roddy McDowell and Tuesday Weld star?
6790. What is peculiar about the lower mandible of a Limpkin?
6791. The abandoned nest of which raptor is used by Gyrfalcons?
6792. Are any of the icterids crested?
6793. Do Ruffed Grouse drum every month of the year?
6794. Do flying terns point their heads and bills downward?
6795. Most members of the mockingbird family have what color eyes?
6796. Does an incubating Dovekie cover its egg with its wing?
6797. Do Cockatiels inhabit Tasmania?
6798. Do Resplendent Quetzals nest more than once a year?
6799. Do Magnificent Frigatebirds mate for life?
6800. During the course of evolution, did flying mammals, such as bats, appear before flying birds?
6801. Which South American seabird was introduced into the Sea of Cortez, Mexico, and why?
6802. What is a poet or singer of great ability called?
6803. Where did the Rev. John Bachman discover the warbler that bears his name?
6804. Which owl has the largest and most perfectly circular facial disc?
6805. Does the female Arctic Loon select the nest site?
6806. How much pressure per square inch can be applied by the specialized bill of a crossbill?
6807. What do Gray Catbirds and American Robins do if a cowbird parasitizes their nest?
6808. Are Rooks highly gregarious year-round?
6809. Which is considered to be the most successful Australian avian family?
6810. What color is the bare skin behind the eye of a Yellow-billed Magpie?
6811. Do Great Blue Herons use their same nest year after year?
6812. Does the aberrant Asian shorebird, the Ibisbill, have a swift, gull-like flight?
6813. Are American Kestrels (sparrowhawks) used in falconry?
6814. Which North American grebe is most likely to swim with only its head above the water?
6815. Why was Wilson's Bird-of-paradise given the specific name *respublica*?

Questions

6816. Do sandgrouse have long, pointed wings?
6817. Are African Swallow-tailed Kites colonial nesters?
6818. Are Canvasback ducks relatively tame and confiding?
6819. Which is the most widespread macaw of Mexico?
6820. Which pheasants have feathered legs?
6821. What is the Horned Lark called in the Old World?
6822. Which American bird is known as a "venison bird?"
6823. How do arboreal monkeys avoid African Crowned Hawk-Eagles?
6824. Why should Purple Martin nesting gourds be painted white?
6825. Prior to the arrival of humans, what was the only predator of the extinct New Zealand moas?
6826. What color are juvenile King Penguin ear-patches?
6827. How many tons of blue-green algae do a million Lesser Flamingos consume annually?
6828. Do the ranges of Eastern and Western Meadowlarks overlap?
6829. Which two cotingas were once considered distinct enough to warrant a separate family?
6830. The tree-cavity nest of a Madagascar Cuckoo-Roller has the strong smell of what substance?
6831. Is the Northern Bobwhite population increasing?
6832. Are the monal pheasants polygamous?
6833. Where the breeding ranges of Red-winged and Yellow-headed Blackbirds overlap, which species is more apt to nest in the deeper water areas of the marsh?
6834. What is distinctive about the head of a Philippine Coleto (Myna)?
6835. What nesting material is used by a Lammergeier?
6836. What color is the rump of a Killdeer?
6837. With respect to plumage color, what do the wings of all flamingos have in common?
6838. Are juvenile Common Raven feathers glossier than adult feathers?
6839. Why is the Indian Jungle Babbler known as *seven sisters*?
6840. Are the edges of an albatross bill blunt?
6841. Is the Australian Wonga Pigeon especially arboreal?
6842. Do birds prey on adult Japanese spider crabs?
6843. Do cranes often fly at great heights?
6844. Is the Blackpoll Warbler one of the more southerly-nesting wood-warblers?
6845. Do Pied-billed Grebes nest in Alaska?
6846. At what age are young Whooping Cranes capable of flight?
6847. Why are birds that hatch in a helpless condition with closed eyes, and are totally dependent on their parents for food and care, known as altricial birds?
6848. Which are the three mergansers (fish-ducks) of Europe?
6849. Do Budgerigars hold food in their feet?
6850. How do running Ring-necked (Common) Pheasants hold their tails?
6851. What is a place where geese are captively-reared called?
6852. Which rails are most apt to construct floating nests, such as those of grebes?
6853. Which larger bird do gnatcatchers superficially resemble?
6854. Why are Ancient, Craveri's, Xantus' and Japanese Murrelets unique among birds?
6855. To which island was the extinct Spectacled Cormorant probably restricted?

Questions

6856. Do White-bellied Sea-Eagles feed extensively on carrion?
6857. Why do owlet-nightjars have dimorphic plumage?
6858. Which Australian bird has a loud scream described as *horrifically reminiscent of a women crying out in terror?*
6859. What percent of the adult's weight does the egg of a murre represent?
6860. What is a casual species?
6861. How did the Australian Gibber Chat obtain its name?
6862. As of 1992, how many New World birds were considered endangered or threatened?
6863. Which insect is favored by a foraging Arctic Warbler?
6864. What is the meaning of the name "brent" or "brant" for the sea-geese?
6865. Which ibis relative formerly nested within what is now the city limits of London?
6866. A number of attempts were made to introduce tinamous to various parts of the world, but where did the only successful introduction occur, and which species was introduced?
6867. With respect to birds as weather prophets, what did people of former times believe would happen if birds became silent?
6868. What color is the crest and chest of a breeding Black-faced Spoonbill?
6869. Is the African Martial Eagle crested?
6870. If the ear-patch of a King Penguin is whitish, what does this indicate?
6871. Are tropicbirds among the most gregarious of seabirds?
6872. Up to 40 percent of a Golden Eagle's winter diet consists of what?
6873. Are most larks gregarious?
6874. Which vertebrate might be consumed by a kiwi?
6875. Which Australian bird is known as a "CWA bird," and why?
6876. How many young per nesting attempt are reared by a pair of Galapagos Hawks?
6877. Which is the largest endemic Hawaiian land bird?
6878. Do Wilson's Storm-Petrels always feed from the surface?
6879. Which wading birds do flying whistling-ducks resemble?
6880. Why do Ostriches have an unusually large number of primaries?
6881. Do duller-plumaged bowerbirds build more elaborate bowers than brightly-colored ones?
6882. What color are Anhinga eggs?
6883. Which bird is synonymous with gluttony in many English-speaking countries?
6884. What color are the horns of a Horned Lark?
6885. Of the approximately 27 species of terrestrial or freshwater birds that became extinct prior to European settlement of New Zealand, the remains of how many species have been found associated with Maori middens?
6886. For what did native Greenlanders use loon skins?
6887. Which is the only sexually dimorphic albatross?
6888. Is the plumage of a rhea loose?
6889. Which is the largest African goose?
6890. Is the Great Gray Owl relatively common in western Europe?
6891. What is the preferred food of a redpoll?
6892. Do all falcons have the same length incubation period?
6893. Which American bird is called a "woodpile quarker?"
6894. Which American pelicans may prey on salamanders?

Questions

6895. Are any of the shrikes colonial nesters?
6896. What color is the inner mouth of a breeding Razorbill?
6897. Is the Australian Brush-turkey sexually dimorphic?
6898. Do Ospreys commence incubation with the first egg laid?
6899. Are juvenile Snow Geese whiter than juvenile Ross' Geese?
6900. Was fish more popular with ancient Romans than poultry?
6901. Do Tufted Puffins have a single incubation patch?
6902. Which heron is considered a hazard at airfields?
6903. Which is the smallest Australian dove?
6904. What is group of choughs called?
6905. How do African natives make use of penduline-tit nests?
6906. Do Rough-winged Swallows breed in colonies?
6907. Does the Great Horned Owl inhabit Cuba?
6908. Do female Black-crowned Night-Herons select the nest site?
6909. Do Ruby-throated Hummingbirds battle with bumblebees over feeding rights to flowers?
6910. Which large native pigeon inhabits the capital of Great Britain?
6911. Which colorful, long-legged wading birds are flightless during the flight feather molt?
6912. Are courting loons prone to call at night?
6913. Do the neotropical potoos have short tails?
6914. Do most fringillids have cone-shaped bills?
6915. When Ruffed Grouse drum on a traditional log, do the birds always drum from the same spot on the log?
6916. What type of habitat is favored by the Brahminy Kite?
6917. Are recently-hatched grebe chicks good divers?
6918. What color is a breeding Cattle Egret's head?
9619. Do African Fish-Eagles only copulate during the breeding season?
6920. Are male skimmers significantly larger than females?
6921. How many of the five species of West Indian todies have a white cheek-stripe?
6922. How does a bee-eater avoid getting stung when stinging bees are captured?
6923. Is the Mountain Plover often found far from water?
6924. In England, what nesting site was formerly favored by the Northern Wheatear?
6925. Which gull does the Iceland Gull most closely resemble?
6926. Which birds rest or sleep in a *pork-pie* posture?
6927. Do all pelicans have short, rounded tails?
6928. How long does it take the blood under normal pressure to make a complete circuit of the body of an adult chicken?
6929. Which American bird is known as a "bullet hawk?"
6930. African Secretary-birds have how many tail feathers?
6931. What is the wing color of a male Pompadour Cotinga?
6932. Which American grouse is most dependent on seeds?
6933. With which birds do Northern Fulmars often nest?
6934. Do Bald Eagles often rob other raptors of their food?
6935. What might a Belted Kingfisher do to escape an attacking Peregrine Falcon?
6936. In the Caribbean, which bird is known as a "Cayenne Tern?"

Questions

6937. In the U.S., which cursorial bird preys on gophers and young rabbits?
6938. Do most cotingas have relatively large feet?
6939. What color is the bare facial skin of a breeding Pelagic Cormorant?
6940. Is the head of a walking Eurasian Hoopoe held stationary?
6941. Are titmice strong fliers?
6942. Who is credited with coining the name "murrelet?"
6943. Do most nuthatches have an eye-stripe?
6944. What is the bill color of a female Bahama (White-cheeked) Pintail?
6945. What is the greatest number of lead shot recovered from the stomach of a Trumpeter Swan?
6946. What physical feature do all honeyeaters share?
6947. Do Great Horned Owls breed within sight of the largest North American city?
6948. Are icterids relatively closely related to tanagers?
6949. Which part of a fish do Bald Eagles consume first?
6950. When were the Pekin ducks that became the founding stock for Long Island ducking first brought to Long Island, and how many ducks were involved?
6951. Does the rare Siberia-nesting Spoonbill Sandpiper winter in Australia?
6952. Which large carnivorous bird was said to be worshipped by the Central American Mayans?
6953. What is the current name for the White-necked Raven?
6954. Which is the largest hummingbird of California?
6955. Does the Louisiana Waterthrush prey on fish?
6956. Which three crested penguins may hatch two chicks?
6957. What is the Snowy (Kentish) Plover called in Australia?
6958. Are all typical falcons in the same genus?
6959. Are Wild Turkeys fond of acorns?
6960. Does the most powerful raptor of Africa continue to feed its young long after fledging?
6961. Have wild Whooping and Sandhill Cranes ever hybridized?
6962. Which U.S. bird with a long, decurved bill may fly in lines as long as 1.2 miles (2 km)?
6963. What is the major predator of Hamerkop chicks?
6964. What is a bamboowren?
6965. Excepting domestic chickens, which bird reportedly has the highest incidence of tumors?
6966. Which is the only bird that has a digestive system with active foregut fermentation?
6967. Do hummingbirds fly in sub-freezing temperatures?
6968. What was the name of the famous Whooping Crane that was maintained in captivity for 24 years and ultimately died at the New Orleans Zoo in 1965, and what was the cause of death?
6969. Does a flying Western Grebe show a conspicuous dorsal wing-stripe?
6970. What is the incubation period of a Barn Owl?
6971. Who directed the 1953 movie *Gentlemen Prefer Blondes* with Marilyn Monroe and Jane Russell?
6972. Is the European population of Corn Crakes declining?
6973. Aside from illegal baiting, how do American waterfowl hunters attract birds?
6974. What was the incubation period of the Passenger Pigeon?

6975. Does a Red-breasted Nuthatch hold food in its feet like a chickadee?
6976. Do New Zealand kiwis line their burrows?
6977. Once a Lammergeier has dropped and smashed open a bone, why must it land without delay?
6978. Which birds have the longest necks relative to their size?
6979. How soon after a Barn Owl's meal are pellets of indigestible parts regurgitated?
6980. Which cormorant relatives continue to feed their chicks after fledging?
6981. In Audubon's *The Birds of America,* which two birds were depicted in the only painting in which the prey species was the focus of the painting?
6982. What are Evening Grosbeaks partial to at backyard feeders?
6983. Which organization publishes *The Passenger Pigeon*?
6984. Are sheathbills solitary birds?
6985. Do any owls engage in noisy courtship flights?
6986. Which of the nine thick-knees is migratory?
6987. Are hornbills noted for sunbathing?
6988. Which two North American grebes subsist almost exclusively on fish?
6989. Is the Gray Catbird known for its ability to mimic?
6990. Are Northern Goshawks heavier than Red-tailed Hawks?
6991. Do fleas transmit disease-causing organisms to wild birds?
6992. Are Red Junglefowl swift fliers?
6993. The legs of which group of birds are splayed?
6994. Why is the Red-eyed Vireo called a "preacher bird?"
6995. How many species of Australian birds are restricted to the rainforest?
6996. The Sage Grouse is often in the company of which large mammal?
6997. Are hawk-owls relatively tame and confiding?
6998. Is the hind toe of an alcid long?
6999. What color are the sides of a male Scissor-tailed Flycatcher?
7000. Why is the scientific name of the Paradise Tanager misleading?
7001. How did the Brambling obtain its name?
7002. Which American bird is called a "crow woodpecker?"
7003. Why do White-tailed Tropicbirds take over Cahow (Bermuda Petrel) nesting sites?
7004. What is an especially delicate and light method of bowing rapid passages on the violin called?
7005. What was the old English name for young cranes?
7006. The New Zealand Parsonbird is better known by what name?
7007. Which bird was called a "rubber turkey" by South African Boers, and why?
7008. In Texas, which icterid nests with herons?
7009. In America, which raptor is known as a "rat hawk" or a "rabbit hawk?"
7010. What is the average lifespan of a hummingbird?
7011. Can Glaucous Gulls capture flying alcids?
7012. Do peafowl generally live in groups?
7013. Which continent has the most species of grebes, and how many occur there?
7014. How much larger is the yolk of a precocial bird egg than the yolk of an altricial bird egg?
7015. Where do Sage Sparrows forage almost entirely?

Questions

7016. Why are the Australian grass parrots so named?
7017. Peregrine Falcons have been known to prey on how many species of birds?
7018. What is *mousing?*
7019. Does the Ibisbill have an unmusical voice?
7020. When a Brown Pelican surfaces after a dive, is it generally facing down wind?
7021. What color are the eyes of an Anhinga?
7022. How do Purple-throated Fruitcrows protect their nest?
7023. Statues of which bird guard the Imperial Throne in Beijing's Forbidden City?
7024. Which birds have the keenest vision of any animal?
7025. What color are the underparts of breeding golden-plovers?
7026. What did the country people of eastern North Carolina formerly believe happened to coots during the summer?
7027. Was Samuel Taylor Coleridge familiar with the albatross when he wrote *The Rhyme of the Ancient Mariner*?
7028. Is the mean body temperature of a passerine higher than that of a non-passerine?
7029. Are the ducklings of *all* northern dabbling-ducks cared for only by the female?
7030. Which is the least wary of the small North American grebes?
7031. The name of which ibis is onomatopoetic in origin?
7032. Do both breeding Horned and Common Puffins have a broad black neck-collar?
7033. What color are the eyes of an Arctic Loon?
7034. Where do Satin Bowerbirds copulate?
7035. Which phalarope is restricted to the New World?
7036. Bald Eagles snatch prey from which mammal?
7037. How far south do wintering American White Pelicans wander?
7038. Which American bird is called "cock-of-the-desert?"
7039. According to legend, what happened to ancient Mexican warriors who were killed in battle?
7040. Clark's Nutcracker has what woodpecker-like habit?
7041. Which are the only monogamous ratites?
7042. Do flocks of feeding seabirds attract other birds?
7043. Are the mosquitos that transmit avian malaria the same species that carry mammalian malaria?
7044. If an Anhinga chick falls from its southern Florida nest, what predator is most likely to get it?
7045. Do American Avocets feed in flocks?
7046. Do Limpkin chicks rather resemble shorebird chicks?
7047. If the beard of a Wild Turkey is cut off, will it re-grow?
7048. Does the Brown Booby prey on parrotfish and flatfish?
7049. Which Atlantic alcid is one of the most social of alcids?
7050. What does *Empidonax*, the genus for a large number of tyrant-flycatchers, mean?
7051. With what do Rosy Finches often line their nests?
7052. Which American bird is known as a "yellow titmouse?"
7053. What is Canton flannel sometimes called?
7054. Do the legs of flying whistling-ducks extend beyond their tails?

7055. What 13th-century European ruler is sometimes referred to as the *father of modern ornithology,* and why?
7056. Were Carolina Parakeets exploited by plume-hunters?
7057. Do grebes prey on amphibians?
7058. At normal walking pace, how many strides a minute does an Ostrich take?
7059. Which owl kills prey as large as Eurasian Capercaillie and young roe deer?
7060. Are diving-petrel eggs significantly larger relative to the size of the female than other petrels?
7061. How are redpolls able to survive very cold winters?
7062. If a Golden Eagle is in *ringtail* plumage, what does this imply?
7063. Why do Great Crested Flycatchers line their nests with bits of cellophane and onion skins?
7064. Which American waterfowl has the most colloquial names?
7065. Which woodpeckers follow Pileated Woodpeckers, and why?
7066. Which American bird is locally called a "copperhead?"
7067. What nesting habit do all birds in the order Coraciiformes share?
7068. Which family of seabirds regurgitates indigestible items, such as fish bones and scales?
7069. How many sperm are released by a domestic rooster during a single copulation?
7070. Do male Northern Harriers migrate earlier than females?
7071. What color are the primaries of an African Saddle-billed Stork?
7072. Do *both* sexes of woodpeckers drum on dead limbs, tree trunks, or other objects producing a loud reverberation?
7073. Was Ross' Goose named after Sir James Clark Ross, the famous Arctic and Antarctic explorer?
7074. In Venezuela, which bird is called a "horseman," and why?
7075. Which North American swift is said to be the "swiftest?"
7076. How do the Green-winged Teal nesting in the Aleutian Islands (Alaska) differ in appearance from those that breed throughout the remainder of North America?
7077. Is the Cuban Screech-Owl a race of the Eastern Screech-Owl?
7078. Do loons engage in courtship feeding?
7079. Which cotinga is the smallest?
7080. What was the name of the talkative Sulphur-crested Cockatoo on the *Beretta* TV series starring Robert Blake?
7081. If parasitized by a cowbird, do Northern Cardinals throw out the intruder's egg?
7082. In America, what is the treecreeper of Britain called?
7083. Which exclusively Pan-American family includes arboreal, terrestrial, territorial, colonial and parasitic birds?
7084. What is the Short-toed Eagle called in southern Europe?
7085. Who wrote the 1977 monograph *Rails of the World*?
7086. Do lyrebirds have strong pair bonds?
7087. Was Frederico Craveri, after whom Craveri's Murrelet is named, an ornithologist?
7088. Do Le Conte's Thrashers feed mostly on the ground?
7089. What well-known American writer penned *Nothing wholly admirable ever happened in this country except the migration of birds?*
7090. What is the mean age of first breeding for a Wandering Albatross?

Questions

7091. Are eagles actually large hawks?
7092. What did Lewis and Clark call the Sage Grouse?
7093. What is the only Southwest Pacific island where both ibises and spoonbills occur?
7094. In what type of bird nest might a Rosy Finch roost?
7095. What do nesting Elegant Terns and Heermann's Gulls have in common?
7096. For what do the natives of some New Guinea tribes use the sharp inner claws of a cassowary?
7097. Which cuckoo follows army ants?
7098. How many bird species occur in the Virgin Islands, and how many are permanent residents?
7099. Do tern chicks consume fish longer than themselves?
7100. How do the underwing linings of Pigeon and Black Guillemots differ?
7101. With respect to neck vertebrae, how do African ground-hornbills differ from arboreal ones?
7102. How long does it take a Lammergeier to dissolve a cow's vertebra in its stomach?
7103. Which is the only breed of "domestic" poultry that had its origin in Africa?
7104. What kind of fried meat is named after a bird?
7105. Which American bird is known as an "orange-chested hobby?"
7106. Do hummingbirds bathe frequently?
7107. When closing their eyes, are diurnal birds more likely to draw up the lower lid than to lower the upper eyelid?
7108. In which European country does the Barn Owl reach its most northerly distribution?
7109. Do Laysan and Black-footed Albatross hybridize?
7110. How does a drumming cock Ruffed Grouse brace himself?
7111. Which is the smallest British corvid?
7112. How many eggs can a domestic Japanese Quail (*Coturnix*) lay annually?
7113. What color is the swollen bare forehead of a courting Old World Great White Pelican?
7114. Do any reptiles lay colorful eggs like those of many birds?
7115. Do Sungrebes and finfoots have large heads?
7116. Is the Asian Shama thrush a bold and curious bird?
7117. Do juvenile Ospreys have a less distinct breast-band than adults?
7118. What is the earliest age at which Budgerigars are capable of laying eggs?
7119. How many exotic bird species have been introduced into Bermuda, and how many have become established?
7120. Which bird is called the Canada Goose's saltwater cousin?
7121. What color is the gular pouch of a breeding American White Pelican?
7122. Do sunbirds have pleasant songs?
7123. In the U.S., which heron has the most restricted range?
7124. Which North American tanager is most likely to hawk insects on the wing?
7125. Which bird did the poet Shelley call *the blithe spirit?*
7126. Is the Eurasian Great Tit gregarious?
7127. Which is the largest stork of Myanmar (formerly Burma)?
7128. Which loon may *fly* back to its chicks carrying a single fish at a time crosswise in the bill?
7129. In southern Ethiopia, what do Ostrich eggs on a grave signify?

7130. At what age can juvenile Caribbean Flamingos fly?
7131. Cranes have how many cervical vertebrae?
7132. How has the peculiar odor of the Hoatzin been described, and what causes the odor?
7133. In the U.S. alone, how many waterfowl perish annually as a result of ingestion of lead shot?
7134. What is the state bird of Idaho?
7135. How large a territory does a breeding male Ostrich defend?
7136. Which two long-legged birds in an exclusively Neotropical family are regarded as gamebirds in South America?
7137. Which is the least gregarious North American grebe?
7138. Are all New World quail monogamous?
7139. What is the Olive-backed Thrush currently called?
7140. During the winter, do Whiskered Auklets lose their facial plumes and crest?
7141. When a flock of European (Common) Starlings is attacked by a Peregrine Falcon, what do the birds normally do?
7142. Is the South American Jabiru stork a scavenger?
7143. What is the origin of the name "godwit?"
7144. Flamingos have how many brood-patches?
7145. Does the Northern Harrier often hunt at dusk?
7146. Do loons mate for life?
7147. In the American tropics, why do kiskadees often nest in a bull-horn acacia tree?
7148. How did the Chinese formerly use live King (Painted) Quail?
7149. When digging nesting burrows, do kingfishers use their bills, or do they use their feet?
7150. How much faster are the swiftest fish than the fastest swimming birds?
7151. What color are the thighs of a Ferruginous Hawk?
7152. How were hummingbird feathers used by the Aztecs?
7153. At six months of age, how much does an Ostrich chick weigh?
7154. How did the name "Wood Duck" originate?
7155. Which bird is called a "flinthead" or "iron head?"
7156. Do shrikes always kill their prey prior to impaling it on a thorn or spike?
7157. When the Dodo became extinct, its loss may have resulted in the near-extinction of what other life form, and why?
7158. Which is the only bird whose survival may be threatened by the obligate brood parasitic-nesting Brown-headed Cowbird?
7159. According to folklore, on what day of the year are Ospreys supposed to return to the eastern shore of Maryland?
7160. What color is the abdomen of the West Indian todies?
7161. Do Burrowing Owl burrows have more than a single entrance?
7162. Which are the only two North American geese with orange or yellow legs and feet?
7163. Is ornithosis a disease restricted to birds?
7164. What is the population size of the African Shoebill?
7165. Are the Palmchats of Hispaniola good songsters?
7166. How were Steller's Sea-Eagle tail feathers used in China?
7167. Where else besides the ground might Black Skimmers nest?

Questions

7168. Of the American waterfowl wintering in the Pacific flyway, which four are the most numerous?
7169. Do feathers obscure the ears of an Ostrich?
7170. How much of its body weight does an American Woodcock consume daily during the summer?
7171. Does the Bateleur eagle have short wings?
7172. Which communal-nesting bird is one of the most detrimental agricultural pests of Argentina?
7173. Are most icterids sexually dimorphic?
7174. Do the young of most North American gulls have a pale terminal tail-band?
7175. How close will flamingos allow an intruder to approach before taking flight?
7176. How much more can a breeding male Sage Grouse distend his esophagus than a non-breeding male?
7177. Do birds use their feet when turning their eggs?
7178. What is the favored prey of the Bald Eagle in the San Juan Islands (Washington State)?
7179. Which American bird is called "everybody's darling?"
7180. Does the Greater Roadrunner have *prominent* eyelashes?
7181. What is the origin of the albatross family name Diomedeidae?
7182. Do Ospreys diving for fish ever break their wings?
7183. In America, do Steller's Eiders winter south of Alaska?
7184. Does lead shot in a bird's flesh cause lead poisoning?
7185. How did Thomas S. Traill, of Traill's Flycatcher fame, aid John James Audubon?
7186. Why are the outer tail feathers of a flying mockingbird conspicuous?
7187. Why was the Golden Eagle so named?
7188. How are rails used by humans in Bangladesh?
7189. What Scrub Jay habit is beneficial in regenerating oak forests?
7190. Does the Marbled Murrelet winter as far south as the Mexican border?
7191. How many times hourly does a female phalarope turn its eggs?
7192. What is the average time interval between copulation and the laying of the first egg in domestic chickens?
7193. Why do Yellow-headed Blackbirds have unusually large feet for a blackbird?
7194. During burrow excavation, can Tufted Puffins remove rocks weighing three times as much as the birds themselves?
7195. Which is the only one of the six families of totipalmate birds (pelicans and their allies) containing birds with large nostrils?
7196. What is the origin of the name "tinamou?"
7197. Are female Rock and Willow Ptarmigan readily distinguishable from one another?
7198. Which long-legged bird commonly disposed of human corpses?
7199. How many owls are noted for soaring?
7200. What is the more familiar name for the "wreathed tern?"
7201. Which is the smallest European owl?
7202. What is the average legal take of quail by U.S. hunters?
7203. Which Australian bird was responsible for the cancellation of a planned townsite?
7204. Which flying birds *never* glide under any circumstances?

7205. What does the Boat-tailed Grackle often use as foundation cement for its nest?
7206. Is the Black-legged Kittiwake darker dorsally than the Red-legged Kittiwake?
7207. Where else besides Ascension Island does the Ascension Island Frigatebird breed?
7208. What color is the nestling down of the Capuchinbird (Calfbird)?
7209. Which bird of India is credited with immortality?
7210. Are the wings of a hovering Ruby-throated Hummingbird beating faster than when the bird is in forward flight?
7211. Which is the largest and darkest North American black-headed gull?
7212. Do rollers construct elegant, domed nests?
7213. Are the two wrynecks more primitive than the typical woodpeckers?
7214. Are all adult male Andean Cocks-of-the-rock bright red?
7215. Are all tinamous lowland birds?
7216. Threatened potoos have what owl-like habit?
7217. What is the *mean* age at first breeding for a giant-petrel?
7218. Was Livingstone's Turaco named after the famous African missionary?
7219. Are the long central tail feathers of any of the three jaegers fully developed by the bird's second year?
7220. Which tropicbird is dorsally barred?
7221. Which displaying birds of New Guinea may turn somersaults and dangle upside down?
7222. When large numbers of live hummingbirds were exported from South America in the 1960's, why did many Sword-billed Hummingbirds die in transit?
7223. What is the origin of the Western Grebe generic name *Aechmophorus*?
7224. Which two states accommodate the largest populations of White-tailed Ptarmigan?
7225. Are African Fish-Eagles generally observed as single birds?
7226. Which American wading bird is known as a "swamp squaggin?"
7227. Are the broad wings of a soaring Bald Eagle held flat, rather than tilted up an angle?
7228. Which sandpiper has the most northerly winter distribution in the American west?
7229. Why do Seaside Sparrows have relatively large feet?
7230. Do cranes have pervious nostrils?
7231. Buffalo-weavers were named because of their association with what animal?
7232. To what animal did the inhabitants of Liberia formerly attribute the bittern-like booming calls of the White-crested Tiger-Heron?
7233. Do any barbets have notched or toothed bills?
7234. What do male Spotted Bowerbirds nesting in sheep country use to decorate their bowers?
7235. Where do Bananaquits typically roost at night?
7236. Are all North American warblers migratory?
7237. Of the 13 exotic birds introduced into Japan, how many became established?
7238. Is a Black-footed Albatross more gentle and placid than a Laysan Albatross?
7239. What is the bill color of a breeding Ancient Murrelet?
7240. Do amphibians prey on warblers?
7241. When spinning on the water for food, how fast do Wilson's Phalaropes rotate?
7242. Do bird claws grow continuously?
7243. Why do obligate brood parasitic-nesting cuckoos usually select insect-eating species as hosts?

Questions

7244. Based on the specific name of the Eared Grebe, where was the type specimen collected?
7245. Why is defecating an effort for tinamous?
7246. Which two of the 14 North American dabbling-ducks are essentially nonmigratory?
7247. How many owls in the genus *Tyto* have ear tufts?
7248. How many species of bird-parasitizing fleas have been described from North American birds?
7249. What does the family name for the Old World flycatchers (Muscicapidae) mean?
7250. What is the more familiar pet-trade name for the Rose-ringed Parakeet?
7251. Why do foraging Hoatzins prefer young fresh leaves, as opposed to mature leaves?
7252. Which bird is most often preyed on by the Bald Eagle?
7253. Do any ratites have four toes?
7254. Which bird was the American ornithologist Arthur Cleveland Bent describing in 1938, when he noted that it was . . . *one of the tamest, or one of the stupidest, of owls.*
7255. What is the incubation period of the Yellow-billed Loon?
7256. Which is the smallest of the eight pelicans, and how much does it weigh?
7257. How much more heat is generated by a passerine in flight than when it is at rest?
7258. In earlier times in the U.S., why did country folk believe it was bad luck to receive wedding presents with designs of birds or bird decorations on them?
7259. Do large birds consume less daily, relative to their weight, than smaller birds?
7260. Which North American heron is most inclined to forage using *foot-raking* or *foot-stirring*?
7261. Were the two collarbones fused in *Archaeopteryx* to form a furcula as in modern birds?
7262. Who is regarded as the *father* of the U.S. federal duck stamp program?
7263. Are female Sora rails duller in color than their mates?
7264. Which alcid may nest in the rocks of boulder beaches *just* above the high tide line?
7265. Which American bird was formerly known as a "fute?"
7266. How did the name "osprey" originate?
7267. How many families of gallinaceous birds inhabit North America?
7268. Do American Oystercatchers prey on sea urchins?
7269. What started a gold rush in western Nebraska in 1911?
7270. Which ibis is the smallest?
7271. Does a juvenile Ross' Gull have a completely white tail?
7272. Does the Bananaquit have an orange rump?
7273. In Britain, what is the Northern Shrike called?
7274. What is the breast color of a male Eurasian Bullfinch?
7275. Based on theoretical calculations, how fast would a 250-pound Ostrich (113 kg) have to fly to remain airborne?
7276. What is the favored prey of Wilson's Plover?
7277. For what is the Muscovy Duck named?
7278. With which bird does the Greater Prairie-chicken often hybridize?
7279. What color is the bill of a breeding drake Masked Duck?
7280. Does the American Black Oystercatcher favor sandy beaches and mud flats?
7281. Which family of pelicaniform birds contains more than half the species in the order?
7282. Which American scoter breeds the farthest south, and how far south does it breed?
7283. Are White-breasted Nuthatches readily attracted to backyard bird feeders?

Questions

7284. Does Steller's Sea-Eagle have a square tail?
7285. Which of the three tropicbirds is the whitest?
7286. During the *winter*, which is the only alcid occurring south of Alaska that shows a conspicuous white wing-patch in flight?
7287. What color is the eye of a Wrentit?
7288. What color is the metallic sheen of a Limpkin?
7289. Which types of birds have been studied using satellite telemetry?
7290. Are three years required for a Red-legged Kittiwake to assume adult plumage?
7291. How many gruiform families occur in North America, and which birds are in those families?
7292. Which is the smallest *native* North American gull, and how much does it weigh?
7293. Is the Crissal Thrasher a common Arizona resident?
7294. Which two birds were named after John Xantus, the 19th-century bird collector?
7295. Which hornbills are the largest, and how much do they weigh?
7296. How many species of raptors occur in China?
7297. How much more does the nest of a Rufous Hornero weigh than the bird that constructed it?
7298. Are the non-gregarious North American arboreal cuckoos generally quiet and shy birds?
7299. Are motmot burrows always straight?
7300. Do Red Phalaropes copulate in the water, or do the birds mate ashore?
7301. What color is the skin of a breeding Greater Vasa Parrot?
7302. At what time of day are megapodes especially vocal?
7303. What color is the bare orbital skin of a breeding American White Pelican?
7304. How are fish a threat to adult Ospreys?
7305. Do hummingbirds prey on aphids?
7306. Do hornbills drink often?
7307. How many species of pheasants have feathered nostrils?
7308. Which American organization publishes *The Chat*?
7309. Why is a hanging gourd more beneficial to nesting Purple Martins than a gourd secured to a tree or post?
7310. What color are juvenile Roseate Spoonbills?
7311. Do incubating meadowlarks flush easily?
7312. Which North American heron *appears* to be the slimmest?
7313. Do Rock (Pigeons) Doves lay eggs every month of the year?
7314. Do Bald Eagles have heavier bills than Golden Eagles?
7315. Is the Louisiana Waterthrush more common than the Northern Waterthrush?
7316. Are both species of brown pelicans strictly coastal in distribution?
7317. Are male bowerbirds responsible for nest construction?
7318. Which is the only green sparrow-like bird of North America?
7319. In what state other than Alaska does the Bristle-thighed Curlew breed?
7320. With respect to nest structure, which New World bird family exhibits the greatest diversity?
7321. How many corvids are native to New Zealand?
7322. Is the Blue Petrel a common Cape Horn-nesting species?

7323. The Brown Towhee uses the burrow of which carnivorous mammal to escape summer heat?
7324. Can a greyhound catch a running Ostrich?
7325. Do roosting swallows of different species ever gather in large heterogeneous flocks?
7326. Where in Mexico do Pinyon Jays breed?
7327. What is significant about the status of the Seychelles Bare-legged Scops-Owl?
7328. In North America, which bird consumes the most ants?
7329. What is a CITES listed bird?
7330. Do the neotropical screamers line their nests with down?
7331. Which bird did ornithologist Frank Chapman describe as *a bit of feathered sunshine?*
7332. In the U.S., is the Mexican Chickadee restricted to Arizona?
7333. Numerically, are the passerines the most dominant land birds of all continents?
7334. Why are the legs of a Harpy Eagle so thick?
7335. When Tree Swallows line their nests with chicken feathers, what colored feathers are preferred?
7336. How many species of nightjars breed in Europe?
7337. Is the head of a Black-necked Crane entirely black?
7338. Do breeding Whiskered Auklets forage close to shore?
7339. What is a group of lapwings called?
7340. Which is the only gannet with an all-black tail?
7341. Are Australian butcherbirds good songsters?
7342. Are Northern Fulmars usually encountered out of sight of land?
7343. What is the bill color of a Black-bellied Whistling-Duck?
7344. Does the Great Kiskadee totally submerge when diving for fish?
7345. Can King and Emperor Penguins, which incubate their single egg on the top of their feet, scratch the back of their head with one foot while balancing the egg on the other?
7346. How much larger in terms of percentage is the avian heart than that of a mammal of comparable size?
7347. Are Secretary-birds normally encountered alone?
7348. Relative to its size, which owl has the longest legs?
7349. Did the name "crake" result because of the bird's call?
7350. Do Florida Scrub Jays breed at one year of age?
7351. What is a colony of Red-cockaded Woodpeckers called?
7352. Henry W. Henshaw, one of the American Ornithologists' Union founders, was an extremely rapid preparator of bird skins, but how quickly could he prepare a museum specimen?
7353. Are Midway Island-breeding albatrosses more numerous today than before the era of exploitation and human habitation?
7354. In those pigeons hatching two chicks, can both chicks be fed at the same time?
7355. Which are the only two of the six crested penguins with orange, rather than yellow, crests?
7356. What are the colorful shoulders of a male Red-winged Blackbird called?
7357. Which American bird is known as a "fool's quail?"
7358. Do Snowy Owls allow a close approach?
7359. Do motmots breed as far south as Argentina?

Questions

7360. The American Ruddy Duck frequently lays its eggs in the nest of which other waterfowl?
7361. How many Old World honeyeaters are ground nesters?
7362. When do skuas assume a hunch-backed appearance?
7363. What did Aristotle believe happened to wintering storks?
7364. Which bird was formerly called a "Bahama honeycreeper?"
7365. Can a female Ruby-throated Hummingbird complete a nest in a single day?
7366. The regurgitated pellets of the Old World Common Kingfisher consist primarily of what?
7367. What color is the forehead of the Thick-billed Parrot?
7368. How many pairs of American Oystercatchers breed in the Galapagos Islands?
7369. Is the Egyptian Vulture a cavity nester?
7370. Do wrens consume bird eggs?
7371. How many white spots does a female Harlequin Duck have on each side of its head?
7372. Female phalaropes have how many brood-patches?
7373. Into what U.S. state was the endangered Masked Bobwhite initially reintroduced?
7374. Do the ranges of the King and Emperor Penguin overlap?
7375. Do alarmed American Jacanas become restless?
7376. Are there any flightless herons?
7377. Where and when were Cattle Egrets first recorded in the New World?
7378. Does the Horned Grebe have a longer, thinner bill than the Eared Grebe?
7379. Where was the type specimen of the Bristle-thighed Curlew collected?
7380. How far beyond the tail do the long tail streamers of an adult Long-tailed Jaeger project?
7381. Are the primaries of a bustard significantly longer that the secondaries?
7382. During the days of unlimited Passenger Pigeon numbers, what popular dish was often served?
7383. Is the neotropical Sharpbill an exclusive frugivore?
7384. How long does it take an Atlantic Puffin to acquire adult bill configuration and color?
7385. What is the only lark native to an oceanic island?
7386. How many shots are fired by American hunters for every duck bagged?
7387. Do pelicans prey on commercially valuable fish?
7388. What color are the large eyes of a Hoatzin?
7389. Which raptor dwells in European castles?
7390. Is the Central American Horned Guan a lowland bird?
7391. Why was the old name for the Sage Thrasher (mountain mockingbird) inappropriate?
7392. Is the Cedar Waxwing larger than a Bohemian Waxwing?
7393. Do fish-eating ducks have smaller livers than ducks that feed on grain or meat?
7394. Do grebes have more neck vertebrae than loons?
7395. Which alcid may use the old nests of other birds?
7396. What are Mallophaga?
7397. Does the Vermilion Flycatcher breed in California?
7398. Where do albatrosses breed in Mexico?
7399. For what are old newspapers commonly used?
7400. Where does Krider's Hawk occur?
7401. When flocks of Passenger Pigeons moving in opposite directions met, did birds ever collide and fall stunned to the ground?

Questions

7402. In birder's jargon, what are tiny warblers, kinglets and gnatcatchers sometimes called?
7403. Which crane has more red on its head, the Sarus Crane or the similar Australian Brolga?
7404. What color is the head of an adult Verdin?
7405. Who starred in the 1942 movie *The Falcon Takes Over*?
7406. Is the shaft of a feather hollow?
7407. Is the plumage of a Barn Owl especially waterproof?
7408. Must sandgrouse lift their heads to drink?
7409. What is the national bird of Nepal?
7410. What is the stride of a running Greater Roadrunner?
7411. Which is the smallest shorebird, and how much does it weigh?
7412. For what is Major Charles Bendire best known?
7413. Do Caribbean flamingos feed extensively on small mollusks?
7414. How long before egg laying do Australian Malleefowl dig their incubator pits?
7415. Why is the Australian population of Silver Gulls increasing?
7416. Which neotropical bird has three long, dull black, unfeathered, worm-like wattles?
7417. Are the horns of a female Horned Lark less prominent than those of a male?
7418. Do both sexes of migrating Ruby-throated Hummingbirds depart together?
7419. Are the long bones of a penguin pneumatic?
7420. Feathers account for what percent of a bird's weight?
7421. How many new species of birds are described annually?
7422. What does the specific name of the Whip-poor-will suggest?
7423. How long does it take a dipper to construct its nest?
7424. Is the wing molt of most birds symmetrical?
7425. What is the maximum number of nuthatches recorded roosting in a single cavity?
7426. How many U.S. states require state duck stamps in addition to a federal duck stamp?
7427. Do Wood Ducks have long, square tails?
7428. Is the New Caledonian Kagu relatively non-vocal?
7429. Which is the only eagle that covers its eggs when it is off the nest?
7430. What color are the flowers of the hawk's beard?
7431. Do Killdeer chicks have two black chest-bands?
7432. When were nesting Black Skimmers first reported in California?
7433. Are the cross-sections of Osprey claws completely round?
7434. Which penguin has the smallest flippers and bill relative to its size, and why?
7435. Do Pinyon Jays gather in flocks numbering in the hundreds?
7436. In Germany, which bird is known as *klapperstorch?*
7437. Do Kelp Gulls prey on penguin eggs and chicks?
7438. Why is the Ring Ouzel unpopular with European gamekeepers?
7439. Why were Egyptian Goose eggs not eaten by ancient Egyptians?
7440. Is the Barn Owl especially numerous in California?
7441. Which is the most gregarious sulid?
7442. Where is the International Crane Foundation located?
7443. In terms of weight, how much zooplankton is taken during the four weeks of Arctic summer by 100,000 breeding pairs of Dovekies (Little Auks)?
7444. Do all three tropicbirds have black eye-stripes?

7445. Why do male ptarmigan retain white wings in the summer?
7446. Are Yellow-bellied Sapsuckers especially vocal when clinging upright to a tree trunk?
7447. Which bright red northern bird has expanded its range since colonial days?
7448. Are American White Pelican chicks more vocal than Brown Pelican chicks?
7449. Why is the western population of Purple Martins declining?
7450. What was the name of the first human-powered aircraft to cross the English Channel?
7451. Is the Andean Pygmy-Owl restricted to the Andes?
7452. Is the Sora rail a common Alaskan bird?
7453. What are Brown Booby livers used for by some fishermen of the Netherlands Antilles?
7454. Do migrating loons ever take *overland* flights of more than 1,000 miles (1,609 km)?
7455. Does the Ivory Gull nest on cliffs?
7456. What does the oystercatcher generic name *Haematopus* imply?
7457. Which American bird is known as a "paisano?"
7458. Do storm-petrels have hooked beaks?
7459. Which is the largest Australia-breeding seabird?
7460. Which feathers are often called *the lining of the wing*?
7461. How did the name "bunting" originate?
7462. Do treeswifts (formerly crested-swifts) have a reversible hind toe?
7463. Does the Curlew Sandpiper breed in Alaska?
7464. Do juvenile Bateleur eagles have shorter tails than adults?
7465. Is the Mississippi Kite expanding its range westward?
7466. Which Arctic bird has the most uniform diet?
7467. What color is the plumage of a breeding Spectacled Guillemot?
7468. Which are the only birds in which the male breeds annually and the female biennially?
7469. Do wrens sing with their tails cocked up?
7470. Do bee-eaters prey on fish?
7471. How many Lammergeiers exist?
7472. Are the wings of an American Kestrel folded when it is diving on prey?
7473. Which type of bird does a flying giant-petrel resemble?
7474. Do American White Pelicans forage at night?
7475. Do combative kiwis fight to the death?
7476. Which South American screamer is the most gregarious?
7477. Do courting female Wilson's Phalaropes drive off other females?
7478. Are the wings of a flying Osprey bent at the wrist?
7479. Does a flying Scissor-tailed Flycatcher open and close its tail like a pair of scissors?
7480. Which hornbills are apparently entirely carnivorous?
7481. In terms of length, what is the size range of birds in the pheasant and quail family?
7482. Are snipe chicks fed by their parents?
7483. Did the extinct Passenger Pigeon prey on earthworms?
7484. Which is the most aggressive African vulture?
7485. What do *all* skuas and jaegers have on their wings?
7486. In terms of percent of the original total, how many species of New Zealand land birds have disappeared since the arrival of humans about 1,000 years ago?
7487. How far east does the Montezuma Quail range in the U.S?

Questions

7488. Does the Common Loon have a heavier beak than the Arctic loon?
7489. Are bowerbirds more accomplished singers than the birds-of-paradise?
7490. According to Australian aborigine legend, how did the first duck-billed platypus egg originate?
7491. Does the Least Bittern have a *distinct* juvenile plumage?
7492. Are any diurnal raptors exclusively restricted to the desert?
7493. Do Summer, Hepatic, Scarlet and Western Tanagers all occur in California?
7494. What does the warning call of a Burrowing Owl imitate?
7495. In the 1977 movie *The Eagle has Landed*, starring Michael Caine and Donald Sutherland, what world leader was to be kidnaped by the Nazis, and what was the code name of the female sleeper German agent?
7496. Do Greater Yellowlegs feed on fish?
7497. What is the major component of a Rufous Hornero nest in southern South America?
7498. Are the outer tail feathers of a bee-eater elongated?
7499. Do Long-eared Owls nest on the ground?
7500. Who first raised the alarm concerning the plight of the Philippine Monkey-eating Eagle?
7501. Which *group* of penguins is noted for mutual preening?
7502. In which state is the Black-whiskered Vireo most numerous?
7503. Which endangered bird was adversely impacted by *Hurricane Hugo* when the storm slammed into South Carolina in 1989?
7504. Where on a bird is the supercilium located?
7505. Is the Snowy Owl an irruptive species?
7506. Is the avian liver larger than the liver of a mammal of equivalent size?
7507. Into what state has Gambel's Quail been successfully introduced?
7508. Aside from scavenging at village dumps, how do Marabou Storks aid African natives?
7509. Why is the population of Yellow-billed Cuckoos declining in the American west?
7510. Are the host species of the obligate brood parasitic-nesting cowbirds generally smaller than the cowbirds?
7511. How do Hoatzins feed their chicks?
7512. Is the bill of a Pyrrhuloxia curved?
7513. After which prominent landmark was the Cape Pigeon (Painted Petrel) named?
7514. In what repugnant trade did the father of John James Audubon engage?
7515. Does the Short-tailed Albatross have a shorter tail than the other 11 albatrosses in the genus *Diomedea*?
7516. After whom was the Zenaida Dove named?
7517. Do Snowy Sheathbills have comparatively large eyes?
7518. Are Red Crossbills shy birds?
7519. Do female Phainopeplas construct their nests without assistance from their mates?
7520. Which of the three tropicbirds is most inclined to breed at less than annual intervals?
7521. Do towhees mate for life?
7522. Do dragonflies prey on birds?
7523. Are finfoots and Sungrebes gregarious?
7524. Is the Fulvous Whistling-Duck the only nocturnal-feeding whistling-duck?
7525. What was the ultimate fate of Henry Palmer, after whom the Laysan Rail was named?

7526. Cave Swallows might nest with which other birds?
7527. Which is the only resident raptor of the subantarctic Auckland Islands?
7528. When approached by a boat, which alcid is the least likely to dive: Horned Puffin, Tufted Puffin, or murre?
7529. What does the gannet generic name *Morus* mean?
7530. Foraging Mountain Bluebirds have what kestrel-like habit?
7531. Frigatebirds were named after what?
7532. How far can a kiwi see in daylight?
7533. Which is the only American gull with an all-white tail during its first winter?
7534. Which woodpecker was once served in American restaurants?
7535. Which is the most abundant wintering Palearctic migrant passerine in Africa?
7536. Did the California Condor inhabit Mexico in historical times?
7537. Do the breeding ranges of the Greater and Lesser Yellowlegs overlap?
7538. Is the Chipping Sparrow one of the least frequent hosts of the obligate brood parasitic-nesting Brown-headed Cowbird?
7539. What does the name "trogon" mean?
7540. When concentrated in compact winter huddles, how many incubating male Emperor Penguins cram onto a single square meter of ice?
7541. Which northern gull was formerly called "burgomaster?"
7542. How did the hermits (hummingbirds) obtain their name?
7543. Which bird imported from China into Australia was formerly sold as a Persian nightingale?
7544. Is the American Linnet a common city dweller?
7545. Which of the four godwits is the largest, and how much does it weigh?
7546. With which camel relatives do rheas often associate?
7547. Why do male Common Moorhens build nests?
7548. What is VVND?
7549. Do roadrunners occur in Australia?
7550. During the winter, which North American dove roosts in large tight clusters?
7551. Who conducted the initial detailed studies on hibernating Common Poorwills?
7552. In Asia, which bird is called a "berkut?"
7553. Is the Horned Lark one of the earliest nesting North American birds?
7554. Which flycatcher was formerly known as Coues' Flycatcher?
7555. Which endangered bird was captively reared in large numbers for reintroduction into Pakistan?
7556. What is the greatest threat facing the Harpy Eagle?
7557. If a bird conforms to Bergmann's Rule, would its body size be smaller if it inhabited a cooler, rather than a warmer, climate?
7558. In American slang, what type of unscrupulous person is called a *hawk?*
7559. Do Wrentits favor dense habitat?
7560. Do female Ocellated Turkeys have warts on their heads?
7561. How did the Asian bleeding-heart pigeons get their name?
7562. Do African crowned-cranes generally forage in solitude?
7563. Are California Thrashers shy birds?

Questions

7564. When skimming, how much of a skimmer's lower mandible is in the water?
7565. Which is more inclined to perch on a telephone pole, a Cooper's or a Sharp-shinned Hawk?
7566. What is the most reliable way to identify the similar-appearing American and Fish Crow?
7567. Which owl has a proportionally longer bill than the bills of other owls?
7568. Do drongos have dull-colored plumage?
7569. Which North American mockingbird relative catches small fish at the water's edge?
7570. Which bird is symbolic of the U.S. Postal Service express mail?
7571. Do American Robins in the west tend to be paler and duller-colored than those of the east?
7572. What is a Blackcap?
7573. Is *The Blue Jay* published by the Virginia Society of Ornithology?
7574. Which *group* of penguins has the greatest longitudinal distribution?
7575. How does a male Bicolored Blackbird of central California differ in appearance from the male Red-winged Blackbird?
7576. What is a hyporhachis?
7577. Which North American gamebird has remarkably developed claws?
7578. Do juvenile Limpkins resemble adults?
7579. Are Old World warblers drabber in color than New World warblers?
7580. What color does the long, iridescent black tail of a Black-billed Magpie reflect?
7581. Most hummingbirds have how many secondary feathers?
7582. Is the bill of a breeding Arctic Tern *entirely* red?
7583. How are gallinule chicks fed?
7584. In which state have most recent sightings of the rare Bachman's Warbler been recorded?
7585. Which bird may take over an occupied Cliff Swallow's nest?
7586. Where is the Canary-winged Parakeet established in the U.S.?
7587. What happened to the library and letters of Rev. John Bachman, the ardent amateur natural historian of the 19th century, and close friend of Audubon?
7588. Why were tattlers (shorebirds) so named?
7589. What is a small, open compartment, such as in a desk, used to file papers, sometimes called?
7590. Which American woodpecker frequently has plumage and eggs soiled with pitch?
7591. Do Australian Royal Spoonbills nest on the ground?
7592. What specialized foraging technique is used by the Indian Black Eagle?
7593. Is the New Zealand Stitchbird a wattlebird?
7594. Which has traditionally been considered the most numerous deciduous forest passerine of eastern North America?
7595. What color are the large facial wattles of a male Bulwer's Pheasant?
7596. Why do Say's Phoebes seldom drink?
7597. Are Lark Buntings solitary for a portion of the year?
7598. Is the Greenland population of the White-tailed Sea-Eagle endangered?
7599. Do Old World orioles often gather in flocks?
7600. Do hummingbirds often copulate while in flight?
7601. In South America, what nest site is often selected by a Brown-chested Martin?
7602. Was the Glaucous Gull named for its call?

7603. Do Pileated Woodpeckers rear more than one brood a season?
7604. Which feathers are sometimes called *hand-quills?*
7605. What color is the crown crest of the Scissor-tailed Flycatcher?
7606. What color is the bill of a winter-plumaged Heermann's Gull?
7607. Is the population of the American Woodcock increasing?
7608. Which American gallinaceous bird is the most popular in terms of the size of overall hunter harvest?
7609. With respect to plumage, what do birds that pair for life have in common?
7610. Are birds nesting in cholla cactus secure from snakes?
7611. Which North American bird is called a "wood wagtail?"
7612. Which birds are favored by fortune-tellers in the Far East, and how are the birds used?
7613. What was the pet-trade name for the Orange-fronted Parakeet?
7614. Do birds lose more water through vaporization via the lungs and air-sacs than through the urine?
7615. In the world of amphibians, what is a mountain chicken?
7616. To which bird is the Cahow (Bermuda Petrel) most closely related?
7617. How large is the crest of a Gray Jay?
7618. Are any of the toucans sexually dimorphic?
7619. When soaring in thermals, how do storks differ from soaring vultures?
7620. What is the greatest recorded age for an Osprey?
7621. Are hornbills especially vocal?
7622. How did the name "macaw" originate?
7623. Are *all* birds in the order Piciformes predominately tree dwellers?
7624. Is the Common Ground-Dove a tame and trusting bird?
7625. What does the ani generic name *Crotophaga* suggest?
7626. What color does the black bill of a male Snow Bunting become during the winter?
7627. What is the population size of the flightless Inaccessible Island Rail?
7628. Black absorbs the sun's heat more readily than white, so why is the Common Raven, a common desert inhabitant, black?
7629. What color are the feathers surrounding the base of a male Northern Cardinal's bill?
7630. Do pratincoles nest in colonies?
7631. What is the average weight of a male Calliope Hummingbird?
7632. Which is the smallest Hawaiian-nesting seabird?
7633. In most gallinaceous birds, how soon before hatching do chicks pierce the air space in the egg?
7634. Do coots consume small birds?
7635. What must a human sprinter be in order to be a winner?
7636. Are most estrildid finches solitary?
7637. How many species of birds of continental Africa have become extinct in historic times?
7638. Does the Clay-colored Robin breed in the North America?
7639. In folklore, what was said would happen to a goat after it had been suckled by a goatsucker?
7640. What color is the bill of an Elegant Trogon?
7641. How can the Crested Francolin of South Africa be accurately separated at a distance from

other southern francolins?
7642. Is the bill of an Eskimo Curlew more curved than the bill of a Whimbrel?
7643. Do Black-billed Magpies reuse their nests annually?
7644. Which hornbills have bare inflatable facial skin?
7645. Do honeyguides have a shorter fledging period than most birds of comparable size?
7646. Do fruit-crows (cotingas) feed extensively on palm fruits?
7647. Is an American Woodcock egg larger and heavier than a Ruffed Grouse egg?
7648. Which cormorant breeds in both Brown and American White Pelican colonies?
7649. After whom is Costa's Hummingbird named?
7650. What is the name of the imaginary creator of a large collection of American nursery rhymes?
7651. Which is the smallest North American goldfinch?
7652. Which is the only North American falcon that breeds within its first year?
7653. What is peculiar about the arrangement of the feathers on the back of an Emu?
7654. Which birds were sketched on cave walls in southern France and Spain by caveman artists 17,000-18,000 years ago?
7655. Do incubating tropicbirds always face the sea?
7656. The feathers of which bird do Verdins lining their nest appear to prefer?
7657. Are mannikins brightly colored?
7658. Which bird is represented in the American Airlines logo?
7659. Does the Loggerhead Shrike have a larger bill than the Northern Shrike?
7660. Is Brewer's Sparrow primarily a marsh species?
7661. How did the Mistle Thrush obtain its name?
7662. What is a loafing bar?
7663. Are *goose*berries sweet?
7664. How great is the wingspan of the cave-dwelling Oilbird?
7665. Do Lewis' Woodpeckers gather in large winter flocks?
7666. Do all American *Empidonax* flycatchers have eye-rings?
7667. Is the long crest of a Hamerkop conspicuous in flight?
7668. What color are the legs and feet of a breeding Bonaparte's Gull?
7669. Prococial young are called chicks, but what are the young of altricial birds typically called?
7670. What color are the chicks of Clark's Grebe?
7671. How do Scrub Jays sometimes kill sparrows and finches?
7672. How many of the 39 trogons inhabit Africa?
7673. Upon hatching, naked Anhinga and darter nestlings have what color skin?
7674. Is post-fledging Brown Pelican mortality especially extensive?
7675. What man-made structure does the Cave Swallow commonly nest under?
7676. How do female Barrow's Goldeneye ducks of the Rocky Mountains differ in appearance from goldeneye hens of the east?
7677. In American slang, what is a stupid, foolish, or talkative person called?
7678. Which birds regurgitate vitamin-rich oil?
7679. Do flamingos require a running start when taking flight?
7680. What is the estimate of the number of birds in Finland?

7681. At what age are young Hoatzins able to fly?
7682. How many Bald Eagles inhabit the lower 48 states?
7683. What bird is pictured on the 50-dollar gold piece that was struck for the Panama-Pacific Expedition of 1915?
7684. Which American bird is called an "Irish snipe?"
7685. Which American cormorant has the most iridescent plumage?
7686. Do flying cranes have rapid upstrokes?
7687. When huddling in a large compact group, how much of a reduction of heat loss is gained by an incubating, fasting male Emperor Penguin?
7688. Is the Hawaiian Hawk the only native Hawaiian raptor?
7689. With which birds do Lapland Longspurs commonly flock?
7690. Is the African Abdim's Stork a marsh resident?
7691. Are male Wild Turkeys darker than females?
7692. What is unusual about a male Asian Fairy-bluebird's plumage?
7693. Do juvenile buteo hawks have breasts that are more streaked than adults?
7694. Is the Green-tailed Towhee the only towhee with a rufous cap?
7695. Which bird was known as a "white dodo?"
7696. Is the egg of an Osprey immaculate?
7697. What is the head color of a juvenile California Condor?
7698. Do pittas migrate at night?
7699. Which North American sparrow's eggs and chicks are preyed on by antelope ground squirrels and leopard lizards?
7700. How many species of birds have been introduced into Europe, and how many of these became established?
7701. The chicks of which birds are called *colts?*
7702. In southern Africa, what are sunbirds often called?
7703. What type of birds are firetails, grassfinches, waxbills and parrotfinches?
7704. What color is the bill of a newly-hatched Anna's Hummingbird chick?
7705. Do sungrebes and finfoots dive often?
7706. Are Secretary-birds especially vocal?
7707. Which is the only hornbill of Southwest Pacific islands?
7708. Is the bill of an African turaco decurved?
7709. What type of tree is a nesting Yellow-bellied Sapsucker prone to select in the northern part of its range?
7710. Which is the only endemic gamebird of Trinidad?
7711. Do Laysan, Black-footed and Short-tailed Albatrosses nest at the same time of year as southern albatrosses?
7712. Male Black Rosy Finches have what unusual courtship habit?
7713. Do juvenile drake Oldsquaw ducks have long tails?
7714. How does the bill color of breeding male and female European Starlings differ?
7715. Do puffins occur in Hawaii?
7716. Why do grebes have such slick, smooth ventral plumage?
7717. What is a parula?
7718. Does the Razorbill occur in Egypt?

Questions

7719. Which is the hardiest North American swallow?
7720. Do cormorants prey on turtles or mammals?
7721. Is the endangered California Least Tern restricted to California as a breeding species?
7722. Are the neotropical jacamars especially gregarious?
7723. Is the Scaled Quail generally encountered in flocks?
7724. Does the flightless New Zealand Kakapo have large wings?
7725. What is the clutch size of the Eurasian Wryneck?
7726. What is the difference between the Common Ringed Plover of the Old World and the Semipalmated Plover?
7727. Do finfoots prey on snakes?
7728. Why is the Little Blue (Fairy) Penguin of Australia and New Zealand a crepuscular feeder?
7729. Do female trogons have metallic coloration?
7730. What is billing?
7731. Which animal did Australian colonials call the *duck mole?*
7732. How can Dusky-legged Guans be sexed?
7733. How many of the 32 extinct penguins identified by fossil remains inhabited New Zealand?
7734. What color are recently-hatched Imperial (Blue-eyed) Shags?
7735. How did the House Sparrow reach the Falkland Islands?
7736. Do frogmouths occur in India?
7737. Is the South African race of the Ostrich common?
7738. What was significant about the artist who won the 1990 federal duck stamp contest?
7739. Do jaegers catch birds in mid-air with their feet?
7740. Do Hairy Woodpeckers feed extensively on winged insects?
7741. What is an eagless?
7742. Were Wood Storks sought by American plume-hunters?
7743. How many states have selected non-native species as the state bird, which states selected them, and which birds are they?
7744. With which aquatic mammal is the King Rail often associated?
7745. In 16th-century Europe, which rare bird was known as a *forest raven?*
7746. In North America, which is the only conspicuously-streaked flycatcher, both dorsally and ventrally?
7747. Do kingfishers use anvil stones to smash open mollusks?
7748. Do Snow Buntings breed in Europe?
7749. Is the Paradise Parrot of Australia a common cagebird?
7750. To which group of raptors are the sea-eagles fairly closely related?
7751. What is a leucistic bird?
7752. What is the Common Loon called in North Carolina?
7753. What commercial fruit trees do American Robins plunder?
7754. What are the only two U.S. states in which Anna's Hummingbird breeds?
7755. Do most of the neotropical curassows have short tails?
7756. Does the African Hooded Vulture have a heavy bill?
7757. Which sex of the Harpy Eagle has a double crest?
7758. When using a barn, do Barn Swallows generally nest inside or outside the barn?

7759. Which American curlew breeds the farthest north?
7760. Does the Anhinga have a terminal tail-band?
7761. Is the red of a Red-tailed Hawk's tail brightest on the dorsal side?
7762. Why do most waterfowl molt all of their wing feathers simultaneously, rather than gradually, like most birds?
7763. Is the Fox Sparrow one of the smaller sparrows?
7764. What color is the face of a male European Goldfinch?
7765. Do juvenile Greater Roadrunners have the blue and red patch of bare skin behind the eye found in adults?
7766. What is the average weight of an Australian Brush-turkey mound?
7767. What color is the bill of a juvenile Cattle Egret?
7768. Which American bird is called a "bluestocking?"
7769. Does the Great Egret have head plumes?
7770. Does the Crested Long-tailed Pigeon of the Solomon Islands have a long, straight bill?
7771. Do Common Ravens tend to travel in pairs?
7772. Which part of a Wild Turkey is called a *leader?*
7773. How do juvenile northern cormorants differ in appearance from adults?
7774. The extinct Mexican Guadalupe Island Wren was a subspecies of which mainland bird?
7775. What does the Hawaiian name *iwa* for the frigatebird mean?
7776. Do African go-away birds (turacos) fly at great heights?
7777. Do snowcock chicks feed primarily on insects?
7778. Does the Everglade (Snail) Kite have a pointed tail?
7779. What were the circumstances surrounding the discovery of the Congo Peacock?
7780. Is the South American Hoatzin sexually dimorphic?
7781. In what 1940 movie did Errol Flynn play a buccaneer?
7782. Do American White Pelicans occur in Alaska?
7783. Are any African hornbills endangered?
7784. To which bird is the Purple Sandpiper most closely related?
7785. Do Northern Gannets fly with their bills pointed downward?
7786. What type of nest site do Black Storks generally select?
7787. Which American plover frequently nests on gravel roads?
7788. Are non-breeding Bushtits gregarious?
7789. Do juveniles of many raptors feed on a wider variety of items than adults?
7790. Which ratite is greatly endangered, and how many remain?
7791. Are Wrentits particularly sedentary?
7792. Which flicker sex selects the nest site?
7793. Is the hearing of most birds superior to that of mammals?
7794. Which California passerine is the most subterranean?
7795. In the American west, which swallow is inclined to fly the highest when foraging?
7796. Was the extinct New Zealand Laughing Owl flightless?
7797. Is the Pinyon Jay nomadic?
7798. If its long trailing tail feathers were absent, which birds would a tropicbird most closely resemble?
7799. Do nesting American Avocets ever dive on intruders?

Questions

7800. Do female Great Horned Owls have lower-pitched voices than males?
7801. Do Mourning Doves ever nest on the ground?
7802. Do swift pairs engaged in aerial copulation ever fall to the ground?
7803. What color are the legs and feet of a Ruddy Turnstone?
7804. Are all estrildid finches sexually dimorphic?
7805. What is the weight of a newly-hatched Adelie Penguin?
7806. What is the origin of the name "troupial?"
7807. When was the first book devoted entirely to birds *published*, and when did it appear?
7808. Can a tiny West Indian tody capture prey as large as a lizard?
7809. Which bird was regarded as the *Thunderbird* by the Indians of interior British Columbia?
7810. Are the African turacos sedentary?
7811. What is a hornet (robber) fly sometimes called?
7812. Has albinism been recorded in New Zealand kiwis?
7813. Are hornbill eggs elongate in shape?
7814. Do the neotropical motmots follow army ants?
7815. Is the bill of an American Woodcock greater than two inches (5 cm) in length?
7816. Why do most honeyguides have white outer tail feathers?
7817. Do juvenile Crab Plovers resemble adults?
7818. Why are feathers white?
7819. Which gull was named for its feline-like vocalizations?
7820. How is bragging commonly described?
7821. Why are there fewer species of American wood-warblers in the west than in the east?
7822. What is unusual about the ovaries of a kiwi?
7823. Which bird was described by a biologist as "This loud and messy bird shares with two other species, *Rattus rattus* and *Homo sapiens*, traits which make the three of them a blight upon the earth: omnivorous food habits, ability to colonize every corner of the globe, and an inordinate ability to procreate in geometric progression?"
7824. Is the Ruddy Quail-Dove one of the most widespread of New World pigeons and doves?
7825. The entire population of the Norfolk Island (Australia) Boobook Owl was reduced to how many birds in 1990?
7826. Are the neotropical tapaculos fruit eaters?
7827. How much does an Ivory-billed Woodpecker weigh?
7828. Are the red-orange feet of a *flying* Tufted Puffin conspicuous?
7829. Who played one of the male leads in the 1976 movie *Network*, for which he won an Oscar?
7830. Is the head and neck of a female Wild Turkey more heavily feathered than the head of a tom?
7831. Are Sora rails more insectivorous than Virginia Rails?
7832. Does a female Black-headed Grosbeak have a black head?
7833. In Florida, which large birds are sometimes called *pier bums?*
7834. What color is the rump of a Northern Flicker?
7835. Is the Boat-billed Heron primarily coastal in distribution?
7836. How many species of raptors breed within the city limits of the capital of Australia?
7837. Which books are commonly known as *Birder's Bibles?*

7838. Do Aleutian Terns winter in the Philippines?
7839. Is a Razorbill egg spotted?
7840. Do any of the American wood-warblers occur in Europe?
7841. Do foraging Northern Shrikes hover?
7842. What color is the female Old World Blue Rock-Thrush?
7843. What color is the rump of the New Zealand Kea?
7844. How do chachalacas feed their chicks?
7845. In most nidicolous birds, are the eggshells of hatched chicks generally eaten or carried away by the adults?
7846. Do ratites nest in hollow logs?
7847. How many species of megapodes are threatened or endangered?
7848. Which is the most abundant nesting seabird of the Snares Islands, south of New Zealand?
7849. When a shrike catches a flying insect, is the prey ever consumed while the bird is in flight?
7850. In terms of percent of the total amount of krill taken by Antarctic seabirds, how much is accounted for by penguins?
7851. In South Africa, which bird is called a "trekduiker?"
7852. Which native raptor was reintroduced into southern California?
7853. Are feeding flocks of Northern Fulmars surprisingly silent?
7854. Is the uropygial gland of a Whip-poor-will functional?
7855. According to Greek mythology, what happened to Procne, the daughter of Pandion?
7856. What is the primary function of the Harderian glands?
7857. With what are Cactus Wren nests thickly lined?
7858. Which raptor robs Hamerkops of their prey?
7859. Which American shorebird is called a "pill-will?"
7860. Are neotropical woodcreeper young fed only by the female?
7861. Which of the New Zealand kiwis is active during the day?
7862. Do any of the neotropical antshrikes have hooked beaks?
7863. Which large, vulturine-appearing pigeon has a conspicuous cape of long, shining hackles that hang over its back and shoulders when the neck is retracted?
7864. Which U.S. president signed the Migratory Bird Treaty Act?
7865. Are Brown-headed Cowbirds of the west larger than those of the east?
7866. Which two weavers are established locally around Miami, Florida?
7867. What shape is the egg of a Secretary-bird?
7868. Was Lawrence's Goldfinch named for Lawrence of Arabia?
7869. Which American bird is known as a "hookum-pake?"
7870. Do Hairy Woodpeckers often reuse the same nest chamber of previous seasons?
7871. What color is the crest of the neotropical Sharpbill?
7872. What was the pet-trade name of the Red-crowned Parrot?
7873. African woodhoopoes have what odor?
7874. What do falconers call primary feathers?
7875. Of the roughly one million geese that winter in Europe, how many winter in Holland?
7876. Excepting the Northern Cardinal, how many other all-red birds of the American southwest have crests?

Questions

7877. Do all pittas have erectile crown feathers?
7878. How much do Great Blue Herons weigh?
7879. In southern Africa, what is a Jacky hangman?
7880. Do female kiwis call more frequently than males?
7881. Which abundant breeding gull of interior North America has pinkish underparts?
7882. Are African rockfowl (bald-crows) primarily terrestrial?
7883. Do American Coots establish lifelong pair bonds?
7884. In Britain, what is the White-winged Crossbill called?
7885. In 1956, the population of Northern Pintails in North America was estimated at about 10 million birds, but what was the population in the early 1990's?
7886. What noise often betrays the presence of a Red Crossbill?
7887. Which of China's 25 species of pheasants is the most endangered?
7888. Which vividly-colored, long-legged aquatic bird do Brown Pelicans roost with in the coastal swamps of Venezuela?
7889. During the winter, is male ptarmigan mortality greater than female mortality?
7890. Do Golden Eagles nest in all 17 of the westernmost continental states?
7891. Is the lovely lavender color of the Pompadour Cotinga caused by prismatic feather structure that refracts light?
7892. Do most chicks hatch early in the morning?
7893. Does a Barn Swallow build an open, cup-shaped mud nest?
7894. Do foraging creepers spiral down a tree trunk?
7895. Which is the only North American grebe lacking a white wing-patch?
7896. Which phalarope breeds in the lower 48 states?
7897. Is the Old World Stonechat a warbler?
7898. When Barn Owls were introduced into the Seychelles Islands in 1951, which native bird was negatively impacted, and why?
7899. What is the bill color of an African White-headed Vulture?
7900. What percent of fledged Emperor Penguin juveniles survive their first year?
7901. Do Trumpeter Swans have rounder heads than Tundra (Whistling) Swans?
7902. Does the white chest of a Common Murre terminate in a V at the base of its black throat?
7903. Do any adult birds have body temperatures as low as humans?
7904. What is the more familiar name for Coues' Petrel?
7905. Are caracaras short-legged?
7906. Which North American woodpecker drinks the most often?
7907. Which organization publishes the *Living Bird*?
7908. Compared to humans, do diurnal birds have large pupils relative to the size of their eyes?
7909. Which is the only white cockatoo with a red vent?
7910. Does the wingspan of a Great Frigatebird exceed that of a Black-footed Albatross?
7911. What is the heart-rate of a resting California Quail?
7912. Does the Gilded Flicker have the head of a Yellow-shafted Flicker with the body of a Red-shafted Flicker?
7913. Does the Inland (Australian) Dotterel have two conspicuous chest-bands?
7914. How long is the nesting burrow of a Belted Kingfisher?
7915. How many species of birds have been recorded in Florida's Everglades National Park?

Questions

7916. Which bird is known as the Alala?
7917. Do the catbirds of Australia and New Guinea prey on birds?
7918. Which nightjar relatives have a "toothed" upper mandible?
7919. Which shorebird is kept as a pet in Latin America, and why?
7920. How many species of gallinaceous birds were successfully introduced into California, and which ones are they?
7921. What color is a juvenile American White Pelican's bill?
7922. Do Clark's Nutcrackers inhabit Mexico?
7923. How did the Australian Pilotbird obtain its name?
7924. Picus was the son of which mythical god, and what happened to him?
7925. Do Cooper's Hawks soar more often than Sharp-shinned Hawks?
7926. Prior to the arrival of humans, what was the size of the New Zealand kiwi population, and what is the current population size?
7927. Are the neotropical trumpeters fast runners?
7928. Are the white wing-patches of a perched Ivory-billed Woodpecker evident?
7929. During the wing molt of most passerines, is the outer primary generally the last to be shed?
7930. Which Eurasian bird is known as a Reed Pheasant?
7931. Which North American hummingbird is the most xerophilous?
7932. What is the length of the walking stride of a Wild Turkey gobbler?
7933. What causes bird's-eye spot, a disease that attacks the leaves of the tea plant?
7934. In New Zealand, which bird is known as Hoiho, and what does the name mean?
7935. Do Heermann's Gulls attain adult plumage at three years of age?
7936. Do shrikes carry heavy prey in their bills, as opposed to their feet?
7937. Do foraging New Caledonian Kagus gather in flocks?
7938. Do American Coots breed in Alaska?
7939. Do male Whooping Cranes incubate at night?
7940. Relative to their size, do plovers have smaller eyes than sandpipers?
7941. Which cavity-nesting American bird is most often victimized by obligate brood parasitic-nesting cowbirds?
7942. Which American waterfowl has a gibbous bill?
7943. Which bird-of-paradise lives at the greatest altitude, and how high does it occur?
7944. Which group of predatory Arctic birds is most numerous?
7945. Excluding kookaburras, do kingfishers prey on birds?
7946. Which state has the most species of woodpeckers, and how many occur there?
7947. In what 1966 movie did James Stewart, Peter Finch and Hardy Kruger star?
7948. Do hornbills commence incubation with the first-laid egg?
7949. Why is the Long-billed Marsh Wren a threat to Least Bitterns?
7950. Is the Cedar Waxwing sexually dimorphic?
7951. Which long-toed aquatic bird is a threat to rice crops in India and Bangladesh?
7952. Were choughs named for their vocalizations?
7953. Do sleeping kiwis tuck their bill under a wing?
7954. Do Sharp-shinned Hawks occur as far north as Alaska?
7955. Where do swallows generally copulate?

7956. Do Lazuli and Indigo Buntings commonly hybridize?
7957. Is the Shore Plover a common New Zealand bird?
7958. How many bowerbirds inhabit the Moluccan Islands?
7959. What are gonoleks and boubous?
7960. According to Allen's Rule, should birds inhabiting warmer climates have longer bills, tails and other extensions of the body than birds in cooler climates?
7961. When European Shelducks parasitize Red-breasted Merganser nests, why do most of the resultant shelduck ducklings perish?
7962. Do female Ring-necked Parakeets have a ringed neck?
7963. Which American longspur is a declining species?
7964. What bird recovered at Poor Knights Islands (New Zealand) in 1936, when introduced pigs were finally eradicated?
7965. Are molting Wild Turkeys flightless?
7966. Is the male Rufous Hummingbird the only North American hummingbird with a rufous-colored back?
7967. Which is the only North American woodpecker with a pinkish-red belly?
7968. What is the camouflaged pattern of many nightjars called?
7969. Does the Palm-nut Vulture nest in palm trees?
7970. Which bird has been seen where the lowest temperature of the planet was recorded?
7971. At what time of day or night does the American Coot generally lay its eggs?
7972. With which birds do American Anhingas commonly nest?
7973. What is a crombec?
7974. Do pipits have stout bills?
7975. Is the crest of a flying Steller's Jay depressed?
7976. Are Black Skimmers on the wintering grounds gregarious?
7977. Which of the two African oxpeckers is most inclined to feed on sparsely-furred mammals, such as the Cape buffalo?
7978. Do flamingos occur in Australia?
7979. How many Old World vultures are inhabitants of the forest?
7980. Is the Yellowhammer one of the least common finches of Great Britain?
7981. Which type of tropical birds do foraging Red Crossbills resemble?
7982. Are avian kidneys relatively larger than those of mammals or reptiles?
7983. Which North American bird is also known as the Siberian Tit?
7984. Which is the most common desert thrasher?
7985. Do Cedar Waxwings catch insects on the wing?
7986. What is an *onagadori?*
7987. Which is the only bird in the family Ibidorhynchidae?
7988. How were duck-billed platypus skins formerly used?
7989. Is the Common Snipe the only snipe found north of Panama in the New World?
7990. What color is the chest-patch of a Gray (Hungarian) Partridge?
7991. Are treeswifts native to Africa?
7992. How many penguins have yellow eyes?
7993. Is the European (Common) Starling known as a mimic?
7994. What caused the demise of the Bonin Islands Thrush?

Questions

7995. How can Bushtits be sexed?
7996. Are fish ever carried in the pouch of a pelican?
7997. What bird is pictured on the Canadian two-dollar bill?
7998. How much greater is the daily energy cost of a molting penguin than an incubating penguin?
7999. Which South American bird is called "chunga?"
8000. Can a Shoebill stork take off without a running start?
8001. How large a territory is defended by a pair of New Zealand Takahe?
8002. Are Marsh Wrens actually marshland inhabitants?
8003. Did the ornithologist Roger Tory Peterson attend art school?
8004. Is the population of Swainson's Hawks declining?
8005. How much does the smallest parrot weigh?
8006. At the beginning of the breeding season, how long do Adelie Penguin males fast?
8007. What color is the breast of a male Mourning Dove?
8008. Which birds are called brownbuls?
8009. In North America, with which other gull does the Little Gull generally associate?
8010. How do terrestrial antbirds differ physically from arboreal antbirds?
8011. Which two pinnipeds prey on Macaroni Penguins at South Georgia Island?
8012. Do Sacred Ibis have black wingtips?
8013. How many species of petrels have populations of over a million breeding pairs?
8014. Which bird is commonly used as a lure when hawks are mist netted?
8015. Where was the type specimen of the Black Rail collected?
8016. Do most Old World white-eyes have melodious songs?
8017. What type of tree does an African ground-hornbill favor as a nesting site?
8018. American hunters average how many shots for each dove bagged?
8019. Do Ospreys have a crest?
8020. Are most babblers sexually dimorphic?
8021. Are the eggs of a European Robin robin-egg blue in color?
8022. Which are the only two western North American waterfowl with black bellies and completely white backs?
8023. Do cockatoo females have darker-colored eyes than males?
8024. Do all barbets nest in tree cavities?
8025. What color is the bill tip of a Greater Flamingo?
8026. Are both peppershrikes stocky birds?
8027. What color are many of the Australian wrens?
8028. Ivory Gulls are often seen with which large predator?
8029. How does a thick-knee make itself inconspicuous?
8030. Which birds of the Bible were referred to as *peacocks?*
8031. Is the White-winged Dove a colonial nester?
8032. What color is the skin surrounding the eye of a Superb Lyrebird?
8033. Which is the only nightjar of Japan?
8034. Does the Lammergeier have a distinct juvenile plumage?
8035. Which American major league baseball team is named after a corvid?
8036. Why was the Christmas bird count initiated?

Questions

8037. How do the Blue Grouse of the northern Rocky Mountains differ in appearance from Blue Grouse occurring elsewhere?
8038. During the early 20th century, on what day did the British traditionally shoot songbirds?
8039. Why is the kiwi incubation period so long?
8040. How many pine cone seeds can a Clark's Nutcracker carry at a time?
8041. Which North American gamebird has the most restricted range?
8042. Are copulating Emus especially vocal?
8043. Which penguin is the most wary around humans?
8044. When and where was the first blue Budgerigar bred?
8045. Which heron has a call that closely resembles a tiger's grunt?
8046. Are stilts more apt to frequent freshwater habitat than avocets?
8047. Which group of seabirds typically soars the highest when foraging?
8048. How did the name "oriole" originate?
8049. Which distinctive group of three seabirds flies in flocks of synchronous movements?
8050. What Spanish dance has the name of a passerine?
8051. What is unique about hornbill cervical vertebrae?
8052. How long is the nestling period of a Marabou Stork?
8053. Do petrels have a juvenile plumage?
8054. When a hungry bird has an empty stomach, is its body temperature likely to rise?
8055. Do shrikes consume impaled prey commencing with the head?
8056. Of all the bird skins deposited in collections around the world, are most specimens housed in Europe?
8057. Which North American rail or rail-relative is most closely associated with fairly deep ponds and relatively open water throughout the year?
8058. Is the temperature of a newly-hatched domestic chicken the same as that of an adult chicken?
8059. Which North American nightjar is most likely to call on the wing?
8060. According to legend, how did the crossbill obtain its crossed beak?
8061. Do flying Mourning Doves ever coast or glide?
8062. Are male Swallow-Tanagers green in color?
8063. How much longer is the upper mandible of a newly-hatched hornbill chick than the lower?
8064. What is unusual about the underwing of an adult Sacred Ibis?
8065. In what plant does the Ladder-backed Woodpecker generally nest?
8066. Do thick-knees (stone-curlews) have webbed feet?
8067. Do both parents of Hawaiian honeycreepers feed chicks?
8068. What is the easternmost state where California Condor bones been discovered?
8069. When foraging, which American egret is most intolerant of other herons?
8070. What is unusual about the nostrils of birds in the turaco family?
8071. Which crane has a voice described as flute-like?
8072. Is the American Goldfinch the state bird of New Hampshire?
8073. Is Townsend's Solitaire a ground nester?
8074. Which birds are generally parasitized by honeyguides?
8075. Does the Kagu of New Caledonia have a long neck?
8076. Which American bird is called a "lettuce bird," and why?

Questions

8077. What ultimately caused the demise of the Heath Hen?
8078. Do shrikes have an undulating, woodpecker-like flight?
8079. How long do the young of neotropical tapaculos remain with their parents?
8080. Do African Pink-backed Pelicans invariably nest near water?
8081. Do all North American chickadees have black bibs?
8082. Which shorebirds have the longest bills relative to their size?
8083. What color are dipper eggs?
8084. The Copper Pheasant is restricted to which country?
8085. Why is the face of the New Guinea Pesquet's Parrot bare?
8086. Are Red-winged Blackbirds colonial nesters?
8087. Who starred in the 1942 movie *The Black Swan*?
8088. Which sex of the extinct Dodo incubated?
8089. Are Greater Roadrunner pair bonds permanent?
8090. How do hartebeest react to oxpeckers?
8091. What is the estimate of the number of petrels and prions killed annually by the 2,500 pairs of skuas on Gough Island (South Atlantic)?
8092. Why do rockfowl (bald-crows) have unusually large eyes?
8093. Do both Sage Grouse sexes have long, stiff, pointed tails?
8094. During its first two months, what percent of a Wild Turkey chick's diet consists of insects?
8095. Is the feral population of British Canada Geese migratory?
8096. How long is the hanging nest of the Central American Montezuma Oropendola?
8097. Do frigatebirds engage in courtship feeding?
8098. What is the *pope's nose?*
8099. After a catch, does the water in the pouch of a pelican weigh more than the bird itself?
8100. Is the Bald Eagle pictured anywhere on the U.S. 20-dollar bill?
8101. When laying its egg in the nest of a host species, how long do obligate brood parasitic-nesting cuckoos remain at the nest?
8102. Are Muscovy Ducks common in West African villages?
8103. Is the Chiffchaff one of the last spring migrant songbirds to return to the forests of Europe?
8104. How many of the 75 species of exotic gamebirds introduced into Hawaii still survive?
8105. Which is the largest bird of the Aru Islands?
8106. Which kingfisher is the most widespread?
8107. Are bowerbirds predominately insectivorous?
8108. Do tern pairs mate for life?
8109. In the early 1950's, rabbit plague (myxomatosis) was deliberately introduced into France to control the rabbit population, and rabbit mortalities of western Europe ran as high as 99 percent. Why did the disease never reach epidemic proportions in southern Spain?
8110. How many species of swifts breed in New Zealand?
8111. Specimens of how many species of extinct parrots are deposited in museum collections?
8112. Which is the most highly-prized songbird of Brazil?
8113. Do both King and Emperor Penguins have circumpolar distributions?
8114. The ground where cock Ring-necked (Common) Pheasants call from early in the spring is sometimes known as what?

Questions

8115. How did Australian hunters capture the duck-billed platypus?
8116. Why does the Palm-nut Vulture not soar in thermals?
8117. Which birds have the most frontally situated eyes?
8118. What are the three main functions of preening?
8119. How high can a Congo Peacock jump up to a branch without using its wings?
8120. Does the gular pouch of the White Stork amplify the sound of bill-clattering?
8121. Why was the introduction of Mallards into New Zealand a major ecological mistake?
8122. Which are the only birds that lay a two-egg clutch in which the *first*-laid egg is always significantly smaller than the second-laid?
8123. What is the Latin name for *Big Bird* of Sesame Street, and how did it acquire the name?
8124. What is the population size of the African Wattled Crane?
8125. With which bird is the Western Sandpiper often confused?
8126. Do Green-winged Teal fly in especially compact flocks?
8127. Do cormorants breed on all continents?
8128. After what is the Australian Spinifex Pigeon named?
8129. Which sex of Bobwhite digs the nest depression?
8130. Is the Black-tailed Gnatcatcher a true desert bird?
8131. Do Australian Brush-turkeys have lateral tails?
8132. How many red blood cells are in a cubic millimeter of hummingbird blood?
8133. Pryer's Woodpecker is endemic to what country?
8134. Are most birds-of-paradise slow fliers?
8135. How many occipital condyles, the point of articulation with the atlas (the first cervical vertebra), are located at the base of the avian skull?
8136. How do American Coots differ in appearance from Eurasian Coots?
8137. Are courting tropicbirds gregarious?
8138. Which is the only northern warbler with a bright red face?
8139. What are secondary feathers often called by falconers?
8140. Which of the three jaegers has the stoutest bill?
8141. Which bird is featured on the coat-of-arms of the Malaysian state of Sarawak?
8142. What does the screech-owl generic name *Otus* suggest?
8143. Are virtually all land birds of the Australasian region endemic?
8144. According to Gloger's Rule, should birds inhabiting more humid areas be lighter in color than birds of arid climates?
8145. What is a *tournament ground?*
8146. What are vultures in the genus *Gyps* often called?
8147. Is the bill of a puffin compressed?
8148. Does the male Northern Cardinal sing only during the breeding season?
8149. What constitutes about 50 percent of a Loggerhead Shrike's winter diet?
8150. When one eats very little, how is this commonly expressed?
8151. What journal is published by the South African Ornithological Society?
8152. In most passerines, what percent of the egg weight does a newly-hatched chick represent?
8153. Are breeding Wild Turkey gobblers territorial?
8154. Why do the Blacks of the West Indies call the Black-capped Petrel "diablotin?"
8155. Are the outer primaries of the North American Broad-winged Hawk notched?

8156. When was the Okinawa Rail discovered, and what is its status?
8157. What is the audible effort to force up phlegm from the throat often called?
8158. Do petrels climb trees?
8159. Is the sense of touch highly developed in birds?
8160. Do storks frequently rest on the ground with the lower part of their legs stretched forward?
8161. Do any of the neotropical antbirds have erectile crests?
8162. Do redstarts hawk flying insects?
8163. Why do the head and breast feathers of many African turacos appear to be hairy?
8164. Is the Carolina Chickadee smaller than the Black-capped Chickadee?
8165. In China, which bird is known as *Fung-whang?*
8166. When did pioneer bird photographers commence work?
8167. The call of which bird was described by Shakespeare as *a word of fear unpleasing to a married ear?*
8168. Is the specific gravity of a loon near that of water?
8169. What position do perching mousebirds generally assume?
8170. Did the Whooping Crane formerly breed in Louisiana?
8171. Which birds have been described as *in many ways the most adaptable and highly evolved of all birds*?
8172. The burrow of which mammal does the Rock Wren commonly select as a nesting site?
8173. Do pigeons nest in tree holes or burrows?
8174. Were all specimens of the Jamaican Petrel, a species not seen since the last century, collected in Jamaica?
8175. Do tropicbirds float low on the water?
8176. Which is the only bright scarlet North American bird with black wings and tail?
8177. Did the extinct Dodo evolve from a flying ancestor?
8178. Do Bell's Vireos of the west tend to be brighter olive than the vireos of the interior?
8179. The average weight of a Sanderling is about 1.77 ounces (50 gms), but how much does it weigh just prior to migration?
8180. The diet of a Killdeer consists almost exclusively of what?
8181. How do the similar calls of the California and Gambel's Quail differ?
8182. Do Long-tailed Jaegers feign injury in an attempt to lead intruders away from their nests?
8183. How does the Australian Emu use its stunted wings?
8184. Are most hornbills sedentary?
8185. Do jacamars use the same nesting burrow of preceding seasons?
8186. In 19th-century Europe, what was a harlot called?
8187. Do male Snowy Owls engage in aerial courtship displays?
8188. At Easter Island, what is the major avian predator of the Chilean Tinamou?
8189. Who introduced the outside world to the Inca's secret of using seabird guano as an organic fertilizer?
8190. Which disease of chickens and their eggs is most feared by the American consumer?
8191. At which time of year do South American Oilbirds nest?
8192. Do all parrots have relatively broad and rounded wings?
8193. Do Sandhill Cranes invariably select a new nest site each year?
8194. Why does a hovering kestrel fan its tail and point it forward?

Questions

8195. When in flight, the primaries of which swan produce a high-pitched, harp-like note on the downstroke?
8196. When excited, what does the Congo Peacock do?
8197. Are Old World rollers aggressive?
8198. Does the lower mandible of a skimmer chick commence to lengthen and exceed the length of the upper mandible within the first week of hatching?
8199. What color is the tail of a breeding adult Roseate Spoonbill?
8200. Does the little Galapagos Lava Heron prey on *adult* Sally lightfoot crabs?
8201. In Great Britain, when was the Night Poaching Act enacted?
8202. What is the origin of the name "Demoiselle" Crane?
8203. Is the head of a Limpkin entirely feathered?
8204. At what age do the eyes of a Short-eared Owl chick open?
8205. Does the Pelagic Cormorant nest on flat ground?
8206. What aquatic plant do aviculturists often feed to stiff-tail ducklings, such as the Ruddy Duck?
8207. Are sandgrouse feathers loosely attached, like those of the closely-related pigeons and doves?
8208. What color are the eye-rings of both North American oystercatchers?
8209. Which "sport" is known as *cocking?*
8210. What color is the rump of Harris' Hawk?
8211. Along the shores of small lakes and streams, which is the most widespread American sandpiper?
8212. Are wagtails swift runners?
8213. What is the greatest number of eggs recorded in a single rhea nest?
8214. Are any of the titmice streaked, barred or spotted?
8215. Do courting Hairy Woodpeckers clap their wings in flight?
8216. When are Marabou Storks most apt to feed on fish?
8217. How many of the more than 150 species of waterfowl dive habitually?
8218. Do male Ostriches outnumber females?
8219. Which American bird is called a "barrel-maker," and why?
8220. Which diurnal raptor is credited with having the widest natural distribution of any bird?
8221. How many species of tinamous inhabit Costa Rica, and which ones are they?
8222. In the Andes, how many flowers does a foraging hummingbird visit daily?
8223. Are birds the only animals with a syrinx?
8224. Do Old World flycatchers have rictal bristles surrounding the nostrils like tyrant-flycatchers?
8225. Do sandpipers and their near-relatives ever catch insects on the wing?
8226. In what country was the mist net invented?
8227. What color are the legs and feet of an African Finfoot?
8228. Domestic pigeons have how many pairs of ribs?
8229. Why are the incubation and fledging periods of the White-throated Swift little known?
8230. Which two boobies are the most abundant?
8231. How was it determined that male Wandering Albatrosses feed in more southerly waters than females?

8232. How many species of kingfishers hover when foraging?
8233. Does the Black Turnstone winter in New Zealand?
8234. How many of the 11 species of coots inhabit the Old World, and which ones are they?
8235. According to aboriginal legend, why does the Laughing Kookaburra have its famous laugh?
8236. Which American bird was known as a "swamp angel?"
8237. Are all egrets of the tropical Pacific white-plumaged?
8238. Where did Barn Swallows nest before Europeans built houses and barns in America?
8239. How many of the 184 species of breeding birds of Madagascar are endemic?
8240. Are birds in the only class of vertebrates that use injury-feigning behavior to lure predators away from their young?
8241. Does the Malleefowl have a high reproductive rate?
8242. Is Townsend's Solitaire migratory?
8243. What is an Elepaio?
8244. Goldfinches have what type of flight style?
8245. How many pounds of food are required to fledge a Wood Stork chick?
8246. When an Anhinga or darter spears a fish, is its bill closed?
8247. Why do Bobolinks have long claws?
8248. How is a dim-witted individual sometimes described?
8249. Which is the most widely-established passerine family on remote islands?
8250. Which American bird is called a "little white crane?"
8251. What is an Anatosaurus?
8252. On what does the endangered Cahow (Bermuda Petrel) primarily feed?
8253. Are Verdins solitary?
8254. What color is the throat of a Mountain Quail?
8255. Is the Australian Ground Parrot a rare, cursorial granivore?
8256. Which bird raids Loggerhead Shrike larders?
8257. What is Rivoli's Hummingbird currently called?
8258. Do King Vultures prey on live reptiles?
8259. Why is the shell of some parasitic cuckoo eggs thicker than would be expected for eggs of such a small size?
8260. Why do Ospreys catch fewer fish when the skies are cloud-covered than during periods of direct sunlight?
8261. How many Emperor Penguin colonies are located north of 65°S?
8262. Which are the most inquisitive and trusting alcids?
8263. Does the Phainopepla have conspicuous rictal bristles?
8264. Do Old World warbler young generally leave the nest before they can fly?
8265. What is the state bird of Vermont?
8266. How do adults of the rare endemic race of Aldabra Island Sacred Ibis differ in appearance from birds of the widespread nominate African subspecies?
8267. Which two starling relatives are totally dependent on mammals for their existence?
8268. Do White (Fairy) Tern pairs allopreen?
8269. With respect to nesting, what habit of the Rufous Hornero and the Firewood-gatherer is disliked by humans?

Questions

8270. Are Ospreys relatively common in Sweden?
8271. In America, which bird is called an "ox-eye?"
8272. Are African mousebirds fond of dust-bathing?
8273. Which is the only one of the neotropical jacamars that does not have four toes?
8274. When sapsuckers drill a series of holes, are the holes evenly spaced?
8275. How far north do rheas occur?
8276. Which North American tanager is the superior singer?
8277. Are jaegers totally carnivorous?
8278. Has the range of the House (English) Sparrow tripled during the past century?
8279. What is the major component of the regurgitated pellets of the European Honey-buzzard?
8280. What is the Plumed Whistling-Duck also called?
8281. Which are the only endemic bird *species* of Cyprus?
8282. Does the White-eyed Vireo actually have white eyes?
8283. Which is the only cavity-nesting flycatcher of eastern North America?
8284. Do all cormorants have black feet?
8285. What color dye is obtained from the various lichens known as canary moss?
8286. Which African bird is called a Cut-throat, and how did the name originate?
8287. Do diving-petrels have hooked bills?
8288. When were Great Bustards extirpated from Great Britain?
8289. Why does the Recurve-billed Bushbird (a neotropical antbird) have a laterally-flattened bill with a semicircular lower mandible?
8290. Do Wild Turkeys in *flocks* frequently vocalize?
8291. Does the dark-color morph of the Northern Fulmar predominate in the Atlantic?
8292. Besides decreasing a bird's size and breaking up its outline, what advantage is there to crouching with respect to concealment?
8293. What was a pelican's pouch formerly used for in southeastern Europe?
8294. Are waxwings especially territorial?
8295. Do Wood Storks feed cooperatively?
8296. According to European folklore, what was soup made from owl eggs said to cure?
8297. How many mounted specimens of the extinct Indian Ocean solitaires exist?
8298. The tiny flightless Inaccessible Island Rail has what type of plumage?
8299. What is the wingspan of a Black Skimmer?
8300. What are anisodactylous birds?
8301. Do Ancient Murrelets dig deep nesting burrows?
8302. Do sleeping penguins hook their bills behind a flipper (wing), like many flying birds?
8303. Do feeding Limpkins ever break open the shell of a snail?
8304. What is unusual about the diet of the Indo-Australian Channel-billed Cuckoo.?
8305. Which avocet relative nests in the Galapagos Islands?
8306. Is the trailing edge of a grebe's leg (tarsus) smooth?
8307. When courting, which large, long-legged birds toss sticks or vegetation high into the air?
8308. What color is the rump of many of the estrildid finches?
8309. Do American Coot chicks fly at an early age?
8310. Including extinct forms, how much larger was the largest bird than the smallest?
8311. Which bird of those examined thus far has the heaviest preen gland relative to its weight?

8312. Is the avian retina unusually thick and well developed?
8313. Are Northern Jacana chicks spotted?
8314. Are New World barbets more sexually dimorphic than barbets of the Old World?
8315. Which New Zealand birds have been described as having a *mole-like* head?
8316. Which bird is known as a "sweetheart owl," and why?
8317. Which birds feed by *hydroplaning?*
8318. Do Anhingas and darters have crests?
8319. What are the particles of stones and sand that many birds ingest to aid in grinding up their food called?
8320. Do Lesser Flamingos inhabit what was formerly the USSR?
8321. Does the Gull-billed Tern have relatively shorter legs than other terns?
8322. Are the elongated crown feathers of the endemic New Caledonian Horned Parakeet movable?
8323. Is the wingbeat of a Red-throated Loon more rapid than the wingbeat of a Common Loon?
8324. Do Glossy Ibis males generally incubate at night?
8325. How many pairs of Red-tailed Tropicbirds breed on Midway Island?
8326. Which American bird is called a "road chippie?"
8327. Which North American icterid has undergone enormous expansion in both range and population size?
8328. Since the inception of the U.S. federal duck stamp program, how much money have stamp sales generated?
8329. Where do Thick-billed Murres breed south of Canada?
8330. What do Eleonora's, Sooty and Red-footed Falcons have in common with respect to breeding strategy?
8331. In Great Britain, is the Common Wood-Pigeon more numerous than the feral Rock Dove?
8332. Why do many vultures have serrated tongues?
8333. Are male Purple Martins actually purple?
8334. In what vegetation does the endemic Antipodes Island Green Parakeet nest?
8335. Do northern hemisphere albatrosses breed only in the Pacific Ocean?
8336. What color is a kiwi egg?
8337. Did the familiar eagle-plumed war bonnet originate with the American Plains Indians?
8338. What happens to a cave swiftlet's salivary glands during the breeding season?
8339. According to superstition, what will happen if a bird, especially a black-as-night corvid, strikes a window?
8340. Was the Jackal Buzzard so named because it associates with jackals?
8341. Can the African mousebirds walk, or must they resort to hopping?
8342. Do African turacos have long claws?
8343. By what nickname was Brother Matthias Newell, after whom Newell's Shearwater is named, known in Texas?
8344. Where is the most important American fall staging and molting area for Eared Grebes located?
8345. Why is a foraging kiwi noisy?
8346. When were Magnificent Frigatebirds discovered nesting in the Florida Keys?
8347. Are nestling Wood Storks vocal?

Questions

8348. What is one forced to do if a major apology is required?
8349. How are the flimsy stick nests of the Red-footed and Abbott's Booby cemented together?
8350. Which organization sponsored the 1985 publication of *A Dictionary of Birds,* edited by Campbell and Lack?
8351. Are most larks brightly colored?
8352. Which African gamebird inhabits Haiti?
8353. Is Wilson's Plover strictly coastal?
8354. How many Hawaiian Mamos were required to supply the feathers for the feather cloak commissioned by the Hawaiian King Kamehameha I in the early 1800's?
8355. What does a soaring Lanner Falcon often do with its tail?
8356. Which northern gull nests in sea caves?
8357. Which two countries supply the most swiftlet nests used in bird's-nest soup?
8358. When a Parasitic Jaeger consumes a small bird, is it swallowed whole?
8359. Do the folded wings of a pratincole extend well beyond its tail?
8360. Which American bird is known as a "Carolina crake?"
8361. Do snake-eagles lay a two-egg clutch?
8362. Which are the only corvids in which the male incubates?
8363. Which is the last bird listed in the sixth edition of the *A.O.U Check-List of North American Birds*?
8364. In which storm-petrel does the male alone excavate the nesting burrow?
8365. How many times an hour are small songbird chicks fed?
8366. Which feathers are called *armpit* feathers?
8367. How do most plovers lure intruders away from their nest or young?
8368. Do swifts live more than two decades?
8369. Do Turkey Vultures weigh as much as five pounds (2.3 kg)?
8370. When do Shoebill storks gather in flocks?
8371. In North America, which was the first bird to have its song recorded, and when did this occur?
8372. Do all parrots have relatively short legs?
8373. Must a coot race over the water to become airborne?
8374. Is the hind toe of a South American screamer elevated?
8375. What is the average harem size of a tom Wild Turkey?
8376. Which bird preys on kiwi eggs?
8377. Why are surface-feeding ducks captured more frequently by snapping-turtles than diving ducks?
8378. Which raptor was successfully introduced to Easter Island?
8379. The juveniles of which jaeger are ventrally barred?
8380. Discounting its long tail, is a Red-billed Tropicbird larger than a California Gull?
8381. Which birds did Columbus take to Hispaniola on his second New World voyage in 1493?
8382. Is the song of a terrestrial wood-warbler, such as the Ovenbird, louder than that of an arboreal warbler?
8383. Do Tufted Titmice store food?
8384. What is the external opening of the preen (uropygial) gland sometimes called?
8385. What is a bird called that seeks insects from leaves in a tree canopy?

8386. Do any passerines have lobed feet?
8387. What was the name of the famous Bald Eagle that was the mascot of the 8th Wisconsin Regiment during the Civil War?
8388. Can a Brown Skua *fly* carrying an egg as large as a King Penguin egg?
8389. Do Ring-necked (Common) Pheasants have erectile crests?
8390. Why do Brazilians refer to the Sharp-tailed Streamcreeper as the *president of filth?*
8391. How does the nesting behavior of the Pied Kingfisher of Africa and Asia differ from most kingfishers?
8392. Which wren nests in the Pribilof Islands, Alaska?
8393. Do older tom Wild Turkeys have a higher-pitched gobble than younger males?
8394. Are male Ruffs *significantly* larger than females?
8395. Do juvenile Lammergeiers (Bearded Vultures) have a beard?
8396. Do the West Indies have a large population of antbirds?
8397. Why do dippers have especially loud songs and call notes?
8398. Do waxwings have a straight-line flight style?
8399. What color is the underwing of an adult Lesser Flamingo?
8400. Do male pelicans generally collect the nesting material?
8401. When prey, such as a shrimp or fish, is encountered by a foraging skimmer, does the prey slide up the inclined edge of the lower mandible?
8402. How many indigenous neotropical families of non-passerine birds contain five species or less, and which birds are in those families?
8403. Which wood-warbler winters the farthest north?
8404. Are juvenile Great Blue Herons paler than adults?
8405. When fledging swifts leave the nest for the first time, is an adult usually in attendance?
8406. Once a female Emperor Penguin relieves the male after his long Antarctic winter incubation shift, how long does she brood the single chick until relieved by her mate?
8407. Which bird is sometimes called a "brotherly-love vireo?"
8408. Which is the only family of birds in which virtually all members are obligate brood parasitic nesters?
8409. Do more than one pair of petrels ever use the same nesting burrow?
8410. Why has the range of the two African oxpeckers (tickbirds) been reduced in the past 30 years?
8411. Which is the only red-breasted woodpecker of South Africa?
8412. The screeching of which bird preceded the murder of Julius Caesar?
8413. How does the behavior of the helmetshrikes differ markedly from typical shrikes?
8414. Is the bright blue of the Bluethroat present all year?
8415. Is the Kagu of New Caledonia completely gray in color?
8416. Does the Caspian Tern range inland?
8417. Do Black-billed Magpies dwell in the capital of California?
8418. Who "discovered" the Western Tanager?
8419. Was Darwin the first to note that the Galapagos Woodpecker Finch was a tool-user?
8420. Which is the rarest Australian parrot?
8421. How are turaco chicks fed?
8422. What hunting dog is known as a griffon?

Questions

8423. In North America, which bird is called a "storm crow?"
8424. What are hybrids between a canary and a goldfinch called?
8425. Which group of wintering Palearctic migrants is the most numerous in Australia?
8426. Are hummingbirds attracted to blue flowers?
8427. Reproductions of which crane adorned the walls of ancient Egyptian tombs?
8428. Aside from the obvious difference of foot and leg color, how does the Yellow-footed Gull differ from the Western Gull?
8429. Are the larger toucans more inclined to use nesting cavities in a living tree than in a dead tree?
8430. What is the mean first breeding age in Northern Fulmars?
8431. Which American bird is known as a "blue canary?"
8432. Secondary feathers are attached to which bone?
8433. Do male Wild Turkeys breed in their first year?
8434. Do male Bobolinks have a distinct winter plumage?
8435. Does the Sandhill Crane occur in Japan?
8436. Birds in which family have the smallest hearts relative to their size?
8437. Do both sexes of titmice have brood-patches?
8438. Are jacana nests substantial?
8439. What causes ringworm, a disease prevalent in chickens, turkeys and pigeons?
8440. Are most pipits virtually the same in appearance?
8441. In German folklore, to which bird did *Adabar* refer?
8442. Do South American Torrent Ducks ever frequent salt water?
8443. Is the Wood Thrush commonly parasitized by cowbirds?
8444. What is an *exodus call?*
8445. Do grebes vigorously defend their nests?
8446. Which four birds were combined into a single species known as the Dark-eyed Junco?
8447. Which large alcid is most inclined to swim with its tail cocked?
8448. Why does a Purple Martin feeding chicks close its eyes?
8449. Do terns feed their offspring after fledging?
8450. Are the flowers of the canary vine yellow?
8451. Is the distribution of the Whimbrel circumpolar?
8452. Do most ground-nesting passerines construct rather substantial nests?
8453. How is something that is done as a joke or on a whim described?
8454. Does the South American Hoatzin ever feed on live prey, such as fish or crabs?
8455. How many woodpeckers have square tails?
8456. Did the Eurasian (Common) Starling reach Jamaica on its own from North America?
8457. Which parrot is endemic to Henderson Island (Pitcairn Island group)?
8458. Do the African mousebirds ever glide?
8459. Do ratites ever sleep on their backs with their feet in the air?
8460. What is cranial kinesis, and in which birds is this most pronounced?
8461. Do skimmers have webbed feet?
8462. Do owls have relatively thick skin?
8463. Are most Hawaiian honeycreepers insectivorous?
8464. Which is the only North American swift with a contrasting black-and-white pattern?

8465. Which is the only migratory upland gamebird of Europe?
8466. Are Inca Terns solitary?
8467. Does the song of Le Conte's Sparrow consist of distinct whistling notes?
8468. In America, which bird is known as a "gull-teaser?"
8469. Exclusive of Antarctica, which is the only continent with no native breeding swans?
8470. Are finfoot or Sungrebe eggs elongate?
8471. Are robin-chats among the best of African avian mimics?
8472. Do African natives make less use of feathers than the Indians of South America?
8473. Do nest-building Black Phoebes use saliva as nest cement?
8474. Which American titmouse is most inclined to pluck hair from woodchucks, squirrels, opossums and humans?
8475. Is the Nocturnal Curassow actually nocturnal?
8476. Has the population of Emus increased since Australia was colonized?
8477. What is distinctive about the nest of the South American White-throated Cacholote?
8478. Are thrashers gifted songsters?
8479. Exactly where was the first nest of the European Starling discovered in North America?
8480. What is the state bird of both Maine and Maryland?
8481. Why are the external nares of a Black Vulture smaller than those of a Turkey Vulture?
8482. Why were stoats and ferrets, both major bird predators, introduced into New Zealand?
8483. In southern California, on what does the Ashy Storm-Petrel primarily feed?
8484. In ancient folklore, which bird was believed to suck the blood of young children?
8485. What are the residents of Kansas called?
8486. During the breeding season, which two alcids have the most cryptic plumage?
8487. Where in Europe can the Red-knobbed Coot be found?
8488. Is the Rosy Thrush-Tanager arboreal?
8489. Is a Crab Plover's egg large relative to its size?
8490. Do waxwings prey on fast-flying dragonflies?
8491. Which is the largest South American woodpecker?
8492. What are the first few swallows to appear in the spring called?
8493. In what country do most ducks that migrate south of the U.S. winter?
8494. Where did the name "drongo" originate?
8495. Have any unbroken eggs of the extinct moas been discovered?
8496. In the Old World, is the Northern Goshawk a bird of the taiga?
8497. Are most grebes essentially tropical?
8498. How did the Hepatic Tanager obtain its name?
8499. When are nuthatches most inclined to feed from the human hand?
8500. What often happens to the color of many captive Green Cissas (Magpie), and why?
8501. Are breeding Carmine Bee-eaters especially gregarious?
8502. Do bitterns regurgitate food for their chicks?
8503. What color are the lores of a breeding Snowy Egret?
8504. Perched kinglets have what nervous habit?
8505. The nest of which bird does the cavity-nesting Lucy's Warbler sometimes use?
8506. Do both flamingo sexes participate in nest construction?
8507. Do boobies and gannets ever hunt by foot in shallow water?

Questions

8508. Are birds more tolerant of daily fluctuations in their body temperature than humans?
8509. What is the raptor pictured on the Mexican flag grasping?
8510. Are the neotropical antbirds noted for courtship feeding?
8511. What animal nest do breeding puffbirds sometimes use?
8512. Why do male sunbirds accompany their mates when they are collecting nesting material?
8513. Which ratite feeds on glow worms?
8514. When did the World Wildlife Fund commence operation?
8515. Which are the most pelagic pelecaniform birds?
8516. How much does a Wandering Albatross egg weigh?
8517. Do the neotropical tinamous have long, pointed wings?
8518. What color is the inner mouth of a Great Cormorant?
8519. Which bird is called a "hackled pigeon" or "vulturine pigeon?"
8520. Do the males of all pelicans select the nest site?
8521. According to a survey by the International Bird Pellet Study Group in 1969, how many species regurgitated pellets?
8522. Which birds have proportionally the widest skulls?
8523. What color are the wingtips of a Dalmatian Pelican?
8524. What type of nest does the Australian Gouldian Finch construct?
8525. Do both sexes of African Wattled Starlings have wattles?
8526. Is the black plumage of an adult frigatebird iridescent?
8527. What is the life expectancy of an Emperor Penguin if it survives the first year?
8528. Which pheasants never leave the nest during incubation?
8529. Do sandpipers perch on utility wires?
8530. Do stilts and avocets breed at one year of age?
8531. Which songbirds inhabit the wooded cays of the Australian Great Barrier reef that are too small to support breeding populations of other passerines?
8532. What is the White-billed Diver of Europe called in North America?
8533. If a falconer indicates that it is *cawking time,* what does this imply?
8534. Which is the only South American country not inhabited by a single antbird?
8535. Which plover breeds exclusively in the Arctic?
8536. Do Ring-necked (Common) Pheasants and Blue Grouse hybridize?
8537. Have models of raptors suspended from balloons or mounted on poles been effective deterrents to birds around airfields?
8538. Is the conspicuous crest of Lady Ross' Turaco erectile?
8539. Which is the only non-altricial obligate brood parasitic-nesting bird?
8540. In the southern states, which bird is known as a "golden swamp warbler?"
8541. Which are the only pelicaniform birds that do not use their feet when incubating?
8542. Is the flight of a nightjar generally silent?
8543. Are male Red-winged Blackbirds late spring arrivals?
8544. Do the merged broods of Ostriches ever exceed 200 chicks?
8545. Which is the palest and grayest North American warbler?
8546. Are any of the true swifts sexually dimorphic?
8547. Is avian pox bacterial?
8548. Does the Ferruginous Hawk prey on bats?

8549. What is the function of an aftershaft?
8550. How much does a Limpkin weigh?
8551. Do woodcocks lay darker-colored eggs than the eggs of most sandpipers?
8552. Are any nocturnal owls actually blind during the day?
8553. What strange nesting technique does the Sulawesi Maleo utilize?
8554. Are sandgrouse vocal in flight?
8555. Where do New Zealand Wrybills winter?
8556. How many species of passerines breed in the Arctic?
8557. Which is the only non-migratory North American vireo?
8558. What services do oxpeckers provide for a rhinoceros?
8559. Who is credited with coining the term oology (the study of bird eggs)?
8560. Compared to North America and Europe, does Australia have relatively few small seedeaters?
8561. What color is the eye-ring of a Glaucous Gull?
8562. Which bird is represented on the military insignia of a U.S. Coast Guard captain?
8563. Is the Broad-tailed Hummingbird a bird of the desert?
8564. Why does so much guano accumulate on the breeding islands of some seabirds, such as boobies and gannets, and some pelicans and cormorants?
8565. Do guineafowl prey on frogs?
8566. Which American bird is called a "flusterer?"
8567. Why do migrating White Storks avoid flying over the Mediterranean Sea?
8568. Do male Spoonbill Sandpipers incubate?
8569. Is cooperative breeding more frequent in passerines than in other groups of birds?
8570. Are paradise-flycatchers noted for fine songs?
8571. Is a Hairy Woodpecker more prone to excavate a nest in a live tree, rather than a dead tree?
8572. How do overheated poorwills cool down?
8573. What type of bird does a Limpkin resemble?
8574. In what African country do most Crab Plovers breed?
8575. When constructing its bower, the Regent Bowerbird often steals items from which bird?
8576. When are sandgrouse most likely to vocalize?
8577. With which mammal do some South American ground-cuckoos associate?
8578. Which country has been the most frequent host of the International Ornithological Congress, and how many times has the congress been hosted there?
8579. Do griffon vultures search for food in groups?
8580. All North American quail have how many primaries?
8581. Are peppershrikes primarily fruit eaters?
8582. Which bird was recorded nesting the greatest distance underground, and how far below the surface did it breed?
8583. With what vegetation and habitat is the Shoebill stork most closely associated?
8584. What is applied ornithology?
8585. Do the filter-feeding flamingos ingest grit?
8586. Are Grandry's corpuscles known only in birds, and what is the function of the corpuscles?
8587. How is a pointless search sometimes described?

8588. Have African mousebirds adapted well to large-scale monoculture agriculture?
8589. The range of the Inca Tern is essentially restricted to what famous ocean current?
8590. After striking a blow at a tree, why does the Hairy Woodpecker often hold the tip of its bill for a moment on the dent it has made?
8591. Do parrots always husk a nut or seed before swallowing it?
8592. If disturbed, do South American Oilbirds become silent?
8593. If a bird is about to lay an egg, but is unable to because of a tumor or an oversized or deformed egg, what is this condition called?
8594. Are Ostriches entirely herbivorous?
8595. The females of which *group* of waterfowl incite their mates to fight?
8596. When obligate brood parasitic-nesting Brown-headed Cowbirds remove host eggs from a nest, do the birds ever consume the eggs?
8597. The feathers of which bird are worn in the hair of Dayak headhunters in Borneo?
8598. Has the Black-billed Magpie population in England decreased the past two decades?
8599. Which megapodes (incubator birds) have prominent casques?
8600. Which aberrant waterfowl constructs a number of nests prior to egg laying?
8601. Do fledged swallows ever return to their nest to spend the night?
8602. Are South American steamer-ducks migratory?
8603. Do bee-eaters consume their prey while in flight?
8604. What does the scientific name of the Common Puffin mean?
8605. Do South American screamers ever fight to the death?
8606. Is the crest of a Congo Peacock permanently erect?
8607. Do jacana chicks grow rapidly?
8608. Why do snake-eagle chicks acquire feathers on their backs and crown while their underparts are still down-covered?
8609. Which cormorant is the rarest, and how many exist?
8610. What do many falconers attach to their birds to help them locate the raptor if it does not return?
8611. Do grouse have long aftershaft feathers?
8612. Is an Andean Flamingo's bill designed for smaller food items than the Puna (James') Flamingo?
8613. Why do rails flick their tails?
8614. How do vocalizing neotropical trumpeters hold their wings?
8615. Are there any nocturnal kingfishers?
8616. Do Yellow-billed Magpies pick ticks off the backs of bighorn sheep?
8617. Which are the rarest swallows and martins?
8618. Are birds more susceptible to tumors than mammals?
8619. Does the tongue of a wryneck have barbs?
8620. Do birds with the highest number of syringeal muscles have the greatest potential for producing different complex songs or calls?
8621. Is the entrance to a bee-eater's burrow oblong in shape?
8622. Do most North American vireos winter in the tropics?
8623. What is the most striking physical feature of the male Wallace's Standardwing?
8624. Which icterid is the only interhemispheric migrant?

8625. Why is the bill of a Helmeted Guineafowl arched?
8626. Is the South American Imperial Snipe a forest bird?
8627. Do wrynecks *often* feed on the ground?
8628. Do all grebes breed on fresh water?
8629. What color is the forehead of a breeding Aleutian Tern?
8630. In which of the six crested penguins does the female, rather than the male, take the initial incubation shift?
8631. Do South American Hoatzins breed throughout the year?
8632. What is the incubation period of a frigatebird?
8633. The crop secretion produced by flamingos feeding chicks is unusually high in what substance?
8634. Was the extinct Rodrigues Solitaire territorial?
8635. What is the more familiar name for the swallow-shrikes?
8636. Do fledging puffin chicks weigh as much as adults?
8637. Do Harpy Eagles range as far north as Mexico?
8638. Are Florida Scrub Jays less bold than western Scrub Jays?
8639. Which birds are said to contract their pupils into the shape of a square?
8640. Do most cotingas lay unusually small eggs?
8641. Which Old World bird is the ecological counterpart of the Lesser Yellowlegs of North America?
8642. What are cuckoo-bees?
8643. Which is the largest mannikin?
8644. Do Australian Green Figbirds build extensive nests?
8645. Aside from size and vocalizations, are Emus sexually dimorphic?
8646. Do the neotropical tinamous call at night?
8647. Do Australian lyrebirds roost high in the forest canopy?
8648. Which endemic New Zealand passerines hold food with their foot like parrots?
8649. What is an individual called who crosses the street disregarding the crosswalk?
8650. Do any of the African turacos prey on snails?
8651. Is the syrinx of an Ostrich more highly developed than that of a rhea?
8652. What objects adhering to leaves and bark do the two neotropical peppershrikes often tear open with their bills?
8653. Which American bird was known as "warnecoutai," and what does the name mean?
8654. How soon after copulation are kiwi eggs laid?
8655. How do juvenile American Coots differ in appearance from adults?
8656. Are mammals more odoriferous than birds?
8657. Which parrot has become established in Great Britain?
8658. Are the most skilled Australian avian mimics primarily arboreal birds?
8659. Is the bill of a Limpkin three times as long as its head?
8660. Is the nominate subspecific name *always* the same as the specific name?
8661. Which phalarope has a white rump?
8662. How many races of the Ruby-throated Hummingbird are described?
8663. How do birds of the Hawaiian race of Black-necked Stilt differ in appearance from stilts of the American mainland?

Questions

8664. Do the West Indies todies capture insects on the ground?
8665. Does the male Black-headed Grosbeak sing while it is *on* the nest?
8666. Are hailstorms more destructive to birds than tornados?
8667. Why do the feathers of an Emu fade?
8668. What was the Pinyon Jay formerly called?
8669. Do boobies and gannets engage in courtship feeding?
8670. Which bird does the Black Skimmer most closely resemble?
8671. Why can seabird guano only be used as nesting material in dry climates?
8672. Which tyrant-flycatcher has the longest and broadest bill?
8673. Do any cracids have aerial courtship displays?
8674. Do the Oilbirds of South America have large crops?
8675. Were the extinct moas swift runners?
8676. Do Red-necked Grebes winter in the Aleutian Islands?
8677. What is at the tip of a kiwi's stubby wing?
8678. Do oystercatchers nest inland?
8679. Which is the only storm-petrel of western North America with a wedge-shaped tail?
8680. Do all species of Darwin's finches sing different songs?
8681. Are breeding female finfoots and Sungrebes more conspicuously marked and brightly colored than the larger males?
8682. Is the bill of the tropical Jabiru stork straight?
8683. When were American Robins ever hunted as table-birds, and what was their market value?
8684. How is complaining commonly described?
8685. Is the leading edge of a flight feather typically narrower than the trailing edge?
8686. What would a human's estimated body temperature be if he had the metabolic rate of a hummingbird, and consumed up to twice his body weight daily, and how many calories would be burned?
8687. Are skimmers victimized by kleptoparasitic gulls?
8688. Do British wagtails with the longest tails engage in the most fly-catching?
8689. In the composition *A Midsummer Night's Dream*, the sounds of what instrument represent the flight of the skylark?
8690. In terms of weight, how much grit may be in the stomach of an adult Ostrich?
8691. In terms of percent, how much of an African mousebird's diet consist of leaves?
8692. Which is the South African counterpart of the American Wood Stork?
8693. How did the koels (Old World cuckoos) obtain their name?
8694. What is the shape of the head of an African White-headed Vulture?
8695. According to mythology, what was Aedon, the Queen of Thebes, changed into by Zeus?
8696. Which insect is responsible for House Wren mortality?
8697. Do drongos rob other birds of food in flight?
8698. What is the status of nesting diving-petrels on subantarctic Marion Island?
8699. What is unusual about African woodhoopoe eggs?
8700. After whom was Lady Amherst's Pheasant named?
8701. When Brown Pelicans arrive on the breeding grounds, are pair bonds already formed?
8702. What is the traditional quarry of Arabian falconers?
8703. What percent of all bird species are monogamous?

8704. After whom was the only Chilean tanager named?
8705. What type of birds are African puffbacks?
8706. Do sunbirds have velvety plumage?
8707. Are hatching chicks assisted out of the shell by their parents?
8708. Which southern South American passerines often feed on offshore floating giant kelp?
8709. Do Ruddy Turnstones nest among breeding jaegers?
8710. What is the dertrum?
8711. Are wild parrots known as mimics?
8712. In old England, which birds were known as storm witches?
8713. Do caracaras lay smaller clutches of eggs than true falcons of comparable size?
8714. Are frogmouths generally encountered as single birds?
8715. Which organization publishes the *Red Data Books*?
8716. What do falconers call a male Old World sparrowhawk?
8717. Do birds suffer from gout?
8718. How do Rooks respond to broadcast alarms or distress calls used to prevent birds from gathering at specific locations?
8719. Is the head of a flying grebe held lower than its back?
8720. What is the percent of annual adult mortality of the Royal Albatross?
8721. Do tropical passerines live longer than temperate passerines?
8722. Do the juveniles of all color morphs of the Red-footed Booby have dark-colored bills?
8723. How can agonistic behavior be described?
8724. What is the function of a juvenile Black-legged Kittiwake's black neck-band?
8725. In what year did the American Ornithologists' Union celebrate its centennial?
8726. Which North American mimid is plain slate-gray with reddish under tail coverts?
8727. What is the wingbeat rate per second of a chickadee?
8728. What might a female kiwi do just prior to laying to facilitate laying?
8729. What is the basic color of the African Openbill stork?
8730. In India, why was nestling pelican fat valued?
8731. Are weavers gregarious throughout the year?
8732. When are thick-knees (stone-curlews) especially vocal?
8733. Do the larger toucans use the same nest cavity year after year?
8734. What color is the bare neck skin of a Helmeted Guineafowl?
8735. Do the chicks of crested penguins gather in crèches?
8736. Do motmots generally hover when plucking fruit?
8737. Do the wings of a resting skimmer project beyond its tail?
8738. What is a *large* group of grouse called?
8739. Woodswallows are partial to what man-made object?
8740. Which endemic New Zealand bird is known as a "roa?"
8741. In falconer's terminology, what is an *intermewed eyass goshawk tiercel?*
8742. Does a stilt have a slightly recurved bill?
8743. How many of the approximately 743 species of Australian birds are native breeders?
8744. Are Hawaiian honeycreeper eggs always unspotted?
8745. On what do Eurasian Jays primarily feed?
8746. Is the lens of the avian eye softer than a human eye lens?

Questions

8747. Are trogons closely related to kingfishers?
8748. With which bird does Abdim's Stork often nest?
8749. When do Secretary-birds feed on carrion?
8750. Do chachalacas generally call in unison?
8751. In what city was the 1990 International Ornithological Congress held?
8752. Is the Bananaquit one of the most abundant birds of the West Indies?
8753. Why don't swallows lay on a thick layer of fat prior to migration?
8754. The flight of an aerial-displaying woodcock is rather like the flight of which raptor?
8755. Do albatrosses come ashore only to breed?
8756. Is the kiwi incubation period about the same as the gestation period of a mammal of comparable size and weight?
8757. Does a feeding flamingo's tongue move up and down?
8758. Which is the only wood-warbler of the Galapagos Islands?
8759. Which wild birds were the first to have their voices recorded, and when and where did this occur?
8760. Which waterfowl relative often grazes with livestock in South America?
8761. What portions of a carcass does a Lappet-faced Vulture favor?
8762. Which diving mammals are smaller than the smallest diving birds?
8763. In what year was the American Birding Association founded?
8764. Which large seabird is depicted on many coats-of-arms?
8765. Do most plover chicks have a dark cap?
8766. Do groups of African oxpeckers often gather on a single mammal?
8767. What type of flight is the most strenuous?
8768. The plumes of which large bird do New Guinea natives often thrust into their pierced nostrils, as well as through the nasal septum?
8769. Do Galapagos Red-footed Boobies breed throughout the year?
8770. In Holland, why has the laying date of Black-tailed Godwits shifted about two weeks in the last half-century?
8771. Does Ross' Gull breed in Greenland?
8772. How high up in trees do American Wood Storks nest?
8773. Is the Snowy Sheathbill a *breeding* species of the subantarctic island of South Georgia?
8774. Why do male rheas sometimes uproot all the vegetation within a radius of 2-3 meters around their nests?
8775. Which two owls of Eurasia are the most migratory?
8776. How many Andean Condors inhabit Colombia, the northern extent of the range?
8777. Do Northern Gannets feed their young after fledging?
8778. What are the two highly-modified, S-shaped, ornamental outer tail feathers of an adult male Superb Lyrebird called?
8779. Do *all* vireos capture prey by flycatching?
8780. With what do tits often line their nests?
8781. Is the Greater Rhea a bird of the campo?
8782. Which is the only freshwater diving-duck of New Zealand?
8783. Which bird-of-paradise may nest near a Black Butcherbird, and why?
8784. Which birds are known as "tree-nightjars?"

Questions

8785. Are buttonquail (hemipodes) terrestrial birds of a forest habitat?
8786. Which familiar North American bird was successfully introduced into Hawaii, Bermuda and southwestern California?
8787. Which African bird is known as "sakabula?"
8788. Which wading bird do Australian aborigines regard as a bird of ill omen?
8789. How did the term *cockpit* originate?
8790. Which type of neotropical bird is called a castle-builder?
8791. Do Northern Gannets gather in clubs?
8792. What color borders the mouth of a nestling Jackdaw?
8793. Why does a roosting or incubating frogmouth close its eyes in the presence of an intruder?
8794. At what prey do British falconers fly Merlins?
8795. Are female phalaropes among the most colorful shorebirds?
8796. Was habitat destruction responsible for the disappearance of the New Zealand Huia?
8797. Do the tiny white-eyes of the Old World pair for life?
8798. Are most South American ovenbirds insectivorous?
8799. Is the brood-patch of a pigeon featherless year-round?
8800. With respect to nesting sites, what does the presence of black lava grit in the stomachs of Pink-footed Geese shot in Scotland indicate?
8801. How long are oystercatcher chicks fed by their parents?
8802. Where do the Old World warblers reach their greatest diversity?
8803. If a penguin is *body-thrashing* in the water, what does this imply?
8804. When and where was the International Ornithological Congress held in the U.S.?
8805. Oil from which ratite was valued, and for what was it used?
8806. What happens to canaries if fed paprika?
8807. Does the Hoatzin of South America commonly glide?
8808. Why do Yellow Warblers nest near breeding Red-winged Blackbirds or Gray Catbirds?
8809. What material does an Oilbird use for its nest?
8810. Which species of African guineafowl has been domesticated?
8811. Do Parakeet Auklets use the same nest site of preceding seasons?
8812. At what age do Arctic Terns commence breeding?
8813. Do tropicbirds copulate on the water?
8814. Which small, endemic family of African birds consumes a high proportion of plants containing toxins or irritants?
8815. Why is the voice of a female painted-snipe deeper than that of a male?
8816. For what were mews originally used?
8817. Do most nightjars have relatively long tails?
8818. Which bird was selected for the logo of the 1990 International Ornithological Congress?
8819. What is a dovecote?
8820. If a kiwi loudly snaps its mandibles together, what does this signify?
8821. Do mammals have coracoid bones like birds?
8822. Is the long tail of a Red-tailed Tropicbird conspicuous?
8823. What is the function of the external nasal tubes of petrels?
8824. Do most flocking birds cry out with a characteristic call upon detecting a predator?
8825. How long do domestic pigeons rearing chicks produce *pigeon milk?*

Questions

8826. How are crane chicks fed?
8827. Are the neotropical jacamars noted for egg neglect?
8828. Do Hooded Crows maintain large territories year-round?
8829. When were the first organized counts of wild waterfowl in Great Britain initiated?
8830. The nesting burrows of most New World kingfishers generally overlook water, but which species may not nest near water?
8831. What is the clutch size of a Black Vulture?
8832. Which Australian bird is known as a "green leek?"
8833. Which bird family has the most species at temperate levels in the Andes?
8834. Aside from size, which are the only sexually dimorphic storks?
8835. Which King Penguin sex takes the first incubation shift, and how long is the shift?
8836. Do male kiwis have bare brood-patches?
8837. Are Reef Herons restricted to the coast?
8838. What is an alar bar?
8839. When do Sharp-tailed Grouse produce tail-rattling noises?
8840. Do female Black Rosy Finches outnumber males?
8841. Why was the Elfin Woods Warbler of Puerto Rico not discovered until 1972?
8842. Which is the only cavity-nesting bird-of-paradise?
8843. What does a Least Flycatcher do every time it calls?
8844. Were the extinct New Zealand moas omnivorous?
8845. What is an oar-winged or angle-winged bird?
8846. How does the toe coloration of a Sungrebe differ from that of the two Old World finfoots?
8847. What is the only North American icterid with a yellow head and black body?
8848. Do pygmy-geese forage underwater?
8849. How many species of owls lay eggs that are not white?
8850. Are Northern Jacanas quarrelsome and combative?
8851. Do skuas ever splash-dive after fish?
8852. What is it called when one avoids an issue?
8853. How many species of sunbirds inhabit Israel, and which ones are they?
8854. Are all Darwin's finches monogamous?
8855. What is duck virus enteritis more commonly called?
8856. Do turnstones probe into the substrate like sandpipers?
8857. Which tinamou is known only from the type specimen collected in 1943 in the foothills of the Sierra de Ocana, Colombia?
8858. Can kingfishers catch more than one fish during a foraging flight?
8859. How might a very dry year impact nesting Mountain Quail?
8860. How many chicks does the Magellanic Plover rear annually?
8861. Do any pigeons breed above the snow line?
8862. Is the Crissal Thrasher a powerful flier?
8863. Do pigeons generally lay the first egg of the clutch early in the morning?
8864. Did Old World vultures formerly inhabit North America?
8865. Is a tchagra a specialized African rail?
8866. When are loons most likely to engage in cooperative feeding?
8867. Was the extinct Dodo a fast runner?

8868. What color is the base of the bill of a Black-footed Albatross?
8869. In terms of percent, how much more do domestic guineafowl weigh than wild guineafowl?
8870. Were the early Greeks fond of cock-fighting?
8871. Which of the three skimmers have black caps?
8872. Do both sexes of the European Robin have red breasts?
8873. Why are the shells of megapode eggs unusually thin?
8874. Do South American seedsnipes regularly dust bathe?
8875. Are most whistling-ducks temperate?
8876. Was the Elephant-bird of Madagascar two-toed?
8877. What is the source of avian eggshell pigments?
8878. Which non-aquatic passerine has the heaviest uropygial gland relative to its body size?
8879. Do tropicbirds ever swallow their prey underwater?
8880. Was the Kakapo population declining prior to the time Europeans arrived in New Zealand?
8881. Do polyandrous female jacanas mate with incubating males or males tending young?
8882. How does a kiwi kidney differ from most avian kidneys?
8883. The Red-footed Falcon of the Old World often uses the nest of which bird?
8884. Is the metabolic rate of a passerine higher than that of a non-passerine of equal size?
8885. How large is the largest-known single colony of Greater Flamingos, and where is it located?
8886. Which is the heaviest seabird?
8887. Which organization publishes *The Zoological Record,* and when was it first published?
8888. Megapode chicks have what type of down?
8889. What was the origin of the generic name *Daption* for the Cape Pigeon (Painted Petrel)?
8890. The feathers of which bird adorn the regimental slouch hats of some Australian regiments?
8891. Is the African Finfoot the smallest of the three finfoots?
8892. How long are females of the large, forest-dwelling hornbills sealed in their nest cavities?
8893. Barred Owls have how many feathers?
8894. Is the Fernwren a New Guinea wren?
8895. Do the South American screamers copulate on the water?
8896. With what do toucans line their nests?
8897. What color is the base of the tail of all lapwing plovers?
8898. Which is the only surviving passerine of Laysan Island?
8899. What are the enlarged, pale-colored margins of the gape of some nidicolous chicks called?
8900. Are bats more maneuverable than birds of similar size?
8901. Are the Old World rollers relatively non-vocal in flight?
8902. Which are the smallest soaring birds?
8903. Which ratites have a laterally compressed bill?
8904. Which woodpecker tail feathers are molted last, and why?
8905. What is the wingspan of an American Woodcock?
8906. At what age do young North Island Brown Kiwis depart the nest for good?
8907. What color is an American Wood Stork's tail?
8908. Are Rose-throated Becards noisy birds?
8909. What nesting site is favored by a Muscovy Duck?

Questions

8910. How many flycatchers occur in the Galapagos Islands, and which ones are they?
8911. What is a precocious flight?
8912. Does the Long-billed Curlew have a long tongue?
8913. Do all South American ovenbirds lay white eggs?
8914. What did Horus, the Hawk God of dynastic Egypt, personify?
8915. What must Ostrich hens do to facilitate copulation?
8916. Do albatrosses feed on offal?
8917. Do skuas have square tails?
8918. When the neotropical White-tailed Trogon nests in a wasp nest, is the invaded nest abandoned by the wasps?
8919. What sound is produced by a disturbed incubating Carolina Chickadee?
8920. What was the fate of the carcass of the last Carolina Parakeet?
8921. Are African White-headed Vultures rather bold?
8922. Are tinamou eggs large relative to the bird's size?
8923. Which mustelid may hoard and store bird eggs?
8924. In which country do Northern Gannets breed well north of the Arctic Circle?
8925. Do frigatebirds have difficulty landing in strong winds?
8926. Why are the eggs of the African Temminck's Courser blackish in color?
8927. Do West Indian todies dig new nesting burrows each year?
8928. What is canary wood?
8929. Do Horned Larks often roost on the ground?
8930. Do nest-building Sunbitterns use mammal dung as foundation cement?
8931. During windless conditions, can a fulmar take off from the water without a running start?
8932. What large predators do African Skimmers mob and chase away?
8933. Can an Ostrich run fast on wet, slimy ground?
8934. If a green Budgerigar is mated with a pure blue Budgie, what color is the resultant offspring?
8935. What term describes the various divisions within the Common Cuckoo population where birds lay eggs of differing colors?
8936. How many of the 28 typical vireos breed as far north as the U.S.?
8937. Which bird is called a "rattlesnake owl?"
8938. Are antbird chicks naked at hatching?
8939. Who is the author of the classic 1936 two-volume work *Oceanic Birds of South America*?
8940. Do cormorants continue nest building throughout the incubation and nestling period?
8941. Are the eyes of a Tawny Frogmouth located in the front of its head, like an owl's?
8942. Is the size and amount of grit ingested by a bird directly related to the coarseness of the food ingested?
8943. Which is the largest of the three New Zealand honeyeaters?
8944. Are most African barbets predominately green in color?
8945. Which American grosbeak occurs the farthest north?
8946. How do Leach's Storm-Petrels from Baja California differ in appearance from those occurring elsewhere?
8947. Do foraging phalaropes dive with ease?
8948. Do most passerine young fledge at weights below those of adults?

Questions

8949. What do falconers call a male Saker Falcon?
8950. Why is the gosling down of Australian Cape Barren Geese unusually long?
8951. Do partridges have larger bills than quail?
8952. Excreta from which group of mammals is called guano?
8953. Which of the three Australian ibises is the most numerous?
8954. What is the origin of the Kookaburra generic name *Dacelo?*
8955. Which stork nests on village huts in Africa?
8956. Are Pheasant-tailed Jacanas restricted to the lowlands?
8957. Who first detected that the distinct hum of flying hummingbirds differed from species to species?
8958. Why are eggshells weaker after incubation than before?
8959. Are male guineafowl significantly larger than females?
8960. Do oystercatchers gather in flocks consisting of thousands of birds?
8961. Do Thayer's Gulls nest on flat rocky beaches?
8962. Is the egg tooth of a penguin chick lost shortly after hatching?
8963. Are woodcock legs relatively short for a shorebird?
8964. Do birds-of-paradise prey on nestling birds?
8965. Does the New Zealand Kokako (Wattled-crow) have pendulous wattles?
8966. What acoustic device is most often used to frighten birds away from specific sites, such as agricultural areas?
8967. Are all Hawaiian honeycreepers sexually dimorphic?
8968. Do Oilbirds occur as far west as Peru?
8969. What color is the cere of an adult Hawaiian Hawk?
8970. How does the nest of a painted-snipe differ from the nests of most other shorebirds?
8971. Do high-altitude birds have larger hearts than low-altitude individuals of the same species?
8972. What is a hern or hernshaw?
8973. What is the average interval between the laying of each egg in a cassowary's clutch?
8974. Do tropicbirds have subcutaneous air-sacs?
8975. Do flying pelicans often scratch their heads?
8976. To which island is the entire population of Narcondam Hornbills restricted, and how large is the population?
8977. Do Long-tailed Jaegers nest in Scandinavia?
8978. Is the crest of a Sandwich Tern erectile?
8979. What is the number of pigeon fanciers in Britain, and how many pigeons do they maintain?
8980. What is the conspicuous location from which male Greater Honeyguides sing commonly called?
8981. Which large *arboreal* hornbill has an extensive amount of bare skin on its head and neck?
8982. Many species of birds migrate, but to what type of mammals is long-distance migration generally limited?
8983. What does a tyrant-flycatcher do if it captures prey too large to swallow?
8984. Do prospecting Purple Martins prefer apartments that have not been cleaned out after a previous occupancy?

Questions

8985. Is the Grenada Dove restricted to the West Indian Island of Grenada?
8986. Do African mousebirds often nest in close proximity to one another?
8987. Are the legs of a standing bird generally situated just in front of its center of gravity?
8988. Do male Ruffed Grouse typically commence drumming just after dawn breaks?
8989. Do all wintering Tundra (Whistling) Swans depart Alaska?
8990. The bill of which seabird does an ani's bill resemble?
8991. What is the main threat facing the Congo Peacock?
8992. How is the refractive effect of moving from air into water minimized in penguins?
8993. The nest of an African Buffalo-Weaver consists of what material?
8994. What does a vocalizing Coppery-tailed (Elegant) Trogon do with its head?
8995. Which Australian ibis is most inclined to feed on a carcass?
8996. When was the Peterson *Field Guide to the Birds of Britain and Europe* first published?
8997. What object do Irish hunters use to attract Corn Crakes?
8998. Do *both* sexes of hummingbirds call in flight?
8999. Why is the bare head adornment of a cock domestic fowl known as a comb?
9000. Which large, white, yellow-eyed bird preys on puffins?
9001. Is the Saddle-billed Stork a ground nester?
9002. Which is the most numerous entirely-black seabird of California?
9003. Do Great Bustards gather in flocks?
9004. Do the neotropical trumpeters lay smooth-shelled eggs?
9005. What is a plumule?
9006. How do New Zealand biologists track down kiwis?
9007. Which is the only resident heron of the Falkland Islands?
9008. Which bird is commonly known by its German name Sprosser?
9009. Which North American jay is the most crow-like in shape?
9010. Do Wrentits defend year-round territories?
9011. What is one of the more interesting theories as to why Purple Martins line their nests with green leaves?
9012. Do Dovekie (Little Auk) chicks go to sea before they are fully fledged?
9013. Have exotic bird introductions generally proved to be less disastrous than mammalian or insect introductions?
9014. Are both finfoots and the Sungrebe in the same genus?
9015. In falconry, which raptor is more likely to be hooded: a falcon or a broad-wing hawk?
9016. Did the extinct New Zealand Huia travel in large groups?
9017. Do boobies typically fly higher than most seabirds?
9018. What type of behavior is known as *jugging?*
9019. Are female Northern Jacanas nearly twice the size of their mates?
9020. Which waxbill finch was successfully introduced into Singapore, Fiji, Sumatra, the Philippines and Mauritius?
9021. Cubital is an obsolescent term for what type of feather?
9022. Do titmice readily nest in nest boxes?
9023. Why do wintering Wilson's Phalaropes in the central Andes feed in small groups around and through the legs of foraging Chilean Flamingos, while nearby feeding Puna and Andean Flamingos are ignored by the shorebirds?

9024. What is the weight of the insects consumed daily by the 4.5 million cave swiftlets of the Niah Cave colony (Borneo)?
9025. What is the most prevalent predator of nesting Florida Sandhill Cranes?
9026. Do all the females of a cock Ostrich's harem incubate?
9027. What is the clutch size of a sandgrouse?
9028. When House Sparrows were first introduced into North America, what pest was it hoped the birds would control?
9029. What color is the dense down of an owlet-frogmouth chick?
9030. Which surviving birds are most closely related to moas?
9031. With what do Northern Shrikes commonly line their nests?
9032. On what do neotropical bellbirds almost exclusively feed?
9033. What is the small, quadrangular board with a handle underneath, used by masons to hold mortar called?
9034. Do herons commonly soar?
9035. Why do most albatrosses inhabit the latitudes of the *Roaring Forties* and *Furious Fifties*?
9036. Do the wings of a drumming Ruffed Grouse strike its breast?
9037. Which African raptor consumes carnivore dung, and even that of humans?
9038. Were bird recordings made in Antarctica earlier than in South America?
9039. South American parrots may use the mud nest of which bird?
9040. Which American-breeding plover with a black-tipped yellow bill is endangered?
9041. In falconry terminology, what is a male Merlin called?
9042. Is the outer pair of tail feathers of most woodpeckers greatly reduced in size?
9043. Are maculate eggs marked, spotted or blotched?
9044. How many U.S. pennies equal the weight of a Black-capped Chickadee?
9045. Are Great Horned Owls often killed on highways?
9046. Which sulid has the longest incubation period, and how long is it?
9047. Are Blue Grouse more apt to inhabit dense coniferous forests than open coniferous forests?
9048. With what do oystercatchers line their nests?
9049. Do African turacos engage in courtship feeding?
9050. Which of the six crested penguins commences nesting earliest in the season?
9051. Are toucan chicks noted for rapid development?
9052. Males of which family of small nectar-feeding Old World passerines assume a female-type eclipse plumage?
9053. In Britain, which are the only two birds that engage in cooperative breeding on a regular basis?
9054. According to Manitoba (Canada) legend, why is the Yellow-bellied Sapsucker so colorful?
9055. What color are the trailing edges of the secondaries of an American Coot?
9056. Why are sulid eggshells exceptionally thick?
9057. When is the South American Whistling Heron most inclined to whistle?
9058. Do Neotropical King Vultures become torpid?
9059. Is nesting highly synchronous in the colonial-breeding Yellow-headed Blackbird?
9060. What is the only country from which birds can be legally imported into New Zealand?
9061. What color is the chick down of a Brazilian Yellow-faced Amazon?

Questions

9062. In the broad taxonomic picture, in what *class* are birds?
9063. Do bee-eaters dig nesting burrows with their feet?
9064. Which especially long-legged bird rides the backs of grazing hippopotamuses, and why?
9065. Why are the claws of an oxpecker extremely sharp?
9066. Is the Flightless Cormorant the only cormorant of the Galapagos Islands?
9067. According to U.S. Fish and Wildlife Service estimates, how many people engage in bird-related recreation in the U.S.
9068. Which is the only frigatebird in which the female has a brown, rather than white, breast?
9069. Does a flamingo's tongue fill the entire cavity of its lower mandible?
9070. Who was President Richard M. Nixon's first Secretary of Health, Education and Welfare?
9071. In Florida, when Glossy Ibises probe into mud holes, what are the birds seeking?
9072. Are the display leks of resident birds traditional sites that are generally used year after year?
9073. Do birds run faster than mammals of comparable size?
9074. Are Australian lyrebirds shy and difficult to approach?
9075. Is the tail of an African Hamerkop proportionally shorter than the tails of most storks?
9076. Which is the only tropical Pacific-nesting cormorant?
9077. Do any tits excavate their own nesting holes?
9078. Do all juvenile frigatebirds have rufous or buff coloring on their head or underparts?
9079. Considering the large number of eggs laid, why is Africa not overrun with Ostriches?
9080. What unique foraging technique do Magellanic Plovers use?
9081. What is the national bird of Belize (formerly British Honduras)?
9082. Are most neotropical manakins about the size of tits?
9083. How do Sunbittern chicks differ from the chicks of most birds?
9084. Why is the flight of a Wood Stork noisy?
9085. Is the Sage Thrasher a bold and inquisitive bird?
9086. Is the Old World Wallcreeper especially vocal?
9087. What is ptilopody?
9088. Which family of tiny land birds has the longest incubation period of any bird of similar size?
9089. Which megapode has the most colorful plumage?
9090. Do Anhingas and darters always nest in trees?
9091. Does the neotropical Great Potoo weigh more than a pound (0.45 kg)?
9092. How many species of birds is a male European Marsh Warbler said to be able to mimic?
9093. Did any of the extinct moas inhabit Australia?
9094. Do American White Pelicans avoid lakes that are still partially covered with ice?
9095. The waxy red drops on a waxwing's secondary feathers are a prolongation of what?
9096. Which tanager was successfully introduced into Tahiti?
9097. Of the seven states with the Northern Cardinal as the state bird, which state first selected the bird, and when was it selected?
9098. Do screamers breed in Patagonia?
9099. Why do many molting passerines become less active and less conspicuous?
9100. Based on fossil remains, have motmots always been restricted to the New World?
9101. Is the Large-billed Tern often observed at sea?

Questions

9102. Are Red-winged Blackbirds polygamous?
9103. Which caprimulgid is cavernicolous?
9104. Which is the most dominant group of seabirds of the Southwest Pacific?
9105. Which northern shorebirds are the most littoral?
9106. Is the bill of an African mousebird downcurved?
9107. Do male junglefowl gather in bachelor groups?
9108. Was the New Zealand Stitchbird named for the type of nest it constructs?
9109. Which are the only two northern wood-warblers in which males are conspicuously blue?
9110. Old nests of which birds are favored by nesting Morning Doves?
9111. Are grebe toenails serrated?
9112. To what mycotic disease are waterfowl particularly susceptible?
9113. Who wrote the delightful line *The bluebird carries the sky on his back*?
9114. Do avian ocular muscles (the small muscles that move the eyeball) differ significantly from those of other vertebrates?
9115. Which bird was featured in Beethoven's Sixth Symphony?
9116. What color are the eyes of most drongos?
9117. What is the size of a Red-tailed Hawk's territory?
9118. To what does nalospi refer?
9119. What does the kiwi generic name *Apteryx* mean?
9120. Do juvenile tropicbirds have serrated bills?
9121. Why are white-eyes sought as cagebirds in Asia?
9122. Is the House Sparrow established in New Caledonia?
9123. Which is the most numerous woodpecker of Texas?
9124. Which North American grebes engage in regular courtship feeding?
9125. What is the heaviest recorded weight of an American White Pelican?
9126. Do male Phainopeplas commonly sing while in flight?
9127. What type of bird is a drepanidid?
9128. Are incubating and brooding Caribbean todies especially attentive?
9129. Is there any ornithological evidence that South America was ever connected to Africa?
9130. Which New World raptor may move its eggs a considerable distance?
9131. What does the Crested Bellbird of Australia customarily place on the rim, or in the nest?
9132. Do Secretary-birds often rear more than one chick?
9133. Which bird has been designated an *honorary mammal?*
9134. Does a juvenile waxwing have a crest?
9135. What is the status of the introduced Canada Goose in Australia?
9136. Do adult Yellow-crowned Night-Herons have yellow eyes?
9137. Which bird was traditionally killed and paraded on St. Stephen's Day in Ireland?
9138. Does the Short-tailed Albatross have an all-white tail?
9139. Do most bowerbirds construct roofed bowers?
9140. How many of the rare Congo Peacocks are in captivity?
9141. How many of the 163 species of land and water birds of the Solomon Islands are endemic?
9142. What is the only bird that when swimming on the surface of the water brings both feet forward simultaneously?
9143. Do cormorant chicks beg for food with open mouths?

9144. Why has the range of the South American plantcutters expanded in recent years?
9145. Do small birds lose water more rapidly than large ones?
9146. What is the name of the cold-water current that sweeps up along the South African coast supporting a huge seabird population, including an endemic penguin?
9147. Do Oilbirds use their same nest year after year?
9148. Which Australian bird is known as a "scythebill?"
9149. Where are cars a threat to ratites?
9150. What is a Koklass?
9151. The 51 members of the *Ptilinopus* genus are generally referred to as *fruit-doves,* but why is this a misnomer in the case of the Atoll Fruit-Dove (*P. coralensis*) of French Polynesia?
9152. Do breeding King Penguins defend territories?
9153. Which shorebird has three annual plumages?
9154. Is the uropygial gland a symmetrical, bi-lobed organ?
9155. What color feathers are worn by the members of the famous Scottish regiment, the Black Watch, and why was that color selected?
9156. Which ratites have flattened bills?
9157. The South American Plush-capped Finch is typically associated with what plant?
9158. Are wagtails tame and confiding birds?
9159. Do lapwings breed in their first year?
9160. Why do Maguari Stork chicks have black, rather than white, down?
9161. Is the heart-rate of a 2-pound mammal (0.9 kg) less rapid than that of a similar-sized bird?
9162. Is the incubation period of the three South American screamers longer than that of any of the waterfowl?
9163. Why do non-breeding sunbirds defend territories?
9164. Is a Peregrine Falcon more maneuverable in flight than a Prairie Falcon?
9165. In the Caribbean, which bird is called a reef heron?
9166. Do Common Snipe prey on vertebrates?
9167. In Great Britain, which organization is known as the IWC?
9168. Which exclusively aquatic birds readily colonize newly-flooded excavated areas, such as gravel pits?
9169. How many of the 36 most-cited journals devoted to birds are specifically named after birds?
9170. What does the German term *ortstreue* mean?
9171. Which two North American warblers are most prone to nest in cavities?
9172. Was the extinct New Zealand Huia a reluctant flier?
9173. When climbing up on the nest to incubate or brood a chick, what do flamingos invariably do with their feet?
9174. Which is the only consistent ground-foraging vireo?
9175. How did the Indian House Crow reach southern Africa and Australia?
9176. What do European mountaineers often call the Wallcreeper?
9177. How many large woodpecker species inhabit Africa?
9178. What color is the terminal half of a Yellow-billed Oxpecker's bill?
9179. Is the wing shape of the South American ovenbirds variable within the family?
9180. Which American bird is locally called a "cut throat?"

Questions

9181. What color is the rump of a Laughing Kookaburra?
9182. Are *all* frigatebirds strictly diurnal?
9183. Do newly-hatched chicks have fewer taste buds than adults?
9184. Why is it more accurate to refer to the American vultures as cathartid vultures rather than New World vultures?
9185. What must one rapidly do if given a dose of caster oil?
9186. Which is the only sexually dimorphic Australian lorikeet?
9187. Which birds do the British sometimes call "yellowshanks?"
9188. Do birds prey on noxious-tasting stink bugs?
9189. What is the Saker Falcon also called?
9190. Are Fork-tailed Storm-Petrels commonly sighted south of the Oregon border?
9191. Do African Wattled Starlings have wattles year-round?
9192. Is the Sarus Crane among the most social of cranes?
9193. When do wagtails congregate on rubbish dumps?
9194. When preening birds nibble at their feathers, what is this sometimes called?
9195. In Argentina, which bird is known as "sleepyhead?"
9196. On which mammal do Peregrine Falcons of the Fiji Islands occasionally feed?
9197. Which eastern U.S. wren is the most reddish in color?
9198. Which American institution maintains the largest collection of recorded bird calls?
9199. Do fledged juvenile skimmers beg food from their parents?
9200. During the summer, which American thrush occurs the farthest north?
9201. Which is the most intensively studied Eurasian avian species?
9202. Are the wattles of a domestic rooster *always* larger than those of a female?
9203. Which is the only passerine with only three toes?
9204. A Paleoxeric bird is an inhabitant of what type of habitat?
9205. Which tern, named after another bird, is one of the most widespread seabirds?
9206. With respect to plumage, how do treeswifts differ from most true swifts?
9207. Do swallow nestlings squirt their guano out of the nest?
9208. Which type of parrot does not have a well-developed crop and a muscular gizzard?
9209. Which is the only North American shorebird that frequents woodland habitat?
9210. In Europe, what birds are favored as hosts by the obligate brood parasitic-nesting Great Spotted Cuckoo?
9211. How do Pheasant-tailed Jacana eggs differ in appearance from those of other jacanas?
9212. Why is the Bohemian Waxwing rarely parasitized by the Brown-headed Cowbird?
9213. What is the Asian Sand Partridge better known as to aviculturists?
9214. Where do the ranges of the Snowy and Black-faced Sheathbill overlap?
9215. What is the favored food of the European Golden Oriole?
9216. Does the European (Common) Starling occur in the Falkland Islands?
9217. Are New World quail more vocal than Old World quail?
9218. Are pelican nests constructed as high as 100 feet (30 m) up in trees?
9219. Regurgitated Rook pellets consist primarily of what material?
9220. Do any bowerbirds have long tails?
9221. Does a Black-footed Albatross have a white head?
9222. The value of which native songbird of Brazil may equal the value of a new car?

Questions

9223. Is the Inaccessible Island Flightless Rail elusive?
9224. Were Eurasian Capercaillie ever introduced into the U.S.?
9225. What type of nests do Laysan and Black-footed Albatross construct?
9226. Do *both* sexes of both North American shrikes have black stripes through their eyes?
9227. Are most babblers frugivorous?
9228. How long is the fledging period of a tropicbird?
9229. Is the Great Blue Heron known to swim?
9230. Do scoters shell mussels prior to consumption?
9231. Do all trogons lay white eggs?
9232. Was Middendorff's Grasshopper Warbler named because of its fondness for grasshoppers?
9233. When flushed, what do Ring-necked (Common) Pheasant cocks generally do?
9234. After an absence of nearly 100 years, when did breeding Pied Avocets reappear in Britain?
9235. In what country can duck ants be found?
9236. How can a breeding-plumaged Japanese Murrelet be distinguished from an Ancient Murrelet?
9237. In South America, which bird is called "potoyunco?"
9238. Which birds probably served as models for the mythical Chinese Phoenix *(Fung-whang)?*
9239. Which of the three phalaropes is the most abundant?
9240. Which corvid is named after an aquatic animal?
9241. Which goose undertakes the longest migration?
9242. Do sleeping hummingbirds tuck their bills into the feathers of their back?
9243. Where in the western U.S. do Gull-billed Terns nest?
9244. Do Greater Roadrunner pairs live in their territories year-round?
9245. Which Hawaiian honeycreeper preys on seabird eggs?
9246. Does the Brambling migrate at night?
9247. What is the primary reason for the decline of the Long-billed Curlew?
9248. What type of individual is known as a bird duffer?
9249. Which feather is called a pinion?
9250. What defense do large Great Blue Heron nestlings have against Turkey Vultures?
9251. Do toucans have coarse plumage?
9252. Is sexual dimorphism evident in juvenile nuthatches?
9253. Has the use of playback bird alarm or distress calls, broadcast to discourage birds from concentrating in certain areas, been effective?
9254. In the most recent taxonomic revision, which birds are in the first-listed passerine family?
9255. In Britain, which bird is known as a Titlark?
9256. In mythology, which birds often accompanied Aphrodite, the Greek goddess of love, and why?
9257. How long is the nestling period of the larger toucans?
9258. How much greater is the volume of the respiratory system of a 2-pound bird (0.9 kg) than that of a mammal of comparable size?
9259. What was Georg W. Steller's (of Steller's Sea-Eagle and Steller's Eider fame) function on the Vitus Bering Arctic expedition?
9260. In what country does the Greenland Wheatear winter?
9261. Is the Wild Turkey currently more widely distributed than during colonial times?

Questions

9262. Is a plovercrest the cranial adornment of a plover?
9263. Aside from thermoregulation, how is body plumage beneficial to flying birds?
9264. Do the West Indian todies have a high metabolic rate?
9265. Is the echo-locating ability of the South American Oilbird superior to that of the Asian Edible-nest Swiftlet?
9266. In Eurasia, which starling often forms mixed flocks with the European (Common) Starling?
9267. Is it more accurate to describe breeding birds at lek display sites as promiscuous rather than polygamous?
9268. Are the neotropical nunbirds more solitary than their larger relatives, the puffbirds?
9269. Is the Pyrrhuloxia migratory?
9270. Which tropicbirds have been recorded in North America?
9271. Where is the center of distribution for Kittlitz's Murrelet?
9272. Which is the only jacana named for its color?
9273. What is a Chowchilla?
9274. Do running Lesser Rheas hold their necks erect?
9275. Do birds that swallow their food whole have small tongues?
9276. Which is the smallest and most numerous stork of India?
9277. How do the tribal natives of Laos attract the bamboo-partridge within weapon range?
9278. In the southern part of its breeding range, what type of nest site does the Northern Parula generally select?
9279. What do some South American natives believe powdered Horned Screamer horn will prevent?
9280. Does the Northern Cardinal construct a large nest relative to its size?
9281. Why is the California condor considered to be a "CMV?"
9282. Which birds are attracted to a neotropical Jabiru nest, and how do the birds benefit the stork?
9283. Do the Old World rollers actually somersault in flight?
9284. What does the generic name *Anastomus* for the openbill storks imply?
9285. How did the term *roosting* originate?
9286. Who wrote the poem *On a Bird Singing in its Sleep*?
9287. Do tyrant-flycatchers have large heads for their size?
9288. What do courting male Superb Lyrebirds do with their long tails?
9289. Are breeding male Hudsonian Godwits more brightly colored than females?
9290. Which chickadee is among the most studied of all birds?
9291. Based on the fossil record, were flamingos once widespread in Australia?
9292. Are the bills of most shorebirds rigid?
9293. In Europe, how does the Winter Wren roost?
9294. Do munias have loud, pleasant voices?
9295. Who introduced the flightless New Zealand Weka rail to Macquarie Island, and why?
9296. What man-made structure might be used by communally-roosting dippers?
9297. The egg of the tiny New Zealand Rifleman accounts for what percent of the female's weight?
9298. Within which biogeographical realm is Hawaii included?

Questions

9299. Is a male canary less likely to sing when its mate is nest building?
9300. For which bird is Cape Kidnappers (New Zealand) noted?
9301. Do Swainson's Hawks have a flat flight profile?
9302. How much do Roseate Spoonbills weigh?
9303. What is a hackle?
9304. Are all three South American screamers larger than ducks?
9305. According to folklore, what colors are *naturally* repugnant to most birds?
9306. How high up in the Himalayas do Red-fronted Rosefinchs occur?
9307. Which bird does the Australian White Ibis most closely resemble?
9308. What is the ghost moth sometimes called?
9309. Do nocturnal birds have color vision?
9310. Do Snow Buntings reuse their nests of previous years?
9311. Do bee-eaters have a blue line through their eye?
9312. How do fish-eating grebes differ in shape from grebes that feed primarily on invertebrates?
9313. What is a Piapiac?
9314. Which 1954 British law extended protection to all wild birds and their eggs?
9315. Which two small endangered Australian passerines have an incredibly long incubation period of 36-38 days?
9316. How much more do male Eurasian Capercaillies weigh than females?
9317. Which ptarmigan is completely circumpolar in distribution?
9318. Is the New Zealand Saddleback a clamorous bird?
9319. In winter plumage, how do phalarope sexes differ?
9320. Are any of the four South American seedsnipe as large as a partridge?
9321. Is the hind toe of a sheathbill generally reduced in size?
9322. Is the Yellow Grosbeak primarily a Costa Rican species?
9323. What shape are the ruff feathers of a California Condor?
9324. Do Gyrfalcons nest in trees?
9325. Which large, long-legged bird was introduced into the Fiji Islands?
9326. Which bird has the longest common name?
9327. Which North American heron is best adapted to salt water?
9328. Are the nostrils of the American shrikes oblong in shape?
9329. Is the down of a Phainopepla nestling especially short?
9330. Rhea chicks cry out with what type of contact calls?
9331. What color are the legs and feet of a breeding Spectacled Guillemot?
9332. Which American bird was known as a "bay-winged bunting?"
9333. Are the offspring of interbreeding domestic Muscovies and Mallards fertile?
9334. Which Asian passerines nest in the side wall of an occupied Changeable Hawk-Eagle or White-bellied Sea-Eagle nest?
9335. What is the estimated weight of *Archaeopteryx*?
9336. When Black Skimmers breed in mixed colonies with Common Terns, which species is most aggressive?
9337. What color is the chin, throat and foreneck of a male Australian Princess Parrot?
9338. How many pairs of Sooty Terns breed on Christmas Island (Pacific Ocean)?

9339. Which gull scavenges in sea lion colonies in southern South America?
9340. Can Andean Condors be sexed by eye color?
9341. The New Zealand ornithological journal *Notornis* is named after which bird?
9342. Do skuas have faster and stiffer wingbeats than gulls?
9343. Which Australian doves are named after precious jewels?
9344. Do Himalayan Monals (Impeyan Pheasants) have crests?
9345. Is a flying frigatebird's wing kinked at the wrist?
9346. Excepting Abbott's Booby, which is the rarest sulid?
9347. What type of plantations do some tropical American trogons frequent?
9348. What is the normal clutch size of a Wattled Crane?
9349. What are *female* Barred Buttonquail used for in eastern Asia?
9350. In terms of angle, to what extent can a skimmer open its mouth?
9351. Are the orbits of the avian skull completely ringed with bone?
9352. What color is the solid casque of the Helmeted Hornbill?
9353. Which British bird is known as a "snake-bird?"
9354. Which U.S. state is nicknamed the Hawkeye State?
9355. Do wagtails defend winter territories?
9356. What type of nests do vireos construct?
9357. About how much farther away can a researcher recording a bird song be when using a parabolic reflector than with an open microphone?
9358. Which bird does Forster's Tern closely resemble?
9359. Which passerine constructs the largest nest, and how large is the nest?
9360. In the popular Christmas carol, *The Twelve Days of Christmas*, in what type of tree was the partridge perched?
9361. Does the stomach of a hungry bird growl?
9362. Is each species of the obligate brood parasitic-nesting whydahs of Africa host specific?
9363. Chlorophonias and euphonias have what unusual chick-rearing habit for passerines?
9364. For what did 19th-century New Zealanders use the feathers of the Australasian Bittern?
9365. Does a Sunbittern have a short, narrow tail?
9366. The nest cavity of which bird is used by a nesting Cactus Wren?
9367. Do all Old World spoonbills have a nuchal crest?
9368. Throstle is the archaic or poetic name for which bird?
9369. Which tropicbird is the most graceful on the wing?
9370. Is the Chaffinch one of the least numerous European birds?
9371. What is "Holboell's Grebe?"
9372. Is the African Fish-Eagle restricted to the lowlands?
9373. Is the Merlin a circumpolar breeder?
9374. Are Rosy Finches solitary?
9375. Which bird was used by southeast Asians to catch fish?
9376. Do storks have webbed feet?
9377. What do most male sandgrouse have on their chests?
9378. What are pardalotes, and what does the name mean?
9379. Why do foraging larks seldom perch in vegetation?
9380. Are Roseate Terns often observed inland?

9381. In terms of days, what is the range of the incubation periods of the 355 parrot species?
9382. In the American tropics, why are Giant Antshrikes troublesome to mist-netting biologists?
9383. What does a Palm Warbler constantly do with its tail?
9384. Do the Old World sunbirds have blunt nails?
9385. Are the wings of the neotropical Sunbittern patterned?
9386. Are female cotingas larger than males?
9387. What is a superspecies?
9388. How many species of land birds inhabit Aldabra Island (Indian Ocean)?
9389. The European (Northern) Lapwing may lay its eggs in the nest of other birds, but which species is favored?
9390. What were the circumstances surrounding the only darter record in New Zealand?
9391. How do newly-hatched chicks of the Laughing Kookaburra and its near-relatives differ from those of other kingfishers?
9392. Which large group of North American passerines is noted for an unusually large number of hybrid combinations?
9393. What is the origin of the vernacular name "verdin?"
9394. Do European White Storks shelter their chicks from driving rain?
9395. How is the tail of a copulating female bird positioned?
9396. Are birds better known taxonomically than any other class of animal?
9397. Which ratite has the largest tail *relative* to its size?
9398. In Holland, what large, round object is often placed on a roof to encourage White Storks to nest?
9399. Who was the first European to describe the pirating feeding behavior of frigatebirds?
9400. The Hawaiian Hawk has both a dark and a pale-colored morph, but which is most numerous?
9401. Do barbets often forage hanging head downwards?
9402. When terns plunge-dive, is underwater propulsion accomplished primarily by the force of the initial impact?
9403. In those birds with a sense of taste, are most taste buds located on the tongue?
9404. Do the neotropical tapaculos have stiff tails?
9405. Is the Eurasian Hoopoe sexually dimorphic?
9406. Is the Smew a highly-regarded gamebird?
9407. Does the Sage Thrasher often sing on the wing?
9408. Do fledged motmots return to the nesting burrow to roost?
9409. Is the inner toe of a bittern longer than the outer?
9410. In English folklore, which birds were called "corpse-hounds?"
9411. Birds have how many jaw muscles?
9412. Which American woodpecker has an unusually high number of alternate vernacular names?
9413. Are most South American ovenbirds small brown birds?
9414. Is a flushed buttonquail (hemipode) more inclined to run than fly?
9415. What is an individual called who works late at night, or does not retire until very late?
9416. When nesting in a cavity, does the European (Common) Starling construct a nest?
9417. Do the neotropical tinamous have thick necks?

9418. Which is the only West Indian island accommodating more than one species of tody?
9419. Which exclusively aquatic birds have hard, compact plumage?
9420. Are all British tits essentially sedentary?
9421. Which animal poses a significant threat to kiwis?
9422. Is the tail of a mousebird three times as long as its body?
9423. How many terns are named after countries, and which ones are they?
9424. Is a Neddicky an African shrike?
9425. Does the Cedar Waxwing winter as far south as Panama?
9426. Do rails ever *nest* in trees?
9427. Are Band-tailed Pigeons solitary?
9428. Which bird is the North American counterpart of the Lanner Falcon of Africa, or the Lagger Falcon of India?
9429. Which American bird is called a "shining fly-snapper?"
9430. Are frogmouths swift fliers?
9431. For what is the roller canary best known?
9432. Which American bird has the same name as a mustelid?
9433. Is the tongue of a penguin split at the tip?
9434. Is a male toucan's bill considerably longer than the bill of a female?
9435. Is the Blue Chaffinch common and widespread?
9436. Which two alcids are most vulnerable to oil pollution, and why?
9437. What color is the bare facial skin of a Rook?
9438. Do sandgrouse land in the water?
9439. What is the bill color of the female Derbyan Parakeet of China and Tibet?
9440. Are Great Horned Owls of the far north migratory?
9441. Sonnerat's Junglefowl of peninsular India is better known by what name?
9442. Do any of the true plovers have specialized bills like some sandpipers?
9443. Was the extinct New Zealand Huia exclusively arboreal?
9444. From claw to claw, what is the span of a California Condor's foot?
9445. Does the Willow Flycatcher breed in southern California?
9446. In French Guiana, artificial flowers made for tourists consist of what?
9447. How long does a young Kagu of New Caledonia remain with its parents?
9448. What is bat-fowling?
9449. Which is the smallest Antarctic-breeding bird?
9450. How many species of trogons inhabit Australia?
9451. Which is the only circumglobal chickadee?
9452. Do the neotropical trumpeters have relatively long necks?
9453. Who was the female star of the 1969 movie *The Sterile Cuckoo?*
9454. Is the type locality of a nominate subspecies *always* the same as that of the species?
9455. Do the large New Zealand Weka rails have short tails?
9456. What color is the eye-ring of an American Robin?
9457. Which vireo is most tolerant of heat?
9458. What is the shape of a Cactus Wren's nest?
9459. Are ibises capable of swimming?
9460. Which bird is known as the *bird of Jove?*

Questions

9461. Are potoo eggs glossy?
9462. Do thrushes occur on all continents except Antarctica?
9463. Which is the only pelican recorded in Japan?
9464. Do Emus brood merge?
9465. If an Evening Grosbeak has a yellow bill, what time of year is it?
9466. On what did migrating Eskimo Curlews feed extensively when on the American prairies?
9467. Do all breeding Blue Grouse cocks have purple inflatable neck-sacs?
9468. Which crane has the longest migration, and how far does it migrate?
9469. The nocturnal singing or ecstatic outpourings of which bird were said to have intrigued Thoreau at Walden Pond?
9470. Is the brightly-colored plumage of a bee-eater iridescent?
9471. Does the Laysan Albatross breed on Wake Island?
9472. What does the name "nightingale" mean?
9473. Which bird did Audubon call the Washington Sea-Eagle?
9474. Do hummingbird chicks defecate over the edge of the nest?
9475. Why do birds in compact, open-country feeding flocks frequently engage in conspicuous *leap-frog* movements, in which the birds at the rear fly over and land in front of the flock?
9476. Which is the only Australian finch that nests exclusively in hollow branches or in holes in termite mounds?
9477. What do the St. Helena Plover and Napoleon Bonaparte have in common?
9478. Which is the only state with a waterfowl species as the state bird?
9479. Do foraging catbirds scratch the ground with one foot?
9480. What type of bird is a Kakariki?
9481. Is the southern population of Wrentits more grayish in color than Wrentits of the north?
9482. Is Alaska the only state where the Surfbird nests?
9483. Which domestic goose has a large knob on its bill?
9484. In the latest taxonomic revision, are penguins listed before grebes?
9485. What is the difference between an avian call and a song?
9486. What was the name of Sir Francis Drake's flagship when he set off in 1577 to attempt the second circumnavigation of the globe?
9487. What color is a vinaceous-colored bird?
9488. Are the changes in pupil diameter of the avian eye strikingly rapid, as well as extensive?
9489. Is a vireo's beak heavier than that of a wood-warbler?
9490. Which bird is said to be the model for the color battleship-gray favored by the U.S. Navy?
9491. Are the long, stiff, pitchfork-like wing quills of a cassowary solid?
9492. Is the Lesser Nighthawk a bird of the mountains?
9493. How do Virginia Rails defend their nests?
9494. Which albatross is sometimes called Carter's Albatross?
9495. What are the two *primary* predators of the Dovekie?
9496. Why is a Loggerhead Shrike called a "cotton-picker?"
9497. Are estrildid finches primarily granivorous?
9498. Aside from pesticides, what important toxic chemicals are potential threats to birds?
9499. Which bird of the Bible was called a *cuckow* or *sea-mew?*
9500. Do bustards dig with their bills?

Questions

9501. Do jaegers have a non-breeding or winter plumage?
9502. Do knots fly in compact flocks?
9503. Is the tail of a juvenile mousebird shorter than that of an adult?
9504. Which bird adorns the national flag of Zimbabwe?
9505. Which vulture has the most powerful grasping feet?
9506. What color are the legs, bill and unfeathered portions of the face of the African Spoonbill?
9507. Do Bobolinks *commonly* nest in hay fields?
9508. Why is the plumage of a cassowary coarse and hair-like?
9509. Are longclaws more slender-bodied than wagtails?
9510. Is the juvenile plumage of many Old World flycatchers streaked?
9511. What does the Village Weaver do with any egg in its nest differing markedly from its own?
9512. Are pipits among the most widely-distributed songbirds?
9513. Is the endemic New Zealand Saddleback known for mutual preening and courtship feeding?
9514. Where was the first Xantus' Hummingbird collected?
9515. The Chinese Egret has what color feet?
9516. Do disturbed yellowlegs immediately cease vocalizing?
9517. Skuas have how many brood-patches?
9518. Is a Sooty Shearwater smaller than a Short-tailed Shearwater?
9519. What does a calling neotropical puffbird often do with its tail?
9520. How much dry weight of guano is deposited by a single Guanay Cormorant per month?
9521. Why is the Hoatzin locally known as "cigana?"
9522. Does Swainson's Thrush sing at night?
9523. Which is the most numerous North American falcon?
9524. Is the name "gull" Scottish in origin?
9525. Which was probably the first bird used in falconry?
9526. At what age do the eyes of a nestling Song Sparrow open?
9527. Which is the largest owl of South Africa?
9528. Which spoonbill has the most restricted range?
9529. Why are Common Ground-Doves known as "tobacco doves?"
9530. Is the Saddle-billed Stork generally found in water and marshlands?
9531. Who first introduced the *father of American ornithology* to the Whooping Crane?
9532. Which American flycatcher is called "José Maria," and why?
9533. Is the white neck of a Chihuahuan (White-naped) Raven conspicuous?
9534. What is the state bird of New York?
9535. What is a Weebill?
9536. Is waterfowl hunting more strictly controlled by law in North America than in Europe?
9537. Are terns more pelagic than gulls?
9538. What is the *ala spuria?*
9539. Do hummingbirds often perch on power lines?
9540. Is a foraging Osprey less successful if the surface of the water is rippled by the wind?
9541. How many claws are on each wing of *Archaeopteryx*?
9542. Are wagtails common Australian birds?

Questions

9543. What is the status of the Tristan da Cunha Moorhen?
9544. Of the 34 species of *Acrocephalus* warblers, most are called swamp or reed-warblers, but why is this a misnomer in the case of the Tuamotu Reed-Warbler of French Polynesia?
9545. Do both Glossy and White-faced Ibis have brown eyes?
9546. The vocalizations of the Brown Noddy are said to resemble the calls of which bird?
9547. Are most bustards Eurasian residents?
9548. Which neotropical raptor is also called a "fishing-hawk?"
9549. Which northern petrel is harvested by humans?
9550. How can one determine from a distance if a Bufflehead or goldeneye duck nest is occupied?
9551. What color is the nape of a breeding Brown Pelican?
9552. Do the Old World white-eyes have long bills?
9553. What do aviculturalists call the Maroon-breasted (Southern) Crowned-Pigeon?
9554. Chicks of which abundant, endemic New Zealand passerine fledge at weights considerably heavier than their parents?
9555. In many parts of Africa, which two large, long-legged birds are known as "grasshopper birds," and why?
9556. Are Antarctic-nesting Snowy Sheathbills exclusively associated with penguin rookeries?
9557. Which gamebirds have translucent, greenish-white toned flesh, and why?
9558. When Anhinga or darter nestlings solicit water, how does their behavior differ from that employed when soliciting food?
9559. How far do the feet of a flying Black-browed Albatross project beyond its tail?
9560. What do fighting rheas do with their necks?
9561. Do swallows mob potential predators?
9562. Which three bird species are endemic to Baja California?
9563. Is the tiny Golden-crowned Kinglet timid?
9564. Do Northern Cardinals rear as many as four broods a year?
9565. Do kingfishers forage in salt water?
9566. Do the seedsnipe of South America roost on the ground?
9567. What is a gull colony called?
9568. What invertebrate avian predator has the name of two birds in its common name?
9569. What is a bird bolt?
9570. Which bird might prompt African villagers to abandon a hut?
9571. Are doves capable of flight before two weeks of age?
9572. Which jaeger is *most* likely to be encountered in South African waters?
9573. Who starred in the 1991 movie *Hudson Hawk*?
9574. How many of the 66 plovers are named after countries, and which ones are they?
9575. Do Hawaiian Geese (Nene) fly in V-formation?
9576. Where is the only place where a flamingo colony may number less than 50 pairs?
9577. Do White-tailed Ptarmigan occur in the far eastern Aleutian Islands?
9578. Which rare trogon occurs sporadically in the Chiricahua Mountains of Arizona?
9579. Which bird do the Japanese call *Tancho?*
9580. What color are the fleshy gape wattles of an Inca Tern?
9581. When did the Spot-breasted Oriole become established in the U.S.?

Questions

9582. In which group of large African gamebirds are the young capable of flight long before achieving adult size?
9583. When were the Common Nighthawk and Whip-poor-will recognized as separate species, and who separated the two?
9584. Do Red-breasted Mergansers often follow winding stream courses?
9585. Which abundant, conspicuous songbird of Hawaii was introduced from South America?
9586. Are Dollarbirds solitary?
9587. What was the Dusky-capped Flycatcher formerly called?
9588. How did the bowling-like game of duckpins get its name?
9589. When crossbill young fledge, are their bills crossed?
9590. In Australia, which bird replaces the Black-crowned Night-Heron?
9591. Which familiar petrel is sometimes known as the Mediterranean Shearwater?
9592. Which *extinct* gamebird was rediscovered in Peru in 1979?
9593. Nesting hanging-parrots have what lovebird habit?
9594. What is the Red-billed Leiothrix called by aviculturists?
9595. How long is the fledgling period of the Nicobar Pigeon?
9596. What color are the legs and bill of a breeding Asian Ibisbill?
9597. After whom was the Narina Trogon named?
9598. Excepting humans, which are the only animals known to build and elaborate upon structures that are considered to be symbolic representations of themselves?
9599. The Saker Falcon is sometimes considered a subspecies of which bird?
9600. Who wrote the novel *Coma*, as well as numerous other medical thrillers?
9601. Is the White-rumped Sandpiper well named?
9602. Do Anhingas and darters often take off from the surface of the water?
9603. What are the South American gold coins stamped with a figure of a condor called?
9604. Where do the terrestrial guineafowl roost at night?
9605. Which two Mexican passerines have the longest tails?
9606. In the latest taxonomic revision, are Old World and New World cuckoos in the same family?
9607. Which of the three tropicbirds is the most pelagic?
9608. Do curlews feed *extensively* on crabs?
9609. Which bird is known as a "white noddy?"
9610. What material do Blue-gray Gnatcatchers often use to strengthen their nests?
9611. What caused the death of Louis Agassiz Fuertes, one of the best-known American bird artists?
9612. Do New Zealand Saddleback pairs remain together throughout the year?
9613. Why was the name of the Monkey-eating Eagle changed?
9614. Which group of large birds has greatly benefitted from exposed city dumps and land fills?
9615. Why do more bird species inhabit a single large rainforest tree than the whole of Antarctica and surrounding seas?
9616. Do many desert birds nest underground?
9617. Were flightless seabirds used by Australian aborigines?
9618. What color are the primaries of the Inca Dove?
9619. Which birds have been recorded preying on the eggs of tree-nesting Marbled Murrelets?

Questions

9620. What color are the inner secondaries of a Short-tailed Albatross?
9621. Which American raptor is known as a "black warrior?"
9622. Which American sparrow breeds the farthest north?
9623. Do California Condors nest in trees?
9624. Do pittas have long tails?
9625. Are peacocks powerful fliers?
9626. Are the wing and tail feathers of an adult Bald Eagle longer than those of a yearling?
9627. Which American bird is known as a "gourdhead?"
9628. How many books did the ornithologist/artist John Gould publish?
9629. Is the Yellow-crowned Night-Heron a solitary feeder?
9630. If a Brown Booby has a yellow-green face, what does this imply?
9631. Which gulls might Common Ravens kill and consume?
9632. Which two pheasants were successfully introduced into France?
9633. In some regions of the tropical Pacific, the appearance of which bird is considered a bad omen?
9634. Which American bird is called "little yelper?"
9635. Which petrels regurgitate pellets of indigestible items?
9636. What percent of the Florida Reddish Egret population consists of the white morph?
9637. Does the tropical American Spectacled Owl have ear tufts?
9638. Are neotropical antbirds sexually dimorphic?
9639. Why are alligators irritating to Louisiana duck hunters?
9640. Is the bill of a European (Common) Starling yellow during the winter?
9641. A Common (Atlantic) Puffin may use which mammal dwelling?
9642. Which bird is known as a cockatoo-parrot?
9643. Do Three-toed Woodpeckers breed in Europe?
9644. Why did the well-known 19th-century American ornithologist, Dr. Elliott Coues, name his mule *Jenny Lind?*
9645. What is a fowling piece?
9646. Which bird has negatively impacted Red-headed Woodpeckers?
9647. Where did the Cincinnati Zoo obtain the breeding stock for its colony of Passenger Pigeons, and how much did the birds cost?
9648. What is the most effective method of capturing the rare Aleutian Canada Geese on rugged Buldir Island?
9649. What is the black crowberry also called?
9650. Who was the first chief justice of the U.S. Supreme Court?
9651. Which bird do Brazilians call the "knocking owl," and why?
9652. Which bird is known as the "large-billed gull?"
9653. Do Ruffed Grouse prey on snakes?
9654. Which is the only cotinga that occurs in the West Indies?
9655. Do Bald Eagles always nest in trees?
9656. To which birds are the accentors most closely related?
9657. Which guillemot in breeding plumage has two white spots at the base of its bill?
9658. Which bird greatly inspired such poets as Shelly, Wordsworth and Tennyson?
9659. Do gannets fly in lines like pelicans?

Questions

9660. What color are the legs and feet of both sexes of South American Kelp Geese?
9661. What color is the bill of an Australian Black Swan?
9662. Which is the only bulbul that breeds in Japan?
9663. On August 3, 1993, what famous neotropical ornithologist perished in a plane crash in the Andes of Ecuador?
9664. During the days of European plume-hunters, which bird-of-paradise was most valuable, and for how much were skins sold?
9665. Is the plumage of a juvenile Limpkin less glossy than that of an adult?
9666. Which group of petrels was named after the pelican, and why?
9667. Which birds are sometimes called "swallow-plovers?"
9668. Which small neotropical passerine, currently in a monotypic family, gathers at leks to display?
9669. Which is the most numerous sparrow of Death Valley, California?
9670. What do Haitian children do that may be one of the greatest threats to the endemic tody?
9671. Must Ostriches drink every day?
9672. Which National Football League team is named after two birds?
9673. Which North American falcon is most inclined to inhabit woodland habitat?
9674. What flower was formerly called a cuckoo bud?
9675. Are Double-banded Dotterel and Two-banded Plover names for the same bird?
9676. Are the flightless Kakapos (New Zealand Owl Parrots) skilled tree climbers?
9677. Do Killdeer call nocturnally?
9678. After whom was Wied's Crested Flycatcher named?
9679. Are Scissor-tailed Flycatchers timid?
9680. On what do sandgrouse feed almost exclusively?
9681. Does the Shoebill feed at night?
9682. Which bird is called a "loper" in Canada, and why?
9683. While flying, do kites consume prey held in their feet?
9684. In Latin America, which birds are called *jacu, camata* and *pava?*
9685. Does the breeding season of the South American Hoatzin coincide with the rainy season?
9686. How long does it take a male American Redstart to assume full adult plumage?
9687. How much of the total volume of a Surf Scoter's stomach is taken up by grit?
9688. Is the spatulate-shaped bill of a Spoonbill Sandpiper evident when viewed in profile?
9689. What is "Cory's Least Bittern?"
9690. Are Crested Treeswifts colonial breeders?
9691. A walking Secretary-bird takes how many paces a minute?
9692. Why are ornithologists concerned about the establishment of the feral population of American Ruddy Ducks in England?
9693. On what animals do Masked Boobies perch?
9694. How many active Osprey nests are in the U.S.?
9695. Which extinct southwest Pacific waterfowl is known from only two specimens?
9696. At what age does a spoonbill chick's bill begin to flatten?
9697. Do Australian butcherbirds have short incubation periods?
9698. What is a place where ravens roost or breed called?
9699. What is a Guaiabero?

Questions

9700. What is the status of the Stewart Island Snipe?
9701. What is a flannel scarf with sleeves worn by people confined to bed called?
9702. What is the origin of the name "motmot?"
9703. What is bull thistle also called?
9704. How many species of penguins breed in Chile, and which ones are they?
9705. What large bird does the Jabiru stork often associate with in the northern part of its range?
9706. What does the term *beccafico* mean?
9707. What color are the underparts of a Lammergeier?
9708. Where does the Eurasian Hoopoe generally copulate?
9709. Do bulbuls inhabit North America?
9710. What is it called when an individual abruptly quits an addictive habit, such as smoking or drinking?
9711. Are male and female Mountain Quail similar in appearance?
9712. Do foraging swifts often fly with swallows?
9713. Which American bird has the local name of "god-damn duck?"
9714. Which bird is called a "New England Jackdaw?"
9715. What is a shaft streak?
9716. Do both sexes of Old World orioles care for chicks?
9717. What is the clutch size of a Lammergeier, and how many chicks are fledged?
9718. Why were female Great Hornbills formerly far more common in zoos than males?
9719. When running, which large, long-legged migratory birds have a bouncing gait?
9720. Which is considered the most numerous and successful raptor in the world, and why?
9721. Which western North American passerine has a number of striking color variations, so much so that the various populations were formerly regarded as distinct species?
9722. When are where was the last Carolina Parakeet collected?
9723. Is the Turkey Vulture native to Puerto Rico?
9724. Do Horned Puffins nest alongside Tufted Puffins?
9725. Which is the only South American seedsnipe not named for its color?
9726. Are the Falkland Islands Ruddy-headed Geese brighter in color than those of South America?
9727. Excluding Heermann's Gull, which is the darkest U.S. Pacific coast gull?
9728. Which of the five loons is said to be the most trusting?
9729. What color is the bill tip of a Purple Gallinule?
9730. What is a fody?
9731. Which American raptor nests in the dwelling of an edentate?
9732. In ancient mythology, what was a swan maiden?
9733. Does the Eared Grebe breed in Japan?
9734. For what talent was the late neotropical ornithologist Theodore A. Parker best remembered?
9735. Do Corn Crakes inhabit swamps and marshes?
9736. Does the South American Rufous Hornero ever build a new nest on top of old ones?
9737. What color are the eyes of all thick-knees?
9738. Do meadowlarks gather in large flocks?
9739. What is the status of the Cuban Sandhill Crane?

Questions

9740. How is the Black Vulture a threat to leatherback turtles?
9741. Which African bird is responsible for the most native legends?
9742. Why was the Summer Tanager so named?
9743. What are people who gather shearwater chicks called in Australia and New Zealand?
9744. What is a punt gun?
9745. How did the slaves of the Old South regard the Blue Jay?
9746. Which bird is almost entirely responsible for the distribution of mistletoe in Australia?
9747. Do all gannets and boobies have bare facial skin?
9748. Do Common Ravens engage in aerial courtship flights?
9749. Which organization annually awards the Brewster Medal, and for what is the medal awarded?
9750. How did the name "scops-owl" originate?
9751. Which ratites are most attached to their territories?
9752. In South Carolina, which bird is called "Judas-bird," and why?
9753. Is Howard Hughes famous seaplane, the *Spruce Goose,* constructed of spruce?
9754. How many exotic species of birds were introduced into Singapore, and how many became established?
9755. What is a brace?
9756. Which American duck may nest on floating logs?
9757. What is the heart-rate of an Ostrich?
9758. Are California Condors susceptible to lead poisoning?
9759. Which neotropical cuckoo is named after a rodent?
9760. Do Northern Fulmars have melodious calls?
9761. Why are the white wing feathers of the Great Hornbill often yellow?
9762. What color is the plumage of a juvenile Red-footed Booby?
9763. Are Alaskan Bald Eagles smaller than those of Florida?
9764. Do female Ruffed Grouse have bright orange combs over their eyes?
9765. Do Surfbirds nest above the timber line?
9766. Which large woodpecker nests within sight of Manhattan?
9767. Which two turacos are rare or endangered?
9768. Are all neotropical tinamous polyandrous?
9769. Do Greater Roadrunners generally nest on the ground?
9770. How do the Sepik River headhunters of New Guinea regard the cormorant?
9771. How long do Laughing Kookaburras live?
9772. Which of the nine species of thick-knees consistently lives near water?
9773. Do most icterids have pointed tails?
9774. What is the average length of time between the laying of the first and second egg in Xantus' Murrelet?
9775. Is the Saffron Finch a common native Jamaican bird?
9776. The northernmost colony of King Penguins is on what group of islands?
9777. How many species of sea-eagles inhabit South America?
9778. Why have Purple Sandpipers extended their wintering range southward?
9779. In American slang, what does the early bird get?
9780. Are the neotropical cotingas among the most diverse of passerine groups?

Questions

9781. Is a Red-footed Booby more apt to feed at night if the moon is full?
9782. How far south does the American Robin range?
9783. Do *all* North American corvids have dark eyes?
9784. How many species of hummingbirds occur in Mexico?
9785. Do noddies at sea typically fly high?
9786. In Scotland, what is cockaleekie?
9787. How many states have selected state birds that are not resident year-round in the state?
9788. Is the Waldrapp ibis crested?
9789. Which large forest raptor has been known to attack wooden duck decoys?
9790. Do babblers lay eggs with glossy shells?
9791. What is the one-dollar Canadian coin called locally?
9792. What does the Hundi name *Hargila* for the Greater Adjutant stork mean?
9793. Does the winter range of the Pine Warbler include a significant portion of its breeding range?
9794. Which Australian dove is named after an African animal?
9795. Is the Brambling a seed eater?
9796. Do Eurasian Hoopoes hawk insects in flight?
9797. Which New Zealand bird was known as an organ bird?
9798. In southern Africa, which bird is called a Mossie?
9799. Do Boat-billed Herons ever feed in broad daylight?
9800. What is the largest and oldest New World organization devoted to the scientific study of birds?
9801. Which birds are sometimes known as woodhewers?
9802. Do alarmed Sunbitterns readily perch in high trees?
9803. What color is the down of a nestling Secretary-bird?
9804. Which of the eight jacanas is named after another bird?
9805. Which woodpecker is named after an American army officer?
9806. Who wrote the foreword to James C. Greenway's classic 1958 book *Extinct and Vanishing Birds of the World*?
9807. Which American bird is called a "cow-frog duck?"
9808. How do songbirds sleep on a perch without falling off?
9809. What is one of the favored nesting sites of the Hook-billed Kingfisher of New Guinea?
9810. Do guillemots ever carry nesting material?
9811. Which eagle is the smallest?
9812. Are Boat-billed Heron chicks fed via regurgitation?
9813. What river is the most important of all staging locations for migrating American cranes?
9814. Which bird do the aborigines of western Australia refer to as *rain brother,* and why?
9815. Do Upland Sandpipers often perch on man-made structures?
9816. How long is the tail of a male Boat-tailed Grackle?
9817. In what state do most American sightings of the Short-tailed Albatross take place?
9818. Which cuckoo was introduced into the Galapagos Islands, and why?
9819. Do Long-billed Marsh Wrens *often* sing at night?
9820. Does a Black-crowned Night-Heron assume full adult plumage at two years of age?
9821. What is the average dive length of the Dovekie?

Questions

9822. Is Pallas' Fish-Eagle of Asia especially vocal?
9823. Which bird is called a firebird in the Falkland Islands?
9824. Which raptors were (and possibly still are) eaten by humans in some Mediterranean countries?
9825. What color is the bare facial skin of the New Zealand Spotted Shag?
9826. How do breeding Australian White-faced Herons differ from most breeding herons?
9827. Which of the pelicaniform birds are the most aquatic?
9828. What is the favored food of the Antillean Euphonia?
9829. Are all avocets lowland birds?
9830. What is a spring chicken or springer?
9831. Which American bird is called a "birch partridge?"
9832. The mythical Phoenix bird is symbolic of what?
9833. What color is the thick chick down of the neotropical screamers?
9834. Which gull is named after a cetacean?
9835. When perched, what do *both* Loggerhead and Northern Shrikes frequently do with their tails?
9836. What legendary English character was called the *prince of thieves?*
9837. Do arboreal hornbills hop, or do they walk?
9838. Which of the six crested penguins, all of which are noted for extreme egg dimorphism, exhibits the greatest size extreme between the alpha (first) egg and beta (second) egg, and how great is the difference?
9839. How many Chinese Black-necked Cranes exist?
9840. Who starred in the 1978 movie *The Wild Geese*?
9841. Which bird call is reproduced by the oboe in Prokofiev's *Peter and the Wolf?*
9842. Which is the largest New Zealand parakeet, and where does it occur?
9843. Do pittas have a strong, direct flight?
9844. Which of the three tropicbirds is most apt to stray inland?
9845. For what was Thomas Traill, after whom Traill's Flycatcher was named, noted?
9846. When were Snowy Owls first recorded breeding in Great Britain?
9847. Are meadowlarks detrimental to crops?
9848. Which is the only insular endemic goose?
9849. What is the greatest cause of Rhinoceros Auklet mortality on Protection Island, Washington, one of the largest-known colonies?
9850. Do wild Red Junglefowl nest in low vegetation?
9851. What does the Maori name *kakapo* for the flightless New Zealand Owl Parrot mean?
9852. Is the Stilt Sandpiper a common North American bird?
9853. At what age do Field Sparrow chicks depart the nest?
9854. Which North American falcon is the most vocal during the breeding season?
9855. How many of the 14 species of albatrosses breed north of the Tropic of Capricorn, and which ones are they?
9856. Are recently-hatched Sunbitterns capable of standing up?
9857. Do grebes and loons often nest on the same body of water?
9858. Why do the people of Bangladesh sometimes cut down trees that may contain over 100 occupied stork nests?

Questions

9859. Why do kiwis not eat snakes?
9860. What is unusual about the migratory behavior of the Boreal Owl?
9861. Do female painted-snipe typically outnumber males?
9862. How can the eggs of the Greater and Lesser (Darwin's) Rhea be differentiated?
9863. With what do both species of choughs often line their nests?
9864. Which wood-warbler is the most frugivorous?
9865. Is the neck skin of the neotropical Jabiru stork especially distendable?
9866. In what country were the fossil remains of the oldest-known parrot discovered?
9867. Which two long-distance migrating birds breed at the highest latitudes?
9868. Which is the only tern with completely white plumage?
9869. What is the only bird endemic to Algeria?
9870. Who starred in the 1930 movie comedy *Horse Feathers*?
9871. How many gallinaceous birds are native to tropical Pacific islands?
9872. How many bones make up each side of the avian jaw?
9873. Do Hooded Orioles have solid-black tails?
9874. Do Shoebills have especially long necks?
9875. When on level ground, are condors obligated to run prior to becoming airborne?
9876. In slang, a person who is easily cheated or misled is sometimes called by the name of which seabird?
9877. When are sandgrouse especially gregarious?
9878. Do spoonbills live longer than 30 years?
9879. Do the terrestrial South American seriemas roost in trees?
9880. What color is the chin of a Great Gray Owl?
9881. Is a cubital band a Latin musical group?
9882. Was the extinct New Zealand Huia a tiny bird?
9883. What is a trainbearer?
9884. Which is the only stiff-tail duck that has never been maintained in long-term captivity?
9885. When the Old World Water Rail is grunting and screaming, what is this sometimes called?
9886. What color are the legs of a Black-necked Stilt?
9887. Is the song of a female Rose-breasted Grosbeak as elaborate as that of a male?
9888. Can the flightless Kagu of New Caledonia glide?
9889. Do male Purple Martins reappear in North America before females in the spring?
9890. Did Marco Polo believe in the existence of the legendary Roc?
9891. Do cotingas typically pluck the fruit while on the wing?
9892. Do titmice have soft, fluffy plumage?
9893. What was the name of the first non-engine powered, helium-filled balloon to successfully cross the Atlantic Ocean?
9894. Aside from being flightless, what do kiwis, penguins and Emus have in common?
9895. What famous naturalist had the same opinion of the Bald Eagle as Benjamin Franklin?
9896. Do Limpkins ever perch atop the tallest trees?
9897. Why do Von der Decken's Hornbills follow packs of dwarf mongooses?
9898. What percent of the indigenous birds of Lord Howe Island has been lost since its discovery by Europeans?
9899. What was Thomas Bewick's main claim to fame?

Questions

9900. Why do Rough-winged Swallows have small, recurved hooks along the outer primary of each wing?
9901. To whom and for what is the Dickin Medal awarded?
9902. In American slang, what are vociferous fans called who commence jeering when their team performs poorly?
9903. What region has the most species of cranes, and which cranes occur there?
9904. How do juvenile ground-hornbills differ in appearance from adults?
9905. Do all grebes regurgitate pellets?
9906. With which bird does the Scissor-tailed Flycatcher hybridize?
9907. What is the second largest city in Guatemala?
9908. Do both sexes of the Everglade (Snail) Kite have a white terminal tail-band?
9909. In American underworld slang, what is an insider of a criminal organization called that informs or spills all?
9910. Is the Sunbittern vocal?
9911. What is an austringer?
9912. What is a cross-country runner sometimes called?
9913. Do skimmers ever feed by standing in shallow water and sifting with their bills?
9914. Is the Copper Pheasant a common Japanese bird?
9915. Which is the most numerous wintering shorebird of northwestern Europe?
9916. Which bird preys on California Condor eggs?
9917. Which is the only sandpiper or sandpiper-relative that breeds in the West Indies?
9918. What state selected a woodpecker as the state bird, and which bird was selected?
9919. Which large southern petrel may construct its deep burrow in saturated ground, often with standing water at the entrance?
9920. Do polygamous northern wrens have longer, more complex songs than monogamous wrens?
9921. Do first-year jaegers return to the breeding grounds?
9922. Is the Burrowing Owl the only subterranean-nesting raptor of North America?
9923. What is unusual about the clutch of a New Zealand Kakapo?
9924. When was the journal of the British Ornithologists' Union first published?
9925. What is the lifespan of a Kakapo (Owl Parrot)?
9926. Which is the only loon that calls out with goose-like cackles?
9927. Is the Dulit Frogmouth one of the more numerous frogmouths?
9928. Does the Five-striped Sparrow of southern Arizona actually have five white head-stripes?
9929. Does the Eurasian Hoopoe occur in Sweden?
9930. Are the external nostrils of loons round in shape?
9931. Which neotropical antbird has a bare face?
9932. The Tawi-Tawi (Sulu) Bleeding-heart pigeon inhabits what country?
9933. Do Barn Owls ever prey on turtles?
9934. Do Wild Turkeys often roost over water?
9935. In American slang, what is a bird in the hand worth?
9936. Are the toes of a Great Gray Owl feathered to the claws?
9937. What percent of adult body weight are day-old Emu chicks?
9938. Which insect may cause a Waved Albatross to abandon its egg?

Questions

9939. Do Sage Grouse and Blue Grouse frequently hybridize?
9940. What plant is completely dependent on the White-cheeked Cotinga of the Peruvian Andes?
9941. Which cuckoo kills rattlesnakes?
9942. Do Shoebills ever perch or roost in trees?
9943. Do rheas prey on venomous reptiles?
9944. What is a rasorial bird?
9945. How do seal-hunting Alaskan Eskimos use swan feathers?
9946. How much larger is the male Rifleman of New Zealand than the female?
9947. In 1989, there was an attempt to change the state bird of Nebraska from the Western Meadowlark to which bird?
9948. Does the Fox Sparrow have a spotted breast?
9949. What color is the bill of a Helmeted Curassow?
9950. Are all hornbills in the same family?
9951. Which aquatic passerines fiercely defend territories?
9952. Do Wild Turkeys always nest on the ground?
9953. When describing the Whooper Swan, Carl Linnaeus confused it with which bird?
9954. Which well-known American ornithologist was the uncle of Louis Agassiz Fuertes, one of the most skilled of bird artists?
9955. How long is the bill of an adult American Wood Stork?
9956. Was Prince Ruspoli's Turaco named after a Hungarian prince?
9957. Do mockingbirds defend fall and winter feeding territories?
9958. Which is the most numerous wild waterfowl of Great Britain?
9959. When laying a replacement or second clutch, do kiwis use the same nesting site?
9960. Is the Lappet-faced Vulture the only vulture that kills adult flamingos?
9961. Must a pelican run over the water to achieve flight?
9962. What pigment is responsible for the color brown?
9963. Who wrote the classic composition *Swan Lake*?
9964. How do male Yellow Warblers differ from females?
9965. In what European country have fossil remains of the Secretary-bird been found?
9966. Do foraging Short-eared Owls hover?
9967. After whom was Burchell's Glossy-Starling named?
9968. Does the Japanese Waxwing breed in Japan?
9969. Are the neotropical jacamars high-altitude birds?
9970. Which bird that occurs in America is called a "tramp?"
9971. Are Magpie-larks common Tasmanian birds?
9972. What Australian subantarctic island is noted for a large number of breeding passerines?
9973. Semper's Warbler is restricted to what island?
9974. Do the Old World broadbills catch insects in flight?
9975. What is silverweed also called?
9976. Who discovered that hummingbirds become torpid, when did this occur, and which species was involved?
9977. What is the population of the American White Pelican, and how many colonies are known?

9978. Do ptarmigans have larger aftershafts (afterfeathers) during the winter than in the summer?
9979. What type of bird is an Ortolan?
9980. Do Mallards have crests?
9981. According to American folklore, what protection was gained if a barn was frequented by swallows?
9982. Are the huge crowned-pigeons of New Guinea ground nesters?
9983. Do gannets make more perpendicular dives than boobies?
9984. Are any muscles located on the lower part of a bird's leg?
9985. What color are the bills of juvenile American Oystercatchers?
9986. Does the nest of a quetzal have two entrances?
9987. During irruption years, is Snowy Owl mortality greater than normal?
9988. Which is the only endemic African sulid?
9989. In what year were the breeding grounds of Kirtland's Warbler discovered?
9990. Who decides on the "official" changes of the common names of birds?
9991. Is the Broad-winged Hawk more migratory than the Red-shouldered Hawk?
9992. Do the neotropical tinamous have noisy wingbeats?
9993. What color is the skin of a Keel-billed Toucan chick?
9994. Do Boat-billed Herons nest in the company of other birds?
9995. Does the Crested Caracara inhabit the rainforest?
9996. Are sunbirds triple-brooded?
9997. When Lammergeiers (Bearded Vultures) drop bones, do they ever catch the bones in the air?
9998. Is the endemic Palmchat of Hispaniola rare?
9999. Are the outer lyre-shaped feathers of a lyrebird's tail notched?
10,000. Is Steller's Jay an abundant breeding bird of Guatemala?
10,001. What company published this book?

Answers

1. The African Palm-nut Vulture. Greatly dependent on palm nuts, the aberrant vulture also feeds on stranded fish, crayfish, freshwater crabs and carrion.
2. The nocturnal South American Oilbird and some Asian cave swiftlets. Loud clicks are produced enabling the birds to navigate through their pitch-black breeding caves, sometimes more than 0.6 of a mile (1 km) from any light. Several swiftlets lack echo-locating ability and breed in the open, or in dim light near the entrance of caves.
3. The adult Kelp Goose gander of southern Chile, Argentina and the Falkland Islands.
4. The Long-billed Curlew, with a bill up to 8.75 inches (22.2 cm) in length.
5. The 23-inch-long (58.4 cm) Mexican Imperial Woodpecker.
6. *The Lone Eagle* (Charles Lindbergh). The flight was accomplished in 1927.
7. The birds of prey (hawks, eagles and owls).
8. The Laysan Duck. In 1930, the entire population was reduced to a single gravid female. An experienced biologist spent 16 days on the tiny island of Laysan, finding but a single pair of the ducks and a nest in which all the eggs were punctured by a Bristle-thighed Curlew. The drake subsequently disappeared, but the female retained sufficient semen in her oviduct to lay a fertile replacement clutch of eggs. Thus, the species owes its existence to that single persistent female. The endemic, insular duck is well-established in captivity, and the wild population on Laysan Island was about 500 birds in the mid-1980's. The Chatham Islands Black Robin was reduced to five individuals in 1979, but by 1987 at least 50 birds existed. In 1974 the Mauritius Kestrel was reduced to four birds, but the population is slowly recovering.
9. Bachman's Warbler. Since 1950, no more than six sightings of the elusive bird have occurred in a single year, and in some years, no sightings at all.
10. The African Ostrich. The reduction to two toes is a running adaptation. All other birds have three or four toes.
11. Gold dust was stored in the hollow shafts (quills) of California Condor flight feathers by the forty-niners. A full-sized quill holds 10 cubic centimeters of gold dust.
12. The extinct Madagascar Elephant-bird. Its eggs weighed up to 27 pounds (12.2 kg), measuring 15 by 12 inches (38 x 30.4 cm).
13. The Emperor Penguin. During the dark Antarctic winter, the temperature averages -50°F (-45.60°C), but it may drop significantly lower, while the winds often howl in excess of 100 mph (161 km/hr).
14. The Wandering Albatross. The record holder had a wingspan of 11.9 feet (3.6 m). Claims of wingspans in excess of 17 feet (5.2 m) have not been substantiated.
15. The young of previous seasons assist in chick rearing.
16. The Snowy Sheathbill.
17. A half-grown swan when it is at a very ungainly stage.
18. The male Australian White Pelican. Its bill length may reach 19 inches (48.3 cm).
19. The Central and South American tinamous. The vividly-colored unmarked eggs are hard, glossy and porcelain-like. The color is variable, including tones of green, blue, turquoise, violet, steel-gray, chocolate and yellow.
20. Probably the Ivory Gull, which nests as far as 85°N. The Snowy Owl has been recorded as far as 82°N during the winter at Ellesmere Island (Canada), and it may *live* farther north than any other bird.
21. Three of the Asian cave swiftlets, especially the Edible-nest Swiftlet.

Answers

22. The Pink-headed Duck of India. Exactly when it vanished in the wild is unknown, but captives in Europe survived until 1936, although rumors persist of one that survived until 1944 or 1945. Sightings are still reported in India, but all invariably turn out to be the superficially similar drake Red-crested Pochard. During the early 1980's, the last-known Marianas Mallard died, but many authorities doubt that this duck was a valid species; it was possibly a hybrid between the Mallard and the Pacific Black Duck.
23. The Ostrich. Its average speed is about 30 mph (48 km/hr), but it can easily run as fast as 45 mph (73 km/hr). Brief sprint speeds of up to 60 mph (97 km/hr) are possibly attainable.
24. Probably the Black-legged Kittiwake.
25. The Cocos Island Finch of Cocos Island, 425 miles (684 km) northeast of the Galapagos Islands. Discovered in 1891, the finch resembles a melanistic Bananaquit.
26. Birds in the family Alcidae, such as the auks, auklets, puffins, murres, murrelets and guillemots.
27. On March 23 ("Buzzard Day"), Turkey Vultures supposedly return to nesting ledges overlooking the town.
28. The Andean Condor. Males weigh up to 31 pounds (14 kg).
29. No, the large, colorful bill sheaths of all three puffins are shed during the non-breeding season.
30. The nearest relatives of the flightless penguins *may* be the petrels and albatrosses, both of which are among the most superb of fliers. The relationship is not especially close.
31. The Oldsquaw (Long-tailed Duck). Oldsquaws have been recovered from nets set at 240 feet (73 m). King Eiders dive nearly as deep.
32. To catch fish and thermoregulation. Over-heated birds cool off by rapidly fluttering their gular pouch. Some pelicans may also carry nesting material *in* the pouch.
33. Apparently none. The remaining large birds are probably semi-flightless Pied-billed/Guatemalan Grebe hybrids, or possibly large Pied-billed Grebes.
34. The huge woodpecker may be extinct. A highland pine resident, no reliable sightings have occurred since 1950.
35. The female Australian Wedge-tailed Eagle has a wingspan of up to 9.3 feet (2.8 m).
36. The Neotropical Region. Neotropical is the usual designation in zoogeography for tropical America and the non-tropical parts of South America, the West Indies and other islands near South America.
37. The Galapagos Woodpecker Finch uses a cactus spine or twig to impale or force an insect or grub out of a hole, and the Egyptian vulture that uses stones to smash open Ostrich eggs. More than 30 species of birds are known to be tool-users, some of which are even tool-makers.
38. To the ancient Egyptians, the ibis was the symbol of Thoth, the god of writing and wisdom, and scribe of the gods, hence it was sacred. Live birds were kept in temples and mummified after death, and were often buried with pharaohs. Huge collections, numbering up to 1.5 million mummified birds, have been found in animal necropolises near the wetland breeding sites. Many of these were possibly nestling birds that died of natural causes and were collected by the Egyptians as a sign of respect.
39. The Emperor Penguin. In recent years humans have also been winter-breeding in the Antarctic.
40. No one knows. Presumably, the sea-duck winters in the Bering Sea along the southern limit of the ice pack.

Answers

41. Possibly the Ivory-billed Woodpecker. No sightings of the continental race have been confirmed since 1951, and the bird may be extinct. Another possibility is the Ooaa (Kauai Honeyeater), which was reportedly reduced to a single pair in 1979.
42. A falconer's term referring to a male falcon that is about one-third smaller than the female.
43. The Australian Magpie Goose. Some, but not all, Ruddy-headed and Magellanic Geese of southern South America and the Falkland Islands molt their flight feathers progressively, bypassing the flightless period.
44. The Wrybill, a bizarre New Zealand plover.
45. No. The extinct *Dromornis stirtoni* of Australia weighed over 1,000 pounds (454 kg) and stood about 10 feet (3 m) tall. The extinct Elephant-bird was nearly as large, but apparently weighed at least 100 pounds (45 kg) less.
46. The 5.5-inch-long (14 cm) Least Storm-Petrel. Scarcely larger than a warbler, it weighs an average of only 0.88 of an ounce (15-24 grams).
47. Their high-pitched call was said to resemble the whine or whistle of a rifle bullet.
48. The Tundra (Whistling) Swan with 25,216 feathers, 80 percent of which were on the head and neck.
49. The Ruby-throated Hummingbird with 940 feathers. Smaller birds have more feathers relative to body size than larger birds. While a Tundra Swan weighs 2,000 times as much as a hummingbird, it has only about 27 times more feathers than the far smaller bird.
50. The Little Spotted Kiwi of New Zealand. Its 9-ounce egg (255 gms) weighs 25 percent of the weight of the 2.6 pound (1.2 kg) female. Wilson's Storm-Petrels and Blue-gray Noddies lay eggs weighing as much as 27.5 percent of their weight. The tiny Puerto Rican Tody also lays an egg equivalent to as much as 25 percent of its weight.
51. The fused right and left avian clavicles (collar bones). The furcula is popularly called the wishbone.
52. A bird obligated to lay its eggs in the nest of another species, such as some cuckoos. The resultant young are reared by the host, often to the detriment of the host's young.
53. The Yokohama chicken (Phoenix fowl). The ornamental chickens are highly prized because rooster upper tail coverts are not molted often (generally only once every six years) and grow continually up to 20 feet (6.1 m) or more in length, with a record of nearly 35 feet (10.7 m). The chicken has been bred in southwestern Japan for over 300 years.
54. The King and Emperor Penguin. In both species, the single white egg is incubated on the top of their feet, which is covered with a muscular fold of abdominal skin.
55. Saliva. Long, thin, glutinous noodles are secreted from a pair of enlarged salivary glands under their tongues that the birds wind into a translucent, rubbery, half-cup nest. The nests are sought for bird's-nest soup.
56. The swifts and nightjars (goatsuckers). The bill of the Glossy Swiftlet is probably the shortest, measuring about 0.157 inches (4.0 mm). Despite their tiny bills, the insect-eating birds have huge gapes.
57. The Antarctic Petrel (80°30'S), Snow Petrel (80°23'S) and South Polar Skua (77°77'S). All three breed farther south than any penguin.
58. Yes, some species do. For example, the 3 to 4-pound (1.5-1.8 kg) piratical frigatebirds with 7-foot (2.1 m) wingspans, have skeletons weighing a mere four ounces (113 gms). Similarly, a Bald Eagle had 7,182 vaned feathers weighing 1.33 pounds (0.6 kg), more than twice the weight of its 9.7-ounce (275 gms) skeleton.
59. The Ruddy Duck. It lays up to 14 large eggs within a 15-day period, a clutch weighing about three pounds (1.4 kg), or approximately three times as much as the bird.

Answers

60. The unfertilized yolk of the egg of the extinct Madagascar Elephant-bird. Of living animals, the largest single cell is the unfertilized yolk of an Ostrich egg.
61. Fat Oilbird nestlings are oil-laden, weighing at least 50 percent more than adults, and possibly twice as much.
62. The avian wrist joint (carpus).
63. The penguins, alcids and diving-petrels. Dippers are also said to "fly" underwater using their wings. Some other birds, such as sea-ducks, may use their wings to assist them in submerging, but only the feet are used once they are underwater.
64. None, but during the molt when all flight feathers are shed simultaneously, some species, such as murres and the Razorbill, are flightless for a short period of time. The extinct Great Auk had degenerate wings and was flightless.
65. One of his initial tasks was to stuff King George IV's pet giraffe. Gould originally gained a reputation as a taxidermist.
66. The Trumpeter Swan. Males weigh up to 28 pounds (12.7 kg) with nearly 9-foot wing-spans (2.7 m). One large male reportedly weighed nearly 38 pounds (17.2 kg).
67. The Turkey Vulture. Its heightened olfactory sense may aid it in locating carrion in forested areas where vision is limited. The Yellow-headed Vulture may also locate carrion by olfactory means, but a functional sense of smell is not confirmed in other vultures.
68. The Asian Helmeted Hornbill. The casques of other hornbills are light and essentially hollow, although supported internally by thin struts.
69. The endangered Kirtland's Warbler. The 62 by 80 mile (100 x 129 km) breeding range in Michigan's north-central lower peninsula is one of the smallest for any bird not restricted to a small island.
70. Like baleen whales, the filter-feeding prions have distinct lamellae hanging from their upper mandibles, hence the name whalebird. The specialized petrels also gathered in large flocks around floating whale carcasses during the southern ocean whaling era to feed on floating oil.
71. The quetzal. The Resplendent Quetzal, the most magnificent of trogons, is the national bird of Guatemala, and is also pictured on the flag.
72. From the Spanish word *bobo* meaning a dunce, stupid, foolish or clownish. They are fearless, and nesting birds can be closely approached, suggesting stupidity to early observers. Boobies often land on boats at sea, a habit that hungry mariners also considered rather stupid.
73. The Ostrich. Its egg is only 1.0-1.5 percent (1/60th) of the weight of the female.
74. The South Polar Skua. The only other vertebrates that have been to the South Pole are humans, sledge dogs and hampsters.
75. The excrement or droppings of the Guanay Cormorant was called *huanu* (dung) by the Peruvian Quechua. The name was later modified to guano by the Spanish.
76. The extinct Great Auk of the North Atlantic. A flightless, penguin-like alcid, it may have numbered in the millions. When European explorers later encountered what are now known as penguins in the late 1400's, they were reminded of the huge alcid and the name was transferred. Even its scientific name *Pinguinus impennis* suggests a penguin.
77. Nineteen, although some species just barely cross the border.
78. A female Northern Royal Albatross banded as a breeding adult in 1937. In great albatrosses, breeding doesn't commence until at least seven years of age (and generally not before 9-11 years) so, as of 1990, she was a minimum of 60 years old. The famous old bird was known as *Grandma,* but she failed to return to the colony at the end of 1990.

79. The three South American screamers all have sharp carpal spurs that attain a length of two inches (5 cm).
80. A southern ocean petrel. Also called Painted Petrel or Pintado, it is among the most conspicuous of southern seabirds.
81. The American Bald Eagle. The largest recorded nest was a huge structure at St. Petersburg, Florida, occupied for many years. It had a depth of 20 feet (6.1 m), was 9.5 feet (2.9 m) across, and weighed nearly three tons (2,722 kg).
82. An Ostrich egg measures up to 6.5 by 5.0 inches (16.5 x 12.7 cm), weighing as much as 4.35 pounds (1.97 kg).
83. The hornbills, Ostrich, Hoatzin, some hawks and owls (especially eagle-owls) and cuckoos. Most birds lack eyelashes.
84. The Arctic Tern breeds in the high Arctic and winters in the Antarctic, necessitating an annual round-trip of 22,000 miles (35,405 km).
85. The Eskimo Curlew. Once extremely numerous, it was virtually wiped out by over-hunting. The last reliable sightings in Texas were in 1987, and two birds were reported in the Northwest Territories. Four birds were reported in Argentina in October, 1990.
86. The Western, Summer, Scarlet and Hepatic Tanagers are the most numerous northern tanagers. The West Indian Stripe-headed Tanager has recently become established in southern Florida. The Blue-gray Tanager is not native, but introduced birds are established in and around Miami, Florida.
87. The pigeons and doves.
88. Yes.
89. The female Common Eider of eider down fame. The thick, heavy down has been used as an insulating material since before the beginning of recorded history. The down is extremely soft, light and cohesive, with the best thermal quality of any known natural substance. No synthetic yet produced is able to match the combined insulating quality and lightness of eider down, which retains its cohesiveness and elasticity for nearly 30 years.
90. The Stephen Island Wren of New Zealand. Living wrens were only observed twice at twilight, and within a year of the semi-nocturnal bird's discovery in 1894, the lighthouse keeper's cat killed every one. The cat brought in the remains of about 20 wrens.
91. The tiny Elf Owl of southwestern U.S. and northern and central Mexico. It is just over five inches (12.7 cm) in length, with a maximum weight of 1.75 ounces (49.6 gms).
92. The Emperor Penguin. Both sexes incubate in the other 16 penguins.
93. No, these are the famous giant clams, the largest of which (*T. gigas*) inhabits Indo-Pacific coral reefs. Weights of 600 pounds (272 kg) have been documented.
94. The economy of Peru was formerly based on guano harvesting. Fish-eating seabirds produce the richest nitrogenous fertilizer known, superior to even the best modern synthetics. Some islands had guano caps exceeding 270 feet (82.3 m) thick, representing 2,000-3,000 years of guano deposition. Liquid seabird excreta desiccates naturally into a crusty, rock-hard material. The Peruvian Booby, Guanay Cormorant and Peruvian Pelican are the major guano producers, and while the population of the three seabirds fluctuates, it often numbers many millions.
95. The Yellow-billed Magpie and California Condor.
96. The Old World megapodes (mound-builders). A Dusky Scrubfowl nest, consisting of nearly 300 tons of forest floor litter, measured 36 feet (11 m) across and was more than 16 feet (4.9 m) high. Orange-footed Scrubfowl mounds approach 40 feet (12 m) in diameter, and reach 10-16 feet (3-5 m) in height.

97. Rictal bristles *might* be used to smack flying insects that foraging birds miss with their beaks. Thus, stunned prey may be more easily captured when the birds rapidly sweep about and make another pass. The bristles may also serve to protect the eyes. Sensory nerve cells are located at the bristles' base, and the modified feathers undoubtedly also function as organs of touch.
98. The South American Hoatzin, but only as chicks. The tiny carpal claws (not to be confused with spurs) on the wrist of the unfeathered wings enable chicks to scamper about the branches. The functional claws are generally lost after about a week or so of age. Considered by some authorities to be the most primitive of birds (living fossils), the Hoatzin is probably not nearly as primitive as formerly believed. Its bizarre physical characteristics may merely reflect a specialized lifestyle.
99. Both are filter-feeders. The whales have horn-like plates (baleen) hanging down from the roof of the mouth used to filter food organisms, and flamingos have distinct filtering lamellae that serve the same purpose.
100. Franklin thought of the Bald Eagle simply as a scavenger, and a bird of bad moral character. He favored the Wild Turkey.
101. The Sword-billed Hummingbird. Its bill exceeds four inches (10 cm) in length, which is longer than the body of the bird. The specialized hummingbird feeds extensively on the climbing passion flower (*Datura*) that has a 4.5-inch-long corolla tube. The flowers are generally located well away from leaves and entangling vegetation so the feeding hummer can hover at the correct angle.
102. Only two: Lewis' Woodpecker and Clark's Nutcracker.
103. The Coppery-tailed Trogon is also known as the Elegant Trogon.
104. The male Bee Hummingbird of Cuba and the adjacent Isle of Pines. Males are slightly smaller than females, measuring only 2.25 inches (5.7 cm) in length (half of which is accounted for by the bill and tail), with a body the size of a large bumblebee. It weighs only 1.6 grams (0.056 oz). Some hawk-moths weigh more than one-and-a-half times as much as the tiny hummingbird.
105. According to the ancient Greeks, a swan sang a *song of death* when its life was about to end. Many people still believe this, including some ornithologists. Possibly a prolonged exhalation of air from the trachea of a dying swan produces a series of musical notes. In general usage, a swan song usually implies *the end.*
106. The Passeriformes. The approximately 5,674 passerines are typically called songbirds or perching birds.
107. The Zapata Wren (*Ferminia cerverai*) was named after Fermin Z. Cervera, a naturalist who was in Cuba in 1926 when the bird was first discovered. The rare wren is confined to a small marsh in southern Cuba.
108. The Amsterdam Albatross, one of the rarest of all birds and, ironically, one of the largest. The recently described species (1983) breeds only on Amsterdam Island.
109. Hawaii. At least 162 exotics were introduced, many of which were detrimental to the local endemics. Of the 162 species, a minimum of 45 definitely became established, with an additional 25 probably established.
110. The smaller southern albatrosses superficially resembling large gulls. The name originated from the Dutch word *mallemok,* and loosely translates to foolish or stupid gull.
111. Wax. Ornithologists formerly believed honeyguides fed only on honey, as well as bees and bee larvae, but it has since been confirmed that it is wax the birds crave. Their primary diet probably consists of insects.

Answers

112. Probably the Alpine Chough. It inhabits the Himalayas between 11,500-20,000 feet (4,572-6,069 m) year-round, but the bird has been recorded as high as 27,000 feet (8,230 m). Some Asian snowcocks may breed nearly as high.
113. A male Sulphur-crested Cockatoo named *Cocky* acquired in 1925 by the London Zoo from an individual who obtained it as an adult at the beginning of the century, died on 28 October 1982 at 80, or possibly 82 years of age. An Andean Condor known as *Kuzya* resided in the Moscow Zoo from 1892 to 1964, remaining outdoors the entire time. Acquired as a full-plumaged adult, *Kuzya* was at least 77 years old when it died. Another Andean Condor obtained in 1902 was possibly in excess of 71 years of age when it died in a Paris zoo in 1973. A Siberian White Crane named *Pops* acquired in June, 1906, died in the National Zoo (Washington, D.C.) in 1967 at the age of at least 62 years. Had the geriatric crane not suffered a compound fracture of its left leg three days previously, it might have lived considerably longer. Despite repeated claims to the contrary, no pet parrot has been documented to have survived a century or more.
114. The birds pile up decaying vegetation, which serves as a nesting mound and natural incubator (microbial heat generation).
115. The Barn Swallow, with a cosmopolitan distribution except for Antarctica and some remote islands.
116. An Elephant-bird egg contains in excess of two gallons (7.6 liters) of material, or the equivalent of 30,000 hummingbird eggs, 140 chicken eggs, or six Ostrich eggs.
117. Kirtland's Warbler.
118. No. Ironically, Sacred Ibis were considered sacred by the ancient Egyptians, but the birds have been extirpated from Egypt for at least a century.
119. The Adelie and Emperor Penguin. Some Adelies breed in the South Sandwich Islands, slightly north of 60°S.
120. The trogons, with skin the consistency of wet tissue paper. One of their generic names (*Apaloderma*) means soft, tender skin.
121. Eldey Island (Fire Island), Iceland, on the afternoon of either June 3rd or 4th, 1844. The incubating pair was discovered by three Icelandic fishermen; Jon Brandsson and Sigourer Isleffson struck the helpless flightless auks down with clubs while Ketil Ketilsson smashed the egg with his boot. A major Iceland breeding site was destroyed 14 years earlier in 1830 by a volcanic eruption. A possibly valid sight record of a Great Auk occurred off the Grand Banks in 1852 by Colonel Drummond-Hay, who later became the first president of the British Ornithologists' Union, so it could be argued that he was a qualified observer.
122. The Snowy (Greater) Sheathbill. The name refers to the fleshy wattles below its eyes, and its rather pigeon-like appearance.
123. The Arctic Tern breeds in the Arctic and winters in the Antarctic, thus the little tern shifts from one region of extended daylight to the other.
124. A guillemot.
125. The Egyptian Plover, that associates with crocodiles, possibly even picking ectoparasites from the large dozing reptiles. Legend has it that the shorebirds also pluck parasites and morsels of food from a crocodile's gaping jaws, but such behavior has never been substantiated by a biologist. The origin of this interesting tale dates back to Herodotus' visit to Egypt in 459 B.C.
126. The South American Black-headed Duck constructs no nest, invariably laying its eggs in the nests of other birds.
127. At least 14 endemics.

Answers

128. The Graylag and Swan Goose.
129. The Wandering Albatross, Royal Albatross and New Zealand Brown Kiwi. All incubate for an average of 78-85 days, but in several instances, 92 days were required for the kiwi.
130. The Cliff Swallow.
131. Probably the endangered Florida Everglade Snail Kite, that depends almost exclusively on the apple snail.
132. Chicks regurgitate and spit stomach contents and oil (projectile vomiting).
133. The Galapagos Penguin.
134. The Cuban Bee Hummingbird is generally credited with laying the smallest egg, but based on surprisingly few measurements, the slightly larger West Indian Vervain Hummingbird lays an even smaller egg. Two Vervain Hummingbird pea-sized eggs measured 0.39 inches (10 mm) in length, and weighed 0.0129 of an ounce (0.365 gms) and 0.0132 of an ounce (0.375 gms) respectively. A single Cuban Bee Hummingbird egg measured 0.45 x 0.32 inches (11.4 x 8.0 mm) and weighed 0.0176 of an ounce (0.5 gm).
135. None. So far as is known, no other vertebrate consumes wax. Microbes or bacteria in the digestive tract enable the unique birds to digest wax. Honeyguides can subsist on wax alone for a time, but it is not an adequate diet and the birds would starve without other food.
136. The South Georgia Pipit.
137. A number of small passerines, such as the Redpoll, Hawfinch, some white-eyes (*Zosterops*), as well as the Black-billed Cuckoo and Great Spotted Woodpecker, hatch after only 10 days of incubation.
138. Megapode means large feet. The huge feet are used to scrape up voluminous amounts of forest litter to create their massive nesting mounds.
139. In the past, the eyesight of a raptor was said to be eight times keener, but it is probably closer to 2-3 times as acute, if that. However, the *visual acuity* of many raptors is far greater than that of mammals, thereby facilitating a much more rapid assimilation of detail in the visual world.
140. The Common (Northern) Raven. Males average 3.7 pounds (1.7 kg) and females 2.4 pounds (1.1 kg).
141. The Common (Eurasian) Starling, House Sparrow and Rock Dove. European immigrants were homesick and the introduced birds reminded them of home.
142. The New Zealand Kea. Sheep ranchers believed the terrestrial parrots killed healthy sheep to eat their kidneys and surrounding fat. While this is probably not true, the birds will feed on a carcass, or possibly even on weakened sheep.
143. The Laughing Kookaburra, a large forest kingfisher from Australia.
144. The Greater Roadrunner can attain speeds of just over 25 mph (40.2 km/hr), but more typically runs at speeds closer to 15 mph (24 km/hr). A running roadrunner can easily outmaneuver pursuing dogs, using its long tail to aid in steering and balance.
145. The especially vocal Arctic duck was named after continually chattering old Eskimo squaws, but it is the male that is the most vocal. In most ducks, females are far more vociferous than their mates.
146. The Red-billed Quelea, an African weaver. Population estimates range between 1.5 and 10 billion birds, and colonies may cover several hundred acres, and contain up to 10 million nests.
147. Birds, and birds alone, have feathers.

148. Powder-down feathers produce a fine powder used during preening, possibly to increase waterproofing or to alter the plumage color so that it becomes more bluish-gray. In fish-eating herons, powder-down may be beneficial in removing fish oil. Parrots, herons, tinamous, Madagascar Cuckoo-roller, Kagu, pigeons, mesities, woodswallows and some of the larger goatsuckers, such as frogmouths and potoos, all have powder-down feathers.
149. The South American Oilbird, that feeds almost exclusively on palm nuts.
150. The Lammergeier.
151. The honeyguides. Burning wax will often attract them and the birds may locate bee hives at least partially by scent.
152. The Least Storm-Petrel weighs less than an ounce (28.4 gms), whereas the two giant-petrels are as large as a small albatross, weighing up to 13 pounds (5.9 kg), with wing-spans of 6.5 feet (2 m).
153. The ankle joint.
154. The Rock Ptarmigan winters as far as 75°N.
155. Scavenging vultures thrust their heads and necks deep into rotting carcasses, and featherless heads and necks are obviously more sanitary and easier to keep clean.
156. Bird's-nest soup, a great delicacy that is one of the most expensive dishes in Asia. The most expensive is that of the Edible-nest Swiftlet.
157. The Kakapo (Owl Parrot) of New Zealand.
158. The male Eurasian Capercaillie. As large as a turkey, the grouse weighs more than 15 pounds (6.8 kg) and exceeds three feet (1 m) in length.
159. Yes, the Common Poorwill sleeps in a torpid condition for up to three months.
160. The male Ribbon-tailed Astrapia with an overall length of 42 inches (1.1 m). The male Australian Superb Lyrebird is probably second with its 2-foot-long (0.6 m) gracefully curved tail.
161. The megapodes (incubator-birds). Chicks are feathered and capable of flight within a day of hatching, and are not cared for by their parents.
162. Invariably about 12 degrees to the right.
163. The male Oldsquaw duck of the high Arctic. Its winter plumage is very distinct from that of the summer, and drakes also assume a dull, female-type plumage for a short time. Other northern ducks have only two annual molts.
164. The King Penguin. Up to 13 months may be required, and during the winter, fasting chicks go without food for weeks, possibly even a month.
165. The Crested Auklet, but only during the breeding season. Nesting burrows can actually be located by the odor. The citrus smell is produced by the bright orange mandibular plates that are shed during the winter.
166. Colombia, with more than 1,700 species. It is possible Peru may actually have more.
167. Most birds have about 14 cervical vertebrae, but swans have up to 25. Most mammals, including giraffes, have only seven.
168. Probably for thermoregulation as birds became warm-blooded. Originally modified scales, feathers became longer and more elaborate over the years and eventually facilitated flight. Feathers possibly also evolved for courtship purposes.
169. The Blue Grouse, if altitudinal movements are considered migration. Some grouse move only 1,000 feet (305 m) between the wintering and nesting grounds, although others may migrate 30 miles (48 km). Other birds that move very short distances include the Mountain Quail, Black-capped Chickadee and Clark's Nutcracker.

Answers

170. Ostriches are noted for their huge size, strength, speed and keen eyesight. Except for the giraffe, Ostriches are the tallest of the keen-sighted grassland animals, and with their superior height and good eyesight, potential danger can be spotted from great distances. If the birds must fight, they can rip a lion open by kicking it with their powerful legs and heavy nails. And if all else fails, Ostriches can flee with great speed.
171. Three: Emperor, King and Royal Penguin.
172. No, but compared to the loud, far-carrying calls of other northern swans, Mute Swans have weak, seldom-used voices. They do utter short grunts, menacing hisses and loud snorts, as well as puppy-like barking notes during the breeding season.
173. The huge, shoe-shaped bill is not used to dig out lungfish as formerly believed. Rather, it is designed for catching and extracting prey from dense aquatic vegetation. Due to its enormous bill equipped with a surprisingly large, sharp hook at the tip, the unique bird is commonly known as the Whale-headed Stork.
174. The steep dive of a raptor. The term generally applies to falcons diving on prey, which they strike in the air.
175. The Crested Caracara.
176. The South American Oilbird. An olfactory sense may play an important role in locating food trees by scent at night, and possibly aids in guiding them back to their nesting caves.
177. The Inaccessible Island Rail of Tristan da Cunha is only five inches (12.7 cm) long and weighs but 1.2 ounces (34 gms).
178. The Cattle Egret of Africa, that currently has a nearly worldwide range.
179. The aberrant vultures are specialized bone and bone-marrow feeders.
180. The purpose is obscure, but the air-sacs may serve as shock absorbers when pelicans are diving. However, of the eight species, only the two brown pelicans are aerial divers. Possibly the air-sacs serve a thermoregulatory function. If pelicans are handled, the prominent air-sacs can be felt under the skin.
181. The Adelie Penguin nests as far south as 77°33'S (Cape Royds, Ross Island), slightly farther south than the winter-nesting Emperor Penguin. Four Adelie pairs bred at a long-abandoned colony at Cape Barne, several miles farther south in 1989, but so few birds do not really constitute a colony.
182. The jacanas. Their extremely long toes and nails are such that the foot imprint may be longer than the body of the bird. The Northern Jacana only weighs 2.8-6.0 ounces (80-170 gms), but its foot covers an area as great as 3.0-5.5 inches (12-14 cm).
183. The endemic, nocturnal, flightless Kakapo (Owl Parrot). In 1976, only 17 Kakapos were known to exist, all of which were males. Subsequently, additional parrots were discovered on Stewart Island. When that population was substantially reduced by feral cats, the remaining birds were moved to small offshore, predator-free islands. As of late 1990, the total population consisted of 46 birds, of which only 13 were females. Two chicks hatched on Little Barrier Island in February of 1991.
184. Europeans formerly believed that the birds obtained food by sucking milk from goats. In reality, the birds were hawking insects stirred up by goats. While their bills are tiny, nightjars have capacious mouths, surrounded by rictal bristles.
185. The solid *ivory* casque is so heavy that the skull constitutes 10-11 percent of the bird's weight.
186. The two wrynecks are closely related to woodpeckers and are in the same family, but they lack strong beaks and stiff tails. The birds have a habit of twisting their necks in peculiar contortions, particularly when disturbed, hence the name wryneck.

187. The penguins, but the often reported speeds of more than 20-30 mph (32-48 km/hr) are inaccurate. Penguins typically swim at 3-6 mph (4.8-9.7 km/hr), but can accelerate to at least 15 mph (24 km/hr) in short bursts.
188. The most recent list (1991) includes about 9,700 species.
189 The little birds feed on bee wax and larvae, but bee hives are often so substantial they cannot get into them without assistance. Therefore, honeyguides pester honey-badgers (ratels) and then "guide" them to the hive where the powerful mammals tear into the nest. Similarly, Black-throated and Scaly-throated Honeyguides attract the attention of African natives and lead them to hives. The superstitious natives believe they must leave part of the hive for the birds because, if not, honeyguides will no longer aid them, or worse yet, may lead them to a poisonous snake or a leopard.
190. The extinct Carolina Parakeet and the endangered Thick-billed Parrot.
191. A thermoregulatory behavior in which the rapid fluttering of the gular pouch dissipates excess heat, thereby cooling down overheated birds. A number of birds use a gular flutter, including pelicans and cormorants.
192. The hummingbirds, whose wings beat between 22 and 78 times per second, but up to 90 beats per second have reportedly been recorded in the Horned Sungem. As a rule, the smaller the bird, the more rapid its wingbeat. Even so, according to some authorities, the wingbeat of a hummingbird is slower than most birds relative to their minuscule weight.
193. The Peregrine Falcon. The fastest *documented* speed is 117 mph (188 km/hr) in a steep dive or stoop, although much higher speeds are commonly reported in the literature.
194. A colony of rooks was called a rookery. The term is currently used for many colonial-breeding birds, as well as for some gregarious marine mammals, such as sea lions or fur seals.
195. Males in flight transport naked chicks under their wings in marsupial-like pouches. The heads of chicks may be visible while the male is flying. The two Old World finfoots lack underwing pouches.
196. Some hummingbirds, swifts and nightjars during cold nights. There is evidence that roadrunners, Smooth-billed Anis, African mousebirds and Turkey Vultures may also become torpid. The core temperature of a Greater Roadrunner may drop as much as 7°F (3.9°C) at night.
197. The New Zealand kiwis.
198. The Black-winged Stilt, not flamingos as is often stated. The length of its legs constitutes about 60 percent of its height. The extremely elongated neck of a flamingo greatly lengthens its overall length, thus the stilt's legs are actually longer relative to its size.
199. The owls. Some owls can rotate their head 280 degrees, and then rapidly swivel it around in the opposite direction in order to maintain visual contact. Accounts of owls being able to twist their heads around 360 degrees are not factual.
200. Countless stories relating to the possible origin of this well-known legend exist, but one of the most plausible is that the spring arrival of European White Storks coincided with a new crop of human babies. Storks were encouraged to nest on the roof in the belief that the birds would bring fertility, good luck and prosperity to the house. In those days there was no thought of birth control because more offspring meant more workers. The legend probably originated in northern Germany, and may have been passed on to children to prevent them from molesting the storks, thus was an early form of conservation.
201. Probably the New Zealand kiwis, with wings reduced to stubs no more than two inches (5 cm) in length.

Answers

202. The Emperor Penguin. It molts and breeds on annual sea-ice (fast-ice), although two small colonies are exceptions. Some hummingbirds never venture to the ground, but the trees the birds nest or roost in are rooted in the ground.
203. Snails.
204. Albatrosses and condors can soar for hours without flapping, possibly even days in some albatrosses. The albatross depends on steady oceanic winds, whereas a condor requires thermals (rising columns of warm air) and can only fly during the day.
205. Their huge, colorful bills may facilitate reaching fruit not accessible to other fruit-eating birds. The exact function is not yet adequately explained.
206. Bones may be dropped from great heights to break them open.
207. Most birds have vestigial tongues, but flamingos, penguins, some alcids, woodcocks, snipe, waterfowl, crossbills, woodpeckers, hummingbirds, sunbirds, parrots, Lammergeier, honeyeaters, white-eyes, larks, swifts, some owls, and some grebes and loons all have functional tongues.
208. A secretion produced in the crop of pigeons and doves fed to chicks for the first few days after hatching.
209. The Sooty Tern. Some authorities indicate the terns *may* remain airborne from 3 to 10 years, depending on when breeding commences. Feeding birds pick food from the surface of the water while hovering, and they undoubtedly doze on the wing, but sightings of resting terns on the ground are becoming increasingly common.
210. The eastern race of Yellow-bellied Sapsucker migrates to the West Indies and south to central Panama.
211. The Kea, a large, alpine, terrestrial New Zealand parrot. The inquisitive birds are partial to rubber and pull to bits windshield wipers, windshield rubber shields, boots, tents, packs, ski roof racks, car aerials, parkas and wiring.
212. Bird temperatures are generally higher than mammals, ranging between 100-113°F (37.7-45°C).
213. Sandgrouse have traditionally been linked to pigeons, but they *may* be more closely related to shorebirds.
214. The true swifts, and even their family name Apodidae means *lacking feet.*
215. No, owls are far more dependent on their acute hearing. Even so, some species can discern prey in 1/100th's the light intensity that humans require.
216. The phalaropes, jacanas, hemipodes, Emu, cassowaries, kiwis, frigatebirds, skuas, jaegers, boobies, as well as many raptors and shorebirds. Some bustards, woodpeckers, cuckoos, hummingbirds, barbets, cotingas, manakins, birds-of-paradise, bowerbirds, and the Madagascar Cuckoo-Roller also exhibit varying degrees of reverse sexual dimorphism.
217. The oil gland. Located just above the tail, the oil from the gland is used during preening to dress the feathers and aids in keeping them waterproofed. Some authorities have challenged the waterproofing aspect, but the glands are generally larger in aquatic species. The oil secretion may have some affect against ectoparasites.
218. Eighty skins, possibly 20 or more skeletons, and about 75 eggs housed in various collections around the world.
219. The Rhinoceros Auklet. The record burrow was 26 feet (7.9 m) in length.
220. Probably the Australian Emu and the Erne, a sea-eagle.
221. The extinct Passenger Pigeon. The population conceivably exceeded ten billion birds, possibly accounting for up to 40 percent of the total population of North American birds.

Some accepted estimates of the number of birds in a single flock alone exceed two billion. Passing flocks obscured the sky, literally blocking out the sun. The pigeon may have been the most numerous bird of all time, and its total elimination within such a short period of time is almost beyond comprehension.

222. The Laughing Kookaburra and Indian Peafowl.
223. The cock Common (Ring-necked) Pheasant with a 21-inch-long tail (53 cm).
224. The Snowy Egret. Its bright yellow feet are very conspicuous because of its black legs.
225. The Guadalupe Caracara. Mexican goat-herders of Guadalupe Island believed the raptors killed and fed on baby goats. The last recorded specimen was shot by the famous bird collector Rollo Beck on December 1, 1900.
226. No, some reptiles also have nictitating membranes. The protective semi-transparent membrane closes across the eye from the bill toward the ear.
227. Probably Zavodovski Island, South Sandwich Islands. The Chinstrap Penguin colony there has been estimated to consist of up to 14 million birds. Such an estimate may not be realistic, but certainly in excess of 500,000 pairs breed there. The remote island is an active volcano where the warm ground precludes snow from accumulating, making it a prime breeding site.
228. The male Scissor-tailed Flycatcher. The 11 to 15-inch-long bird (28-39 cm) has a 9-inch-long (23 cm) tail. The Fork-tailed Flycatcher, a casual visitor to the eastern U.S., has an even longer tail, and the 10.75-inch-tail (27 cm) represents about 77 percent of its total length of 14 inches (35.6 cm).
229. Probably the New Zealand kiwis.
230. Females of polyandrous species breed with more than one male. Cassowaries, rheas, most tinamous, Harris' and Galapagos Hawk, Pale Chanting Goshawk, Red-legged Partridge, buttonquail (hemipodes), Tasmanian Native-hen, jacanas, painted snipe, some shorebirds, phalaropes, Black Coucal, and at least one honeyeater (Noisy Miner) are all polyandrous.
231. The Laysan Albatross, but the name is also applied to the Black-footed Albatross. They are sometimes known as the White Gooney and Black Gooney respectively.
232. The Mute Swan. Large males weigh nearly 50 pounds (22.7 kg), but many are essentially domesticated, and such abnormally heavy swans are probably incapable of flight.
233. No. Between 1987 and 1991 all California Condors were in captivity. Shortly after the last California Condor was removed from the wild in 1987 for captive propagation, female Andean Condors were released to function as a surrogate species.
234. The Gray Fantail, Rifleman, Tui, Kea, New Zealand Pigeon, Morepork Owl and Takahe respectively.
235. When boiled down, Oilbird chicks yield an odorless and durable oil used in cooking and lighting. The oil was especially valuable because it kept for months without turning rancid. The extremely fatty condition results from the chick's principal item of diet, the palm nut, from which commercial palm oil is obtained.
236. A huge Bald Eagle nest at Vermilion, Ohio was occupied for 35 consecutive seasons. Had the tree not ultimately collapsed because of the weight, the nest undoubtedly would have been used longer. When the 4000-pound (1814 kg) nest crashed to the ground during a storm, the eaglets within were killed. Some Osprey nests are said to have been continuously occupied for more than 40 years, and some Peregrine Falcon *sites* have been occupied for at least 50 years (falcons don't construct nests).
237. The California Condor at 55-60 days, and the Northern Fulmar at 51-55 days.

238. Possibly some of the arboreal hornbills. The female is sealed in a cavity while the male feeds her and the chicks through a slit. The incubation and rearing periods are very long and during this time, the female may undergo a complete molt. When the imprisoned female breaks out, the provisioning male is tattered and exhausted and reportedly, some diligent males have actually died as a result of overwork. Hornbills are sealed in to protect the chicks and the vulnerable female that is possibly flightless during the molt from predators, especially roving bands of monkeys.
239. Turning assures that the egg is incubated evenly throughout. It may also prevent the embryo from sticking to the side of shell. The Old World palm swifts do not turn their eggs because the eggs are glued firmly to the nest. The nest consists of a few feathers glued to the inner side of a broad leaf and even if the leaf turns upside down in blowing wind, the glued eggs remain secure. The megapodes, possibly the White (Fairy) Tern, potoos and kiwis don't turn their eggs either.
240. Birds that are hatched as helpless, blind, naked chicks. Therefore, they require much parental care, and remain in the nest for some time after hatching.
241. Yes, but the primaries have been reduced to 5-6 glossy, black, vaneless, rigid, hollow primary quills up to 11 inches long (27.7 cm). The bare quills curve down and under the ratite's body, apparently protecting its flanks as it crashes through the dense jungle vegetation. Cassowaries lack tail feathers.
242. No. The long trailing feathers that exceed five feet (1.5 m) in length in adult males are not tail feathers, but rather are upper tail coverts (the feathers that cover the actual tail).
243. A gannet or a booby (family Sulidae).
244. The extinct Giant Moa of New Zealand stood approximately 12 feet (3.7 m) tall and weighed about 500 pounds (227 kg). Other extinct birds weighed more, but were not as tall.
245. About 100 billion.
246. The male Crested Argus Pheasant, with its tail feathers that may reach seven feet (2.1 m) in length. Some authorities state that the cock Reeve's Pheasant has the longest tail, up to eight feet (2.4 m) long, but I am not aware of any *documented* records to support this.
247. If the mound temperature becomes too warm or cool, some megapodes (generally the male) scrape off or add material to adjust the temperature. Exactly how the birds are able to determine the correct temperature is not fully understood. Nesting crocodilians also maintain a constant incubation mound temperature, and how they accomplish this is not known either.
248. The storm-petrels. Foraging birds dangle their long, thin legs, giving the appearance of walking on water. The name petrel itself was derived from St. Peter who, according to the Bible, walked on water.
249. Divers.
250. The Galapagos Sharp-beaked Ground-Finch pokes a small hole in the base of a developing flight feather of a molting Masked or Red-footed Booby to obtain the blood.
251. The Anhinga. The name is of South American Tupi-Guarani Indian origin meaning *snake-bird.* The bird frequently swims submerged with only its long, thin neck and small, snake-like head visible, rather resembling a snake.
252. The Wood Stork.
253. The jacanas, because of their ability to walk on floating vegetation.
254. The Asian Red Junglefowl was domesticated over 5,000 years ago, and geese over 4,000 years ago.

Answers

255. The African Secretary-bird.
256. The nearly flightless Kagu, of which only 500-1,000 remain. The unique bird is currently restricted to the large forests of the southern third of the island. Kagus often fall prey to marauding dogs.
257. The Eurasian Great Bustard. Males average about 37 pounds (16.9 kg), with wingspans of eight feet (2.4 m). Exceptionally heavy males weigh as much as 46 pounds (20.8 kg), possibly even over 50 pounds (22.7 kg). Some authorities suggest that the African Kori Bustard is heavier.
258. Jaeger is a German word for hunter, but the name may have originated even earlier from the Norse word *jaga.*
259. Turacin, a water-soluble copper pigment. Some people believe the red coloration of a turaco's wing will run in the rain, but this is doubtful.
260. A terrestrial cuckoo.
261. The hawksbill turtle.
262. Mousebirds are peculiar African birds that scurry about like mice in trees, and have remarkably fine hair-like feathers.
263. Most neotropical motmots have prominent racket tails. The tail feathers develop normally, but preening birds reportedly pick out the weakened vanes of the central feathers, causing the distinctive racket tail.
264. The sunbirds. This group of generally colorful, nectar-feeding passerines occupies the same ecological niche, but sunbirds are passerines that are not even remotely related to hummingbirds.
265. A thrush.
266. Isla Rasa in the midriff region of the Sea of Cortez (Gulf of California), between Baja California and the west coast of mainland Mexico. Up to 95 percent of the population breeds on the tiny island.
267. The hummingbirds. At least twice their body weight in food, primarily nectar and tiny insects, is required daily. With the possible exception of shrews, the tiny birds have the highest metabolic rate of any animal.
268. The female of the nominate race of the Cotton Pygmy-goose weighs only 6.5-9.0 ounces (185-255 gms). The female African Pygmy-goose is second at 9.1 ounces (260 gms).
269. The casque might make it easier for the huge flightless birds to force their way through the dense jungle because it acts as a shock absorber. The casque is also used like a shovel to turn over loose sand and debris when the birds are foraging.
270. Birds have no teeth, and their hollow bones are fragile. Only teeth and solid bones undergo fossilization.
271. The vulture has a long, thin tongue shaped like a narrow trowel used to scoop marrow out from the bone core.
272. The Golden-crowned Kinglet is only 3.25-4.25 inches (8.3-10.8 cm) long, weighing one-fifth of an ounce (5.7 gms).
273. The Egyptian Vulture.
274. The ducks inhabit noisy, rushing waters, and their high-pitched calls are clearly audible over the tremendous background noise of the rapids.
275. A flightless bird with a flat (non-keeled) breastbone. The only surviving ratites are the Ostrich, rheas, cassowaries, Emu, and kiwis. Ratite comes from the Latin *ratis,* meaning a "raft," implying flat.

Answers

276. A cormorant. The name refers to the shaggy head of breeding adults, and is generally applied to a totally marine cormorant that does not nest in trees.
277. Hall's (Northern) and the Antarctic (Southern) Giant-Petrel. When disturbed on the nest, agitated petrels may vomit up copious amounts of stinking stomach contents, hence the descriptive name stinker. The birds are both scavengers and predators.
278. New Zealanders are typically referred to as *Kiwis.*
279. Old World vultures are more closely related to hawks and eagles than to American vultures, and they have a greater ability to grasp with their feet. The less powerful feet of New World vultures are not designed for grasping, and they have considerably weaker bills with perforated nostrils.
280. The Australian Duck-billed Platypus, a bizarre mammal with a duck-like bill that lays eggs requiring incubation.
281. The Peregrine Falcon.
282. The European (Common) Swift may spend up to three years in the air prior to breeding for the first time. It eats, sleeps, drinks and sometimes even copulates on the wing.
283. The name came from a copy of *The Birds of the West Indies* by James Bond. Fleming noted the field guide on his bookshelf as he was wrestling with the name for his famous fictional British superspy.
284. The Little Blue (Fairy) Penguin in Australia, especially Tasmania. Some penguins are also road-kill victims in the Chatham Islands, east of New Zealand.
285. *Greater* Antarctica with only 11 species.
286. Eggs are covered with their large, highly vascularized feet because the birds have poorly defined brood-patches (or lack them altogether). Pelicans also position eggs on top of their feet.
287. Incubating male Emperor Penguins fast for 110-115 days, but may occasionally extend this to 134 days (nearly 4.5 months).
288. The huge Hyacinth Macaw of South America and the Papua-Australian Palm Cockatoo.
289. The popular belief that its name originates from its resemblance to an old-time secretary with a quill pen stuck behind his ear has recently been convincingly challenged. The name is possibly derived from the Arabic *saqr et-tair* (*saqr* = hunter or hawk; *tair* = flight, or is a collective term for bird). This argument is even more plausible if, as seems probable, the name was first corrupted into French as *secretaire*.
290. The Whooping Crane stands about five feet tall (1.5 m).
291. The enlargement on the upper mandible may function as a shock absorber when a hornbill pounds on a tree to create a nesting cavity.
292. Dumpy little ducks with very stiff tail feathers. The most familiar American stiff-tail is the Ruddy Duck.
293. The entire population was vaporized when Isla San Benedicto (Mexico) erupted in 1952.
294. Indiscriminate logging of southern hardwood forests and large dead trees. The bird might be considered a victim of over-specialization, whereas the similar Pileated Woodpecker, a far more adaptable close relative has survived, as it thrives in softwood forests, such as pine.
295. It *reportedly* has an incredibly short incubation period of only 10.5-11.0 days. Chicks hatch in a very underdeveloped state with closed eyes, almost naked pink skin, and rudimentary feet.

Answers

296. The tips of an owl's flight and contour feathers lack the structures that hold the barbs together, thus sound is deadened. This modification is greatly reduced, or even absent, in some diurnal owls that have no need for silent flight.
297. The spectacular birds-of-paradise.
298. Whistling-ducks. While many do perch in trees, the ducks often call with loud sibilant whistles.
299. The megapodes (mound-builders), in which the well-developed chicks are completely independent at hatching. The young of obligate brood parasitic-nesting birds, such as some cuckoos, don't see their biological parents either because the chicks are reared by foster parents.
300. Up to 50 pounds (22.7 kg), and it stood up to four feet (1.2 m) tall. Some researchers argue that the Dodo was much more slender than the traditional image and probably weighed closer to 26.5-33.0 pounds (12-15 kg).
301. Probably the African Social-Weaver. Its huge communal nest contains up to 300 separate nest chambers.
302. Falconry has traditionally been called the sport of kings, but in current usage it generally refers to horse racing.
303. The flightless, nocturnal Kakapo (Owl Parrot) of New Zealand. Males weigh as much as 7.5 pounds (3.4 kg).
304. *Archaeopteryx* has long been accepted as the earliest avian fossil, a species that existed 130-140 million years ago. However, in 1984, in the badlands of western Texas, a fossil proto-type bird believed to be 75 million years older than *Archaeopteryx* was discovered in 225-million-year-old rocks. Named *Protoavis* ("first bird"), it, unlike *Archaeopteryx*, apparently had hollow bones, and only the distal portion of its jaws were armed with teeth.
305. The Osprey.
306. Greater albatrosses do not begin nesting prior to their seventh year, and generally not before their 9th-11th year.
307. The three skimmers. The birds skim the surface of the water with their blade-like lower mandible, snapping down on any fish or crustacean that comes in contact with it.
308. The Marbled Murrelet nests as high as 148 feet (45.1 m) in large coastal coniferous trees, as far inland as 29 miles (47 km) in Oregon. Very few species, much less a seabird, nests as high in trees.
309. At least 90.
310. The African Crowned Hawk-Eagle.
311. The three New Guinea crowned-pigeons. All are the size of a large chicken or small hen turkey.
312. A name imitating a bird's call. The Indian name of *chachalaca,* for example, is an imitation of the cracid's reverberating vocalizations.
313. The Greater Roadrunner.
314. Hawks in the genus *Buteo*. The raptors are not closely related to the seven American vultures. New World vultures are in a separate family, and it is incorrect to call them buzzards.
315. The drake Ruddy Duck with its brilliant powder-blue bill. The bill color of the male Masked Duck is equally bright, but the duck is not a North American resident, although it is an occasional vagrant.
316. The South American Coscoroba Swan (*Coscoroba coscoroba*).

Answers

317. The Black-billed Magpie.
318. The South African Jackass Penguin's loud, braying vocalizations are donkey-like.
319. The birds sit on exposed branches with their neck feathers puffed out, projecting a very dumpy or puffy appearance.
320. The extinct Elephant-bird of Madagascar.
321. The South American Monk (Quaker) Parakeet.
322. The name alludes to their spectacular aerial acrobatics.
323. The Canvasback.
324. A thrush.
325. The South American Oilbird feeds almost exclusively on palm nuts gathered at night.
326. The Upland (Magellanic) Goose. Considered a competitor for grass by South American and Falkland Islands sheep ranchers, bounties were offered and the geese were slaughtered by the hundreds of thousands. Despite this, the birds thrived, and may even have increased because of improved habitat that resulted when virgin southern beech (*Nothofagus*) forests were cut for pasture.
327. A fish-eagle.
328. All feathers are molted simultaneously, thus the birds are not waterproof and cannot enter the sea to feed.
329. Swallows have 12, rather than 10 tail feathers, nine, rather than 10 primaries, and swallows have rictal bristles that swifts lack.
330. The storks.
331. As the most aquatic of swans, the birds are barely capable of terrestrial locomotion. Carrying young enables adults to simultaneously brood offspring and feed in the water.
332. Until recently, the record holder was a New World condor-like vulture with a wingspan of 16-17 feet (4.9-5.2 m). In 1979, a new species was unearthed in Argentina with a wingspan of possibly 25 feet (7.6 m). The bird could have weighed 250 pounds (113 kg), with a length of 11 feet (3.4 m). Both of these extinct birds were soaring species and were probably not capable of sustained flapping flight.
333. A highly specialized behavior of some birds, mainly passerines, where parts of the plumage are treated with ant body fluids, which includes formic acid. The exact function of anting is poorly understood, but formic acid may be beneficial in combating ectoparasites. Anting may also assist in feather maintenance, especially of the wings.
334. Loons are practically helpless ashore. When ashore, a loon rests its weight entirely on its breast and "frog-leaps" for a limited distance, kicking its legs out backwards. Swifts and hummingbirds are also essentially incapable of walking.
335. Ostrich males stand eight feet (2.4 m) tall and may weigh in excess of 300 pounds (136 kg). The record holder was a 9-foot tall (2.7 m) domesticated cock that weighed nearly 350 pounds (159 kg).
336. Sprig is the American hunter's name for the Northern Pintail.
337. Only two: Ocellated and Wild Turkey.
338. The name alludes to the Limpkin's distinctive halting, limping gait.
339. Some of the smaller sandpipers.
340. On April 19, 1987, the last remaining California Condor was captured for captive propagation. The wild population dwindled from 50-60 birds in 1971, to a mere six by 1986. In October of 1991, two captive-reared condors were hacked back to the wild and released in January, 1992. Additional condors were subsequently released.

Answers

341. Shrikes typically impale prey on a suitable spike, such as on a barbed wire fence, twig or thorn.
342. The Ruby-throated Hummingbird.
343. The Slender-billed (Short-tailed) Shearwater in Australia and the Sooty Shearwater in New Zealand, where downy chicks are harvested in huge numbers for food.
344. The Waved Albatross (Galapagos Islands).
345. The Masked (Blue-faced) Booby weighs up to 4.8 pounds (2.2 kg).
346. Possibly the Red-winged Blackbird. Some winter roosts consist of more than a million birds. The overall population undoubtedly exceeds 100 million, but the Common Grackle may be as numerous.
347. The Galapagos Penguin lives astride the equator.
348. Hawaii and mid-Pacific islands, necessitating a non-stop flight of nearly 2,500 miles (4,023 km), one of the longest, uninterrupted flights for a land bird.
349. Coots, ibises, screamers, Limpkin, gulls, rails, herons, and even caracaras and Snail Kites serve as host species.
350. The neotropical raptor favors snakes.
351. Yes, both are in the same order (Piciformes).
352. The male Emperor Penguin incubates the single egg on the top of its feet continuously without relief for 64-67 days. The male kiwi incubates longer, but departs the nest to feed.
353. About 1790, from England, not Asia where it is native.
354. A chisel.
355. The Chinstrap Penguin, because its loud, raucous calls sound like stones being smacked together.
356. Large plovers with bony carpal spurs or knobs, crests and facial wattles.
357. The birds spin in tight circles on the water to stir up aquatic life and bring food particles to the surface.
358. The skua. While related to gulls, the extremely aggressive predatory birds are behaviorally much more like raptors.
359. A dark tropical tern with a rounded tail. Most terns are essentially white with forked tails.
360. Neither sex. Megapodes (incubator-birds) bury their eggs in a mound of rotting vegetation to be incubated by natural fermentation (microbial heat generation).
361. Cuckold: a man whose wife has committed adultery.
362. The frogmouths and the fruit-eating Oilbird.
363. Its bill is so long and heavy that the bird is obligated to hold it upward at about a 45-degree angle.
364. Florida in the New World and Spain in the Old World. Based on midden remains, the species was widespread several thousand years ago, occurring off the coasts of temperate North America, the British Isles and Scandinavia.
365. Small 2-3 mm long projections that grow out from the sides of each toe of some grouse during the winter. The pectinations function as snowshoes by increasing the foot surface area, enabling the birds to walk more effectively over snow.
366. The rollers. The name came from the pale, circular, silver dollar-sized patches on the wings of the Broad-billed Roller, that in flight appear almost transparent.
367. The frigatebirds.
368. The Horned Puffin.

369. The Emperor Penguin can dive to at least 1,772 feet (540 m), and one bird remained submerged for 18 minutes. King Penguins may dive to depths of at least 1,000 feet (305 m). Mammals, however, dive much deeper than any bird; the elephant seal can dive to depths exceeding 5,000 feet (1,524 m) and the sperm whale to depths approaching 10,000 feet (3,048 m).
370. The Anhinga, because when fanned, its long, stiff tail resembles a turkey's tail.
371. The protective, semi-transparent nictitating membrane.
372. The petrels have a distinctive flight style of skimming just above the water on set wings.
373. From the weird calls of a loon, especially its cacophonous, maniacal laughter.
374. The circumpolar Rockhopper Penguin.
375. The conspicuous, separated, external nasal tubes of an albatross are not joined on top of the bill as in petrels.
376. The three New Zealand kiwis.
377. The ratites (Ostrich, Emu, cassowaries, rheas and kiwis). Their closest relatives are probably the two rheas. Tinamous resemble gamebirds and a close relationship with the ratites appears far-fetched. However, appearances can be deceiving, and some taxonomists have placed tinamous and ratites together in the same order.
378. Between 30 and 40: approximately 2 vs. 90 pounds (0.9-41.0 kg).
379. Female Southern (Double-wattled) Cassowaries weigh up to 130 pounds (59 kg). The closely-related Emu is often credited with being the second largest bird, but female Emus average about 100 pounds (45.4 kg), although they may reach 120 pounds (54.4 kg).
380. The dippers.
381. The Northern Cardinal: Illinois, Indiana, Kentucky, North Carolina, Ohio, Virginia and West Virginia.
382. The Cattle Egret and the wide-ranging Arctic Tern. The egret has been sighted on a number of occasions in the Antarctic, but it has never became established.
383. The cranes. Courting cranes perform some of the most beautiful and elaborate of avian dances. Dancing cranes are given to sudden outbursts of flapping, bowing, leaping and prancing on the ground.
384. A communal display area or arena where males gather to attract and court females that come there to be bred. Birds using leks include some shorebirds, of which the Ruff is the most notable, some grouse, such as the Sage Grouse and Prairie-chicken, Great Bustard, Owl Parrot, cocks-of-the-rock, some neotropical manakins, some birds-of-paradise and hummingbirds and Jackson's Widowbird.
385. Their distinctive crests recall an umbrella.
386. The Marbled Murrelet.
387. Birds with a keeled breast bone or sternum. All surviving birds, except for the flightless ratites, are carinates. Carinate is from the Latin *carina,* meaning a "keel."
388. No, not since the Rose-throated Becard, formerly considered a cotinga, was shifted into the tyrant-flycatcher complex.
389. The Chubut (White-headed) Steamer-Duck, in 1981. The flightless duck is restricted to coastal Chubut, Argentina.
390. About 2,000 B.C., in China.
391. Europeans believed the birds lacked feet. Skins prepared by New Guinea natives lacked feet, so it was assumed the birds must be from paradise.
392. A dark-color morph of the Lesser Snow Goose.

Answers

393. Clark's Nutcracker and the Gray Jay.
394. Indiscriminate logging of virgin forests.
395. Due to their feeding style, calmer waters are required. In high winds or if the water is choppy, Lesser Flamingos gather along the shore passing the day resting or preening.
396. Swan young when still in down.
397. The legendary Cahow that disappeared in the early 1600's, not to be seen again until 1916, following an absence of nearly three centuries. In 1951, 18 (some accounts indicate six) pairs were discovered nesting on five small islands (Castle Rocks) off Bermuda.
398. The European race of the Common Merganser. Its name refers to its goose-like size.
399. A bird that feeds primarily on seeds or grain.
400. The nest of a raptor, usually that of an eagle or falcon.
401. Yes, most species do, but the Australian Malleefowl often builds a new mound annually.
402. The Oilbird. The name is of Indian origin, referring to the bird's weird, deafening vocalizations. The name loosely translates to *howler* or *one who cries and laments.* The loud calls are entirely different from the echo-locating clicks emitted by the cave-dwelling birds.
403. Four: Black-chinned, Calliope, Ruby-throated and Rufous Hummingbird.
404. The Canada Goose. The current estimate of the total number of individuals in the various races is approximately 4.8 million birds. Until recently, the Lesser Snow Goose was more numerous with a population of 3.5 to 4.0 million geese in good years.
405. The hornbills and cassowaries. Other birds, such as some cracids and megapodes, also have casques, but large casques are not typical of the families.
406. The Reddish Egret.
407. The fleshy covering of the proximal portion of the upper mandible of some birds that contains the nasal openings, such as in raptors and parrots.
408. The Pied-billed Grebe primarily, but hell-diver is also a local name for the Common Loon, Horned Grebe and Bufflehead duck. When shot at, these birds dove so quickly, and remained underwater so long, that hunters concluded that they had dived to hell.
409. The Common Eider, Lesser Snow Goose and Bar-headed Goose. The latter is the most gregarious nesting goose. On one island of 3,000 square meters, 40-50 pairs bred, with nests only 1.3-4.9 feet (0.4-1.5 m) apart. Ross' Goose and the Barnacle Goose are also quite colonial in nesting.
410. Only two: the widespread Black-legged Kittiwake and the much less abundant and more localized Red-legged Kittiwake.
411. To increase reproductive potential. A laparotomy is a minor surgical procedure used to sex birds where both sexes are identical in appearance. The procedure consists of a 1-2 mm incision between the ribs and the use of a powerful light to peer into the body cavity to examine the gonads.
412. An assemblage of downy young from more than one pair, still dependent on their parents. Adults only feed their own chicks (except in waterfowl where young feed themselves). The young of some penguins, Common Eiders, some shelducks, sheldgeese, northern geese, white pelicans (as well as ground-nesting brown pelicans), flamingos, Pied Stilts, and some crested terns and Kelp Gulls all gather in crèches.
413. The South American Coscoroba Swan.
414. The incubation period of this Australian megapode *may* range from 50 to 96 days. If so, it would have the longest avian incubation period.

Answers

415. An Old World term referring to a rookery of murres.
416. Small Old World grebes.
417. The Anhinga. The vernacular name darter describes its darting or stabbing behavior, and the Old World species are known as darters.
418. The Eclectus Parrot of Australia, New Guinea and the Solomon Islands. Males are green, but females are reddish-maroon. It is among the few parrots where females are more beautiful than males.
419. The dull, female-like plumage stage of many northern duck drakes during the summer. It is strikingly different from the colorful breeding (winter) plumage.
420. Seventy-five condors, most of which were maintained in captivity at the San Diego Wild Animal Park and the Los Angeles Zoo. Captive propagation commenced at the Wild Animal Park in 1989 when four chicks were successfully reared. Young were subsequently reared in Los Angeles, and by 1993 second-generation chicks were being produced. A pair of captive-reared condors was released into the wild in January of 1992, but the male was found dead on October 8, following ingestion of a toxic substance (ethylene glycil, a fluid used in anti-freeze). Six additional condors were released in October of 1992. Two of these were killed within three weeks of one another in June of 1993, after colliding with high-voltage power lines near Fillmore, California. Another male was found dead on October 30, 1993, apparently also a victim of a power line collision. Consequently, the four remaining condors in the wild were captured and along with five more captive-hatched birds were moved to Lion Canyon in the Los Padres National Forest, about 110 miles (177 km) northwest of Los Angeles. The site is isolated and no power lines are within reasonable reach. The nine condors were released in December, 1993. Additional condor captive-breeding facilities are available at the World Center for Birds of Prey in Boise, Idaho, and another is planned in Bartlesville, Oklahoma. In September of 1993 six pairs were sent to Idaho, and the young produced there are scheduled for release in Arizona and New Mexico. If the population ever reaches 450 birds, the California Condor will be removed from the endangered species list and upgraded to a threatened species.
421. Sandgrouse carry water in their breast feathers, presumably to increase the humidity on their eggs when incubating. Males, and to a lesser extent females, have specially modified abdominal feathers that have a much greater capacity to hold water than the remainder of the plumage. Chicks drink by squeezing and stripping the water-laden breast feathers.
422. The last day of June, 1940, Perry River Delta, Northwest Territories, Canada.
423. Despite complete protection, the population still probably does not number more than 50 breeding pairs. In 1991, 41 pairs nested on the small islands off Bermuda (Castle Rocks).
424. To equally distribute their weight, enabling the birds to walk on floating vegetation.
425. Supposedly on foot, both uphill and downhill.
426. The Common Poorwill, the first of which was discovered in a rock crevice in the Chuckwalla Mountains of California's Colorado Desert in December, 1946. Its temperature was only 64.4-68.0°F (18-20°C), well below the normal temperature of 104-106°F (40-41°C). The nightjar was banded and found in the same location in subsequent winters. It has been calculated that normal fat deposits can sustain a torpid poorwill for 100 days at a body temperature of 50°F (10°C), a drop of well in excess of 50°F (27.7°C) from its normal temperature.
427. In 1988, the breeding population was estimated at 21 pairs, with an overall total of about 65 birds.

Answers

428. The American Elf Owl.
429. According to some authors only the skimmers, but most experts indicate that both sexes incubate. The only other seabird in which a single sex incubates is the Emperor Penguin, but it is the male that incubates.
430. Five: Tennessee, Texas, Florida, Mississippi and Arkansas.
431. The Galapagos Flightless Cormorant.
432. Nesting was not documented until 1989 (Red-billed Tropicbird).
433. *Hawkeye,* but the last of the Mohicans was *Chingachgook.*
434. Yes, but true diving is generally restricted to the four diving-petrels. Most petrels feed from the surface, occasionally dipping under, but only the diving-petrels make a living underwater, swimming with their wings rather than their feet. A South Georgia Diving-Petrel dove to a depth of 160 feet (48.6 m).
435. Yes.
436. Over a year is required to fledge a King Penguin chick in the wild, but hand-reared captive young grow rapidly, fledging in five months or less if fed daily. The obligatory fasts of wild chicks obviously greatly prolongs the fledging period.
437. Yes.
438. Yes.
439. No, its narrow, graduated, stiff tail is quite long.
440. Probably the Sacred Ibis by the ancient Egyptians.
441. A small specialized woodpecker. It seasonally feeds on sap obtained from special pits excavated and maintained in certain trees, as well as on insects attracted to the sap.
442. Yes, the cocks can.
443. Some swifts and hummingbirds.
444. Stone curlews.
445. The female Steller's Sea-Eagle.
446. Terns, probably because of their forked tails.
447. Up to 50 percent of their weight. Approximately 22 pounds (10 kg) of an 83-pound (37.5 kg) Emperor Penguin consists of fat, its sole source of fuel during the extended fast.
448. The jacanas.
449. Bonxie is an Old World name, specifically Scottish, for the Great Skua. It apparently means an *old fat woman.*
450. A patch of distinctive, usually metallic, color on the wing, especially in some dabbling-ducks.
451. The shrike.
452. The little New Zealand ratites have a more highly developed olfactory sense than most birds, and nostrils at the bill tip presumably aids them in locating food at night. Kiwis probe for their food, thus it is advantageous to have their nostrils at the tip of the bill.
453. The two rheas. The name originated from one of their distinctive calls. Courting males call out with a deep, resounding *nan-du* sound, a vocalization that is more like the roar of a large mammal than the call of a bird.
454. Ascension Island, a tiny volcanic island in the mid-Atlantic. The loud, continuous vocalizations of nesting Sooty Terns, (also known as the Wideawake Tern) kept everyone wide awake all night long. In Hawaii, the native name *ewa'ewa* translates to "to make uncomfortable," referring to the incessant screeching cries made by flocks of Sooties at their rookeries.

Answers

455. A large number of seabirds swept ashore or inland as a result of a severe storm.
456. The Carolina Duck.
457. Hornbill ivory is highly valued for carving.
458. When a number of generally smaller birds engage in a social attack and dive continuously on a larger bird considered a threat, the behavior is known as mobbing. Mobbing birds typically swerve away without striking the larger bird, but not always. A Snowy Owl was killed by mobbing Arctic Terns, and an Osprey was killed by mobbing frigatebirds.
459. Yes, some cormorants apparently do.
460. A term describing the behavior of a raptor flying in search or pursuit of prey. The term also applies to insectivorous birds that sally out from a perch to capture flying insects. The practice of flying trained raptors at living prey is also known as hawking by falconers.
461. The two oxpeckers in the starling family. Also known as tickbirds, both feed almost exclusively on the ectoparasites of large African mammals.
462. The toucans, but the two groups are not related.
463. A species active at twilight, such as a nightjar.
464. As great as 1,000 beats per minute, but it could double if a hummingbird is excited or engaged in sustained flight.
465. It feeds extensively on wild celery buds, a fact reflected in its scientific name: *Aythya valisineria.*
466. The Snow and Antarctic Petrel. Both may nest several hundred miles (322 kilometers) inland, and the Snow Petrel may nest as high as 6,500 feet (2,000 m) at some inland sites.
467. The Harlequin Duck and Steller's Eider.
468. Mutual preening, generally by a pair of birds.
469. The mergansers.
470. Australia.
471. No, it is a New World cuckoo. Three of the four species are black, long-tailed birds with distinctly compressed bills.
472. Mother Carey's Chickens is an old sailor's name for storm-petrels, possibly first applied to Wilson's Storm-Petrel. The name refers to the Virgin Mary (Mater Cara), to whom the protection of sailors was commended.
473. The tropical American birds frequently swing their tails from side to side, like a pendulum.
474. The right avian ovary is vestigial.
475. A downy King Penguin chick, because its long, dense, brown down resembles hair or wool. The down may exceed three inches (8 cm) in length. The huge chicks are so unlike adults in appearance that early seafarers considered them a separate species.
476. A small gregarious toucan.
477. The Baldpate, because from a distance, the creamy-white stripe on the drake's forehead creates an illusion of a balding pate.
478. The Emu and the three cassowaries.
479. The adhesive saliva coating the tongues of some, if not all, ant-eating woodpeckers is alkaline, possibly neutralizing the potent formic acid.
480. A bird that digs a nesting burrow, such as a puffin.
481. Hialeah Racetrack (Miami, Florida).
482. The New Zealand Kea. Between 1890 and 1970, when partial protection was provided, more than 150,000 parrots were killed. Sheep ranchers believed the alpine birds killed stock.

Answers

483. *Wild Turkey, The Famous Grouse*, or *Old Crow*.
484. The Garganey Teal, because the drake utters cricket-like calls.
485. Brown Pelicans are saltwater inhabitants that dive for fish, whereas all white pelicans are partial to fresh water and fish from the surface.
486. The Andean Condor.
487. The mockingbird. Courting birds commence singing loudly long before dawn.
488. The Prairie Falcon.
489. The serrated feather comb is beneficial in preening. Most nightjars, pratincoles, frigatebirds, Crab Plover, barn owls *(Tyto)* and, to a lesser extent, dippers, all possess such serrations.
490. The woodpecker collects acorns and stuffs them into individual holes it has drilled. Some entire trees appear riveted, and these storehouse trees are defended.
491. The parrots.
492. Large, downy Short-tailed (Slender-billed) Shearwater chicks are harvested and sometimes canned in Australia, where the birds are marketed as Tasmanian squab.
493. The Horned Screamer and Horned Guan.
494. The Yellow-billed Spoonbill of Australia.
495. The Fulvous Whistling-Duck.
496. Probably the Osprey, but the cosmopolitan Barn Owl is also a major contender.
497. They were named for their peculiar oven-shaped mud nests.
498. Robert Stroud, a lifer in Alcatraz Prison, who wrote a book on bird diseases. Burt Lancaster played him in the movie.
499. Terrestrial carnivores, such as Polar Bears and Arctic Foxes, constitute a major threat to flightless birds. All penguins inhabit the Southern Hemisphere, and it is unlikely that they could cross the warm, relatively unproductive waters of the tropical doldrums to reach the Arctic.
500. The Harlequin Duck.
501. About 11 mph (17.7 km/hr).
502. Flamingos are behaviorally similar to waterfowl. They share some anatomical similarities with storks and ibises, but their egg-white protein places them closer to herons. Based on recent work, a possible relationship with the shorebirds, particularly the stilts, is possible. In the most recent taxonomic revision, flamingos have been elevated to a separate order (Phoenicopteriformes).
503. The California Condor. In the wild, it typically breeds only every other year. By removing eggs, captive birds have been induced to lay up to three eggs a season.
504. Males are known as cobs, whereas females are called pens.
505. Seven.
506. It is basically gray in color and tends to lag behind other geese during northern migration.
507. An Old World sandpiper.
508. The Willow and Rock Ptarmigan.
509. A small, long-winged Old World falcon that feeds mainly on insects.
510. Blue. The blue color results when seemingly colorless light is broken apart as it strikes the feather structure that acts as a prism. Some colors are absorbed while others are reflected. Absorption or reflection depends on feather structure, and most avian colors result from a combination of both pigments and feather structure.
511. Ringing.

Answers

512. The male incubates, an unusual galliform trait.
513. The last *official* record was of a specimen shot at Long Island, New York, in 1875. Another duck was shot by a small boy at the overflow of the Chemung River, Elmira, New York, on December 12, 1878, but the record is disputed because the specimen was lost.
514. The parrots.
515. The five todies in the family Todidae.
516. The Brown Noddy. It breeds on Bird Key, Dry Tortugas, Florida, building nests up to 12 feet (3.7 m) above the ground in vegetation.
517. Colorful Asian pheasants.
518. All have extremely long central tail feathers.
519. The petrels and albatrosses, both of which have conspicuous external nasal tubes.
520. Yes, but diurnal feeding is more typical.
521. The alcids: auks, auklets, murres, guillemots, puffins, etc. All 22 surviving species can fly (the extinct Great Auk was flightless) and are not closely related to penguins, but they occupy a similar ecological niche. The two widely separated geographical groups illustrate a classic example of parallel or convergent evolution.
522. The Velvet Scoter.
523. Possibly because the Arctic owls feed primarily on lemmings, and the small rodents do not inhabit Iceland.
524. The female Calliope Hummingbird is the smallest bird *nesting* north of the Rio Grande River. The smaller Bumblebee Hummingbird occasionally strays into Arizona.
525. The American Woodcock.
526. Macquarie Island, south of Australia. In recent years, a few breeding pairs have been discovered in the Falkland Islands.
527. The flightless Greater Rhea of South America. Large cocks weigh in excess of 55 pounds (25 kg).
528. As a result of mounting concern over the excesses of the plume-hunters, President Teddy Roosevelt designated Pelican Island on the east coast of Florida as the first National Wildlife Refuge in 1903. Before the Civil War, the four acres (1.6 ha) of Pelican Island were covered by thick mangroves supporting thriving colonies of herons, White Ibises and Roseate Spoonbills. The multitude of nesting birds and a severe frost in 1886 killed the mangroves, and plume-hunters subsequently eliminated the birds. The island then became a treeless mudflat taken over by nesting Brown Pelicans. Roosevelt was keenly interested in birds, and by the time he left office, he had created 51 bird reservations on public lands.
529. The drake Oldsquaw may be the most vocal of all waterfowl.
530. The Brown Pelican.
531. A bare area of highly vascularized skin on the lower abdomen where eggs come in contact with the warm skin. If both sexes incubate, both have brood-patches. During the non-breeding season, brood-patches are generally, but not always, absent.
532. The Christmas Bird Count. Participants count as many species and individuals as possible during a 24-hour period within an area of 15 miles (24 km) in diameter.
533. The Bean Goose, because it formerly fed on bean crops.
534. Cock-fighting.
535. A small projection on the tip of the upper bill of chicks that aids them in breaking out of the shell. The egg tooth generally drops off or is absorbed within a few days of hatching.
536. The starling family.

Answers

537. A modified inner secondary. The inner vane of the 12th secondary is greatly enlarged to produce the sail-like structure. When the duck is in flight, the peculiar feathers are flattened against the body.
538. A growing feather filled with blood. The term is generally applied to (but not restricted to) a flight feather. Mature feathers are not living structures, and can be cut without inflicting pain.
539. The Mallard and the neotropical Muscovy Duck.
540. Chipping or starring.
541. About 2.7 percent: 260 out of approximately 9,700 species.
542. A nest into which a number of females dump eggs that are seldom incubated. Whistling-duck dump nests containing more than 100 eggs have been recorded.
543. The Mute Swan. A condor egg was the ultimate for any collector and sold for as much as $2,000 to $3,000 each, an astronomical sum at the turn of the century.
544. The 15 pochards are aquatic ducks and are accomplished divers. Five species nest in North America: Canvasback, Redhead, Ring-necked Duck and Greater and Lesser Scaup.
545. The three South American screamers. Practically all of their bones are hollow, including even the end of the spine and the distal digits of the wings and toes.
546. Shining a powerful light through an egg to determine fertility or the state of incubation.
547. A running or walking species, such as an Ostrich.
548. A yoke-toed foot where two toes are directed forward and two point backward. Parrots, woodpeckers, toucans, barbets and cuckoos all have zygodactyl feet.
549. A domestic bird gone wild, or an escaped exotic.
550. An aberrant African stork, also known as a hammerhead or anvil-head.
551. The alula. Located on the avian thumb, it consists of 2-4 small, stiff quill feathers that can be moved and separated. These are used to control the leading edge of the wing in flight, and help prevent stalling, particularly in raptors.
552. Some of the savanna-dwelling Old World barbets in the genus *Pogoniulus*.
553. When a pair of birds is copulating, the male is said to be treading. The term is most often used for waterfowl mating on the surface of the water.
554. Its conspicuous white wing-patches reminded early sailors of the familiar Common Wood-Pigeon back home, and the petrel was commonly encountered around the Cape of Good Hope and Cape Horn. Its chick even resembles a pigeon squab.
555. Waders.
556. The female Bufflehead averages only 11.5 ounces (325 gms) in weight.
557. No, but there are exceptions. Birds with a fairly highly developed sense of smell include the kiwis, petrels, Oilbird, Turkey Vulture (and possibly the Yellow-headed Vulture), honeyguides, and perhaps even some grebes, cranes and albatrosses, and the Black-billed Magpie.
558. Emperor Penguins are annual breeders. If breeding commenced in the austral spring, the limited time available would be insufficient to rear a chick because of the long incubation and fledging periods. Ideally, chicks should be fledged at the optimum time of the year when the weather is best, food is most abundant and the annual fast-ice edge has receded close to the rookery, thus eliminating a long trek to open water.
559. The Dusky Seaside Sparrow (a dark race of the Seaside Sparrow) of the east coast of Florida. The last surviving male, a captive, died in 1987. In 1975, five females with young and 47 males were counted, but in December of that year, fire destroyed the limited

remaining habitat, and a female has not been seen since. In 1980, the population dropped to just six males, five of which were captured with the intention of breeding them with the closely related Scott's Seaside Sparrow. Unfortunately, all died before the project really got underway.

560. A chat.
561. Although regarded as a great delicacy in Asia, the saliva nests have no nutritional value whatsoever, but the Chinese believe that because the cave swiftlets fly continuously, exhibiting great strength and vitality, their nests must have medical or aphrodisiac qualities (i.e. the fountain of youth).
562. When swimming penguins leap out of the water in a graceful arc to breathe, they are porpoising, and they superficially resemble small, leaping porpoises.
563. Seabirds, as well as some other birds, have supra-orbital salt glands that remove excess salt from the blood, enabling them to ingest salt water.
564. The Wood Duck.
565. No, they are vocal shorebirds.
566. The Black-legged Kittiwake.
567. On September 1, 1914, at about 1:00 p.m. at the Cincinnati Zoo, Ohio.
568. The owls.
569. The hummingbirds, but the relationship has been the subject of much debate and may not be settled.
570. The raptor is associated with honey. It rips open wasp and bee nests in quest of bees and their larvae.
571. Technically the United States. At least 162 exotics were introduced into Hawaii, but the state was a territory at the time of most introductions. New Zealand is second with no less than 133 species.
572. The name is Spanish, meaning *cover your posterior (ass).* Tapaculos typically hold their tails tilted well forward toward their heads like wrens. However, the name may have originated because of their vocalizations
573. The American Robin.
574. The Kelp Goose of South America and the Falkland Islands.
575. The large New Zealand Weka rail.
576. Only 23. In 1937, a mere 15 wintered along the Texas coast, along with lesser numbers in Louisiana, but the Louisiana cranes disappeared by 1949. Through careful management and the employment of avicultural techniques, the population slowly rebounded and numbered 267 cranes by December of 1993.
577. The male Ruffed Grouse.
578. Either a vulture or an eagle, but it is also credited with being the largest of the kites. Its position within the raptor complex remains undetermined, but the Lammergeier is generally considered an aberrant vulture.
579. The five frigatebirds.
580. Reportedly in 1776, about the time the mission was built.
581. Allan O. Hume, who was also less affectionately called the *Pope of Indian Ornithology* because of his dogmatic zeal. Hume, along with his protégés, collected over 60,000 bird skins and large numbers of eggs and nests, primarily between 1870 and 1885. The impressive collection was ultimately deposited in the British Natural History Museum.
582. The grebes.

Answers

583. The Japanese and Chinese use cormorants to catch fish. The practice dates back to the late 6th century A.D. in Japan, and as far back as at least 317 B.C. in China. The tradition is still carried on today in Japan, primarily for tourists, and is even subsidized by the Japanese Imperial household because of its cultural value. Exhibitions are performed at sunset from a boat with a metal basket dangling from its prow containing burning wood, which produces light to attract fish. Several cormorants are tossed into the water, but are controlled from the boat by a rein attached to a leather collar around each bird's neck that prevents them from swallowing the catch.
584. Snails and freshwater mussels.
585. The tiny falconets, the Philippine and Bornean Falconets probably being the smallest. The sparrow-sized falconet of northwestern Borneo is only 5.5-6.0 inches (14.0-15.2 cm) long and weighs 1.24 ounces (35.2 gms).
586. In 1936, in Zaïre, then known as the Belgian Congo.
587. A gam. As many as 10 or more birds may engage in gamming.
588. The phalaropes.
589. Emperor and King Penguins produce melodious, resonant, trumpeting calls.
590. Bright pink.
591. On the ground.
592. A small English thrush with a red breast (European Robin). The British have always regarded *their* robin so highly that they carried the name wherever they went, giving it to red-breasted foreign species that reminded them of their familiar homeland bird. Hence robins, related and unrelated, are found the world over in America, Australia, India and Africa, wherever English is spoken.
593. The guillemots.
594. The White (Fairy) Tern's single egg is balanced on a horizontal tree branch. The "nest" of a Marbled Murrelet is not much more substantial, and in most cases, no nest at all is built.
595. The pigeons and doves are best known for this ability, but some of the waxbill finches, mousebirds and birds in a few other groups, including shorebirds, may also drink without lifting their heads. Most birds must lift their heads to drink.
596. The New Zealand Kea is an alpine species.
597. The Sandwich Tern.
598. The Peregrine Falcon. The raptors readily adapt to nesting on ledges of tall buildings and thrive on Rock Doves (Pigeons) that occur in limitless numbers in most large cities.
599. The South American Torrent Duck (43-44 days).
600. They have a toothed upper mandible (tomial tooth), a structure that typifies the falcon family.
601. When birds collide with an aircraft, the event is known as a bird strike. On 23 November 1962 a flock of Tundra (Whistling) Swans shattered the tail of a Vicker's Viscount flying at 6,000 feet (2,000 m) over Maryland, forcing the crippled plane into an uncontrolled dive, killing 17 people. Other extreme examples include a 4 October 1960 event involving a Lockheed LI 88 Electra that took off from Boston and sucked a flock of Common Starlings into its engines. The Electra went down and of the 72 people aboard, only 10 survived, of whom nine were seriously injured. On 26 February 1973, a Lear jet flying over Atlanta, Georgia, ingested Brown-headed Cowbirds into its engines, and all seven people aboard perished in the resulting crash, and a bystander on the ground was seriously injured.
602. The Little Gull weighs about 4.4 ounces (125 gms).

Answers

603. Clarice *Starling.*
604. The partridges, but the two groups are not related.
605. A sprinkling device used to water lawns. Some cuckoos are also known as rainbirds, and in Britain, the Green Woodpecker is called a rainbird, as is the Gray Currawong in Australia, and Peale's Petrel in New Zealand.
606. The Ruby-throated Hummingbird. It winters in Central America and must cross the Gulf, a non-stop flight over water of 620 miles (998 km), a remarkable feat for a tiny bird weighing only 3-4 grams (0.14 oz).
607. All have beaks.
608. One under par.
609. The domestic chicken.
610. Possibly because of the crude similarity of the bird's head and neck pattern to the stalk and shell of the goose barnacle. Legend has it that the geese arose from barnacles, a myth that may have been initiated by those of the Catholic faith, because if the geese were barnacle progeny, they were considered seafood, not meat, and could be consumed on meatless Fridays. The barnacles rather resemble an upside-down legless goose. These were believed to be the embryos of geese, which hatched in trees and dropped onto the water, later to metamorphose into adult geese.
611. MacGillivray's (Fiji) Petrel. A petrel flew ashore at Gua Island and literally struck the head of an ornithologist in 1983. The adult petrel was captured and subsequently released. In 1985, a fledgling was discovered on Gua Island.
612. The hummingbirds. The minuscule, cup-shaped, thimble-sized nests of the Cuban Bee and Vervain Hummingbirds can be covered with a pop bottle lid, measuring no more than 0.78 inches (1.98 cm) across and 0.78-1.2 inches (1.98-3.0 cm) deep.
613. The Masked Booby, that is well known for sibling murder.
614. It stood 5.5 feet (1.7 m) tall, and *may* have weighed as much as 300 pounds (136 kg). It was one of the first penguin fossils found and was discovered in 1903 on Seymour Island, Antarctica, by a member of Otto Nordenskjold's expedition.
615. The cassowary. The innermost toe of a female Double-wattled Cassowary is armed with a sharp spike-like claw up to five inches (12.7 cm) long and 1.25 inches (3.2 cm) wide at the base. The bird's massive legs are extremely powerful and an irate attacking cassowary leaps into the air, striking out with both legs.
616. The Pompadour Cotinga, presumably because of the male's brilliant, ornamental purple-colored plumage.
617. The Western Foundation of Vertebrate Zoology in Camarillo, California. Their extensive collection consists of approximately 800,000 eggs.
618. Only one, the Kelp Gull.
619. Steller's Eider. Both sexes average slightly less than two pounds (0.9 kg).
620. Yes, Greater Flamingos nest in southern France and Spain.
621. The woodpeckers and wrynecks. The long, extensible barbed tongue enables the birds to extract ants and other insects from their tunnels or under tree bark. The tongue extends up and over the top of the skull and coils in a loop behind the head and is joined to the extremely long horns of the hyoid bone. The hyoid bone may be five times the length of the upper mandible and the tongue apparatus is anchored in the right nostril. The wryneck's tongue and hyoid processes are nearly two-thirds the length of its body, excluding the tail. Therefore, it may have the longest tongue relative to its size. Some hummingbirds also have very long tongues, especially the Sword-billed Hummingbird.

622. The Ivory-billed Woodpecker, California Condor, Carolina Parakeet, Passenger Pigeon, Heath Hen, Bachman's Warbler, Great Auk, Labrador Duck, Dusky Seaside Sparrow and Eskimo Curlew.
623. Piscivorous birds, such as pelicans, are fish-eaters.
624. Pigmented feathers are more resistant to wear than unpigmented feathers.
625. A very fine, almost invisible net, resembling a huge hairnet, used to catch birds.
626. The Short-tailed Albatross.
627. The Magpie Goose and the Musk Duck, both of which are Australian. Adults offer food to the young from their bills, but all other waterfowl young feed themselves.
628. White, but in most populations, the white morph is relatively rare.
629. The two rheas.
630. An immature female domestic fowl.
631. The penguins, primarily Adelie and Emperor Penguins.
632. Stint is a British name for some of the small Old World sandpipers, specifically those in the genus *Calidris*.
633. The origin is obscure, but the murre's distinct purring and murmuring vocalizations might be responsible.
634. The Takahe. The large gallinule-like bird was rediscovered in the Murchison Mountains of Fiordland in 1948, near Lake Te Anau. Prior to that time, it had not been seen since 1898. In 1990, the population consisted of about 200 birds, some of which are captive.
635. The Sooty Albatross and the Light-mantled Sooty Albatross.
636. 1894.
637. The name alludes to the relatively slow wingbeat of the large plovers.
638. Ornithosis. A viral disease communicable to humans, it is not restricted to parrots and is carried and transmitted by a number of birds. It was initially described from pet parrots, hence the name psittacosis.
639. The noise created by a hornbill's powerful wings recalls a chugging steam engine. The sound is caused by the absence of feathers covering the bases of the flight feathers, enabling air to pass through more easily.
640. Apparently never. Some species hatch two chicks, but one invariably dies by the end of two weeks. Eudyptid penguins lay a clutch of dissimilar-sized eggs, and the second egg laid is always significantly larger than the first. Usually the chick from the first egg perishes, but not always.
641. The American Ruddy Duck. A courting drake continuously slaps its bright blue bill against his inflated chest producing a drumming sound. Air is forced out from beneath the feathers, creating a ring of bubbles around his chest.
642. The penguins.
643. The Thick-billed Murre.
644. A wattle, generally located at the gape.
645. Only three. The two most common are the widespread Belted Kingfisher and the Green Kingfisher. While not generally considered a resident, the Ringed Kingfisher is a rare vagrant to Texas, but in 1970 a few were discovered nesting along the Rio Grande River.
646. No, it is a large stork.
647. A wisp of snipe.
648. Yes, Woolly-necked Storks in India, for example, have been observed flying through rising swarms of winged termites, snapping up the insects in their bills.

649. Addled is a term applied to an egg (generally rotten) in which the developing embryo has died (not to be confused with a clear or infertile egg). In common usage, the term is often applied to any egg that has gone bad.
650. Aransas National Wildlife Refuge (Texas), with 389 species.
651. Yes.
652. The avocets.
653. The Long-eared Owl.
654. A large endemic Australian crane.
655. A gathering of non-breeding seabirds, generally formed up on the edge of a colony.
656. The tropical American curassows.
657. The American Bittern, because of its guttural gulping, belching, braying or booming calls.
658. The three volume set *The Viviparous Quadrupeds of North America*, published between 1846 and 1854. The work was a cooperative effort with the Rev. John Bachman.
659. In 1932, war was declared on 20,000 crop-damaging Emus in western Australia. A military detachment of machine-gunners engaged the birds, but failed to inflict much damage because the Emus were far too elusive. After about a month of fruitless pursuit, only 12 Emus had been killed and the war was called off. Subsequently, Emu-proof fences were erected that proved to be effective.
660. At least 26 species.
661. To prey on invertebrates flushed by the foraging terrestrial birds. Bee-eaters also ride Secretary-birds, goats, camels, antelope, zebra and warthogs, and may follow tractors. Captured prey may be pounded against the horns of an antelope or on the back of an Ostrich.
662. A gander.
663. According to legend, if the Ravens disappeared, serious disaster would befall the realm.
664. The Andean and Puna (James') Flamingo.
665. The extinct flightless Great Auk of the North Atlantic.
666. Heermann's Gull. It nests in Mexico in the Sea of Cortez, but winters along the Pacific coast as far north as Washington and British Columbia. Some may move as far south as Guatemala.
667. Crustaceans.
668. Moorfowl is an old English name for the Red Grouse (Willow Ptarmigan).
669. The African Hamerkop.
670. A grafting procedure, generally used in falconry, that mends broken flight feathers. The broken feather is cut in the middle and a similar-sized piece of a molted feather is joined by gluing.
671. The South American Ringed Teal and Chilean Teal.
672. Aspergillosis and avian malaria.
673. The Rhinoceros Auklet. It is probably a puffin lacking the colorful bill sheath.
674. A hawk in the genus *Accipiter*.
675. Their loud, explosive, bell-like peals carry a half-mile (0.8 km) or more through the jungle.
676. The social display ground of the Greater Prairie-chicken is known as a *booming ground*.
677. The aggressive display of a Mute Swan cob when he rapidly advances over the water toward an intruder with his neck drawn back on its shoulders and its wings arched. His movements are very jerky because both feet are paddled in unison.
678. A castrated domestic fowl.

679. The Demoiselle Crane, that weighs a maximum of 6.7 pounds (3 kg).
680. The male Eurasian Black Grouse.
681. The fossilized remains of a bird.
682. Birds are unable to produce carotenoid pigments, but flamingos derive the required carotenoids from their food of blue-green algae and aquatic invertebrates.
683. The Least Auklet. One of the smallest seabirds, it weighs only 3.5 ounces (99 gms).
684. The Killdeer.
685. The Little Blue Heron, because it forages for crayfish along rice field levees.
686. About 80. Island forms tend to be specialized and are incapable of adapting rapidly to change or moving to safer areas.
687. The Neotropic (Olivaceous) Cormorant.
688. Both collected bird eggs.
689. The stiff tail functions as a brace when the birds cling to vertical tree trunks.
690. The three ptarmigans.
691. Australia and Antarctica.
692. The Common and Thick-billed Murre. Their eggs are cone-shaped with a sharp end and, in theory, the eggs roll in a tight circle, preventing them from rolling off the nesting ledge.
693. The Greater Prairie-chicken.
694. Princess Eleonora d'Arborea, a 14th-century war leader, regent and judge of a large part of the island of Sardinia, who issued a law protecting hawks and falcons.
695. Rails have narrow, strongly compressed bodies, enabling them to run through thick marsh grass, weeds and underbrush.
696. The grebes, phalaropes, coots, Sungrebe and finfoots.
697. The five loons.
698. The three distinctive seabirds rarely stray beyond tropical seas.
699. Legend has it that the birds return on March 19.
700. The 21 storm-petrels. Wilson's Storm-Petrel has an incubation period of 39-48 days, but this may be prolonged to 52 days by absence from the nest (nest neglect).
701. The Vulturine Guineafowl of tropical East Africa. Males weigh up to 3.6 pounds (1.6 kg).
702. A shorebird (a short-billed curlew).
703. Biologists formerly believed that because frigatebird feathers quickly become water-logged, it was very difficult for the birds to achieve flight. Frigatebirds apparently never voluntarily enter the sea and probably don't land on the water because of a structural inability to take off, not because of the lack of an oil gland that would result in water-logging. In addition, their tiny feet make them inefficient swimmers.
704. The domestic chicken.
705. The last remaining rails were eaten by starving Japanese troops during World War II.
706. The loons. Although strong fliers, loons have difficulty becoming airborne, and cannot take off from land.
707. The diving-petrels.
708. The huge nest may weigh up to 500 pounds (227 kg). Each parakeet pair constructs and maintains a separate nest chamber within the structure.
709. The 8-10 pound (3.6-4.5 kg) South American Coscoroba Swan.
710. The tropicbirds, pelicans, cormorants, Anhinga and darters, boobies and gannets, and frigatebirds are known as totipalmate swimmers. All are in the order Pelecaniformes.
711. The sparrow-sized painted-quails of Africa, Asia and Australia.

712. The Imperial (Blue-eyed) Shag.
713. The ptarmigans. Males assume a female-like plumage during the summer, and both sexes turn white during the winter.
714. The Least Bittern, that weighs only 1.5-4.0 ounces (43-113 gms).
715. The Korean Crested (Kuroda's) Shelduck; three were collected in Korea and Japan, and one in Russia. The last, a female, was collected in December 1916. Only three specimens remain, two of which are in Tokyo and one in Copenhagen.
716. A flying crow.
717. Probably to minimize the effects of insect bites and stings.
718. The Great Gray Owl, but both the Snowy Owl and Great Horned Owl weigh more.
719. Three of the four are flightless, and when racing over the surface of the water, the ducks furiously thrash the water with their reduced wings, producing much spray. This "steaming" reminded early sailors of a sidewheel paddle-boat.
720. The Congo Peacock.
721. The Sage Grouse. Cocks weigh nearly eight pounds (3.6 kg).
722. The syrinx. The mammalian voice box is called a larynx.
723. The Spotted Sandpiper, from its habit of continually wagging its rump and tail up and down.
724. "Quoth the raven, *Nevermore*!"
725. The Mourning Dove.
726. Alcatraz Prison in San Francisco Bay. The facility is no longer operational, and is currently a historical site. *Alcatraz* is a Spanish and Portuguese name for pelican.
727. The Double-crested Cormorant.
728. The cock Ruffed Grouse during the breeding season.
729. In the late 1980's, between 2,500 and 5,000 macaws, with the actual number probably about 3,000.
730. Unlike many birds, incubating waterfowl lack brood-patches. Down is plucked from the region of the breast where the eggs come in contact with the warm skin. The down keeps eggs warm and conceals them when the female is off the nest.
731. Australia.
732. Two under par.
733. None since the Wrentit family (Chamaeidae) was eliminated. The Wrentit is now included in the Old World warbler complex.
734. The American Woodcock, a large, nocturnal, interior woodland shorebird.
735. No, but there formerly were when skuas of all populations were considered races of the northern Great Skua.
736. A Pigeon Hawk.
737. The albatross was practically exterminated by Japanese plume-hunters early in the century, and for a time, it was feared extinct. In the early 1880's, at least 100,000 albatrosses bred on Toroshima alone, the southernmost of the "Seven Isles of Izu," some 400 miles (644 km) south of Tokyo. In 1887, feather-hunters took up permanent residence on Toroshima, and over the ensuing 17 years slaughtered either 500,000 or five million birds, depending on one's source. The volcanic eruption of 1903 not only killed all the feather-hunters, but many albatrosses as well. The respite was temporary, because more feather-hunters returned and were active until the late 1930's, but by then the entire albatross breeding population had been essentially exterminated. Following World War II, the island was

declared a sanctuary. Toroshima erupted again in 1965, which no doubt adversely impacted the albatrosses.

738. A gannetry.
739. The naked chicks are black or dark purple.
740. Australia.
741. The frigatebirds, but virtually all other seabirds are not dimorphic.
742. African crowned-cranes often perch in trees.
743. The Caspian Tern, that weighs nearly 1.5 pounds (0.7 kg).
744. Probably not. The gulls were possibly not Antarctic residents prior to the days of sealing and whaling, and may have reached the ice by following whalers or sealers south, but this is unlikely because most gulls seldom stray far from land.
745. The Great Egret.
746. The New Zealand kiwis.
747. The Phoenix. According to legend, there could only be one Phoenix in the world at a time, and it lived in paradise, a land of infinite beauty lying beyond the eastern horizon. Every 1,000 years the Phoenix left paradise and flew westward to die. It collected aromatic plants on the way from the spice groves of Arabia, and built itself a nest in the top of a tall palm tree. Here, on the first dawn after its journey, it sang a song of such surpassing beauty that the Sun God stopped his chariot to listen. When the Sun God whipped up his horses again, sparks from their hooves set fire to the nest of the Phoenix, and it died on a blazing aromatic funeral pyre. The new Phoenix grew from a worm which was found in the ashes of the nest, and flew back to paradise for its own allotted life-span of 1,000 years.
748. The covering or hiding of prey with the wings by a feeding raptor.
749. A thrush.
750. The Osprey. The spiny projections aid in holding slippery fish.
751. Females have a broad chestnut-colored chest-band that males lack.
752. The Barn Owl.
753. No, the Downy Woodpecker is the smallest, weighing only about an ounce (28.6 gms).
754. A kettle.
755. The white morph of the Reddish Egret.
756. The Black-billed Magpie.
757. Lemmings and Arctic hares.
758. The feathers function as snowshoes, and may also have some thermoregulatory value.
759. Sabine's Gull.
760. The Greater Roadrunner, because of its fondness for lizards.
761. The male Andean Condor.
762. An exaltation of larks.
763. The stamp must be signed across the face.
764. Heermann's Gull.
765. The Brent (Brant) goose. The goose is so dependent on eel-grass that following a massive grass die-off in the 1930's and 1940's, the Atlantic Brent population crashed drastically.
766. The Bobwhite.
767. The 15.0-16.5-inch-long (5.9-6.5 cm) Laughing Kookaburra weighs 1.1 pounds (.05 kg) and is generally credited as the largest kingfisher. Some authorities suggest that the 18-inch-long (46 cm) African Giant Kingfisher is larger, but it definitely weighs less than the Laughing Kookaburra.

768. The Flying Steamer-Duck.
769. The tropicbirds, gannets and frigatebirds.
770. The swallows.
771. Common Eiders. On the Farne Islands north of Scotland, St. Cuthbert, who died in 687 A.D., protected nesting eiders around his hermitage, thus creating the first duck sanctuary. To this day, eiders are known locally as St. Cuthbert's Doves, or Cuddy Doves.
772. No, the aberrant shorebird is native to coastal east Africa and southwest Asia.
773. The Swallow-tailed Gull (Galapagos Islands) and Sabine's Gull.
774. The Tricolored Heron.
775. The Inca Tern of Chile and Peru typically nests in rock crevices or holes in hard guano.
776. Early sailors provisioning ships killed many for food. The introduction of dogs, pigs and pet monkeys (crab-eating macaques) ultimately eliminated the flightless Dodos. Feral dogs preyed on adults, while the pigs and monkeys ate eggs and young.
777. The three tropicbirds.
778. Yes, on the Dry Tortugas, just off the Florida Keys, in the Gulf of Mexico.
779. The Thick-billed Murre.
780. In the early evening hours or even during the darkest nights, when the water tends to be calmer and small fish and crustaceans rise to the surface.
781. The African Gray Parrot, but some Amazon parrots are nearly as proficient.
782. The turacos of Africa.
783. Grayson's Dove is the avicultural name for the Socorro Island Dove. The species is extinct in the wild, but approximately 300 birds survive in captivity.
784. The Razorbill.
785. Yes, but only sporadically. Vast numbers were sold as cagebirds, even in Europe. Carolina Parakeets could have been easily captively bred, but because the supply was seemingly inexhaustible, there was no reason to. Captive birds had the reputation of being inattentive parents, and nesting failures were high.
786. The Brown-headed Cowbird has been recorded parasitizing the nests of more than 200 species.
787. The Dodo: *Dead as a Dodo.*
788. Probably to keep the eggs cool on hot days.
789. Its cheeks are naked, rather than feathered.
790. The nocturnal caprimulgids (nightjars and allies).
791. A nestling pigeon.
792. The three puffins.
793. Tyrant-flycatchers.
794. The little owlet-nightjars of Australia and New Guinea, a group of eight species formerly known as owlet-frogmouths.
795. The three American anis. The birds gather in flocks and even nest colonially.
796. The Crested Auklet.
797. There is no technical difference. The term *pigeon* is generally, but not always, applied to the larger species.
798. From its long crest that is held horizontally, rather like the nail-pulling end of a claw hammer.
799. The brilliant male Vermilion Flycatcher. It is one of the few tyrant-flycatchers to exhibit extreme sexual dimorphism. Females are dull brownish-gray.

Answers

800. The Spruce Grouse, because it is remarkably tame, failing to recognize humans as potential enemies. The Blue Grouse is sometimes known as a fool hen for the same reason.
801. The three New Guinea crowned-pigeons.
802. An abandoned woodpecker hole in saguaro cactus.
803. The Mourning Dove. It was introduced into Hawaii in 1964.
804. A kiskadee may splash into the water after a small fish, like a kingfisher.
805. *Coot-footed.* The name alludes to the development of the feet, which are rather like those of the much larger coots, because a phalarope's toes are distinctly lobed.
806. Xantus' and Craveri's Murrelet and Cassin's Auklet breed as far south as Baja California.
807. Potoos typically lay their single, spotted egg atop a broken stump. Both sexes incubate in a very erect position so that the cryptic-colored birds resemble a continuation of the stump.
808. The birds have unfeathered legs, and their feet have rough scutellations on the bottom to aid in grasping slippery prey.
809. Not normally, but during the austral summer of 1988/89, at least five Black-necked Swans were repeatedly observed in various locations along the Antarctic Peninsula and surrounding islands, 500-600 miles (805-966 kilometers) south of the tip of South America. Swans reappeared in the Antarctic during the 1993/94 summer. During the 1916/17 austral season, an emaciated swan was found in the South Shetland Islands.
810. The owls don't nest in trees. They nest on the ground, generally in open, inland habitat around both fresh and saltwater marshes.
811. The little birds clip off leaves, fruits and buds, and even sever small plants at the base.
812. Both birds often line their nests with shed snake skins.
813. The entire population of the Chatham Islands Black Robin was restricted to a tiny islet of no more than an acre (0.4 hectare). It has since been established on another small predator-free island.
814. The long tail feathers of a displaying male Superb Lyrebird resemble a lyre.
815. The three puffins.
816. A swallow.
817. No.
818. Wildfowl.
819. Approximately 650 species nest in North America, with another hundred or so occurring infrequently or accidentally.
820. The islands of Tristan da Cunha in the south Atlantic, where about five million pairs nest.
821. The Yellow-billed Loon, that weighs as much as 14 pounds (6.4 kg).
822. The Whiskered Auklet, that is only slightly larger than the Least Auklet.
823. The sexes are similar, but males have a very long tracheal loop extending over the chest before returning and entering the chest cavity.
824. The Emperor Goose.
825. Yes, the Black-footed, Amsterdam and the juvenile Wandering Albatross.
826. The Wood Duck.
827. The Mountain Quail, that weighs up to 10.25 ounces (291 gms), but averages 8.35 ounces (234 gms).
828. No, Red-throated and Yellow-billed Loons have slightly upturned bills.
829. In the Komandorskiye Islands, as well as the Nemuro Peninsula of Hokkaido, Japan.
830. Four: Tennessee, Connecticut and Kentucky Warbler, and Louisiana Waterthrush.
831. Females have white eyes, whereas males have red eyes.

832. The Sharp-shinned Hawk.
833. The Plain Chachalaca, but it barely reaches southern Texas.
834. Approximately 20 billion.
835. It forages on the ground in quest of ants.
836. Adult males have blue, rather than rusty-colored, throats.
837. A molting, blotched blue-and-white, year-old Little Blue Heron.
838. Yellow, but adults have dark bills.
839. The Sora rail.
840. The male Pennant-winged Nightjar of Africa.
841. The Resplendent Quetzal of Central America, but females and immatures are crestless.
842. The Pacific Eider, a race of the Common Eider. Drakes weigh more than six pounds (2.7 kg), larger even than a small goose.
843. Turkey Vultures soar with their wings held at a V-shaped angle, but the flight profile of a condor is flat.
844. Dabbling-ducks (dabblers) feed from the surface, often tipping up (up-ending), as opposed to diving-ducks that forage underwater. Dabbling-ducks dive on occasion, but generally not very efficiently.
845. The Black Oystercatcher.
846. Possibly, but this is considered doubtful by some authorities. Its specific gravity is much lower than that of water, hence the bird cannot stay on the bottom without exerting some force to keep it from bobbing to the surface. If a dipper really does walk on the bottom of fast-moving streams, it must tilt forward facing the current so the resultant pressure stabilizes it.
847. The Pomarine Jaeger. Females average about 1.6 pounds (0.7 kg), and males 1.4 pounds (0.6 kg).
848. Ross' Gull, but it is primarily a Siberian-breeding species.
849. Basically white.
850. The gum exuded by some *Acacia* trees.
851. The plumage becomes whiter.
852. Black, whereas that of a male Red-shafted Flicker is red.
853. Facial discs enhance hearing by collecting acoustic energy and directing it toward the ears.
854. The Gull-billed Tern.
855. In 1953, 20 years after the last one was reported. It was formerly regarded as extinct.
856. The numerous nondescript *Empidonax* tyrant-flycatchers.
857. No. The New World orioles are related to icterids (blackbirds, meadowlarks, grackles, oropendolas, etc.), whereas the 29 Old World orioles are in the family Oriolidae.
858. To prey on insects stirred up by the active primates.
859. A greatly enlarged oil gland, an apparent necessity for aquatic birds.
860. Their stubby tails are cocked bolt upright.
861. Steller's Jay.
862. The Tufted Titmouse, that weighs up to an ounce (28 gms).
863. The Vikings believed swans possessed supernatural powers and flew to their ancestral home of Valhalla, or to the moon.
864. No, the bird has mewing, cat-like calls.
865. A rail.
866. The aberrant Crab Plover.

Answers

867. The dorsal ridge of the upper mandible, from the forehead to the tip.
868. The birds *spear* fish while swimming underwater. Most fish-eating birds grasp their prey.
869. The 8.5-inch-long (22 cm) Giant Hummingbird weighs about 0.7 of an ounce (20 gms).
870. The trogons may dig out a nesting cavity in large paper nests of tropical wasps. The birds first eat the wasps and then consume the larvae as they dig.
871. A foot where two of the three front toes are joined for part of their length. Kingfishers, hornbills, todies, motmots, bee-eaters, rollers, hoopoes and cocks-of-the-rock all have syndactyl feet.
872. The nuthatches.
873. Bright red.
874. The Limpkin. Its varied wailing and screaming calls are heard mainly at night.
875. Birds with a highly developed sense of smell, such as a kiwi, Oilbird or Turkey Vulture.
876. The eggs were sometimes splotched with brown shoe polish and sold as Snail Kite eggs.
877. Tarantula hawks. Inhabitants of tropical South America, these are the largest wasps in the world, with a body length of up to 2.64 inches (6.7 cm) and a wingspan reaching 4.17 inches (10.6 cm).
878. The Australian Magpie Goose, whistling-ducks, and some Black Swans and African White-backed Ducks. In other waterfowl, only the female incubates.
879. The American Woodcock. Chick-carrying also has been reported, but not confirmed, in Spotted Sandpipers, Common Moorhens, Clapper Rails and even whistling-ducks.
880. Wood Buffalo National Park, Alberta, Canada.
881. An Atlantic mutant of the Common Murre with a distinct white eye-ring and a white line extending down the postorbital crease.
882. Cygnets are patterned like the ducklings of whistling-ducks. The cygnets of all other swans are mono-colored with no pattern.
883. Fruit eaters.
884. Cover the eggs with down to conceal them and keep them warm.
885. The oil gland.
886. A twitcher.
887. The Laughing Kookaburra.
888. The Rufous Hornero, an ovenbird whose mud nest resembles an oven.
889. Some of the smaller neotropical puffbirds. Their drab coloration is reminiscent of the dull-colored clothing worn by nuns.
890. The South American tapaculos seldom flutter more than a few meters.
891. The South American Toco Toucan, although some authorities suggest that the Black-mandibled and Red-billed Toucans may be larger.
892. Some of the bellbirds.
893. Probably the Wedge-tailed Eagle. It had the unwarranted reputation of being a lamb killer, and tens of thousands were killed in West Australia and Queensland. The Golden Eagle of the U.S. did not fare much better. Both raptors are now completely protected.
894. The Australian Brolga crane.
895. The Common Raven.
896. The bee-eaters.
897. The white spot on the otherwise black wing tip of some gulls. A mirror formerly referred to the brightly-colored area of the wing, especially the metallic speculum of dabbling-ducks.

898. The last *confirmed* Dodo vanished in 1662, but a few may have survived until 1681.
899. The Eurasian Blackbird, with a population estimated at 15 million birds. The more elusive Winter Wren is possibly more abundant, and during peak years the population may number 15-20 million.
900. Nineteen, compared to 25 in swans.
901. The aracaris, a distinct group of small toucans.
902. The Gray Phalarope. The phalarope is red during the breeding season, but becomes gray during the winter. As a result, the bird is far better known as the Gray Phalarope in Europe, where it is commonly observed during migration.
903. South America, where one species has been introduced. When plume-hunters threatened many forms, Sir William Ingram, a noted newspaperman and aviculturist, imported into Trinidad a number of immature Greater Birds-of-paradise from New Guinea. In September of 1909, about 48 birds were released on Little Tobago Island (also known as Bird-of-paradise Island), directly adjacent to Trinidad. In 1959, a maximum of 29 birds inhabited the island, but in 1966, only seven were seen.
904. The Galapagos Swallow-tailed Gull. Its huge eyes reflect a nocturnal lifestyle, and the gulls feed mainly on squid and fish caught on the surface. The birds may focus on phosphorescence created by moving prey.
905. The preen gland of the drake produces a strong musky odor during the breeding season.
906. The elaborate courtship structures of male bowerbirds. The 9-inch-long (23 cm) male Golden Bowerbird may construct a roofed gazebo that, in extreme cases, towers nine feet (2.7 m) in height.
907. The Northern Gannet, that weighs up to seven pounds (3.2 kg).
908. The Ground Woodpecker of South Africa, that even excavates its nest in the ground in the manner of a kingfisher.
909. *The Pelican Brief.*
910. The Aleutian Tern of Alaska.
911. Green. Selective captive breeding has resulted in a wide variety of colors, such as blue, white, yellow, gray and violet.
912. Seven: Whooping, Hooded, Siberian, Manchurian, White-naped, Black-necked and Wattled Crane.
913. The Ivory Gull.
914. At least 265 feet (81m). A Common Loon was entangled in a net set at that depth.
915. Yellow, but in flight, the webs are not always visible.
916. These Old World warblers construct nests in a cup formed by sewing the edges of one or two leaves together. To sew the leaves, the birds first pierce holes in the leaf with their beaks and draw plant fibers or silk from insect cocoons or spider webs through the hole. Each stitch is separate and the "thread" is knotted on the outside to prevent its slipping back. This might be considered tool use.
917. Five.
918. The loss of virginity.
919. The thrush complex, that includes such notable songsters as the Nightingale, Song Thrush, solitaires, Shama and Dyal Thrush.
920. Dark purplish-chocolate or even glossy black, a very unusual color for a bird egg.
921. Great Curassow males weigh up to 10 pounds (4.5 kg).
922. Yes, the Northern Jacana barely ranges into southern Texas.

Answers

923. The distal portion of one wing at the carpal joint is removed to render a bird permanently flightless. The shortening of one wing makes the bird aerodynamically unstable. Large birds, such as waterfowl, cranes and flamingos not maintained in an aviary are generally pinioned.
924. A Rueppell's Griffon was sucked into a jet engine at 37,000 feet (11,278 m) on November 29, 1973, over Abidjan, Ivory Coast, West Africa. The engine was damaged so severely it had to be shut down.
925. The Yellow-eyed Penguin of New Zealand and its subantarctic islands. While never numerous, possibly no more than 3,400 to 4,500 birds remain, an estimated reduction of about 80 percent during the last half-century.
926. A Reeve.
927. No. Loons nest on solid ground, but grebes typically construct large floating nests of aquatic vegetation. Their nests are either anchored to the shallow bottom or float free in dense vegetation.
928. Flamingo tongues were considered a great delicacy.
929. The Snow Bunting, that even breeds up to the northern extreme of Greenland.
930. The Common Raven.
931. The Cattle Egret.
932. The Northern Flicker.
933. The Dovekie, Razorbill and Common Puffin.
934. The Phainopepla of the southwest.
935. Three: Brown (Common), Little Spotted and Great Spotted Kiwi.
936. Only one, the Yellow-breasted Sunbird.
937. Pigeon-hearted or chicken-hearted.
938. Canaries were kept in mines for the early detection of toxic methane gas, to which the birds were particularly sensitive. If the canaries dropped dead, there was real cause for alarm. During the 1990-91 Persian Gulf War, sentinel chickens were used for the early detection of gas in the event that Iraqi troops resorted to chemical warfare.
939. No, but the male does.
940. Birds related to one another subspecifically.
941. Rookeries may be located 100 miles (161 kilometers) or more from open water on annual fast-ice. The travel time required to reach open water is such that it is not practical for both sexes to travel back and forth to relieve one another during the long incubation period. When the chicks hatch, the recently-arrived female is very corpulent and food-laden, a situation superior to having both parents in marginal condition, which might be the case if both incubated. The male is the logical candidate to incubate because he is larger and more resistent to the cold, and the female has burned up some of her reserves laying the 1.0-pound (0.45 kg) egg.
942. Yes, as long as both partners are alive. Contrary to persistent popular belief, if one dies, the surviving partner will not pine away and die, and sooner or later will probably take another mate.
943. An aberrant African Snake-Eagle, that, as an adult, has an exceptionally short tail.
944. Yes.
945. The Greater and Lesser Scaup.
946. Sometime during the 16th century.

947. The Greater Roadrunner. In the 1920's, a bounty was placed on them because it was believed they ate gamebirds, especially quail. Roadrunners do take some quail chicks, but such prey does not form a significant portion of their varied diet.
948. The Lammergeier (Bearded Vulture). Ossifrage translates to *bone-breaker,* no doubt alluding to the vulture's habit of dropping bones to smash them open or crush them.
949. The quills of the large flight feathers of the Trumpeter Swan. Audubon wrote *The quills are so hard and yet so elastic that the best steel pen of the present day might have blushed, if it could, to be compared to them.*
950. Bob *Crane,* who was murdered on June 29, 1978, in Scottsdale, Arizona.
951. The Flightless Atitlan Grebe of Lake Atitlan (Guatemala), in 1984.
952. The burrow-nesting Crab Plover.
953. The Guadalupe Storm-Petrel was last sighted in 1912, and may already be extinct.
954. The Blue or Paradise Crane.
955. The crows and jays (corvids).
956. The name presumably refers to the bobbing behavior typical of dippers. The American dipper may bob up and down as often as 40 to 60 times a minute.
957. No, the diurnal hawk-owls are more dependent on sight.
958. Between 1851 and 1900, no less than 27 full species vanished along with 19 subspecies. The most disastrous period was the 23 years between 1885 and 1907, when 24 species and 14 races, 38 forms in all, were lost. From 1901 to 1950, 19 species and 19 races disappeared.
959. Small tropical tanagers. The name "euphonia" is from the Greek meaning *goodness of voice*. Euphonias sing better than the other tanagers, most of which are not good songsters.
960. No.
961. A maximum of 66 birds.
962. Possibly. Paleontologists have long believed that birds evolved from reptiles, and some herpetologists still refer to them as *glorified reptiles.* Birds may have actually evolved from dinosaurs, which were possibly warm-blooded, although this theory is controversial. If true, some dinosaurs may have evolved into birds. In April, 1993, a new dinosaur from the Mongolian Gobi desert was described; *Mononychus* (one claw). This turkey-sized animal looks like a modern flightless bird, complete with feathers, but has bone structures of both birds and dinosaurs, and some of its bone structure suggests that it may have evolved from a flying ancestor. Its discovery cements the bird-dinosaur link, but some researchers remain skeptical.
963. A nestling falcon or hawk, especially of those species favored by falconers.
964. The increase is probably correlated with the abundant refuse of the whaling and fishing industries. Fulmars are opportunistic scavengers.
965. A hair-like feather.
966. The bill tips cross at oblique angles to facilitate extraction of seeds from pine cones.
967. A neotropical ovenbird.
968. The owls, which have extremely large wings for their weight and can fly very slowly, with a slow flapping rate.
969. The name wheatear has nothing to do with wheat or ear, but is a euphemism for the Anglo-Saxon *white arse,* referring to the small thrush's white rump.
970. The Puna (James') Flamingo. Confined to the Andean highlands, it generally does not venture below 10,500 feet (3,200 m).

Answers

971. The flightless, nocturnal Kakapo (Owl Parrot).
972. The five frigatebirds.
973. Ross' Goose, that incubates for only 19-25 days, with a mean of 22 days, and young are fledged at 40 days of age. The compressed breeding cycle is a necessity because the little geese breed in the far north, where spring arrives late and winter arrives early.
974. The three puffins.
975. The hawks hunt at dusk and catch swallows, martins and swifts on the wing.
976. An Indian bustard.
977. South American Hoatzins nest over water and if threatened, chicks allow themselves to fall into the water, where they dive and swim, using both wings and legs. Once the danger has passed, the chicks climb out and scramble back up through the vegetation.
978. The long, flowing plumes grown by some egrets during the breeding season. The beautiful feathers have barbs that are long and free. "Aigrettes" were highly sought by plume-hunters.
979. The neotropical flycatcher has a huge fan-shaped crest of long orange-vermilion feathers, tipped with violet. The crest can be erected perpendicularly to the axis of the head. The flower-like crests of both sexes *might* be useful in attracting insects within reach, but this is unlikely. The crest is probably related to courtship, but possibly also signals aggression (both intraspecific and interspecific), as well as functioning as a predator deterrent.
980. The Northern Gannet.
981. The social display ground (lek) of Lesser Prairie-chickens.
982. A band of color in some plumage patterns, located on the throat or upper breast. Males of many hummingbirds have gorgeous, colorful gorgets.
983. The three enormous crowned-pigeons of New Guinea.
984. Steller's Sea-Eagle. Females weigh nearly 20 pounds (9.1 kg).
985. At one time considered a separate species, it is currently regarded as a white morph of the Great Blue Heron. The bird is essentially restricted to southern Florida and the Keys.
986. The 1,000 copies went on sale on New Year's Day, 1941, and were sold out by the end of the year. Most copies were purchased by Europeans living in Burma, and the books were left behind when they evacuated prior to the 1942 Japanese invasion. The Japanese collected as many copies as they could and shipped them to Tokyo, where they were housed in the library of the Royal Veterinary College, which was subsequently destroyed in an air raid. The revised second edition did not appear until 1953.
987. The Common Raven and Old World woodswallows.
988. An especially noisy, olive-brown African ibis, the Hadada in the south constructs a flimsy nest structure, and many eggs and young fall from the nest.
989. The Yellow Wagtail nests in northern Alaska and Canada. The White Wagtail occasionally straggles from Asia to Alaska, but is not a North American breeder.
990. The Indian Black Eagle. It feeds extensively on nestlings, and even eggs, and the eagle probably finds it advantageous to fly very slowly over forests and hillsides to locate such prey.
991. No native swans breed in the eastern U.S., but the introduced Mute Swan is well established in some regions.
992. At about a month of age.
993. They continually pump their long tails up and down.
994. No, honker is an American name for the Canada Goose.

995. The South American Inca Tern.
996. The cutting edge of the bill.
997. Abbott's Booby.
998. The shrikes. They feed primarily on large insects, but also readily take frogs, lizards, small rodents and even birds almost as large as themselves.
999. The name means *little star,* referring to the spangled appearance of the Eurasian Common Starling in its fresh autumn plumage.
1000. The Crested Ibis. In 1989, the total number in China, and possibly North Korea and Russia, was estimated at 46 birds. A few are maintained in captivity, and a pair in the Beijing Zoo produced two young in 1989, but both unfortunately died.
1001. The woodswallows of Australia and Asia.
1002. Upon entering the hole, the male turns around facing out with its long plumes doubled forward over his back with the tips still outside the hole. By end of the nesting season the plumes are generally frayed or broken.
1003. The lorys and lorikeets. The parrots are nectar-feeders and a brush tongue is useful for lapping up nectar, as well as for feeding on very soft fruit.
1004. The seven peacock-pheasants.
1005. No, it is an icterid related to blackbirds and orioles.
1006. A kiwi is pear-shaped.
1007. Yes, oxpeckers may drink blood from wounds caused by ectoparasites, but this is not typical behavior.
1008. The Glossy Ibis is the only cosmopolitan ibis, occurring over much of the Americas, southern Europe, Australia and Africa.
1009. The Helmeted Curassow.
1010. Both names apply to the same bird. Kentish Plover is its British name.
1011. The Village (Spot-backed) Weaver was probably initially introduced from West Africa when the slave ships were plying between this region and the West Indies, and the birds could have been established as early as 1783 in Haiti. Some authorities suggest the weavers were intentionally introduced by French colonists. During the 1960's, the Hispaniola weaver population underwent an explosion and the birds became a serious threat to rice crops, sometimes accounting for up to 20 percent crop loss. Although shot and poisoned, weavers are still as abundant as ever in all parts of lowland country, wherever there is water.
1012. The Song Thrush.
1013. None.
1014. The drastic decline was apparently caused by the accidental introduction of bird-eating and egg-eating snakes from Australia and the Philippines. The snakes appeared during, or immediately following, the Second World War. The density of brown tree-snakes is estimated at a staggering 12,000 snakes per square mile (2.6 sq. km).
1015. Two.
1016. Yes, the crepuscular raptor swallows bats and small birds whole while in flight.
1017. Lewis' Woodpecker.
1018. Only two, the Indian and Javan Peacock. The Congo Peacock is not really a peacock.
1019. The 3-4 pound (1.4-1.8 kg) frigatebirds with wingspans of 7-8 feet (2.1-2.4 m).
1020. Very loud, large, polygamous, colonial-nesting neotropical icterids.

Answers

1021. The African elephant.
1022. The oystercatchers.
1023. The grooves probably serve to counteract the tendency for the bill to be pushed downward in the water when the birds are skimming (foraging). The elongated mandible is immersed for about two-thirds of its length, cutting the water at a 45-degree angle, so that the striations along the side of the bill are parallel to the surface.
1024. Passerines share a number of similarities, but the most striking is that all have four toes joined at the same level. The hind toe is directed backwards and is not reversible.
1025. Many do, but others do not.
1026. The seven peacock-pheasants and the two extant argus pheasants.
1027. Vaned (contour), down, semi-plume, filoplume and powder-down feathers.
1028. A neotropical nightjar that ranges north to southern Texas, and as far south as northern Argentina and southern Brazil.
1029. Blinks. Each time a dipper dips, it generally blinks its silvery-white nictitating membrane.
1030. The Cinnamon Teal.
1031. The Snares Islands Crested Penguin, that essentially breeds only on the Snares Islands.
1032. The magnificent South American Hyacinth Macaw.
1033. The Australian Laughing Kookaburra.
1034. The downy natal plumage of a chick.
1035. The grebes, but exactly why remains a mystery. Grebes are unique in the habit of eating their own feathers, sometimes in large quantities, especially in fish-eating species. The ingested feathers accumulate in the main compartment of the stomach where they decompose and are transformed into a green, spongy, felt-like material, which is mixed with the food, making a characteristic feather-ball that lines the gizzard. The feather-balls, whole or in bits, are regurgitated periodically in the form of pellets, along with the parts of the prey that are the most difficult to digest. In piscivorous grebes, the action of the gizzard may be insufficient to crush fish bones. The feather-balls protect the stomach, wrapping up the bones and delay their digestion, so that the sharp, pointed fish bones are fairly well dissolved by the time they pass into the intestine. A feather-pad, formed by a small, more or less digested mass of feathers in the lower chamber of the stomach, blocks the entrance to the intestine, like a pyloric plug, protecting it from the indigestible remains. Just why grebes require such extreme protection from fish bones is puzzling because many species of birds consume fish and are not bothered by bones in the gut.
1036. The Orinoco Goose and at times, the Ashy-headed Goose (in the hollows of burnt trees).
1037. Krill (small, swarming, shrimp-like crustaceans).
1038. Steller's Sea-Eagle, named in honor of Georg W. Steller (1709-1769).
1039. The horny covering of the unfeathered parts of the avian leg and foot.
1040. An Ostrich may seek to escape detection by remaining immobile and flattening its body, neck and head out on the ground, assuming the appearance of just another hillock on bare, scrub-covered ground, a posture that may deceive predators. This is a common tactic of incubating Ostriches, and such behavior could be the origin of the belief that alarmed Ostriches bury their heads in the sand.
1041. The four diving-petrels.
1042. The southern part of the Southern Hemisphere.
1043. The Indian Pond-Heron.
1044. The four eared-pheasants.

Answers

1045. A tropical American oriole.
1046. Only a single chick out of 552 nesting attempts hatched on Anacapa Island (southern California), the sole western U.S. colony at the time. In 1969, a maximum of five chicks hatched out of 1,272 nests. Highly toxic residues from organochlorine pesticides, mainly DDT, caused eggshell thinning, and virtually all eggs were crushed during incubation.
1047. The woodpecker ceased to exist, along with the Yellow-shafted and Gilded Flickers. All three were merged into a single species collectively called the Northern Flicker.
1048. The Australian Cape Barren Goose. Its short, thick black bill is almost completely obscured by an extensive waxy, pale greenish-yellow cere. Its generic name *Cereopsis* ("wax-like") refers to the waxy cere.
1049. The Ruff. The distinctive neck ruff of the male occurs in a variety of colors, and no two individuals are exactly alike in color or pattern.
1050. No, it is an Old World falcon.
1051. *To cook one's goose.*
1052. The membranous fold of skin along the anterior margin of a bird's wing.
1053. The Penguin, who was played by Burgess Meredith. In the 1992 Batman movie, the role was played by Danny DeVito.
1054. 1916.
1055. Six to nine acres (2.5-3.5 hectares).
1056. Yes.
1057. The Asian Crested Partridge. It was formerly known as a wood partridge, but all wood-partridges are of New World origin.
1058. Laysan Albatrosses have strayed into Arizona, probably drifting up the Colorado River from the Sea of Cortez, Mexico.
1059. Between 5,000 and 7,000 of the aberrant shorebirds.
1060. The ibises.
1061. An African glossy staring.
1062. A desert. Xerophilous species are adapted to arid conditions.
1063. The Whimbrel.
1064. The screamers, pelicans, and gannets and boobies.
1065. Not currently, but the extinct Stephen Island Wren was believed to be flightless, although some authorities suggest that limited flight was possible. Live wrens were only seen on two occasions at twilight, and both times the bird reportedly scurried about among the rocks like a mouse and did not fly.
1066. The African Secretary-bird.
1067. Not always, and floating nests may be moved about by the wind.
1068. No, males have two rusty-brown wingbars, but are otherwise entirely deep, rich blue.
1069. The Dusky Seaside Sparrow. When the NASA Space Center was established at Cape Carnaveral (later Cape Kennedy), Merritt Island was extensively diked to control mosquitos. The dikes prevented the tidal flooding of the marshes, sealing the doom of the specialized sparrow, the last of which succumbed in 1987.
1070. The total number of eggs laid by one female during a single nesting attempt and incubated simultaneously.
1071. It is restricted to an island.
1072. Sálim Ali.

Answers

1073. Food poisoning affecting aquatic birds, caused by the anaerobic bacterium *Clostridium botulinum*. Also called western duck sickness, botulism in some years causes the deaths of millions of ducks and other aquatic birds.
1074. One whose newly-hatched young are well developed, covered with down, and are able to move about and feed themselves. Galliform birds, waterfowl and shorebirds are precocial birds.
1075. A flight of birds organized in V or wedge-formation.
1076. A hollow, thin-walled, bubble-like, bony prominence on the syrinx of many drake ducks.
1077. The Bonneville Salt Flats in Utah, a region practically devoid of birds.
1078. A toucan.
1079. Between 12 and 14 pounds (5.4-6.4 kg).
1080. The line of juncture between the closed mandibles when viewed laterally.
1081. A projection of land jutting into the sea, over which migrating eiders pass. Such sites are favored Eskimo hunting locations, and one of the most famous is at Point Barrow, Alaska.
1082. A specific region on the skin where feathers grow.
1083. Wilson's Storm-Petrel is often cited as the most numerous, and while hundreds of millions have been estimated, no counts have ever been made. Several shearwaters, and possibly the Sooty Tern, may be equally, or more numerous.
1084. The Little Blue Penguin of New Zealand and Australia.
1085. More than 350.
1086. Ford Motor Company.
1087. The much larger female will usually kill the male. The jungle-dwelling cassowaries are solitary, except during the breeding season, when fruit is most plentiful in the forest.
1088. The shorebirds turn over stones and seaweed in search of small crustaceans and other invertebrates. Both species have specialized, slightly upturned bills and especially strong neck muscles that aid them in turning stones over.
1089. The Galapagos Sharp-beaked Ground-Finch. The name alludes to its habit of drinking blood from developing flight feathers of molting Masked or Red-footed Boobies.
1090. 1961.
1091. Blue.
1092. The Marbled Murrelet. On August 7, 1974, administrators at Big Basin Park, California, were worried that a large Douglas fir near campsite J-1 would drop a branch on a tourist. A tree surgeon was sent up to cut the limb and while doing so, a chick was discovered. He caught it, but the bird escaped and fell 130 feet (40 m) to the ground. The chick survived the fall, but died two days later.
1093. None, although a few were introduced.
1094. About 60 mph (97 km/hr).
1095. About 45.
1096. The Wandering Albatross averages 278 days.
1097. The Dwarf Emus of Tasmania, King and Kangaroo Islands. On Kangaroo Island, the endemic species died out prior to 1836, probably because it was heavily hunted by the Flinders, the first settlers on the island, who arrived in 1802, and also because the habitat was irreversibly altered by extensive burning. The King Island Emu disappeared at the beginning of the 19th century, before arrival of the settlers, and while it was no doubt hunted to some extent by the natives, just why it disappeared is not known. In Tasmania, over-hunting and habitat alteration doomed the local Emu.

Answers

1098. Wandering and Royal Albatrosses have up to 40 secondaries, but like all petrels and albatrosses, they have only 11 primaries.
1099. Admiral Richard E. *Byrd.* He and three others flew over the pole in a Ford monoplane in November, 1929.
1100. 1941 or 1942.
1101. No, adult pelicans are essentially mute.
1102. The beautiful pink gull was discovered by and named after Admiral Sir James Clark Ross (1800-1862), the noted Arctic and Antarctic explorer.
1103. The name is an imitation of their distinctive call.
1104. The estimate is only five mph (8 km/hr).
1105. Nearly 100,000.
1106. The specific specimen on which a biologist bases his or her description of a new species or subspecies.
1107. Penguins.
1108. Mandarin Ducks. The pair symbolized marital fidelity, and in old Japanese painting and embroidery, the striking, colorful drake is traditionally pictured with his rather drab mate.
1109. A bad conduct discharge.
1110. Yes.
1111. The male Huia had a sharp, straight bill used for tunneling into dead wood. When an insect or grub was located, the female took over, and with her long, thin, decurved bill, was able to extract the prey which the male could not reach. Not all ornithologists subscribe to this theory and the extreme bill dimorphism may have evolved for courtship purposes.
1112. Mythical creatures with the head of a woman, but the wings, tail, legs and claws of a bird. The ghoulish creatures were said to carry off the souls of the dead.
1113. The pigeons. There are exceptions and some species construct delicate nests of finely woven material.
1114. Parrots were maintained in French forts and on the Eiffel Tower where the screeching birds generally gave warning of approaching German aircraft long before the planes could be detected by the French defenders.
1115. The petrels: Wilson's Storm-Petrel, Sooty Shearwater, Short-tailed Shearwater, Great Shearwater, etc. The 21,000-mile (33,800 km) annual migration of the Short-tailed Shearwater, for example, assumes a figure-of-eight course from southern Australia to the North Pacific and back again.
1116. No, birds have no skin sweat glands.
1117. *Houston, Tranquillity Base here. The Eagle has landed.*
1118. Yes, as a rule.
1119. Yes, Common (Brown) Noddy chicks may be uniformly dark, or mostly white, or intermediate in color.
1120. This is a widespread belief, but based on my own extensive experience, I don't believe the birds are particularly near-sighted. When ashore, most penguins can certainly detect movement several hundred feet (60 m) away.
1121. The American Woodcock.
1122. Chicks peck at the conspicuous terminal bill spot to stimulate the parent bird to regurgitate food.
1123. Korhaans.
1124. The Great Tit and Blue Tit.

Answers

1125. As incubation progresses, there is an evaporative water loss as water passes through the porous shell causing the eggs to become lighter. Fresh eggs sink if placed in water because the specific gravity is high, but float after the eggs have been incubated for a time. A 10-20 percent egg weight loss from laying to hatching is typical.
1126. No, most feed extensively on insects as well.
1127. Five: asities (false sunbirds), vanga shrikes, roatelos (mesites), ground-rollers and Cuckoo-Roller.
1128. Many insects have an unpleasant taste and some are poisonous if consumed, and a discriminating taste might serve as a safeguard by alerting a bird that a captured insect could be poisonous.
1129. Only one: Struthioniformes (Ostrich).
1130. No.
1131. Four to five feet (1.2-1.5 m) in length.
1132. The Mute Swan. Its egg averages 4.53 x 2.89 inches (11.5 x 7.3 cm).
1133. No, but hatching chicks have a small projection on their bills called an egg tooth that assists them in breaking through the shell. Some extinct proto-birds had teeth.
1134. The Wild Turkey, with a tail up to 15.75 inches (40 cm) long.
1135. Bey Al-Arnaut Abdim (1780-1827), at one time the governor of the Wadi Halfa area in northern Sudan.
1136. The hallux.
1137. Cape Royds, Ross Island (77°33'S), the site of a small Adelie Penguin rookery, although a few pairs recently bred several miles farther south at Cape Barne.
1138. The Heath Hen was last seen on the night of March 11, 1932, at Martha's Vineyard, Massachusetts.
1139. Between 500 and 700 years ago.
1140. Approximately 14 feet (4.3 m) of worms. This was determined experimentally when a large robin chick was fed all it would eat.
1141. Only the Ostrich. In most birds, the liquid waste is almost completely re-absorbed through the walls of the cloaca. The re-absorption of water provides birds an almost perfect system of water usage and conservation.
1142. The South American Orinoco Goose.
1143. 1905. State societies existed much earlier (36 by 1901 alone), but became united when the National Association of Audubon Societies was formed. The name was subsequently changed to the National Audubon Society.
1144. The birds and turtles.
1145. The feather, one of the lightest and, at the same time, one of the strongest materials produced by any animal.
1146. The Red-breasted Goose. It breeds in Siberia, but many winter in Rumania.
1147. The lore.
1148. From a distance, the white head of an adult eagle apparently appeared bald to early observers.
1149. The Northern Jacana, but it just barely reaches southern Texas.
1150. The primary and secondary flight feathers.
1151. Feeding cranes dig their bills into iron-rich mud and rotting vegetation, and when preening the birds "paint" themselves. Once heavily stained or pigmented, the gray cranes become brownish. Juveniles are also brown.

Answers

1152. Pegasus.
1153. Forty-two.
1154. The horn is shed.
1155. Haiti, on April 26, 1785.
1156. General Motors Corporation, Pontiac division.
1157. Squid, generally taken at night when the deep-dwelling cephalopods migrate to the surface.
1158. The Royal Penguin.
1159. The Puna Flamingo of the Andes, with a population of about 50,000 birds.
1160. The Galapagos Islands. The main breeding grounds of the much rarer Hawaiian race are on Maui, in the walls of Haleakala Crater.
1161. The Blood Pheasant. It lives near the snow line, usually between the 9,000-15,000 foot-level (2,743-4,572 m) in the Himalayas. Monal pheasants have been observed higher, but not consistently.
1162. The goose is endemic to Hawaii, formerly known as the Sandwich Islands.
1163. The boobies. Frigatebirds often grab boobies by the tail or wing, forcing them to disgorge their fish, which the kleptoparasitic frigatebirds skillfully catch in the air.
1164. Madame Adelie, the wife of the French Admiral Jules-Sebastian-Cesar Dumont d'Urville, a well-known 19th-century Antarctic explorer.
1165. Iceland, with a population of 8-10 million birds.
1166. Turkey grass.
1167. The Dalmatian Pelican. The birds have declined dramatically since the last century when "millions" were said to inhabit Rumania alone. In 1979, the population was estimated at 665-1,000 pairs, about half of which were breeding in what was formerly the USSR. Between 1,926 and 2,710 pairs were estimated in 1991 in 21-22 colonies.
1168. The five tragopans.
1169. The cormorant is still as numerous as when discovered, but is rare and vulnerable.
1170. Tiny voles and shrews, but the huge owl may occasionally take frogs and small birds.
1171. The gonads may enlarge several hundred times (or more). Following the breeding season, they shrink to such an extent that in many cases testes are scarcely larger than a pinhead.
1172. Probably to reach past the fragile wings of butterflies and dragonflies to grasp their bodies firmly and to keep their fluttering wings, as well as the stingers of bees and wasps, away from the bird's face while the bird smacks its prey against a perch. Jacamars feed extensively on large morphos and swallow-tailed butterflies, and once the brilliant wings are knocked off, the body is swallowed whole. Such prey is generally shunned by other insect-eating birds.
1173. The Magellanic Plover of southern Argentina and Chile. It has a crop used to carry food to the young.
1174. A bird of the open sea, generally remaining out of sight of land.
1175. Slightly more than 81 pounds (37 kg). The huge turkey was sold at auction in England in December, 1986 for £3,600. The money raised went to the Save the Children Fund and the bird itself was donated to the children's ward of a London hospital.
1176. An instrumented Thick-billed Murre dove 689 feet (210 m).
1177. The chachalacas.
1178. The South American Hyacinth Macaw.
1179. Yes, but not by much. In 1987, the global chicken population was about eight billion birds and, as of mid-1993 nearly 5.5 billion humans inhabited the planet.

Answers

1180. Prion is from the Greek *prion*, meaning "a saw," referring to their bills that have serrated edges used for filtering food from the sea. Prions are a distinct group of subantarctic petrels.
1181. No. During the breeding season, the unfeathered skin surrounding the eyes of both sexes becomes a deep brilliant blue, but the eye color itself is dark.
1182. At least twenty.
1183. The penguins. Some species spend up to 75 percent of their lives *in* the sea.
1184. Bright yellow.
1185. The Ostrich. On a more limited scale, Emus have been commercially raised on farms in Western Australia since 1970, primarily for skin, which makes a fine leather. The meat has a beef-like flavor.
1186. Panama, where pelicans commonly fly across the narrow isthmus of the republic. Frigatebirds also commonly fly across the country.
1187. The Magellanic Penguin. The Argentine population alone exceeds a million birds.
1188. Humphrey Bogart, who played the famous private-eye Sam Spade.
1189. A small plover.
1190. Pennsylvania.
1191. The Mauritius Parakeet (Echo Parrot). As of 1992, probably no more than 14 existed, most of which were males.
1192. Shakespeare referred to London as *a city of crows and kites*. Centuries ago, kites scavenged in central London, and while in much of the literature the raptors are called Red Kites, they were more likely Black Kites. For many years it was a capital crime to kill a kite in London because the birds were said to keep the streets clean of filth. Ironically, kites had a price on their heads in the countryside.
1193. To be given an insult by an obscene gesture where the middle finger of one hand plays a major role. The gesture is also commonly known as *the finger*.
1194. Defensive birds face a predator and fan out their wings and tails, presenting a body area four times their normal size. They may also hiss like a snake or growl like a dog.
1195. The flightless Takahe of New Zealand weighs more than seven pounds (3.2 kg).
1196. No, but the endemic Brolga was considered the only native crane until 1966, when Sarus Cranes were discovered. How they got there from Asia is not known, but the cranes were probably in northern Australia all along and simply overlooked because of the similarity in appearance to the Brolga.
1197. As of 1990, the Guam Kingfisher, Socorro Dove and Edward's Pheasant. The last remaining Guam Kingfishers were captured in 1986, and captive propagation has commenced. Based on recent surveys in Vietnam, it appears that Edwards' Pheasant is extinct in the wild. However, from seven birds exported by Jean Delacour in 1925, a captive population exists, despite some inbreeding problems. The last surviving Guam Rails were also removed for captive propagation, many of which were subsequently reared. In 1990 some were released on snake-free Rota Island, 30 miles (48 km) north of Guam. For several years, all California Condors were in captivity.
1198. Owl ear openings are asymmetrically placed. The left ear is higher with the opening tilted downwards, so that it is more sensitive to sounds from below. An owl's head is relatively wide, so sounds reach one ear a fraction of a second before the other and this, coupled with the placement of the ears, permits a precise location of the sound.
1199. The Surf Scoter.

Answers

1200. The loons and grebes

1201. The San Clemente Loggerhead Shrike, a race of the mainland form. The population is unstable and possibly no more than seven birds remained on the island in 1991. Predators such as feral cats, native foxes and Common Ravens, as well as the elimination of nest sites by browsing feral goats, are responsible for the decline. The exotic predators are being hunted and trapped and as of late 1991, only three goats, possibly 50 pigs, and hundreds of cats still inhabited the island. It is doubtful that cats can ever be eradicated. In 1991, 12 eggs were collected and taken to the San Diego Zoo where seven chicks were reared. A rearing station was also established on the island, and a number of captive-reared shrikes have been released onto the island, more than doubling the population in less than two years; nearly 50 as of March, 1993.

1202. There is no difference.

1203. The Marabou Stork, with a *documented* wingspan of 10.4 feet (3.2 m).

1204. Yes, the Northern (American) Jacana and the Black Coucal. Simultaneous polyandry was also reported at least once in a population of Spotted Sandpipers.

1205. Noddies are more pelagic, settling on the water for hours without becoming water-logged.

1206. The New Zealand Scaup.

1207. The Philadelphia Academy of Sciences, Philadelphia, Pennsylvania.

1208. Woodpeckers constantly pound on trees and their brains must be cushioned to withstand the repeated shocks.

1209. Yes, but only once. The only *documented* incident occurred on June 5, 1932, at Leka, Norway, when a White-tailed Sea-Eagle picked up 4-year old Svanhild Hansen, carrying her for nearly a mile (1.6 km) before dropping her on a narrow ledge only 50 feet (15 m) from its eyrie, 800 feet (240 m) up the side of a mountain. Aside from some scratches, the child was unharmed. The little girl was particularly small for her age, and the eagle probably hit a powerful updraft of air at precisely the right moment to provide the required lift.

1210. The Emperor Penguin. Territories are not defended because the egg is incubated on top of the male's feet. For males to successfully incubate during the middle of the brutal Antarctic winter, the penguins must gather in large, compact huddles of thousands of birds, which greatly reduces individual heat loss. The birds must be in physical contact with one another, thus aggressive territorial behavior would be counterproductive. In other words, Emperor Penguins depend on each other for their success as a species, and huddling in such a manner represents an extraordinary act of cooperation in the face of a common hardship.

1211. As early as 16,000 B.C.

1212. *Martha.* The captive-bred bird was named after Martha Washington.

1213. From its distinctive, often repeated call of *kill-dee.*

1214. The marine geese rarely stray out of Alaska, but females occasionally dump eggs in the nests of Pacific White-fronted Geese. The resultant young may follow their foster parents down the Pacific coast as far south as northern California.

1215. The Northern Flicker's bill is slightly curved.

1216. Light blue, but the eggs turn chalky-white upon drying and subsequently become nest-stained.

1217. The thin frontal spike may be up to six inches (15.2 cm) long.

1218. No. Avian color vision, at least in diurnal species, is apparently about equal to that of humans, although many species can also see colors in the near ultraviolet range.

Answers

1219. On the dried dung of large mammals, such as an hippopotamus or elephant.
1220. No, but wheat was formerly called corn in Britain, and the rail does consume wheat.
1221. The waxbill finches of Africa.
1222. Kingbirds are especially pugnacious and have been known to attack hawks, crows, vultures, humans and in one instance, reportedly pursued an airplane.
1223. The Western Grebe.
1224. The avian and reptilian ear bone. The single bone corresponds to the three small bones of the mammalian middle ear.
1225. The woodcocks, that feed by thrusting their long bills into the soil, and their eyes are located high on the head so that the field of vision of each eye overlaps fore and aft. As a result, the birds have binocular vision to the rear as well as to the front, and can spot danger while feeding.
1226. As of the late 1980's, more than 562 billion.
1227. John James Audubon. Audubon later regretted this, as well as the loss of his hearing through so much firing of his gun. He often collected far more birds than were required for specimens or for the pot, although he commonly consumed his specimens.
1228. Yes. Most molting cranes are probably flightless for a brief period, with the possible exception of the Demoiselle Crane.
1229. The tail feathers.
1230. The Yellow Wattlebird of Tasmania. The nearly 20-inch-long (50 cm) bird weighs 8.8 ounces (250 gms) and has a pair of long, pendulous, orange-yellow wattles dangling from the side of its face.
1231. No, even hair-like structures such as down, eyelashes and rictal bristles are modified feathers.
1232. The Hawaiian Goose (Nene).
1233. The Spectacled Eider.
1234. The Wood Stork.
1235. Not really. While the silky-flycatcher preys on some insects, the berries of the mistletoe growing on mesquite bushes are far more important. It also feeds extensively on elderberries.
1236. The Common Wood-Pigeon. The pigeons cause millions of pounds worth of crop damage each year, even though hunters may harvest as many as 12-16 million annually.
1237. Skuas.
1238. The Oldsquaw duck and King Eider.
1239. Both groups forage on the wing, but swifts speed through the air with bursts of flickering wingbeats interspersed with glides, often at considerable heights. Swallows are slower with frequent wingbeats, but are more maneuverable. The swallow's flapping flight uses more energy, but the rewards are greater as the bird can twist and turn after large insects. Conversely, the swift's rapid glides are economical in terms of energy used, but it catches mostly small, weakly-flying insects.
1240. A large, horny knob grows near the middle of the upper mandible. During or following the breeding season, it drops off. Defensive pelicans snap at one another with their bills, and the horny growths may function as targets, thus averting damage to the delicate skin of the pouch or head.
1241. Pigeons and doves, flamingos and male Emperor Penguins.
1242. Population estimates range as high as 42 million birds.

Answers

1243. Yes, non-unison flapping would be aerodynamically unstable.
1244. None.
1245. The Striated Caracara.
1246. Only one, the Roseate Spoonbill.
1247. The Smew, a petite Eurasian merganser. Drakes in breeding plumage are basically white.
1248. The Lammergeier (Bearded Vulture).
1249. The birds need to make maximum use of low tides to obtain sufficient food because feeding opportunities are limited during high tides.
1250. The Dominican or Southern Black-backed Gull.
1251. The Brahminy Kite.
1252. The unique bird has a huge, conical bill used for digging up grubs, centipedes, snails and other small invertebrates, and even vertebrates such as lizards, but its favored prey is the earthworm. In mangrove habitats, it also preys on small crabs. The kingfisher thrusts its specialized bill up to three inches (8 cm) into the soil, working it from side to side for up to a minute at a time, while plowing through a square meter of leaf-littered soil.
1253. The turkey. A courting albatross gobbles loudly, squeals, brays and bill-clatters.
1254. No, but the two sheathbills consume small amounts of algae. During the winter, as much as 50 percent of the diet of some Black-faced Sheathbills consists of algae.
1255. *Lagopus* is from the Greek and means "hare-foot," a reference to the feathered feet and toes of the ptarmigan.
1256. Yes, the raptor has a distinctive two-note, laugh-like call, usually repeated many times.
1257. Yes, but this is unusual in shorebirds. Food is brought to small oystercatcher chicks, but they are led to feeding grounds by their parents when older.
1258. About 20 beats a second.
1259. Not really, but as of early 1994, it was still on the endangered species list. In the late 1960's and early 1970's, it was certainly at risk. While currently numerous, Brown Pelicans remain extremely vulnerable to reduction of food fish, disturbance on the breeding grounds, or resumption of use of certain types of pesticides.
1260. Probably the gannets and boobies. The birds plunge into the sea from a height of 100 feet (30 m) or more, striking the water at about 60 mph (97 km/hr). Most dives are relatively shallow, seldom exceeding 30 feet (9 m).
1261. The Snowy Sheathbill, that steals food from penguins feeding chicks. When a penguin passes a food bolus to its chick, a sheathbill may suddenly fly up and frighten them and, as the penguins jerk apart, food is spilled that the sheathbill gobbles up. A few Kelp Gulls have learned to do the same thing.
1262. The Green Heron has been known to drop bait in the water to attract fish into striking range. This might be considered tool use, but it has only been noted in a few individuals.
1263. The Impeyan Pheasant was named after Sir and Lady Impey, who initially introduced the bird to England. The first captive monals were maintained by Lady Impey in India, but the pheasants never reached England alive, having died during the journey.
1264. No, avocets have recurved bills. Flamingos have bent bills.
1265. The pigeon berry.
1266. Nearly 80 percent migrate to some extent.
1267. Following birds conserve energy by using the lift of disturbed air created by the wingbeats of the bird ahead. The lead bird works the hardest and, at least in northern geese and probably swans and cranes, the leader is replaced periodically.

Answers

1268. The Eskimo Curlew was called a dough bird because during its fall migration the shorebird was extremely plump and its skinned breast revealed a thick layer of fat resembling dough.
1269. Its compressed bill is unusually sharp.
1270. White winter plumage conceals the grouse in a world of snow and ice, but black in the tails may serve as *flags* to help keep winter flocks together in flight.
1271. The Eurasian Hoopoe. Its extremely foul-smelling cavity nest may function as a predator deterrent. The penetrating stench is produced by the large preen gland.
1272. The Northern Harrier.
1273. The megapodes (incubator-birds). Eggs are buried deep within nesting mounds or pits, thus emerging feet first is advantageous. Kiwi chicks may exit the egg feet first as well.
1274. Yes, as adults, but the newly-hatched chicks of many species are poikilothermic (unable to control their body temperature) and must be brooded by adults. Birds that hibernate are also periodically poikilothermic.
1275. The Snowy Sheathbill.
1276. No, most are insectivorous.
1277. Dr. Jared P. Kirtland (1793-1877), a zoologist from Ohio.
1278. No, but Magnificent Frigatebirds breed along the central coast of Texas, and on Marquesas Key in Florida.
1279. Until recently, it was believed that both Emperor and King Penguins fed primarily on squid, but fish constitute a greater portion of their diet. Squid beaks are indigestible, and when large numbers of beaks were found in penguin stomachs, the assumption was made that squid was the primary dietary item. The penguins also feed on krill.
1280. Above the eyes.
1281. A small, drab, difficult to identify bird is often called a *little brown job* (LBJ).
1282. The Old World Black Woodpecker pounds between 8,000 and 12,000 times a day. Pounding rates no doubt vary from species to species.
1283. No native geese inhabit New Zealand, but the Canada Goose was introduced and is currently numerous and widespread.
1284. Yes, turkeys sometimes gang up on a rattlesnake and kill it, and the vanquished snake may be consumed.
1285. The Peacock Throne. The grand throne is crusted with rubies and diamonds and in 1960, was said to be valued at more than $100 million.
1286. The Short-tailed Albatross.
1287. A Mallard. On July 9, 1963, the duck struck a Western Airlines L-188 Electra above Elko, Nevada, at an altitude of 21,000 feet (6,400 m).
1288. *The Auk.*
1289. The Red-footed and Abbott's Booby.
1290. The bullfrog, that preys on ducklings.
1291. Birds such as the Sage Grouse, where males engage in communal courtship on traditional display grounds, such as a lek.
1292. The White-chinned Petrel. The strange sounds the birds make in their underground burrows recalled the noise produced by cobblers (shoemakers).
1293. A total of 489.
1294. Auklet remains have been discovered in the stomachs of *bottom-dwelling* codfish caught 200 feet (61 m) below the surface.
1295. Behind waterfalls.

Answers

1296. Harry *Robbins* Haldeman. Haldeman died of cancer in 1993.
1297. The intentional killing of a sibling by its nest-mate, also known as sibling murder or siblicide.
1298. The Black Guillemot, a name that apparently originated from its twittering notes.
1299. The Wild Turkey, but the beard is on its chest, not its face.
1300. Strangely enough, no one seems to know.
1301. A small, neotropical honeycreeper.
1302. The Whooping Crane, because of its large size and whiteness.
1303. The American Kestrel, that is only 9-12 inches long (23-30 cm), with a wingspan of 20-24.5 inches (51-62 cm). Females average 4.25 ounces (120.5 gms), whereas males weigh 3.4-4.0 ounces (96.4-113 gms).
1304. A mass or ball of food swallowed by a bird. Bolus generally applies to a ball of food regurgitated by an adult to feed its chicks.
1305. The Pileated Woodpecker.
1306. An eye-shaped spot (ocellus) on the feathers of certain birds, such as on the tail of an Ocellated Turkey or the train (upper tail coverts) of a male Indian Peacock.
1307. The North American anis.
1308. A thermoregulatory behavior where a bird defecates on its legs in order to cool off via evaporative cooling. This is a common habit of storks and New World vultures, as well as some seabirds, such as cormorants and boobies.
1309. Yes, the vultures may attack, en masse, skunks, killing and consuming them.
1310. If penguins molted gradually, as do most birds, they would not be waterproof and would perish in the frigid sea. During the catastrophic molt, the non-waterproof penguins must fast ashore or on the ice as they cannot enter the sea to feed.
1311. The tropicbirds. Their long central tail feathers reminded early sailors of marlin spikes, and the shrill, trilling calls of the Red-billed Tropicbird resembled the whistles of a boatswain's pipe. Jaegers are also known as bosun birds because of their long tails.
1312. Bird-loving, a term applied to plants pollinated by birds.
1313. Two, both of which inhabit Australia: Superb Lyrebird and Albert's Lyrebird.
1314. A thrush.
1315. The Gray (Black-bellied) Plover, that weighs up to 10.5 ounces (298 gms).
1316. The Great Cormorant.
1317. Yes, most species do.
1318. The flamingo.
1319. The California Condor. Some American Indians also regarded eagles and hawks, and other types of birds as well, as thunderbirds. The birds were believed to control rainfall.
1320. The four South American seedsnipes are distant relatives of the shorebirds. They feed on seeds and other vegetable matter, and fly like snipes.
1321. The lipstick line, but unless viewed closely, the red is not visible.
1322. *The Johnny Penguin,* but the origin of the name is not known. Striated Caracaras often feed on Gentoo Penguin carcasses, bringing about the kelper name *Johnny Rook* for the raptors.
1323. Courting noddies have an elaborate head-nodding ceremony, and when greeting one another, the terns also frequently nod; presumably the name originated from this behavior.
1324. It is not known, but probably 22-24 days.
1325. White.

Answers

1326. Robber crabs of tropical Indo-Pacific islands and atolls are not only the heaviest land crab, but also the largest land crustacean. These formidable creatures are actually shell-less hermit crabs with leg spans of 3.0 feet (1 m), weighing as much as 5-6 pounds (2.3-2.7 kg).
1327. Small birds that are swept out of the air into the nightjar's large open mouth, but the birds are probably taken accidentally.
1328. A dated American slang expression referring to loose, under-age girls on the prowl.
1329. Yes.
1330. The Rancho La Brea Tar Pits at Hancock Park. No less than 133 fossil bird species have been discovered in the tar pits, of which at least 20 are extinct, including the Passenger Pigeon.
1331. The Long-tailed Duck.
1332. Even at rest, at least 50 times greater.
1333. Yes, males are responsible for most nocturnal incubating and brooding.
1334. The Gray Jay.
1335. The Wild Turkey.
1336. Probably the openbill storks. Like the Limpkin, the specialized birds feed extensively on snails.
1337. The caracaras. Typical falcons build no nests.
1338. The two murres. Their eggs are among the most variable in color and patterning, theoretically enabling the birds to locate their own single egg on packed nesting ledges.
1339. The Yellow-billed Loon.
1340. The Crab Plover.
1341. The Dark-rumped (Hawaiian) Petrel.
1342. When birds alter their normal coloration by "painting" themselves while preening after feeding in mineral-rich mud or decaying vegetation, it is referred to as adventitious coloring. The term is used for color in feathers caused by a chemical or other matter in the environment that discolors or soils plumage.
1343. An albinistic form of the Mute Swan in which the legs and feet are pinkish-gray rather than brownish-gray, and the down of newly-hatched cygnets is white, rather than grayish.
1344. The Ruddy Duck. Hunters thought the duck's feathers were impervious to shotgun pellets.
1345. The parrots, with more than 350 species.
1346. The hummingbirds, with more than 320 species.
1347. A dish-shaped clay target used in skeet and trap shooting.
1348. The Lesser Flamingo.
1349. The 16 to 20-inch-long (40.6-51 cm) American Black Skimmer.
1350. Limpets are swallowed whole and the shells regurgitated later, often in extensive middens.
1351. Mosquitos were introduced, allowing exotic diseases, such as avian malaria, to be transmitted to endemic birds. As a result, many were eliminated. Most of the survivors occur above 2,000 feet (610 m), where mosquitos are less numerous.
1352. An African thick-knee (stone-curlew).
1353. *Robin.*
1354. From the inferred sadness of its call.
1355. Yes, the large, heavy birds are surprisingly capable swimmers, and can cross lakes and wide rivers without difficulty.
1356. There is little sexual dimorphism in northern geese except for size, but the South American Kelp Goose and the Magellanic (Upland) Goose exhibit extreme dimorphism.

Answers

1357. The Ruddy Turnstone, because of its fondness for horseshoe (king) crab eggs.
1358. The tinamous, that rank among the most tasty of all birds. As a youth growing up in Panama, one of my assigned tasks was to roam the jungle several times a month to bag a brace of tinamous for my father, who considered them a gastronomic delight.
1359. An American birding term describing small, generally difficult to identify, nondescript passerines.
1360. The White-tailed Tropicbird. The larger, more aggressive bird usurps nesting cavities, sometimes even tossing Cahow chicks out.
1361. The air space.
1362. He is credited with bringing the statue Venus de Milo to the Louvre in Paris.
1363. The King Vulture, a tropical raptor that is mostly white.
1364. Land is near.
1365. 1903, at Kitty *Hawk*, North Carolina.
1366. Some of the manakins, especially the Red-capped Manakin.
1367. The fish-eating mergansers, because their serrated bills resemble saws.
1368. Parrot-fish.
1369. Grays Lake, Idaho, where Whooping Crane eggs were cross-fostered under incubating Sandhill Cranes. The resultant young winter in central New Mexico with the Sandhill Cranes. As of 1993, no Whooping Cranes had bred in Idaho and until this occurs, the noble experiment cannot be considered successful. In February, 1993, 14 captive-reared Whooping Cranes were released in south-central Florida in an attempt to reintroduce the species to the state, but within a month of release, five were killed by bobcats. Five more cranes were shipped to Florida in November of 1993, and 14 more are destined for Kissimmee in early 1994.
1370. The Rockhopper Penguin.
1371. The Waldrapp, or Northern Bald Ibis.
1372. *Many-tongued mimic,* referring to the classic mimicking ability of the mockingbird.
1373. The *Kondor* Legion.
1374. Bright blue.
1375. Harris' Sparrow, that weighs up to 1.75 ounces (50 gms).
1376. Domestic (Graylag) Geese.
1377. The Pinyon Jay.
1378. The Great Cormorant, that weighs 5-11 pounds (2.3-5 kg).
1379. *To talk turkey.*
1380. Rod Taylor and Tippi Hedren.
1381. Males may construct a number of dummy or *cock* nests.
1382. The east coast of Australia and throughout the lowlands of New Guinea and adjacent Papuan islands. The pigeon was named for its booming, far-carrying calls.
1383. The Ostrich.
1384. Yes, bluish-gray.
1385. The Bald Eagle.
1386. Pekin Duck.
1387. The Common Loon.
1388. The ripe old age of 29 years.
1389. The main complaint was that steel shot lacked the killing power of lead, and more cripples would result.

Answers

1390. The Fox Sparrow, because of the rufous, fox-like color of birds of the eastern race.
1391. *Legal eagles* or *vultures*.
1392. Yes, killer whales, the largest of the dolphins, prey on penguins, especially in the subantarctic. Penguins typically do not constitute a major portion of their diet because it is far more economical for orcas to prey on seals weighing 600-1,000 pounds (272-454 kg) than birds only weighing 10-90 pounds (4.5-41 kg).
1393. None. The pole is 800 miles (1,287 km) from the nearest coast, high atop more than 9,000 feet (2,743 m) of ice.
1394. A covey or bevy.
1395. A bird that primarily feeds on fruit. Soft-billed is essentially an aviculturist's term.
1396. Crow's feet appear at the corners of the eye.
1397. The African mousebirds.
1398. Yes, the owls feed on bats.
1399. The Snares Islands Crested Penguin roosts on low, horizontal, wind-swept branches of the daisywood tree (*Olearia*). Rockhopper penguins inhabiting some southern Chilean Fuegian islands may also roost on the branches of southern beech trees.
1400. The Masked Duck.
1401. The Australian Gouldian Finch. Three light-reflecting gape tubercles or *reflection pearls* are located at each angle of the nestling's jaw. The bead-like bodies are brilliantly opalescent in emerald-green and blue. Chicks of many species have a brightly-colored lining to the mouth that serves as a beacon or target for provisioning adults. The colors are usually orange, yellow or white, and the mouth lining may be patterned.
1402. The extinct Steller's sea cow and Steller's sea lion. The only description of the sea cow was made by Steller when he was shipwrecked on Bering Island in 1741. The population probably did not exceed 2,000, and by 1768 fur-hunters had eliminated every one of the 24-foot-long (7.4 m), 10-ton sea mammals. Until recently, the sea lion was numerous, but its population dropped to about 40,000 in 1991, down from 170,000 in the late 1950's. The specific cause of the continuing decline is unknown, although incidental net kills, shooting by fishermen, and reduction of important prey species by commercial fishing operations have contributed significantly to the decline, and the sea lion is currently considered threatened.
1403. No, although in North America it only inhabits Florida. While once widely distributed in Florida, the kite is currently more or less restricted to the Loxahatchee Wildlife Refuge in a corner of Lake Okeechobee.
1404. The Northern Gannet.
1405. The whitish guano of seabirds and cliff-nesting raptors.
1406. The Peruvian Plantcutter. All three South American plantcutters feed extensively on foliage.
1407. The endemic Galapagos Lava Gull. The entire population consists only of about 400 pairs, but the gull is not considered endangered.
1408. The common chamber in birds and reptiles that receives the rectum, ureters and gonaducts.
1409. Kirtland's Warbler.
1410. A captive pigeon used to lure flocks of wild pigeons to the ground where the birds were caught by trappers with nets. The stool (lure) pigeon was firmly attached to a stool or platform. Stool pigeons were commonly used to trap Passenger Pigeons. Stool pigeon is an American slang term for an informer.

Answers

1411. None.
1412. The Rockhopper and Macaroni Penguin, on some islands off southern Chile.
1413. The Australian Emu. Both are in the order Casuariiformes.
1414. Yes.
1415. Robert *Hawke* and Andrew *Peacock,* with Hawke prevailing.
1416. It preys extensively on lamprey eels, as well as squid.
1417. The two threatened African bald-crows (*Picathartes*) are taxonomically puzzling and were long classified with the babblers, although the birds are unlike any other babblers. The unusual birds are so distinct that they were placed in a separate family (Picathartidae).
1418. The drake Musk Duck of Australia and Tasmania.
1419. The toma*hawk.*
1420. Almost. With the exception of several pairs nesting on Isla Malpelo, Colombia, all 10,000-15,000 pairs breed in the Galapagos Islands.
1421. The Jackass (Black-footed) Penguin, that breeds on islands off the coast of southern South Africa, and locally on the mainland in areas generally protected from terrestrial predators.
1422. There is no nail on an Ostrich's outer toe. The thick inner toe is armed with a 4-inch-long (10 cm) sturdy, flattened claw.
1423. The Cackling Goose.
1424. A highly venomous fish more commonly known as a lion fish.
1425. The Red-tailed Hawk.
1426. The Long-eared Owl.
1427. The two magpies. The 17.5-22.0-inch-long (44.5-55.9 cm) Black-billed Magpie has a tail length of 9.5-12.0 inches (24.0-30.5 cm), or slightly more than half the total length of the bird. The 16-18-inch-long (40.6-45.7 cm) Yellow-billed Magpie has a tail 9.5-10.25 inches (24-26 cm) in length.
1428. An average of about 10 pounds (4.5 kg), but up to 16 pounds (7.3 kg).
1429. Tower Island (Isla Genovesa, Galapagos Islands), a boobery consisting of approximately 140,000 pairs. About a third of the world's population of Red-footed Boobies nest in the Galapagos Islands.
1430. No, the Common Gallinule was recently officially renamed the Common Moorhen. However, the Purple Gallinule is still known as the Purple Gallinule.
1431. Birds.
1432. Most cranes construct substantial nests, but the Demoiselle and Stanley (Paradise or Blue) Cranes use little or no nest material.
1433. A watch of Nightingales.
1434. The Band-tailed Pigeon. It does not weigh more than a pound, attaining a maximum weight of 12 ounces (340 gms).
1435. Parrots are cavity-nesters, thus there is no need to camouflage eggs with color or pattern.
1436. In 1903, the plumes were valued at $32.00 an ounce (28.4 gms). An ornithologist of the time sagely noted that as long as egret plumes are worth their weight in gold, there will be someone to supply them.
1437. Yes. While sometimes observed in small groups, the sandpiper is generally found alone.
1438. Yes, crows swallow golf balls but later regurgitate them. Ostriches commonly pick up and ingest foreign objects, including golf balls, and often perish as a result. An Ostrich cock in the London Zoo had in its stomach an alarm clock, three feet (91 cm) of rope, a pencil, three gloves, a roll of film, a cycle valve, a handkerchief, a comb, a collar stud and assorted

coins of various nations. Another bird at a South African Ostrich farm ingested 484 coins weighing 8.25 pounds (3.74 kg) in its stomach.

1439. Probably the Wood Duck.
1440. Mercury, but in Greek he was known as Hermes. His wings were feathered.
1441. Florida.
1442. The two rheas.
1443. Kiwi fruit.
1444. The nest of a specific termite, and the parakeet's Central American range corresponds with that of the termite.
1445. Yes, but the marine ducks are so tough that they are generally only eaten once.
1446. Wisdom.
1447. The Cactus Wren, that attains a length of 8.75 inches (22 cm).
1448. Not really, although some biologists claim that the Red-throated Loon can, but only with great difficulty.
1449. *Opus* the penguin.
1450. June 20, 1782.
1451. The population of the Siberian-breeding sandpipers is estimated at 2,000-2,800 pairs, but even these low numbers are possibly optimistic.
1452. None.
1453. No.
1454. Canada and Mexico.
1455. The Eurasian (Common) Starling.
1456. Icarus flew too close to the sun and his flight feathers, which were attached to his arms by wax, dropped off when the wax melted. Icarus then plunged to his death in the Aegean Sea. To this day, an Icarian flight refers to one who soars too high for safety.
1457. It is not a bird at all. The lengthy scientific name refers to the unique duck-billed platypus.
1458. The Nene (Hawaiian Goose), that inhabits highland lava fields on the big island of Hawaii and Maui.
1459. The Red-footed Booby weighs about 2.2 pounds (1 kg).
1460. Hawaii.
1461. The Flamingo, built by the playboy mobster "Bugsy" Siegel in 1946.
1462. Yes, some nest on the land closest to the North Pole, at times within 600 miles (966 km) of the pole.
1463. Isla Rasa, in the Sea of Cortez, Mexico, where 120,000 pairs breed.
1464. Not normally, but on September 16, 1980, Emperor Penguins hatched for the first time outside of the Antarctic at Sea World in San Diego, where the birds have bred sporadically ever since.
1465. The *Osprey* is a special tilt-roter aircraft built by the Boeing Aircraft Company and Bell Helicopter using two movable rotors to take off and land like a helicopter, but flies like an airplane.
1466. The Waved Albatross.
1467. When both partners of a bird pair sing simultaneously as part of the courtship display or to maintain the pair bond, they are duetting. A special type of duetting is known as antiphonal singing where the partners may alternate with extraordinarily accurate timing, often singing different phrases, so that unless one is actually watching, it may not be possible to determine that the song is not coming from a single bird.

Answers

1468. The frigatebirds, in which the gular pouches of displaying males are inflated to grotesque proportions, resembling bright red balloons.
1469. The hummingbirds, with their flight muscles constituting about 30 percent of their overall weight.
1470. No.
1471. *Snowbirds.*
1472. Three: American, Northwestern and Fish Crow.
1473. The Scarlet Ibis.
1474. Boobies frequently perch on boats, often selecting a vantage point near the bow from which they dive on flying fish. A hatch is often located near the bow, and the term *booby hatch* almost certainly was derived from the booby's habit of perching on or near the bow hatch.
1475. The Swallow-tailed Gull of the Galapagos Islands.
1476. Craveri's Murrelet, in the Sea of Cortez, Mexico.
1477. No.
1478. The African Goliath Heron. Standing 4.9 feet (1.5 m) tall, it weighs up to 9.5 pounds (4.3 kg), with a bill length exceeding eight inches (20.3 cm).
1479. 1861, in a limestone quarry at Solenhofen, Bavaria, southern Germany.
1480. Chicks hiss like snakes, and if a predator persists, they let loose a stream of pressurized excrement into its face.
1481. No, if a chick is fledged, the albatross does not nest the following season.
1482. Long Beach, California, in 1910, when a plane struck a gull and crashed. The pilot did not survive; the fate of the gull was not recorded.
1483. Based on some studies, about 2,000 fish.
1484. The Lammergeier (Bearded Vulture), with a wingspan of up to nine feet (2.7 m) and a maximum weight of 16.5 pounds (7.5 kg). The female White-tailed Sea-Eagle is equally as heavy.
1485. Adults feed primarily on seeds, grain and fruit, but up to 95 percent of a nestling's diet consists of insects.
1486. More than 4,000.
1487. Duck legs are placed far back on the body as a swimming adaptation. This no doubt increases the efficiency of aquatic propulsion, but forces them to move with a waddling gait because they must position their center of gravity above the supporting leg to maintain balance. Because of their wide bodies and relatively short legs, this positioning can only be accomplished by a rotation and shifting to the side, which in turn brings about the duck's characteristic waddle.
1488. Darwin's finches of the Galapagos Islands.
1489. The Purple Martin, with a fledgling period of 28-35 days.
1490. The waterfowl (ducks, geese and swans), from the family name Anatidae.
1491. Yes. Hummingbirds, House Finches, Purple Finches, Bushtits and small birds of the Old World tropics become entangled in spider webs and perish.
1492. Audubon's Warbler (now known as the Yellow-rumped Warbler) and Audubon's Shearwater.
1493. Probably about 44 days.
1494. The three South American screamers. While not similar in appearance, screamers and waterfowl share the same order.

Answers

1495. Clams or oysters occasionally snap shut on the feet of small shorebirds that subsequently drown as the tide rises, or a mussel may clamp down on the tongue of a foraging scoter, usually with fatal results.
1496. The Kelp Gull, that nests along the Antarctic Peninsula as far south as 65°.
1497. No, owls hatch with their eyes closed, but hawk chicks hatch with open eyes.
1498. Artificial burrows have been provided for Cahows. Wooden baffles across the entrance of burrows permit access by the endangered petrels, but the holes are too small for the slightly larger tropicbirds.
1499. The Parakeet Auklet.
1500. Yes, at least 15 species of birds, including Common Crows and Mallards, and a number of hawks feed on toads, but the often poisonous toads are shunned by most birds.
1501. The African Jacana scoops up its chicks, and holding them firmly against its body with its wings, walks away with the long legs of the chicks dangling.
1502. Eleven, the first of which is always small.
1503. Yes, some species do. During hot weather, water is carried in their gular pouches and crops and poured over chicks and into their opened bills.
1504. Asparagus.
1505 *Wilee E. Coyote.*
1506. No. King Penguin young average about 80 percent of adult size, whereas Emperor Penguin juveniles are only about 60 percent (or less) of adult size.
1507. No, but it formerly did. The last Ostrich was reportedly eaten by Arabs in Saudi Arabia during World War II. A few survived until 1968, but the Arabian Ostrich is now extinct.
1508. The long hairs shed from horses' tails.
1509. Its soft, downy plumage is responsible for its specific name *pubescens*: from the Latin meaning "having hair-like feathers" or "with hair of puberty."
1510. No, the female's bill is more upturned.
1511. The long, narrow wing is only nine inches (23 cm) wide.
1512. Up to two months, and the burrow may extend back four feet (1.2 m).
1513. The Northern Shoveler. Discriminating hunters often gave the less tasty ducks to their neighbors while retaining the Mallards for themselves. The unpleasant taste of the shoveler is a result of its diet of invertebrates.
1514. Yes, probably so.
1515. The American Robin, a common urban resident, thus it is observed with great frequency. Other birds not observed as often, such as the Eared Grebe, may be just as prone to albinism.
1516. The Red-footed Booby, and its larger eyes reflects this.
1517. The breast beard is not feathered, but rather consists of solid, horny filaments.
1518. *One Flew Over The Cuckoo's Nest.*
1519. Yes. The first record was in July, 1967, when one was discovered grounded 60 miles (97 km) north of San Francisco, California.
1520. The Bar-headed Goose.
1521. No, it was named in honor of Charles Lucien Jules Laurent Bonaparte, the nephew of the French emperor.
1522. Habitat destruction and introduced species. Only one percent of the forest present in the mid-1700's remains, and the endemic parrot must compete for nest sites with introduced parakeets and mynas. Introduced monkeys and black rats also prey on eggs and nestlings.

Answers

1523. The raptors mimic Turkey Vultures in flight by soaring with wings held in a V-shaped angle, thus gaining an advantage in getting close to and surprising prey. The hawks may even fly with Turkey Vultures. The similarity to the innocuous Turkey Vulture may have originated as an aggressive disguise, but may now also serve as a means of avoiding being mobbed by passerines.
1524. The albatross doesn't build a nest; its egg is laid on the bare ground.
1525. The woodcocks and snipe.
1526. Williamson's Sapsucker. The male has a distinct red throat with two white stripes on a solid glossy green-black head. Its neck, back, upper breast, wings and tail are black. The female is brown-headed and "zebra-backed."
1527. In 1892, McIlhenny created Bird City at Avery Island, Louisiana. Egrets were extremely scarce at the time, and he brought in eight Snowy Egrets, and from this stock developed an enormous nesting colony, one of the largest American rookeries.
1528. No, Yellow-billed Magpies are more social.
1529. The Tufted Puffin weighs about 1.6 pounds (0.73 kg). A Horned Puffin weighs just under 1.5 pounds (0.7 kg), whereas a Common Puffin weighs just over a pound (0.45 kg).
1530. In June of 1989, a set of John James Audubon's *The Birds of America* was sold for a record $3.96 million at Sotheby's Auction House in New York. Only about 134 complete sets remain: 94 in the U.S., 17 in Great Britain and 12 elsewhere.
1531. Arizona, with 14 species.
1532. 1937.
1533. No, but the Common Snipe is.
1534. Antarctica, South America and Australia.
1535. Yes, as a rule, and only mammals have proportionally larger brains.
1536. The phalaropes, from their association with bowhead whales, and eating the same food as the enormous cetaceans.
1537. Yes. Swallows, flickers, orioles and a California Gull have been struck down by golf balls, sometimes while in flight.
1538. The American Dipper.
1539. A common South American caracara.
1540. The California Condor is the largest with a wingspan approaching 10 feet (3 m), followed by the American White Pelican with a wingspan of 9.5 feet (2.8 m), and the Trumpeter Swan with a wingspan of nearly nine feet (2.7 m).
1541. The crane has an especially long windpipe of which at least two feet (0.6 m) is looped within a concavity in the sternum. The length, in conjunction with the peculiar configuration of the vocal apparatus, is responsible for its loud calls.
1542. Yes.
1543. An ornithological term coined by the Germans referring to avian pre-migratory restlessness.
1544. Yes, although Ostrich thighs are essentially bare.
1545. Probably the waterfowl (wildfowl).
1546. 1850, in Brooklyn, New York.
1547. Between 10 and 20 percent.
1548. China, followed by the former USSR and the U.S.
1549. The Australian Magpie Goose, Hawaiian Goose (Nene) and at least some Cape Barren Geese copulate ashore. If water is available, the Nene will use it.

1550. Yes, both in flight and while at rest.
1551. Rich chicken stock.
1552. No, but bees are. Birds are kept in aviaries.
1553. A thin-walled pouch extending from the ventral wall of the esophagus. It enables birds to rapidly swallow large amounts of food, at a faster rate than their stomachs can accommodate it. Pigeons, gallinaceous birds, parrots and many birds of prey have well-developed crops.
1554. The Parasitic Jaeger.
1555. Twenty-one.
1556. The adult Heermann's Gull.
1557. Black, but an adult's head is red.
1558. *Falcon.*
1559. No, its population may number in the hundreds.
1560. Yes, very rarely at Churchill, Canada, and possibly a few other northern locations.
1561. The Sanderling.
1562. The Pelagic Cormorant, in breeding plumage.
1563. Four folio-size volumes.
1564. Probably the Swallow-tailed Kite.
1565. Typically within 24 hours of hatching. Waterfowl can be accurately vent-sexed and are usually pinioned according to sex: males on the right wing and females on the left.
1566. Yes, but only on recently-hatched alligators or crocodiles.
1567. An inter-grade colored heron resulting from a mating between a Great Blue Heron and a Great White Heron.
1568. Franklin's Gull.
1569. Pigeon Guillemots have two black bars in their white wing-patch, whereas the wing-patch of a Black Guillemot is immaculate.
1570. Greater Yellowlegs average slightly more than two inches (5 cm) longer than Lesser Yellowlegs: 11.0 versus 8.75 inches (28 vs. 22 cm).
1571. Yes, if a bird lacks an olfactory sense, it is anosmatic.
1572. Yes. Dovekies (Little Auks) have been found in the stomachs of beluga whales, but the birds were probably ingested incidentally.
1573. A turkey.
1574. The American (Common) Goldeneye, and less commonly, Barrow's Goldeneye, a name alluding to its speed or the whistling sound of its wings in flight.
1575. The Ruffed Grouse has a red and a gray-color phase, and the White-tailed Ptarmigan has a winter and a summer plumage.
1576. The Common Snipe.
1577. Danforth *Quayle* (pronounced quail).
1578. The estimates range from 154,000 to 1.5 million.
1579. Nowhere.
1580. Between 34 and 35 days.
1581. Between 33 and 41 days.
1582. Not normally, but a Golden Eagle appeared at Kauai, where it lived for 17 years. It survived on wild pigs and goats, but was killed in 1984 when it attacked a helicopter.
1583. The Yellow-billed Cuckoo.
1584. The California Gull. In 1848, gull flocks arrived to prey on locusts consuming crops.

Answers

1585. Wilson's Phalarope.
1586. Six. Three adorn each side of the face, and the birds have a dark, quail-like crest.
1587. Reddish, but the air-sacs of male Greater Prairie-chickens are orange.
1588. *To Kill A Mockingbird.* Peck won the Academy Award for best actor that year.
1589. The *domestic* turkey.
1590. The Ostrich. The trisyllabic booms are repeated several times at short intervals and are typically used to proclaim territory, or are uttered during displays. Booms are audible from nearly two miles (3 km) away.
1591. The Swallow-tailed Kite, that averages just under a pound (0.45 kg).
1592. No, but the Smooth-billed Ani does.
1593. The Terek Sandpiper, that occasionally feeds in shallow water, sweeping the water like an avocet.
1594. A person who collects stamps featuring birds.
1595. The Crested Caracara.
1596. Yes, but this has rarely been documented. In one instance, a Loggerhead Shrike carried off a 16.5-inch-long (42 cm) green snake that weighed more than the bird, and a Bald Eagle was observed flying downhill carrying a 15-pound (6.8 kg) mule deer fawn, that it ultimately dropped. Ospreys may carry fish that weigh as much as themselves.
1597. Black, but the crest of the closely-related female Pileated Woodpecker is red.
1598. The name refers to auk's small size, rather like a dove.
1599. The Wild Turkey. The turkey was initially introduced into Europe via Spain, and was probably imported into Great Britain between 1525 and 1532 by William Strickland of Boynton-on-the-Wold, as he was permitted to incorporate a turkey cock in his family crest.
1600. Egyptian Plover eggs are buried in the sand and the chicks hatch *underground.* Crab Plover chicks also hatch underground because the birds nest in burrows.
1601. The Burrowing Owl.
1602. Sexually dimorphic rails of Africa and Madagascar.
1603. A falcon or falconet.
1604. Yes, both are in the same family (Mimidae).
1605. The Bobwhite, with a clutch of 12-24 eggs.
1606. Probably the Bahamas.
1607. The Black Phoebe.
1608. It was named after Savannah, Georgia, by Alexander Wilson.
1609. Almost. Virtually the entire population of 12,000 pairs nests on Isla Espanola (Hood Island), but several pairs breed on tiny Isla de la Plata off the coast of Ecuador.
1610. John Graham Bell, a naturalist/taxidermist (1812-1889), in 1875. The Sage Sparrow *(Amphispiza belli)* is named in his honor.
1611. Its recurved bill is nearly four inches (10 cm) in length.
1612. Harris' Hawk.
1613. To protect the nostrils from wood dust raised by the woodpecker's bill when drilling into a tree.
1614. The Imperial Woodpecker. The rare Mexican parrot depended on the woodpecker for its nesting holes.
1615. The gulls are nocturnal feeders and feed their chicks at night. The white-tipped bill presumably provides the chicks with a highly visible target to peck at to elicit a feeding response.

Answers

1616. Turkey.
1617. The Green Heron, from its habit of excreting a stream or line of white guano when startled.
1618. Yes.
1619. Flying fish.
1620. Tragopan chicks are reportedly capable of flight within a few days of hatching.
1621. Snow is a good insulator, enabling the birds to remain relatively warm.
1622. No. The weight of a Common (Eurasian) Starling's eyes, for example, constitute 15 percent of its head weight, compared to only one percent in a human's head.
1623. Anna's Hummingbird.
1624. The male South American Long-wattled Umbrellabird. The extendible feathered wattle measures 13 inches (33 cm), whereas the bird itself is 18 inches (45.7 cm) long.
1625. June 30, 1954, just south of the Great Slave Lake, on the border of Alberta and the Northwest Territories. After a decade of intense aerial searching, the breeding grounds in the northern portion of the 17,300 acre (7,000 hectares) Wood Buffalo National Park were finally discovered.
1626. Probably the Semipalmated Sandpiper.
1627. The Least Auklet. Based on recent research, the lighter colored the plumage, the more dominant the bird, and the more likely it is to obtain a nesting site. In other species, birds with darker feathers are generally dominant.
1628. Probably the Black-crowned Night-Heron.
1629. The cracids (curassows, guans and chachalacas), African Congo Peacock and the peacock-pheasants. Feeding chicks is an unusual trait, unknown in other galliforms.
1630. Three: Cape (South African), Northern and Australian Gannet.
1631. The eggs are used as canteens.
1632. 1948, in Alaska. The curlew was among the last North American birds whose nest and eggs went undiscovered until well into the 20th century.
1633. The anis. Several pairs may construct a communal nest with each female laying in the nest. Up to 29 eggs have been recorded, but 10-15 eggs is more typical. Eggs at the bottom of the nest often fail to hatch, even if fertile. All participants share in incubation and chick-rearing.
1634. In excess of 900.
1635. The dark (red) meat is working muscle requiring a large blood supply because much oxygen is necessary. Conversely, white meat consists of muscle not used as much and less oxygenated blood is needed. The white breast muscle of a domestic chicken is evidence that the bird is a poor flier, while the dark leg meat is evidence that it is a powerful runner.
1636. The Northern Hawk-Owl and the Northern Pygmy-Owl.
1637. No, the sparrow was named for Thomas Lincoln, a companion of Audubon.
1638. Kori Bustard males, that weigh up to 30-40 pounds (13.6-18.1 kg).
1639. While analyzing Togo frog calls, naturalists heard the voice of the rail in the background.
1640. Probably the Snowy Owl and the three ptarmigans.
1641. The Giant Moa. The Maoris may have been at least partially responsible for its extinction, as well as for the extinction of a number of the smaller moas. Within 600 years of the arrival of the Maoris, the endemic moas were essentially eliminated due to a combination of habitat modification and overkill.
1642. The Ostrich.
1643. About 250.

Answers

1644. The study of bird eggs.
1645. No, the Utah state bird is the California Gull.
1646. The Gyrfalcon, but the Peregrine Falcon is a close second.
1647. For centuries Polynesians collected young frigatebirds from the nest and raised them. When setting out on a long fishing trip, and upon arriving safely at their destination, notes were attached to the birds' leg, because the birds invariably returned to the house where they were raised.
1648. Just over 110 mph (177 km/hr), but the pigeon involved was assisted by a tail wind. Under normal conditions, it is doubtful that speeds much in excess of 60 mph (97 km/hr) are achievable.
1649. Between 17 and 18 inches (43-46 cm).
1650. The hummingbirds.
1651. No one really knows because cormorants, Anhingas and darters have well-developed preen glands. Their feathers do become water-logged, making them less buoyant. This presumably is more efficient for a diving bird, although other diving birds are adequately waterproofed and are still efficient divers.
1652. As of 1993, three have been documented. In July, 1992, I sighted a Black-legged Kittiwake flying over the pole, and in May, 1987, a Royal Navy submarine surfaced at the pole, whered a Snow Bunting was observed. In August, 1993, a Northern Fulmar was seen at the pole.
1653. Just over 20 mph (32 km/hr).
1654. The Solitary Sandpiper. Thrush nests are favored.
1655. No, but the American Kestrel (Sparrowhawk) is.
1656. Yes, but some others, such as the Common Raven, may occasionally prey on vertebrates.
1657. The woodcocks.
1658. The single chick is cared for by both parents at the "nest site" for 28-36 days, after which time the chick flies to the sea.
1659. More than 10,000 sticks in some instances.
1660. Probably the European Blue Tit. It typically lays 11-12 eggs, but as many as 19 eggs have been recorded.
1661. Yes, but there are some exceptions.
1662. Talons.
1663. The two kittiwakes.
1664. The Chimney Swift.
1665. Yes.
1666. The Purple Sandpiper.
1667. The Spectacled Eider.
1668. *Killing two birds with one stone.*
1669. The huge, four-volume set weighs 56 pounds (25.4 kg).
1670. The woodpecker relative has an incubation period of only 12 days.
1671. Up to a half million, but generally about 60,000.
1672. Yes.
1673. The bird is non-migratory.
1674. The Blue-footed Booby.
1675. The Spectacled Guillemot and Japanese Murrelet.
1676. A molt in which all feathers are molted simultaneously, such as in penguins.

Answers

1677. The heavily-armored goliath beetle is the largest-known insect in terms of weight and bulk. Adult males of the 4.33-inch-long (11 cm) beetle weigh up to 3.53 ounces (100 gms), or about 63 times as much as the diminutive 0.056-ounce (1.6 gm) male Bee Hummingbird.
1678. If receptive, the land iguanas may rear up on all four legs in a posture called *stilting,* to facilitate parasite and loose skin removal by the birds. Marine iguanas, on the other hand, do not solicit the attention of the passerines.
1679. Doves; pro-military demonstrators were called hawks.
1680. The Rufous Hummingbird breeds as far north as coastal southern Alaska.
1681. 1934.
1682. The first specimens sent to Europe were shipped from the West Indian island of Curaçao, hence the name curassow. Ironically, the large cracids do not occur in Curaçao.
1683. The Australian Jabiru (Black-necked) Stork.
1684. Cassin's Auklet.
1685. No, two species have only three toes.
1686. Probably the House Wren. A male that lost its mate when the chicks were 12 days of age returned to the nest 1,217 times within a 16.25- hour period of time, an average of one nest visit every 47 seconds.
1687. Four. Only the Bald and Golden Eagle are American breeders, whereas Steller's and White-tailed Sea-Eagles are casual visitors.
1688. It travels about in small groups of about a dozen birds, like the twelve apostles.
1689. Chuck-will's-widow reaches a length of 13 inches (33 cm), with a wingspan of up to 25.5 inches (65 cm).
1690. As many as 50 million, out of a population of about 500 million.
1691. The New Zealand Kea, but some authors suggest the alpine terrestrial parrot is monogamous.
1692. Aransas National Wildlife Refuge, on the south coast of Texas.
1693. Two. The Horned Lark is the only native species, but the Eurasian Skylark was introduced into Vancouver Island, British Columbia.
1694. Lady *Bird* Johnson.
1695. Yes, Tufted Puffins winter off northern Baja California, Mexico.
1696. Less than six percent.
1697. The heaviest fledgling of record weighed nearly 35.5 pounds (16 kg).
1698. No, grebes have no close relatives. The superficial physical similarity of loons and grebes is a result of convergent evolution.
1699. The Indian Peacock.
1700. The two murres. Males of the Bear Island race of Thick-billed Murre have bills nearly 2.5 inches (6.4 cm) in length.
1701. The name is said to have originated with hunters because the birds called loudly while a stalk was in progress, alerting game that danger was near, enabling them to "go away."
1702. The Black-necked Stilt.
1703. The hummingbirds. Penguin wings also move essentially only from the shoulders, but the flippers can be slightly bent.
1704. By carrier pigeon.
1705. Growing feathers push out the old feathers, resulting in a very tattered appearance because of the double coat of feathers. The bioenergetic cost of molting is high, and the double coat of feathers may serve a thermoregulatory function.

Answers

1706. At least 85 percent.
1707. The Short-tailed Hawk, that has two distinct color phases.
1708. A Chihuahuan Raven's nest was constructed entirely of barbed-wire strands. A pair of crows constructed a nest out of gold spectacle frames stolen from an open shop window in Bombay, India.
1709. The elephant eye is *not* larger than that of an Ostrich. The Ostrich eye, at about two inches (5 cm) in diameter, is the largest of any terrestrial animal. However, the eye of the giant squid is the largest of any animal, living or extinct, at up to 15.75 inches (40 cm) in diameter. The eye of a large blue whale, the heaviest animal that ever lived, is up to 4.72 inches (12 cm) in diameter, with that of the southern elephant seal spanning 2.75 inches (7 cm). The eye of a human is about 0.94 inches (2.4 cm) in diameter.
1710. The Band-tailed Pigeon, but the population has recovered.
1711. The Wattled Crane.
1712. Rivoli's Hummingbird has traditionally been considered the heaviest, but the Blue-throated Hummingbird is slightly heavier. Male Blue-throated Hummingbirds weigh up to 0.3 ounces (8.4 gms), whereas a male Rivoli's Hummingbird attains a maximum weight of 0.27 ounces (7.7 gms).
1713. A hanger-on at racetracks, or an individual who secretly clocks race horses during private trials.
1714. Grebe claws are flattened and incorporated into the foot.
1715. The flightless Kakapo, the Night Parrot and the rare Ground Parrot.
1716. The former USSR. In 1985, the estimated size of the flock was 66 million birds.
1717. Short straps attached to the legs of falconer's birds.
1718. The eight quail in the genus *Coturnix*. Most do not assume adult plumage for 10-12 weeks, but reproduction may commence as early as 38 days.
1719. No, he favored water colors.
1720. The name refers to their huge, heavy, "gross" beaks.
1721. Nearly 12.5 inches (32 cm).
1722. The curlews winter on tiny mid-Pacific islands, often feeding on tern and other seabird eggs, including albatross eggs. Curlews may steal eggs from beneath incubating birds.
1723. No, prey is speared.
1724. The similar appearing, but larger, Hairy Woodpecker.
1725. Yes, despite their relatively short legs, seedsnipes run quickly and easily.
1726. The Romans were aware of the homing abilities of migratory swallows, and when they went off to sporting events, such as chariot races, they took adult birds from the nest. The practice may have originated about 1 A.D. when Pliny, a noble fond of chariot racing, took swallows to Rome from his home in Volterra, some 134 miles (216 km) away. The swallows were color-marked or dyed to indicate the race winners and released to carry the news back home to his friends.
1727. *Stray Feathers: A Journal of Ornithology for India and Dependencies*. Published between 1872 and 1888, its 11 volumes are a veritable gold mine for the ornithologist.
1728. The angler fish (*Lophius piscatorius*) has reportedly swallowed geese, as well as cormorants, loons, ducks, grebes, alcids and other seabirds.
1729. The Parakeet Auklet.
1730. Between 20 and 30 minutes.
1731. The hummingbirds, but other birds may be blown backwards in high winds.

Answers

1732. No, but most grebe chicks are conspicuously striped.
1733. In 1840, Audubon banded an Eastern Phoebe by tying a silver wire around its leg.
1734. The drake Black Scoter in breeding plumage.
1735. The cormorants. The typical clutch size averages 2-4 eggs, with extremes of 1-7.
1736. The interior of the casque consists of a very tough, elastic, foam-like substance, and is not supported by any protuberance from the skull.
1737. The ducks are obligate brood parasitic nesters.
1738. Yes.
1739. The four longest primaries are black tipped.
1740. The Red-throated Loon, that becomes especially noisy before foul weather.
1741. Stuttering.
1742. Up to four pounds (1.8 kg), but generally less.
1743. No. Bald Eagle males weigh 8-9 pounds (3.6-4.1 kg) and females about 10-14 pounds (4.5-6.4 kg), compared to 8-10 pounds (3.6-4.5 kg) for a male Golden Eagle and 9-13 pounds (4.0-5.9 kg) for a female.
1744. No, the kittiwakes also nest in the Komandorskiye Islands, Russia.
1745. A gosling.
1746. The Waved Albatross (Galapagos Islands). Nests are not constructed and the birds frequently move their single egg, sometimes for considerable distances. Exactly why is not clear, but moving the egg may be a means of escaping "nest" parasites.
1747. An enormous ponderosa pine tree in the San Jacinto Mountains, California contained up to 50,000 stored acorns.
1748. The Egyptian Goose, Comb Duck, Spur-winged Goose, Hartlaub's Duck and African Pygmy-goose. Gray Kestrels, Speckled Pigeons and even Black Storks may also use a Hamerkop nest, as well as honey bees, genets, mongooses, monitor lizards and a variety of snakes.
1749. Nearly black. Soft, black, downy feathers on the back absorb heat and form an insulating layer over the black skin that prevents heat loss while the bird is sunning. The roadrunner maintains a sunning position throughout the cold hours of the morning when the ground temperature is not high enough for insects and reptiles to become active. When sunning, the bird's back is toward the sun, its wings partially extended, and the feathers ruffled to expose the black skin.
1750. The Old World Tawny Eagle, with a population possibly exceeding 200,000 birds.
1751. The Roseate Spoonbill.
1752. The coucals, because of their long straight hind-claws.
1753. If grouse walked to their snow-insulated roosting site, their tracks could be followed by foxes or other predators.
1754. No, a flamingo chick's bill is only slightly curved, but is essentially straight.
1755. The 1.49-ounce (42 gms) brain accounts for only about 0.03 percent of an Ostrich's weight. Despite its small brain, the Ostrich is not regarded as stupid.
1756. The nocturnal Little Blue Penguin. It nests under houses in New Zealand and Australia where its continual, loud, night-time calling is not looked upon with favor.
1757. An owl.
1758. The Senegal Thick-knee.
1759. The Gyrfalcon. Females weigh about 4.5 pounds (2 kg), and males 3.0 pounds (1.4 kg).
1760. None. The Giant Hummingbird is the largest, but it weighs only 0.7 of an ounce (20 gms).

Answers

1761. The White-tailed Ptarmigan, that weighs slightly more than a pound (0.45 kg).
1762. Henry *Robin*son "Birdie" Bowers.
1763. The Black Rail, that weighs a maximum of 2.75 ounces (78 gms).
1764. The Purple Finch.
1765. The corvids (crows, ravens, etc.).
1766. No, some huff, grunt, whistle, bark, squeak, cluck, yodel and coo.
1767. Bitterns have shorter legs, a shorter and usually heavier body, are more secrective, and tend to seek concealment in cattail and sedge marshes.
1768. Some megapodes lay eggs in pits dug into sand heated by subterranean vulcanism. The Polynesian (Pritchard's) Megapode of northern Tonga buries its eggs in hot ash from the still active volcano on the island of Niuafo'ou.
1769. The Pacific Gull of southern Australia and Tasmania.
1770. The Peregrine Falcon.
1771. No, it generally uses old raptor nests.
1772. About 46.
1773. Lengthwise, not crosswise, as is typical of most birds.
1774. Steller's Eider, from its occasional habit of flocking birds swimming in single file.
1775. Pods.
1776. No, apparently not. Virtually all prey is taken on the wing.
1777. Man originated in Africa about three million years ago, but habitat was probably not greatly altered until the use of fire was established some 350,000 years ago. Even then, a small human population of hunter/gatherers could set fire to enormous areas. Such fires were started deliberately, either to smoke out beehives, or simply to clear long grass so that people could move about with greater ease.
1778. The male kiwi may sit for 75-85 days, although in some instances the female may incubate as well.
1779. Four, but the hind toe is tiny.
1780. Yes, dormitory cavities are used by many woodpeckers.
1781. The 30 diverse Hawaiian honeycreepers.
1782. They have ceres.
1783. The Peregrine Falcon and Rough-legged Hawk. The sharp-eyed raptors do not molest the geese and they keep predators away.
1784. No, not in most instances.
1785. Yes, but it is an introduced species.
1786. From the mistaken belief that the grosbeak sang mainly in the evening. Its Latin name *C. vespertinus* reflects this belief.
1787. Chicks crouch in a hollow while their parents bury them completely by throwing sand over them with their bills. To prevent overheating, buried chicks may be wetted with river water carried in the adult's belly feathers.
1788. The Willow and Rock Ptarmigan.
1789. The eagle rays.
1790. None. Most albatrosses breed in the subantarctic.
1791. Marshal *Rooster* Cogburn.
1792. The greater albatrosses (Wandering, Royal and Amsterdam) and the King and Emperor Penguins. Exceptionally heavy white pelicans may exceed the weight of a small greater albatross.

Answers

1793. The Hispaniolan Woodpecker. As many as 26 pairs have been recorded nesting in a single tree.
1794. No.
1795. The Elf Owl.
1796. Aristotle, in the 4th century B.C.
1797. Usually just one, but rarely up to five.
1798. The Adelie Penguin, but the predatory seal preys on other species of penguins as well.
1799. Small birds.
1800. Yes, but not all herons are egrets.
1801. The aborigines recognized that Emus were extremely curious, so they climbed trees and twirled balls of Emu feathers and rags rapidly in the air. This fascinated the Emus, and when the birds gathered around the tree, they were speared from above. The struggling victims in turn attracted more inquisitive Emus.
1802. The South American Black-headed Duck, because of its obligate brood parasitic-nesting habits.
1803. An American game biologist's term describing the breakup and autumn dispersal of family groups of gamebirds, especially Ruffed Grouse.
1804. Cowardice.
1805. The Swallow-tailed Kite. It even gathers nesting material while in flight, and drinks by skimming over the surface of a pond like a swift or swallow.
1806. The Galapagos Hawk. As many as four males may breed with a single female, and all aid in caring for the eggs and chicks. Harris' Hawk and the pale morph of the Chanting-Goshawk are also known for this, but to a lesser extent.
1807. The South American Hoatzin.
1808. A rada, that is blessed and said to be imbued with magical powers. If something other than a rada is used to collect the delicate nests, the superstitious collectors fear that the cave spirits will be angered. Collecting nests is extremely risky, requiring skilled climbers to scale ropes and rickety bamboo ladders to reach the nearly inaccessible nests, and many have perished in the course of nest collecting.
1809. Crane flies.
1810. It bears a striking resemblance to a dark-plumaged juvenile Wandering Albatross. It is somewhat smaller with dark cutting edges and tips on both mandibles, and its vocalizations and courtship displays differ.
1811. The *duck*tail.
1812. A sticky substance smeared on twigs or branches used to catch small birds.
1813. The Spotted Sandpiper.
1814. The House (English) Sparrow.
1815. The Cape Barren Goose of Australia, that was named for its pig-like grunts.
1816. The Asian Green Magpie.
1817. The Ringed Kingfisher weighs about 10.25 ounces (291 gms).
1818. None.
1819. The Brown-headed Nuthatch uncovers insects by flaking off tree bark, using a separate piece of bark held in its bill as a tool.
1820. The Sanderling. Probably no beach in the world is without Sanderlings at some time of the year.
1821. Calling every second, a record number of 1,088 calls.

Answers

1822. European terms referring to the swallowing of young, usually dead or weak, by the parent birds.
1823. About three gallons (11.4 liters).
1824. Black.
1825. Formerly believed to be a distinct species, the bird was subsequently discovered to be a hybrid between the Blue-winged and Golden-winged Warbler.
1826. Probably because their feet are too small to be effective paddles, and their plumage is not particularly waterproof.
1827. The first record of a band recovery occurred in 1710, in Germany, when a Gray Heron was found wearing metal rings placed on it several years earlier in Turkey. The earliest known bird-marking for message transmission (not to be confused with scientific banding) occurred between 218-201 B.C. when Quintus Fabius Pictor, a Roman officer, tied a thread to the leg of a swallow and released it to inform a besieged Roman garrison when relief could be expected.
1828. The bald-headed adults have no head feathers.
1829. The arrival of great herds of cattle that destroyed the native tall grass essential to the quail's survival.
1830. No. The Emu incubation period is 56-60 days, whereas that of an Ostrich averages 42-46 days.
1831. The Wood Stork, because of its black-and-white coloration and solemn demeanor.
1832. About 100, but relatively few became established.
1833. Not really. Drakes often accompany females with young, but little attention is paid to the ducklings. The males are probably still sexually interested in the hens.
1834. The extinct Dodo, both of which were probably related to pigeons and doves. The Reunion Solitaire was gone by 1746, but the Rodrigues Solitaire survived until about 1791.
1835. A number of snakes consume eggs, but the African egg-eating snakes in the genus *Dasypeltis* are exceptionally well adapted for the task. The mouth and neck are astonishingly distensible, and a surprisingly large egg can be swallowed as far as the neck. The ventral projections on the snake's neck vertebrae are elongate and sharp, and the tips actually pierce the dorsal wall of the gullet, creating a saw in the form of a row of up to eight points in the gullet. The swallowing motions bring the egg against the saw and a cut is made in the eggshell. Finally, the egg collapses from the pressure and the expelled contents swallowed and the shell is regurgitated.
1836. The Least Sandpiper, that weighs about an ounce (28 gms).
1837. The eagle.
1838. The alert vocal geese are quick to give alarm if trespassers intrude.
1839. The Bald Eagle. Eight pairs nested in Anchorage in 1993.
1840. Yes, but only once in the Lubumbashi Zoo (Zaïre).
1841. Anna's Hummingbird.
1842. Exposed external nares, as opposed to those covered by frontal feathers, such as in grouse.
1843. None.
1844. No, it is a rare straggler from the West Indies, with four documented records this century.
1845. The Snowy and Great Gray Owl.
1846. Yes, as a rule.
1847. Australian cockatoos with long, thin upper mandibles that are ideal for digging up roots and bulbs.

Answers

1848. All have barred plumage.
1849. No.
1850. The average is about 240.
1851. No.
1852. The Common Nighthawk and Barn Swallow migrate between Alaska and Argentina.
1853. No, the state bird is the California Quail.
1854. Taste. Humans have about 9,000 taste buds, compared to 27-59 in the domestic chicken, 50-75 in pigeons, 200 in starlings and ducks, and 300-400 in parrots.
1855. A bird with a varied diet.
1856. Yes.
1857. No, antbirds rarely forage on the army ants themselves. The approximately 253 antbirds are mainly lowland neotropical forest birds, some of which feed primarily on invertebrates, lizards and frogs flushed by army ants on the march.
1858. A number of insects have significantly faster wingbeats. A species of a tiny midge may achieve the unbelievable rate of 1,046 wingbeats a second, although such a high figure is challenged by some entomologists. Even so, the 250 wingbeats per second of a honeybee far exceeds a hummingbird's 90 wingbeats per second.
1859. The Red-cockaded Woodpecker is almost exclusively dependent on old southern pine trees infested with red heart fungus. The trees are soft enough to provide for easy excavation, and because they are still living, are less susceptible to fire damage than dead ones.
1860. The King Rail male weighs just over a pound (0.45 kg).
1861. Twelve: Arizona Woodpecker, California Gull, California Quail, California Thrasher, Carolina Chickadee, Connecticut Warbler, Kentucky Warbler, Louisiana (now known as Tricolored) Heron, Louisiana Waterthrush, Mississippi Kite, Oregon (now known as Dark-eyed) Junco and Tennessee Warbler.
1862. The two finfoots and the neotropical Sungrebe. The thorn-like claw probably assists young birds in scrambling through vegetation, rather like young Hoatzins. Unlike the Hoatzin, adults often retain the claw.
1863. The West Indian todies. *Pedorrera* means flatulence, a reference to their buzzing flight.
1864. A nest attached to a vertical surface, such as a hollow tree or the inside of a chimney or cave wall, by means of mud plaster or, in some swifts, saliva.
1865. Yes, some races of the New Zealand Snipe inhabit the Auckland, Antipodes and Snares Islands.
1866. The name originated with the nocturnal European Nightjar because of its churring or jarring notes.
1867. Waterfowl generally don't travel at all, using only material gathered as far as the birds can reach from the nest site. However, Magpie Goose ganders occasionally bring material from as far away as 60 feet (20 m).
1868. According to some authorities, the King Vulture. However, I have been bitten many times by California and Andean Condors, as well as King Vultures, and I don't agree. The condor bites were far more serious than those of a King Vulture, particularly Andean Condor bites.
1869. When raked slowly through the water, the egret's bright yellow toes are believed to attract fish.
1870. Descending nuthatches may be in a position to locate food in bark crevices overlooked by such birds as treecreepers and small woodpeckers that forage by moving up a tree trunk.

Answers

1871. About 150 million.
1872. *The Ibis*.
1873. Franklin's Gull migrates south to Chile, and occasionally even to South Africa.
1874. The Osprey. The reversible outer toe enables the raptor to grasp fish with two toes in front and two in back.
1875. South America and Antarctica.
1876. The Lesser Flamingo numbers between 2.5 and six million birds, more than all other flamingo species combined.
1877. The social, ground-nesting Andean Flicker inhabits the high, cold puna of the Andes.
1878. The two lyrebirds.
1879. Only 47 birds were accounted for in 1963, but by the late 1980's, the population consisted of between 300 and 400. Most breed on Toroshima Islet, a tiny volcanic island 400 miles (644 km) south of Tokyo. In 1993 ten pairs nested in the Senkaku Islands (southern Ryukyus).
1880. Eleven. The Canary Islands Oystercatcher disappeared about 1913.
1881. The Carolina Parakeet reached northern U.S., but not Canada. The parakeets were non-migratory, and even those that ranged as far north as New York or Wisconsin did not move south during the winter.
1882. To attract birds. A squeak lure is created by loudly kissing the back of the hand. *Pshing* (an extended *spshspsh* sound) imitates natural scolds or distress calls.
1883. The Zapata Wren is only found in a single swamp of less than five square miles (13 sq km) in southern Cuba. In 1990, at least four of the rare wrens still survived.
1884. The Papuan Frogmouth.
1885. Stuffing mattresses and pillows. The Funk Island Great Auk feather trade commenced about 1785 when a base was established on the little breeding island. Auks were driven into pens, beaten to death with clubs, and carcasses tossed into cauldrons of boiling water to loosen their feathers. Because no other fuel was available, their fat-laden bodies were used as fuel. By 1840, the Great Auk was exterminated from Funk Island, a period of only 55 years after serious feather trade began.
1886. To quail.
1887. About 13 percent.
1888. Merrit was the writer who composed the famous limerick that begins with: *A wonderful bird is a pelican, His bill can hold more than his belican.*
1889. The two painted-snipe.
1890. The African Marabou Stork and the Greater Adjutant stork. The grotesque, naked, pink, bulbous, pendent throat pouch or air-sac may attain a length of 18 inches (46 cm).
1891. Yes. Monophagous describes a species limited to feeding on a single type of food, and the kite depends almost exclusively on the apple snail.
1892. About 78-80, less than one percent of all bird species.
1893. No.
1894. Jazz saxophone legend Charlie "Yardbird" Parker.
1895. The Royal Tern, with a wingspan of up to 44 inches (1.1 m), is exceeded in size only by the Caspian Tern.
1896. Graminivorous birds feed mainly on grasses, whereas granivorous birds favor seeds.
1897. A newly-growing feather just emerging from the skin still enclosed for most of its length in a horny sheath.

Answers

1898. Yes, but such behavior is not typical. Both sexes of the widespread Violet-eared Hummingbird are nearly identical, and males do assist with the domestic chores.
1899. Port Royal, Nova Scotia, 1606, by the French.
1900. The Baltimore Oriole. In 1973, much to the disgust of bird listers, the American Ornithologists' Union combined the Baltimore, Bullock's and Black-backed Orioles (all of which were distinct species) into a single species known as the Northern Oriole.
1901. Yes, but the penguins are not successful during one of those years. It takes 10-13 months to fledge a chick, therefore it is impossible to fledge a chick annually, and only two chicks can be reared within a three-year period. If eggs are laid late in the season, the resulting chicks perish during the winter and the parents become early breeders the following season.
1902. No, Black Guillemots are whiter in winter plumage.
1903. No, its long trailing feathers are actually elongated inner secondaries.
1904. Yes, as a rule.
1905. About 46 feet (14 m).
1906. The Japanese (Manchurian) Crane and Japanese Green Pheasant respectively.
1907. Harris' Hawk. A trio (two males and a female) may nest cooperatively, with the extra male helping feed chicks, or bringing food to the nest.
1908. The South American Hoatzin, especially chicks when they still have the carpal claws.
1909. The Black Swift.
1910. The Northern Gannet is approximately four times heavier than a Red-footed Booby.
1911. The avian tail bone consisting of several fused vertebra.
1912. Yes. Coots dive surprisingly well and reportedly can dive as deep as 25 feet (7.6 m), remaining submerged for as long as 16 seconds.
1913. The three skimmers. The vertical pupils apparently enable the birds to tolerate glaring sunlight. The unique slit-like pupil may be correlated with their twilight and nocturnal fishing habits, partially closing to protect the retina when the skimmers are loafing or nesting on glaring white sand, and opening by night.
1914. No, but grebes do.
1915. Because the stork appeared to be wearing a black coat and cape, like a bishop.
1916. Some of the smaller moas were possibly still present when Captain Cook first visited New Zealand in 1769, but were probably gone shortly thereafter.
1917. The African Saddle-billed Stork, that stands nearly 5.5 feet (1.7 m) tall.
1918. The Australian lyrebirds. Their single-egg clutch is incubated for an average of 50 days.
1919. Generally not until their sixth year.
1920. About 232.
1921. Between 28 and 32 days.
1922. Three.
1923. The highly-pelagic Sooty Tern has been observed 1,250 miles (2,000 km) from the nearest land.
1924. Yes, on rare occasions. A northern pike was observed leaping out of the water to grab a Black Tern. In another instance, a black bass jumped up and gulped down a hummingbird. An American Dipper was found in the stomach of a Dolly Varden trout, but it could have been taken underwater.
1925. The 20-cent coin.

Answers

1926. Sometime prior to the 12th century, Mute Swans were domesticated in Great Britain, and for hundreds of years the birds were considered property of the Crown and could only be raised by persons holding a permit from the Royal Swan Master. The Crown later granted "royalties" enabling certain corporate bodies to own swans and to mark their bills with registered symbols known as swan marks. Swan marks consisted of a pattern of notches cut into the upper mandible, and any unmarked swans on the River Thames were considered property of the Crown. Young swans were captured annually to be marked in a colorful traditional ceremony known as swan-upping. Much of this ancient tradition is still carried on today, generally on the Monday of the third week in July.

1927. Many, but not all, do. Some, such as flickers and the Pileated Woodpecker, feed chicks by regurgitation, but other species carry food to their nestlings in the bill.

1928. Cranes fly with both their legs and neck extended, but herons retract their necks.

1929. Funk Island, off Newfoundland, was named *Isla de Pitigoen* by the Portuguese in about 1520. The name alluded to the enormous number of Great Auks breeding there, perhaps as many as 100,000 pairs. The Great Auk was the original penguin.

1930. A bird adapted to climbing, such as a woodpecker.

1931. No, but the Eurasian Oystercatcher does.

1932. The Eastern Great White Pelican. The chicks of all other species have white down.

1933. The Hooded Merganser.

1934. No, chicks are fed partially digested seeds.

1935. Only rarely, but even then, a larger clutch may be the result of more than a single female laying in the nest.

1936. Alexander the Great in 300 B.C. The parrot retains the great conqueror's name: Alexandrine Parakeet.

1937. The most ornate feathers are probably those of the male King of Saxony Bird-of-paradise. Adult males are only seven inches (17.8 cm) long, but have two 18-inch-long (46 cm) plumes trailing backward from their heads. Each plume bears a series of 30-40 miniature flags along one side of the vane that are brilliantly enameled blue outside and brown inside.

1938. The Emu and the cassowaries.

1939. No. Speared fish are brought to the surface, tossed into the air, caught in the bill head first, and then swallowed.

1940. The fleshy projection extending from the forehead of a Wild Turkey.

1941. No, it is a tall North American fern that grows in swamps and moist places.

1942. Abbott's Booby.

1943. No, the birds are diurnal.

1944. An overlapping rib projection that strengthens the rib cage by serving as a supporting cross-strut without adding much weight.

1945. No, it is slightly downcurved.

1946. *Drumstick* is an American term for the leg of a cooked chicken or turkey: the tibia or tibiotarsus (avian foreleg), the long bone between the knee and the ankle.

1947. The Ruddy Duck.

1948. The crane. It is also symbolic of longevity.

1949. The Phainopepla. After rearing a brood in the southwest, the birds may move westward in search of fruit and cross over the Coast Range in California, where some subsequently rear a second brood.

Answers

1950. Not generally, but there are reports of octopus grabbing terns and Little Blue Penguins.
1951. A Red-eyed Vireo sang 22,197 songs in a single day.
1952. No, although larger females typify the family. In some of the larger hawk-owls, males exceed females in size.
1953. The Colima Warbler, in 1889.
1954. Large, South American, tree-dwelling tarantulas, some of which have a leg span of 10 inches (25 cm), that prey on small birds.
1955. The average is about 1,200 trips.
1956. No, the Thick-billed Murre is more abundant.
1957. The Common Nightingale.
1958. A race of the Willow Ptarmigan.
1959. Yes, mainly Tasmania.
1960. No, all are insectivorous birds of the Old World tropics.
1961. Georg Wilhelm Steller of Steller's Sea-Eagle fame. The cormorant was numerous at the time, and some were eaten by the marooned explorers.
1962. Smoky blue.
1963. No.
1964. No, but swifts do. Swallows perch like typical perching birds, but swifts apparently are incapable of this.
1965. Yes, stenophagous birds are limited to a restricted diet.
1966. Paul Newman.
1967. Penguins typically use only their flippers.
1968. Darwin didn't. He never saw the unique bird because it was not described until 1897.
1969. Yes, on occasion.
1970. Males have rich blue ceres and bluish legs and feet, whereas females have pale to dark brown ceres with pinkish legs and feet.
1971. Only six.
1972. No. Based on several studies, a southern Bald Eagle territory averages 2-3 square miles (5.2-7.8 sq km), whereas that of the Golden Eagle ranges between 19 and 59 square miles (49-153 sq km). Territory sizes vary, but there is little question that Golden Eagles require much larger territories.
1973. The Scissor-tailed Flycatcher.
1974. Only a single live pair was ever collected. The pair was imported into France in 1924, from what was then known as French Indochina (Annam), by the famous ornithologist/aviculturist Dr. Jean Delacour. The few Imperial Pheasants in collections all descended from that original pair. The existing birds are the result of crossing a pure-blood cock back to a Silver Pheasant hen, and over the years birds resembling pure wild birds were produced. Recent surveys in Vietnam have confirmed that the species still survives there, but it surely is at great risk.
1975. A large, New Zealand forest parrot.
1976. No, cracid chicks instinctively climb, seeking an elevated roost shortly after hatching.
1977. No, it is much smaller and less colorful.
1978. The birds were originally named for Guinea, West Africa.
1979. The Maleo, a megapode of Sulawesi (formerly Celebes). The 6 to 8-egg-clutch is laid at 10-12 day intervals, taking 2-3 months to complete. Chicks are hatching before the clutch is complete.

Answers

1980. No, prey is swallowed immediately after capture.
1981. The Tricolored Blackbird.
1982. Plantcutters. The birds waste far more than they eat.
1983. By their distinct vocalizations.
1984. The Bristle-thighed Curlew. In 1991, it was documented that at least some wintering curlews in the northwestern Hawaiian Islands picked up small pieces of coral and repeatedly threw them at abandoned Laysan or Black-footed Albatross eggs. Once a hole is made, the curlews insert their long bills into the hole and enlarge it, and then feed on the egg contents.
1985. The owls.
1986. Pekin Ducks.
1987. No. Ironically, crows already have tongues that are slightly split.
1988. *Sleeping one*, alluding to the poorwill's hibernating ability.
1989. The Brazilian Merganser. The population of the elusive duck is not known, but some experts fear it may number less than 250 birds. Its nest was not discovered until 1954.
1990. *Wamp* is an American Indian name meaning white and was applied to the Common Eider. The King Eider was known as *wamp's cousin.*
1991. The owls.
1992. The King Vulture weighs up to 8.25 pounds (3.7 kg).
1993. The only specimen (from Sangihe Island north of Sulawesi) was destroyed in a bombing raid over Germany during World War II.
1994. No, but this is a common belief. The seal smacks a penguin to pieces on the surface of the water, dismembering it.
1995. Yes, particularly in the waters off California.
1996. The Wild Turkey.
1997. The Band-tailed Pigeon is of equal size and weight.
1998. The tiny African Pygmy Falcon. The 2-ounce-raptor (57 gms) may evict the owners and take over a completed nest, especially that of the Social Weaver or White-headed Buffalo-Weaver, both of which are communal nesters. The weavers benefit because the falcons may keep snakes away, and the raptors need the weavers because the passerines maintain the communal structures in good repair. The falcons may occasionally prey on a young weaver bird in the nest, but the symbiotic relationship would be destroyed if this were a common practice.
1999. The Surf Scoter.
2000. Snakes. They kill far fewer snakes than generally believed and apparently prefer rodents.
2001. The prospects are good.
2002. Discrepancy exists with respect to the actual date. Most accounts state that the last survivor died in late September 1914, at the Cincinnati Zoo. Other accounts state that a pair of parakeets named *Lady Jane* and *Incas* resided at the zoo for about 32 years, and during the late summer of 1917, the female died. On February 21, 1918 (or possibly a week earlier), *Incas* died in his cage surrounded by his keepers.
2003. The guineafowl's head is bare, superficially resembling a vulture.
2004. Yes.
2005. Yes, as a rule, but many fish are also vividly colored.
2006. The four diving-petrels of the Southern Hemisphere.
2007. Both the upstroke and the downstroke provide thrust.

Answers

2008. *Father Goose.*
2009. The bills of the sicklebills are curved almost into a half-moon.
2010. Unlike most plovers that slip away from the nest and run a short distance before flying to conceal the nest site, the Mountain Plover waits until the last possible moment, and then "explodes" into the air. This behavior possibly evolved from the days when large bison herds roamed the plains and late explosive flight could have startled the bison, preventing them from trampling nests containing eggs.
2011. A half-mile (0.8 km) or more.
2012. Not really, although the Narina Trogon is partially migratory in southern Africa.
2013. No, the Sarus Crane is taller.
2014. The Pied-billed Grebe.
2015. The Crab Plover.
2016. The cuckoo of cuckoo-clock fame.
2017. Hatchling crocodiles and turtles.
2018. No.
2019. No, its neck is extended.
2020. Unlike typical ptarmigan, the Red Grouse does not assume white winter plumage.
2021. A thrush.
2022. Buldir Island, Aleutian Islands, Alaska.
2023. Brightly-colored birds with broken patterns harmonize with their surroundings. The vivid splashes of color may cause a bird to be mistaken for a flower or fruit among the foliage.
2024. No, nightjars have extremely thin skin.
2025. Two captives reportedly attained ages of 62 and 67 years.
2026. Basically white.
2027. The Emu.
2028. The titmice.
2029. No, pair bonds are not formed at all.
2030. *Rump foot*, referring to the fact that grebe legs and feet are located well to the rear of their body, near the rump.
2031. The Brolga crane. A dewlap is a throat wattle hanging from the lower part of the bill.
2032. No, an Ostrich eggshell is pitted.
2033. The Double-banded Argus. A portion of an 1871 primary feather of unknown origin is deposited in the British Museum. The pheasant might be from Java, and is probably extinct.
2034. The Green Heron. The descriptive name alludes to its habit of excreting if frightened or when in flight.
2035. It was flightless, or nearly so.
2036. The neotropical Harpy Eagle. One particularly heavy female named *Jezebel* weighed 27 pounds (12.2 kg). A large female at the Los Angeles Zoo in the late 1960's weighed nearly 24 pounds (10.9 kg).
2037. The Golden and Lady Amherst's Pheasant.
2038. The Satin Bowerbird. Males apply colored fruit pulp and masticated charcoal to the walls of their bowers using a wad of sponge-like fibrous bark as a brush.
2039. Some gallinules and swamphens.
2040. Yes, a domestic duck once laid 363 eggs in 365 days.
2041. The Bald Eagle.

Answers

2042. The eyelid. In hummingbirds, the feathers are only 0.4 mm in length.
2043. *The Ornithology of Francis Willughby,* edited by John Ray, was published in London in 1678.
2044. Mannikins are Old World finches, whereas manakins are small neotropical birds related to cotingas.
2045. About one-fifth heavier.
2046. An endemic Hawaiian honeycreeper.
2047. The six monotremes: the platypus and the five echidnas (spiny anteaters) of Australia and New Guinea.
2048. No, all woodswallows are Australian or Asian.
2049. The Corsican Nuthatch.
2050. In 1986, a racing pigeon named *Peter Pau* was sold for £41,000, an equivalent of more than 10 times the bird's weight in gold at the time. In October of 1978 a racing pigeon named *De Wittslager* was sold to a Japanese fancier for £25,000.
2051. The Least Bittern seldom flies more than 100 feet (30 m) per flight.
2052. As many as 50.
2053. White-tailed Tropicbirds nesting on Christmas Island (Indian Ocean) use tree cavities, although ground nesting is typical elsewhere.
2054. No.
2055. No, Turkey Vultures have longer wings.
2056. In most instances, probably not. Some shrikes are known to return to food stored months previously.
2057. Feathers covering the bases of other feathers that are usually named for the feathers they cover. For example, secondary coverts cover the bases of the secondaries.
2058. The Spotted Redshank. Its head, neck and underparts are totally black with the remainder of the plumage spotted with white, and its legs are red. This combination of black underparts and conspicuous red legs is unique in shorebirds except for the rare New Zealand Black Stilt.
2059. The Snowy Sheathbill.
2060. The extinct Huia.
2061. Its large beak.
2062. No, their nostrils are concealed by feathers.
2063. The Torrent Duck.
2064. The upper mandible is not longer. The lower mandible is conspicuously longer and broader.
2065. Skimmers probably work back over their feeding tracks seeking prey attracted to the disturbance of the surface of the water during the first passage.
2066. A pigeon.
2067. Yes, but only rarely. The first nest was not discovered until 1973, in Massachusetts.
2068. *Turtle Dove.*
2069. At least three: Short-winged (Lake Titicaca) Grebe, Atitlan Grebe, and Puna (Lake Junin) Grebe. All are confined to large neotropical lakes. The Atitlan Grebe of Guatemala is extinct.
2070. Both species have only three toes, but all other flamingos have four toes.
2071. The sexes are identical except for the slightly smaller size of the female.
2072. Nearctic species inhabit the temperate and Arctic regions of the New World.

Answers

2073. No, but it is often more active at dusk and some activities are semi-nocturnal.
2074. *Buffalo wings* consist of a portion of a chicken wing often offered as hors d'oeuvres in American bars. The idea apparently originated in Buffalo, New York.
2075. The Herring Gull.
2076. The Emperor Penguin. During the Antarctic winter of 1911, three members of Scott's South Polar expedition (Wilson, Bowers and Cherry-Garrard) sledged from Cape Evans (Ross Island) to Cape Crozier to collect eggs from the winter-nesting Emperor Penguin. At the time, the penguin was considered very primitive, and it was believed that the embryos might shed light on the origin of all birds by establishing a conclusive link between reptilian scales and feathers. The hardships endured to obtain the eggs were such that the saga was subsequently dubbed *the worst journey in the world.*
2077. Swainson's Hawk.
2078. The region surrounding the cloaca, often including the under tail coverts.
2079. A regurgitated pellet consisting of an agglomeration of the undigested portions of a bird's food, such as feathers, bones, claws, beaks and beetle wings.
2080. *Big Bird.*
2081. The Rifleman, a tiny endemic with a very short tail.
2082. The Marabou Stork. The scavengers take advantage of dump sites and abattoirs, and are quite abundant around some villages.
2083. The courting male Bare-necked Umbrellabird of southern Central America.
2084. Yes.
2085. Yes. Some crested penguins and Sooty Terns are not successful breeders until 10 years of age.
2086. According to aboriginal legend, *Buralga* was a beautiful and famous dancer. She spurned the attention of an evil magician who changed her into a tall, graceful, wonderful bird that became known as the Brolga crane.
2087. April 17, 1915, in Nebraska.
2088. Typically two eggs, with a range of 1-3.
2089. Large crustaceans, as its name implies.
2090. The large eiders. Steamer-ducks and eiders are diving sea-ducks that feed extensively on marine mollusks.
2091. The Thick-billed Parrot.
2092. No.
2093. Four pairs, plus a single unpaired sac for a total of nine air-sacs.
2094. The Marbled Murrelet. It is immortalized in song by the Tlingit *Murrelet Clan.*
2095. About 4,000 are considered migratory to some extent.
2096. The Australian Banded Stilt has a chestnut-colored chest-band, and the New Zealand Black Stilt is completely black.
2097. Subantarctic Campbell Island where no less than five species breed: Wandering, Southern Royal, Black-browed, Gray-headed and Light-mantled Sooty Albatross.
2098. The white morph predominates by as much as 89 percent.
2099. Active termite mounds (nests).
2100. In 1990, the delicate nests commanded up to U.S. $1,000 per pound in Hong Kong, and the price was increasing. The nests of other cave swiftlets are less expensive.
2101. No, it favors pine-oak scrublands and mangrove swamps.

Answers

2102. Crows do not inhabit Central or South America, but four species breed in the Caribbean: Cuban, White-necked, Jamaican and Palm Crow. Conversely, jays are absent from the West Indies, but abound in Central and northern South America.

2103. The grebes.

2104. No, but early naturalists *suggest* that a closely related species did. This little-known bird, the Painted Vulture (*Sarcoramphus sacra*), closely resembled the King Vulture, but was smaller with a white, black-tipped tail, rather than an entirely black tail. The mysterious vulture supposedly disappeared about 1800. Its existence has been debated, and it is not even mentioned in Greenway's 1958 book on extinct birds.

2105. No one really knows, but navigation is probably a result of a combination of factors, such as the position of the sun, moon, stars, constellations, and infrasonic waves, slight changes in barometric pressure, polarized light, as well as landmarks, geomagnetic force, coriolis force, and possibly even wind direction.

2106. None.

2107. Heterodactyl feet resemble zagodactyl feet, but the *second* toe, not the forth toe, has shifted to the rear. The only birds with such feet are the trogons.

2108. None.

2109. Up to seven days. Extended absence from the nest (egg neglect) prolongs the incubation period.

2110. Adult plumage in those birds with juvenile plumage.

2111. Based on some studies, approximately 85 percent.

2112. Probably Sabine's Gull. It breeds in the Arctic, but winters off southwest Africa, Peru and northern Chile.

2113. The spines of a hedgehog. The downy feathers have also been described as resembling a pile of straw.

2114. The inside quarters for raptors maintained by falconers.

2115. This causes a flow of sap, forming a sticky resinous mess that may deter potential predators, such as snakes.

2116. Overgrown. The bills of some captive birds, particularly raptors, require periodic trimming or coping.

2117. Yes, although some may hiss.

2118. No. Ospreys, fish-owls and fish-eagles use their talons.

2119. The hot, inland, nitrate deserts of Chile and Peru.

2120. The Wild Serin (Island Canary) of the Canary Islands.

2121. The rock trout.

2122. Based on some studies, about every eight minutes.

2123. The Red-crowned Parakeet. The birds disappeared shortly after the arrival of man and the subsequent introduction of exotic animals.

2124. The name implies that the penguin inhabits New Guinea. Forster applied the name to Gentoos collected in the Falkland Islands because the penguin had previously been erroneously reported from New Guinea. The word *papua* is Malayan for curly, but the Gentoo Penguin has no feature to which this could reasonably be applied.

2125. Eggs were initially seen by outsiders when natives brought two shells to a trading post to have them filled with rum.

2126. The rare Queen Alexandra birdwing, a butterfly restricted to the lowland forests of Papua New Guinea. The larger, less colorful females average 8.27 inches (21 cm) across the wings, and possibly up to 11 inches (28 cm). Females are also the heaviest known butterflies, weighing up to 0.42 of an ounce (12 gms), or 7.5 times heavier than a male Cuban Bee Hummingbird.

2127. The endangered ibis is restricted to no more than a half-dozen colonies in Morocco. Until the late 1980's, there was a single colony in eastern Turkey, but it has since disappeared.
2128. Curassows whistle loudly, but also produce deep booming or croaking sounds.
2129. The Eurasian Hoopoe.
2130. The Falkland Islands with no less than six: King, Magellanic, Gentoo, Rockhopper and Macaroni Penguin. Recently, several pairs of breeding Royal Penguins were discovered.
2131. Almost exclusively on blue-green algae and planktonic organisms strained from the water by filter-feeding.
2132. The unfeathered external parts, such as the eyes, bill, feet, eyelids, wattles and bare patches of skin.
2133. The woodpeckers. They have a distinctive roller-coaster type flight style, diving with their wings almost closed and alternating with strong wingbeats which regain the lost height.
2134. *Lady-of-the-waters*, because of its grace and elegance.
2135. *Ladyhawke.*
2136. None.
2137. The Dalmatian Pelican, that weighs up to 28.7 pounds (13 kg).
2138. The larks, that pipits resemble both in appearance and behavior.
2139. The Dovekie preys primarily on zooplankton.
2140. The Ostrich, megapodes and possibly kiwis.
2141. No, but the raptor migrates to South America.
2142. Salvadori's Teal.
2143. No, the terns breed every nine months.
2144. Brilliant blue.
2145. The Black-crowned Night-Heron, that breeds as high as 15,750 feet (4,800 m).
2146. At least 1,000 as of 1992, a four-fold increase over the number a decade before. The greater number also reflects greater knowledge in that a number currently considered endangered or threatened were probably at risk earlier, but their plight was not recognized. Another 5,000 species are declining.
2147. The mythical bird was so huge that it bore off elephants to feed its young.
2148. About 20 days.
2149. The Hooded Merganser.
2150. The loons.
2151. Yes, even newly-hatched ducklings.
2152. Marabous are mainly scavengers, but are also efficient predators.
2153. Some of the Old World barbets. The birds may sit for hours calling monotonously, and their monosyllabic clangings become annoying and exasperating, especially during hot weather. The persistent singers became known as brain-fever birds by those who were forced to endure the noise in the heat. The name is also applied to some cuckoos.
2154. Yes.
2155. Up to 21 days, based on a torpid brood monitored in 1907, a highly unusual situation in altricial young. The chicks depend on stored fat and become torpid, losing as much as 50 percent of their weight. During periods of inclement weather, flying insects are rare and parents have difficulty in obtaining sufficient food.

2156. The Bateleur, that incubates its single egg for about 55 days.
2157. Not really. The birds typically nest in the open or under protective overhangs, but most other petrels are cavity-nesters.
2158. The largely terrestrial pigeon abandoned its traditional habit of nesting on the ground and shifted to tree-nesting. Even so, the birds are scarce and locally distributed in the mountains of Savaii and Upolu.
2159. The Caribbean Flamingo, a species generally considered a race of the Greater Flamingo.
2160. At night, whereas the more cryptically-colored females incubate during the day.
2161. Florida and Georgia. Limpkins are also rare visitors to South Carolina.
2162. Yes, the huge parrots regularly descend to the ground to feed on fallen fruits or palm nuts.
2163. Yes, as a rule.
2164. The name was given by the English naturalist John Ray (1627-1705) who erroneously thought the birds were puffins.
2165. Yes, generally so. A Carolina Chickadee collected in the winter had 1,704 feathers, while one taken in the summer had only 1,140 feathers. Conversely, an American Goldfinch collected during the early summer had 1,901 feathers, compared to 2,107 on a winter-collected bird.
2166. Some of the albatrosses.
2167. No.
2168. The White-crested Hornbill follows monkeys that stir up insects and other potential prey.
2169. Miniature adults.
2170. John Gould named the gorgeous finch after his wife, who illustrated many of his books.
2171. The birds often nest in giant saguaros, and when holes are bored, the cactus secretes a sap that rapidly builds up to a hard scar tissue, preventing evaporation in the dry desert heat. The hard, gourd-like cavities remain intact long after a fallen saguaro cactus has disintegrated, and the Apaches used these indestructible shells as water jars.
2172. The Asian Pheasant-tailed Jacana.
2173. Yes, some crabs prey on seabird chicks, such as terns.
2174. Yes, but hole-nesting is rather unusual in corvids.
2175. No, the mollusk is extracted from its shell.
2176. The Green-backed Firecrown breeds as far south as the Straits of Magellan.
2177. Small Asian estrildid finches. The name is a corruption of Ahmadabad, the Indian city from which the first of the finches were sent to Europe.
2178. *The Spruce Goose.*
2179. The lark-sized Hume's (Tibetan) Ground-Jay of central Asia.
2180. Birds have no teeth, but many species "chew" food in their muscular stomach or gizzard.
2181. At least 52.
2182. Yes, Lesser Noddy chicks resemble adults, even to the presence of a white patch on their forehead and crown.
2183. Yes, but it never became established.
2184. No, a pair may construct 3-5 nests a year.
2185. No, but many of the white pelicans do.
2186. The Martial Eagle. Females may weigh 13.6 pounds (6.2 kg).
2187. The seriema.
2188. An Australian megapode.
2189. Longevity.

Answers

2190. The Icelandic name means *foul gull.* Fulmars are petrels and the name undoubtedly was given because of their superficial appearance to gulls, and from the rank odor of the musky oil disturbed birds regurgitate.
2191. Two eggs.
2192. The name may have been given because the initial introductions into Europe were from the Turkish Empire. At the time, it was customary to refer to foreign traders as turks, therefore, turkey might have merely meant a foreign bird.
2193. An albatross requires many years to mature, and some immatures were still at sea when the last breeding adults were killed. Were it not for this, the Short-tailed Albatross would undoubtedly have become extinct.
2194. No. Both the Himalayan Griffon and the Cinereous Vulture are larger.
2195. Pelicans were said to pierce their breast and fed their chicks on their own blood. Pelicans frequently rest with their bill tip touching their breast, and the legend was probably based on the fact that breeding Dalmatian Pelicans have a reddish-golden spot on their upper breast superficially resembling a wound.
2196. The Streamertail, an endemic hummingbird.
2197. No, piculets lack stiffened tail feathers, and their tails are short and soft.
2198. The nearly naked chick has black skin.
2199. No.
2200. Shirley Temple.
2201. The Northern Jacana has a three-lobed yellow frontal wattle, whereas the South American Wattled Jacana has a dull red, two-lobed frontal wattle and a lateral (rictal) lappet overhanging the side of the bill. Nevertheless, some taxonomists have combined both jacanas into a single species.
2202. The endangered Kakapo (New Zealand Owl Parrot).
2203. No. The pygmy shrew is often cited as the smallest mammal, weighing about a tenth of an ounce (2.0-3.35 gms), or about the same as the male Cuban Bee Hummingbird. However, Kitti's hog-nosed bat is even smaller, weighing only 0.053 of an ounce (1.5 gms).
2204. When the falcons declined because of chlorinated hydrocarbon pesticide poisoning, it possibly impacted the geese that subsequently underwent a population decline. The high Arctic-nesting geese may depend on the breeding falcons and nest near them because the raptors function as predator deterrents.
2205. Two eggs, but at times, only a single egg.
2206. Stirrups.
2207. The opening of the Scottish grouse season. The *glorious twelfth* is as much a social occasion as it is a hunting event.
2208. The Hemlock Warbler, because it often nests in hemlock trees.
2209. No, most loon bones are solid rather than filled with air spaces.
2210. About 20 pairs per square meter.
2211. Yes, the breast and belly of a breeding adult takes on a distinct rosy wash.
2212. The Red-footed Booby can be either white or dark, with some birds intermediate.
2213. The grebes, whose legs are well designed to reduce water resistance.
2214. Montagu's Harrier. The low point was in 1974 when there were no breeding pairs. In 1986, seven pairs bred, of which six pairs reared 13 young.
2215. The bowerbirds. Males construct specialized structures, which may be very elaborate, that serve no function but display, and are entirely distinct from the nest.

2216. An adult may periodically incubate during the cooler part of the day, but generally only at night. Its eggs are covered with warm sand for most of the day. Adults sit above the buried eggs to provide shade and periodically soak the sand with water transported in their belly feathers.
2217. Yes, even on perches that are nearly vertical.
2218. Nowhere else.
2219. The proliferation of golf courses.
2220. Loon feet project beyond the tail and are held together sole to sole.
2221. The Pectoral Sandpiper, from its habit of squatting.
2222. A major portion of its head is devoid of feathers revealing bare skin of brilliant turquoise-blue grading to cobalt-blue below. Immatures have bare greenish-yellow skin, whereas the bare facial skin of a juvenile is brownish.
2223. Yes.
2224. No, but females do. Both sexes of guans and chachalacas brood chicks under their wings.
2225. Yes.
2226. The huge grouse feeds mainly on a low grade diet of conifer needles and reportedly, the bird excretes daily sufficient fecal material to form a line exceeding nine feet (2.7 m) in length.
2227. Yes.
2228. No. Snails are held under water and the blade-like lower mandible cuts the columeeller muscle, releasing the mollusk from its almost intact shell. At least in the Asian species, a narcotic secretion is produced in the saliva that flows down the lower mandible onto the operculum, relaxing the snail's muscle, allowing for an easier entry.
2229. Yes.
2230. In 1893, in honor of Dr. James G. Cooper, recognizing his pioneering ornithological work in California.
2231. No known dinosaur egg is larger. An Elephant-bird egg is three times as large as the largest dinosaur egg.
2232. Yes, apparently so.
2233. In Australia, the nocturnal birds fly down to seize prey illuminated on the road and are struck by motor vehicles.
2234. The Parasitic Jaeger. It pirates fish from the terns on the long ocean journey.
2235. The Hottentot Teal is sometimes cited as the smallest, but the female African Pygmy-goose is actually smaller, weighing only nine ounces (260 gms).
2236. At least 172.
2237. To rail.
2238. The theoretical potential is 19,531,000 descendants.
2239. No, tropicbirds often catch airborne flying fish, although plunge-diving is more typical.
2240. About 10 times as large.
2241. Lewis' Woodpecker is the most efficient at foraging on flying insects. Accordingly, its wings are broader and its mouth opens wider than the mouths of other American woodpeckers. Foraging woodpeckers may remain aloft for more than a half-hour, circling over open country, often in the company of swallows.
2242. Sable Island, Nova Scotia, Canada.
2243. Its legs are dark but the conspicuous ankle joints ("knees") are red.
2244. Gobblers or toms.

Answers

2245. The petrels have a very exact laying schedule with 85 percent of their eggs laid between November 24 and 26. Laying generally starts on November 20-21 and finishes by December 1-3.
2246. Yes, flamingos often up-end in deep water to feed like dabbling-ducks.
2247. In 1973, the population was estimated at between 309 and 580 birds, but possibly less than 200 currently exist.
2248. *Sitting in the catbird seat.*
2249. No, the large African lake is almost devoid of birds on open water, and birds are not even common around its shores. Deep lakes are typically less productive than shallow lakes.
2250. No, its breeding cycle exceeds a year, thus the booby breeds only in alternate years.
2251. The Austral Parakeet breeds as far south as Tierra del Fuego.
2252. A ring of small shingle-like bones encircling the optic cavity in front. The sclerotic ring is especially prominent in owls, but is also a feature of some reptiles.
2253. A jerkin.
2254. Two: Greater and Lesser (Darwin's) Rhea.
2255. Not as a rule, but a few instances have been reported where a walrus swam up under eider ducklings and alcids and sucked them down. Fulmars have also occasionally been taken on the surface.
2256. The Whistling, Bewick's and Jankowski's Swan.
2257. The American Museum of Natural History, with over 850,000 specimens.
2258. The ratites (Ostrich, Emu, cassowary, rhea and kiwi) and the waterfowl (ducks, geese and swans). Most birds merely require cloacal contact for fertilization to occur. In some of the Ruddy and blue-billed ducks, the "pseudopenis" may be nearly four inches long (10 cm) and spirally twisted. After copulation, the organ may trail in the water and more than a minute may be required for it to be withdrawn.
2259. California Condors cannot be sexed externally. Both sexes are identical in appearance.
2260. A chicken.
2261. Yes. After having been eliminated by 1760, the huge grouse was successfully reintroduced into Scotland.
2262. Yes.
2263. The cormorant. Its feathers are not very waterproof, and following diving and prior to flight, it often extends its wings, either to dry them or to warm up after diving.
2264. No, but they have longer bills and legs.
2265. The Adelie Penguin numbers many millions, but the Chinstrap Penguin is possibly as abundant.
2266. Chambers carved out in active termitaries on trees.
2267. A crane.
2268. Some eagles crack open tortoises by dropping them on rocks below, and Aeschylus met his death in 456 B.C. in Sicily when an eagle, probably a Lammergeier, carrying a tortoise mistook his bald head for a smooth rock (he was nearly 70).
2269. The Crab Plover.
2270. The South Georgia Pintail and Eaton's Pintail, the latter of which is subdivided into the Kerguelen and Crozet Pintail.
2271. A wren pair once nested on a human skull.
2272. Males of most species are brightly colored and adorned with fancy, ornate feathers, thus elaborate songs are not required, although a number have exceptionally loud voices.

Answers

2273. The ptarmigans.
2274. The Yellow Warbler.
2275. A sparrow-bill.
2276. Yes, sometimes even in flight, and in Barn Swallows, this may continue for several weeks.
2277. Female skuas are larger than their mates.
2278. Pâté de foie gras.
2279. Stone-fly larvae, but the specialized ducks may also take a few fish and mollusks.
2280. No, it is nearly flightless.
2281. Yes.
2282. Grebes, along with some other aquatic birds, expel air trapped between their feathers, and probably also from their air-sacs, enabling them to sink out of sight, scarcely leaving a ripple behind.
2283. The Green Sandpiper.
2284. The Black-necked Swan.
2285. Nineteen.
2286. Some live above the 15,000 foot (4,572 m) level.
2287. Yes, African crowned-cranes may nest in trees, but all other cranes nest on the ground.
2288. Yes.
2289. The Little Wattlebird doesn't have wattles.
2290. Yes.
2291. On March 24, 1989, the supertanker *Exxon Valdez* ran aground in Prince William Sound, Alaska. It spilled nearly 11 million gallons (41,639,400 liters) of Alaskan crude oil that ultimately drifted over 10,000 square miles (25,898 sq km) of coastal and off-shore waters. The spill killed an estimated 375,000 seabirds, of which more than 70 percent were murres. This figure represents the greatest toll of marine birds from oil pollution ever documented. At least 146 Bald Eagles and roughly 1,000 sea otters perished as well.
2292. Yes, even the African Square-tailed Drongo has a slightly forked tail.
2293. The New Zealand moas didn't have wings. Even the humerus was lacking, and their shoulder girdles lacked sockets into which wing bones fit.
2294. Primarily seed eaters, the parakeets quickly modified their feeding habits to take advantage of new sources of food that became available with the spread of organized cultivation, and the birds soon were regarded as pests in the eyes of the colonists. Their preference for the seeds of various fruits, combined with their communal feeding habits, resulted in the ruination of orchards. Long before ripening, fruits would be torn from the stalks and ripped open, and a flock could destroy the produce of an apple or pear tree in short order. Corn fields were also sought, and in 1831, Audubon indicated that flocks attacking grain crops *cover them so entirely, that they* (parakeets) *present to the eyes the same effect as if a brilliantly coloured carpet had been thrown over them.*
2295. Possibly less than 1,000 birds. The entire population of the aberrant shorebird is restricted to southern Chile and Argentina.
2296. No. A Wandering Albatross may follow a ship for days, but a Royal Albatross seldom spends much time behind a ship.
2297. About 1,030 species.
2298. No. Diving-petrels brood their chicks for several weeks, an unusually long time. Several days of brooding is far more typical for most small petrels. Diving-petrels lack stomach oil to feed their young, possibly resulting in a slower chick growth-rate than the other petrels.

Answers

2299. The Red-breasted Nuthatch.
2300. The screamers, penguins and ratites, and some authorities also include the African mousebirds. The skin of most birds appears to be fully and evenly covered by feathers, but in reality the feathers grow only from distinct feather tracts.
2301. Yes, in most instances.
2302. Gannets generally roost at sea, but boobies tend to return to islands to roost.
2303. The pictorial representation of Quetzalcoatl, the principal god of the Toltecs of Central America.
2304. The cracids (chachalacas, guans and curassows).
2305. The mergansers. They are primarily fish eaters, although eiders and scoters are nearly as carnivorous.
2306. The tropical antpittas.
2307. No.
2308. The neotropical jacamars, in which chicks hatch covered with copious amounts of long whitish down.
2309. The famous nocturnal penguin parade that takes place when Little Blue (Fairy) Penguins venture ashore after dark to visit their burrows. The event is a major tourist attraction.
2310. Yes, the New Zealand kiwis utilize nesting burrows.
2311. Yes, about 25 percent heavier.
2312. About two-thirds.
2313. Approximately 2.5 million metric tons.
2314. The White-tailed Ptarmigan.
2315. No, fish are snatched from the surface of the water.
2316. The Highland Guan.
2317. No one knows. Gentoo in various spellings means a pagan inhabitant of Hindustan.
2318. Four: Wood, Milky, Yellow-billed and Painted Stork.
2319. A term applied to nest-mates when the first-hatched and largest chick attacks and kills the smaller, weaker chick. This often occurs with eaglet siblings.
2320. The Sarus Crane.
2321. The Northern Fulmar.
2322. The White-headed Duck. While the current world population is about 19,000 birds, it is unlikely that more than 1,000 remain in Europe. An estimated 12,000 winter in Turkey and neighboring countries, 1,000 in Israel, and at least 5,000 in southwest Asia.
2323. The British Museum of Natural History, with a collection of more than 1.25 million skins.
2324. A zero on the score board.
2325. Totally black except for a conspicuous white V on its back, a white rump and upper tail coverts, and some white scapulars. It is also known as the African Black Eagle.
2326. The Indian Peacock.
2327. An Australian and New Guinea robin that is sometimes called the Brown Flycatcher.
2328. Someone in the house will die.
2329. Well camouflaged, either by color or pattern, or both.
2330. Yes, the territories are defended, and over 90 percent of the birds occupy the same site annually.
2331. The Chaffinch. The name was apparently derived from one of its calls.
2332. Yes and no. Cavity-nesting species hatch with closed eyes, but the chicks of surface-nesting penguins hatch with their eyes open.

Answers

2333. The Cattle Egret.
2334. The netting of Common Puffins, murres and Northern Fulmars with a long-poled, hand-held net. In Iceland alone, 150,000-200,000 puffins are netted annually. Icelanders are not permitted to take puffins in burrows.
2335. Nuptial plumage.
2336. The tiny 5.5-inch-long (14 cm) Bushtit constructs an elaborate nest consisting of a long, pendent bag up to a foot (30 cm) in length.
2337. Bach wrote the book *Jonathan Livingston Seagull.*
2338. When line becomes snarled on the reel, it is called a *bird's nest,* a term that is not complimentary.
2339. Northeast Bolivia.
2340. The Secretary-bird. The central tail feathers of a male may reach 34 inches (85.4 cm) in length.
2341. Zero: the parrot is probably extinct. None has been seen since 1915.
2342. On islands such as Macquarie Island, where the Weka rails were introduced, the birds consume petrels.
2343. Steller's Sea-Eagle females weigh up to 20 pounds (9.1 kg).
2344. The adult males of the five tragopans.
2345. As of 989, Indonesia with 126, but Brazil was a close second with 121.
2346. No.
2347. Probably the rails.
2348. The nocturnal Kakapo. Courting males inflate their thoracic air-sacs to gross proportions and produce loud booming vocalizations. The unique New Zealand parrots have been described as "feathered footballs."
2349. Yes.
2350. The Red-crowned (Manchurian or Japanese) Crane.
2351. Yes, at times. Emperor Penguins breed on fast-ice, and the capability of entering the water while still down-covered may be an adaptation to early ice break-up.
2352. The Barn Owl.
2353. The flightless steamer-ducks The name refers to their huge heads.
2354. At least 5,000. Following Gould's death in 1881, it was sold to the British Museum, where most of the impressive collection remains today.
2355. Yes.
2356. No, only the American Black Skimmer.
2357. Yes, a larger gland may be required because Ospreys dive into the water after fish.
2358. A stilt.
2359. The African Yellow-throated Longclaw bears a striking resemblance to the meadowlark, but it is not remotely related.
2360. Although not documented, the vultures probably kill such prey as hyraxes, monitor lizards and young dogs.
2361. The chicks are down-covered and hatch with their eyes open, as opposed to naked and blind. In contrast to the Sungrebe, only females incubate in the two Old World finfoots.
2362. No.
2363. No, the warbler feeds primarily on moth caterpillars, especially spanworms.
2364. Push or roll its eggs as much as 50 feet (15 m) from the original nest site. Eggs may even be floated over areas of open water to a safer location.

Answers

2365. The ovenbird generally nests in the burrow of a vizcacha, a large gregarious rodent.
2366. A feral population of approximately 1,000 pairs (perhaps as many as 2,500) Mandarin Duck is established in England. As a rule, introductions are not beneficial, but in this instance the secondary population should be encouraged, because the Asian population numbers less than 6,000 pairs.
2367. The normal frequency may exceed 20 blows per second.
2368. Yes, such as pre-fledging Wilson's Storm-Petrels.
2369. A foot where all the toes are directed forward, or at least *capable* of being so directed, as in many swift species. Four forward-pointing toes may improve a swift's ability to cling to vertical surfaces.
2370. The Bluethroat, in northwestern Alaska.
2371. Eggs are placed on each point of a 7-pointed star. Church members believe this will protect the church from lightening strikes. The seven eggs represent truth, benevolence, brotherly love, harmony, spirit (soul), justice and peace.
2372. Yes.
2373. No, fish are grasped with the bill.
2374. Based on one study, 2,600 lemmings.
2375. The three tropicbirds.
2376. Yes. Between 36,000 and 50,000 pigeons were maintained by U.S. forces during the war, and 5,000 during World War I.
2377. The Hungarian (Gray) Partridge.
2378. Very rarely, but the closely-related Anhingas and darters commonly soar.
2379. No, it has the longest legs.
2380. Africans regard the stork as having magical powers, and as a result it is never persecuted.
2381. The North Pacific (mainly the Bering Sea) where 16 of the surviving 22 alcids occur, 12 of which are endemic to the region.
2382. Either 14 or 15.
2383. The Akiapolaau, an Hawaiian honeycreeper. The straight lower mandible is used to chip away bark to reveal beetle larvae, which are apparently then extracted with the curved, probe-like upper mandible. Approximately 1,500 Akiapolaaus survive in the Koa and Mamame-Naio forests of the big island.
2384. About 10 percent of their body weight daily.
2385. The Eurasian Red-crested Pochard.
2386. Yes. While its center of abundance is in the western Mato Grosso region of Brazil, small numbers have recently been discovered in northeastern Bolivia and northern Paraguay.
2387. A 150-ton adult blue whale weighs in excess of 48 million times as much as a Kitti's hog-nosed bat weighing less than a 0.053 of an ounce (1.5 gms).
2388. The Wedge-tailed Eagle.
2389. Yes. Unlike their closest relatives, the waterfowl, screamers frequently fly to considerable heights and may soar for hours with their long legs trailing, especially the Southern Screamer.
2390. The adult African Palm-nut Vulture.
2391. In June of 1964, a White-tailed Tropicbird circled around remote-controlled model airplanes at Newport Beach, California. After displaying to model planes in the air, the bird landed next to a model on the ground and attempted to copulate with it at least five times.

Answers

2392. About 20 percent: two percent by the lungs and 18 percent by the air-sacs. In humans, about five percent of the total volume of the body is taken up by the lungs.
2393. Yes, but only the Spot-winged Falconet of the semi-deserts and savannas of western and northern Argentina. It is in a different genus than the Old World falconets.
2394. The Inca Tern.
2395. The Blue or Paradise Crane.
2396. The feathers grow continuously and are never molted.
2397. Only three: Wood, Jabiru and Maguari Stork.
2398. John Clark, a 19th-century ornithologist. Many people incorrectly assume that the grebe was named after one of the leaders of the Lewis and Clark expedition.
2399. The American Wigeon. It "steals" floating water plants uprooted by other waterfowl.
2400. No, iridescence is a result of light being refracted as it passes through a feather's structure.
2401. Possibly because insects and parasites have become established in the first nest, prompting the building of a new nest. In addition, a new nest would be clean. In regions where the breeding season is short, the same nest may be used for more than one brood.
2402. Yes.
2403. Black.
2404. Happiness.
2405. The woodcocks. Such movement enables the bird to seize an earthworm deep in the ground without expending the energy required to open its entire beak to push aside the soil through the entire beak penetration depth.
2406. *The Partridge Family.*
2407. Ostriches don't have scapulas.
2408. The Short-tailed Albatross.
2409. A bird's footprint preserved in stone.
2410. About three hours.
2411. A large kite.
2412. The Limpkin and the three South American trumpeters.
2413. No, the Aleutian Tern is an Arctic-breeder as well. Nesting Common Terns also reach the Canadian Arctic, but generally well south of breeding Arctic Terns.
2414. A bird distributed around the world in the tropics.
2415. Bark fungi, lichens and possibly termites.
2416. Both have been domesticated.
2417. The Dunlin.
2418. Possibly because the birds consume large numbers of ants, they acquire a distinctive ant odor and are not recognized as intruders. The ants fiercely attack other intruders, but run over the woodpecker's eggs and nestlings without harming them.
2419. No.
2420. Yes, as a rule.
2421. No, the Kagu is only nocturnal when incubating. The endangered bird was believed to be nocturnal because of its large red eyes, but the large eyes are probably an adaptation to low light levels in thick undergrowth.
2422. The Snow Bunting, because a snow burrow is relatively warm.
2423. Indeterminate species continue to lay eggs until a specified number has accumulated in the nest. Conversely, a determinate layer produces a certain number of eggs and then stops, regardless of the fate of the eggs. In one well-documented study of the Yellow-shafted

Flicker (an indeterminate layer), eggs were experimentally removed, and the persistent female laid 71 eggs in 73 days.

2424. The colorful, ground-dwelling Old World pittas.
2425. Not until their fourth or fifth year, and possibly longer. For a passerine, this is remarkably long.
2426. Most have only three toes, but the aberrant Australian Plains-wanderer has four. Buttonquail are not related to quail and appear to be most closely related to the cranes and their allies.
2427. Possibly because the Ostrich, with an incubation period averaging 42 days, inhabits regions swarming with large terrestrial carnivores, a shorter incubation period is advantageous. The Australian Emu is not faced with such threats, and can afford to safely incubate longer (56 days average).
2428. The bill evidently enables the hummingbird to capture and swallow insects, which form a larger part of its diet than in most other hummingbirds.
2429. The Christmas Island Frigatebird. Possibly no more than 1,600 pairs breeding in three colonies remain. Disturbance and habitat destruction caused by phosphate mining, hitting electric lines, illegal capture of birds for food by the island residents, and cyclones knocking down trees with nests are serious threats.
2430. The eagle reportedly perches concealed in the foliage near a band of monkeys and whistles softly. This supposedly attracts the old males and, when in range, the monkeys are grabbed.
2431. Cowbirds commonly feed near the feet of ruminant mammals, and may have followed the great bison herds on their northward treks in the spring and back again in the fall. The herds moved continuously, thus the cowbirds had no time in a given area to build nests and raise young. Therefore, birds that were sedentary during the breeding season were parasitized by the continually moving cowbirds.
2432. About 45.
2433. No, temperate birds typically produce larger clutches.
2434. At least 400.
2435. Jouanin's Petrel. Its breeding grounds remain undiscovered, but it is suspected to breed in Oman, either inland in desert mountains or on coastal islands.
2436. *The Sea Swallow*.
2437. The dove.
2438. The South American Coscoroba Swan.
2439. Yes, very much so.
2440. The Tree Swallow feeds on some seeds and berries, primarily the bayberry. Such food is generally taken early in the breeding season or during periods when insects are scarce.
2441. Yes, the two head plumes are not only erectile, but can be moved in all directions.
2442. Yes.
2443. Yes, but only once from Dusky Sound, New Zealand in July of 1884.
2444. Probably because in lightly-wooded areas where Sulphur-crested Cockatoos are found, white plumage is an advantage when the goshawk hunts the white-plumaged parrots.
2445. The three tropicbirds.
2446. No, Nicaragua is as far south as ravens range in the New World.
2447. The elusive West African forest passerines inhabit bat caves. They apparently feed on invertebrates attracted to the accumulation of bat guano on the cave floor.

Answers

2448. Probably the Whooper Swan. On 9 December 1967, about 30 swans were spotted by an airline pilot flying over the Inner Hebrides at an altitude of just over 27,000 feet (8,230 m).
2449. The European Honey-buzzard, one of the most variable in plumage of all raptors.
2450. Turtle hatchlings are preyed on.
2451. No, the Black Vulture is more abundant.
2452. The last remaining rail was eliminated in late 1990. Wekas were a threat to burrowing seabirds.
2453. Small wren-like Australian passerines with erect tails.
2454. No. The Northern Shrike is larger, weighing more than two ounces (57 gms), whereas the Loggerhead typically weighs less than two ounces.
2455. Iceland. Nesting eiders are vigorously protected and the approximately 200 eider duck farms farms yield around 8,000 pounds (3,629 kg) of processed down annually. Eiders are also farmed in Norway and Russia.
2456. Within two days of hatching.
2457. No, the storks pant and defecate on their legs.
2458. Based on much of the literature, a typical clutch consists of a single egg, but two eggs are actually more common, and three rare. The interval between laying is about 3-4 weeks. An egg takes about 30 days to form, the greatest amount of time for any bird. In most birds, eggs are formed in a day or two, or possibly a week. In both the Great Spotted and Little Spotted Kiwi, the clutch size is only one egg.
2459. Blakiston's Fish-Owl is the only fish-owl with fully feathered legs. The bird apparently wades in shallow water seeking crayfish.
2460. Yes, but not always. Even when nesting in the "open," the auk generally seeks a protective rock overhang, although some sites are quite exposed.
2461. The bustards.
2462. At least 31, of which no less than 14 are in the order Charadriiformes.
2463. The flightless Dodo.
2464. Flocks exceeding 400 cocks sometimes gather to display.
2465. Ambient air temperatures may exceed 100°F (38°C), but the tropical penguin's range is characterized by the presence of cold-water currents. The Antarctic-spawned currents cool the penguins, and the chilled nutrient-rich waters abound with food fish.
2466. The successful 1979 rescue mission mounted by H. Ross Perot, the colorful Texas businessman, billionaire, and 1992 U.S. presidential candidate. Several of Perot's executives were taken hostage by militant Iranians.
2467. The jacana has the curious habit of elevating its wings, displaying the conspicuous yellow undersides.
2468. The White-tailed Ptarmigan.
2469. Fifteen, including the aberrant Lammergeier.
2470. The Parasitic Jaeger.
2471. Owls typically swallow prey whole and regurgitate pellets. Hawks, falcons and eagles tear food apart, usually rejecting fur, feathers and larger indigestible parts.
2472. Oscines are in the suborder Passeres, encompassing the true songbirds with a complex syrinx; at least 80 percent of the passerines are in this huge, diverse division. Suboscines have a simpler syrinx but have a more advanced middle ear apparatus. All passerines are known as perching birds, but technically only the oscines are considered songbirds.
2473. Quail. Quail eggs are even available in modern supermarkets.

Answers

2474. Yes. On rare occasions, the birds nest in underground cavities, especially on offshore islands lacking terrestrial predators.
2475. The tinamous. The purity and softness of the tones of many tinamous are attained by few other birds.
2476. The Tussock-bird (Blackish Cinclodes).
2477. No. The American Sparrowhawk is a small falcon more correctly called the American Kestrel, whereas Old World sparrowhawks are small accipiters.
2478. The green iguana. The large arboreal lizard is highly prized because its flesh tastes like chicken. The reptile has disappeared in many areas, and in 1989 the black-market price was as high as $20 each.
2479. Pairs may incubate simultaneously, but in different nests. The female builds two nests and then lays in the first nest. The eggs remain unincubated for several weeks while the female lays another 10 eggs or so in the second nest. She then returns to the first nest to incubate while the male incubates the second clutch. Both clutches often hatch and each parent independently cares for its own brood of young.
2480. Balance. Hoatzins are top-heavy and unstable among the branches.
2481. No, moonlight inhibits them, as does a strong wind.
2482. Young cassowaries are kept until large enough to be eaten or traded, or until the bird kills a villager. A cassowary is very valuable and can be exchanged for up to eight pigs, or even a wife.
2483. The White-capped Albatross.
2484. A charm: a charm of goldfinches.
2485. No, the owl is known only from a single specimen collected in 1952.
2486. It would be hazardous for roof-top or tree-hatching storks to walk too early in life, but it is advantageous for a ground-nesting species. The legs of the terrestrial precocial cranes grow most rapidly between the 8th and 16th day, whereas the legs of a stork chick grow most rapidly between the 20th and 38th day.
2487. About 5.7 beats per second.
2488. Its bill is thrust down the gullet of the young, and then the adult regurgitates.
2489. The Eastern Rosella.
2490. Yes, praying mantises occasionally catch small birds.
2491. A type of pincers used by surgeons.
2492. The large raptor has a very short tail, and must cant from side to side in flight to steer. The name *bateleur* is from the French referring to a tightrope walker using a horizontal balancing pole.
2493. A dinosaur. The footprints of some species are similar in size and shape.
2494. Blue-green algae.
2495. The highly venomous snakes eat both eggs and small young. Burrows are also shared with large chicks and adults, and possibly the snakes are warmed by the birds. How the birds benefit, if at all, is not known.
2496. Steller's Eider.
2497. The Northern Gannet.
2498. Nesting sites are scarce and the endangered coots may pile up small stones until the rock pile breaks the surface of the water. Some often-used nesting "islands" reportedly consist of 1.5 tons of piled rocks with a base of 13 feet (4 m).
2499. The Lammergeier. In 1980, possibly no more than 75-90 pairs remained.

Answers

2500. The largest-known dinosaur eggs are those of *Hypselosaurus*, a 30-foot-long (9.1 m) sauropod of the late Cretaceous of Europe. Its 10 by 12-inch (25 x 30.5 cm) egg with a capacity of 5.77 pints (3.3 liters) probably represents the largest size possible for a reptile egg, because reptile egg shells are more fragile than bird egg shells, and if the egg were any larger, the shell could probably not withstand the great internal pressure.
2501. The cock Eurasian Black Grouse.
2502. The three species of bluebirds.
2503. Yes, but the feathers are very small and scale-like.
2504. Yes. Most rails are extremely vocal.
2505. Robert Redford and Faye Dunaway.
2506. The ptarmigan.
2507. Leach's Storm-Petrel.
2508. The false bird-of-paradise plant. The true bird-of-paradise plant is the *Strelitzia.*
2509. No, bustards rarely, if ever, rest with one leg drawn up.
2510. Yes.
2511. Molted penguin tail feathers. Other material is utilized as well, including lichens, limpet shells, bones, eggshells and shed elephant seal skin.
2512. No, but the sole exception is the Red-crowned (Japanese) Crane, which is mostly resident, even when winter conditions are severe. The Japanese population is the most sedentary. The three southern *races* of the Sandhill Crane are also resident.
2513. The Whiskered Auklet has been observed by very few ornithologists.
2514. Probably the African Crowned Hawk-Eagle. It rears only one young every two years in Kenya, but in South Africa it may rear a chick every year.
2515. Large male King Penguins may weigh up to 40 pounds (18 kg).
2516. The frog duck, because a calling merganser sounds rather like a pickerel frog.
2517. Bright red.
2518. The bitterns. They direct their eyes under the bill when they *freeze* with the bill pointing upwards, enabling the birds to keep close watch on a potential predator. Most birds can scarcely move their eyes within the socket, and head movements compensate for this lack of mobility.
2519. Only about 80.
2520. No, nor can the skimmers walk except for very short distances.
2521. When birds are copulating, the cloacal contact is sometimes called a *cloacal kiss*.
2522. Yes.
2523. The Black-crowned Night-Heron.
2424. The Turkey Vulture.
2525. Large, skulking, weak-flying, terrestrial cuckoos. All 30 species are Old World birds, some of which are known as crow-pheasants.
2526. Yes, but their ceres are generally obscured by bristles projecting laterally from the base of the bill.
2527. No, unlike most ibises, it is a solitary nester.
2528. Three days usually, but up to six days.
2529. Yes, the large shorebirds are nocturnal or crepuscular.
2530. Hartlaub's Gull.
2531. The magnificent Bulwer's Pheasant.
2532. Abdim's Stork.

Answers

2533. Four eggs of such a shape can be closely packed together by arranging the narrow ends to point inwards, thus minimizing the area to be covered by the incubating adult. A compact group also dissipates heat rather slowly.
2534. Both the Black and Turkey Vulture.
2535. Texas, with 320 species.
2536. Yes.
2537. No, it is flexible.
2538. Traditionally two, the Laughing and Blue-winged Kookaburra, but two others are in the same genus *(Dacelo)*: Aru Giant Kingfisher and Rufous-bellied Giant Kingfisher. The Shovel-billed Kingfisher *(Clytoceyx rex)* is also currently called a kookaburra.
2539. Probably no more than 500. The 265 singing males counted in 1990 reflect the highest number since 1961.
2540. The Harpy Eagle and the cassowary. The eagle has the largest and most powerful feet and claws of any raptor with the sharp talons of a female exceeding 1.5 inches (3.8 cm) in length. The 5-inch-long (12.5 cm) nail of a female Double-wattled Cassowary is equally as fearsome. More humans have been killed by cassowaries than by eagles.
2541. Only two inches (5 cm).
2542. According to the legend of Leda and the Swan, on the eve of her wedding to King Tyndareus of Sparta, Leda was visited by the great god Zeus (Jupiter), who assumed the form of a swan. Leda was seduced and subsequently bore two sets of twins, each enclosed in an egg. One set of twins was said to be the progeny of Zeus, while the other twins were the mortal children of Tyndareus. Countless works of art have been inspired by this classic legend, including a magnificent painting by the master Michelangelo, illustrating the most intimate phase of the liason between the god and his mortal mistress, that is regarded as one of the world's greatest art treasures.
2543. No, but albatrosses, many petrels and frigatebirds are exceptions.
2544. The Winter Wren. In Europe and Asia, it is simply known as *the wren.*
2545. The banded penguin complex consists of four tropical or temperate species (Magellanic, Humboldt, Galapagos and African Black-footed or Jackass Penguin). Of these, the Humboldt and the Black-footed Penguin have a single chest-band.
2546. Four: Trumpet Manucode, Paradise, Victoria's and Magnificent Riflebird.
2547. No, the endangered duck is restricted to Madagascar.
2548. A crane.
2549. The name is probably onomatopoetic in origin.
2550. The Peregrine Falcon, Bald Eagle and Osprey.
2551. Hornby's (Ringed) Storm-Petrel nests in the coastal deserts of Chile and Peru.
2552. The names are phonetic renderings of the owl's clear, disyllabic repeated calls.
2553. About 10 inches (25 cm).
2554. No, the males do.
2555. Yes.
2556. The nearly extinct Kakapo (Owl Parrot) of New Zealand.
2557. No, but the similar-appearing Tundra Swan has varying amounts of yellow on its bill.
2558. Yellow.
2559. No. They have large wings, but the birds fly poorly.
2560. Fourteen, of which the Eurasian Dotterel, Northern Lapwing and Oriental Plover are visitors.

Answers

2561. About 96.5°F (35.8°C).
2562. Approximately 400.
2563. The Siberian Red-breasted Goose.
2564. An Australian bell-magpie.
2565. The Whooper Swan.
2566. The Wild Turkey. In the spring, a thick mass of fibrous tissue arises and overlays the breast and crop of tom turkeys. Just before the breeding season, this material may become so fatty that it weighs up to two pounds (0.9 kg). It serves as a reservoir of energy during the breeding season when gobblers are so occupied with displaying that they eat very little.
2567. Louries.
2568. Yes, sparrow and quail chicks may be included in its varied diet. It feeds primarily on fish, usually brook trout, freshwater sculpin, stickleback and salmon up to 5.5 inches long (14 cm). Other fish are also preyed on, as well as crayfish, frogs, lizards, water-shrews, insects, and in winter, even berries.
2569. Probably the buttonquail (hemipodes), that incubate for an average of only 12-13 days.
2570. The Sacred Ibis.
2571. A form of aerial-pursuit used by food-pirates. The bird repeatedly flies in a hostile manner at its intended victim, with or without contact, in an attempt to steal food.
2572. Possibly from two Welsh words: *pen* meaning head and *gwyn* meaning white, alluding to the large, white oval spot on the head of the Great Auk, the original penguin. The name may have also come from the Latin word *pinguis*, referring to the fat or blubber of the Great Auk.
2573. The Palm-nut Vulture.
2574. Yes, especially Gull-billed, Black and Whiskered Terns. Most terns are fish eaters.
2575. The Mauritius Kestrel that was down to only 4-6 birds in 1973 when the first pair was captured for captive propagation. Large numbers were shot when it was formerly known as *mangeur-de-poules* (chicken-eater), even though the kestrel's favorite food appears to be geckos. By late 1991, the 200th captive-reared kestrel was released into the wild, and the population is currently regarded as almost self-sustaining.
2576. A flock of black Rooks, traditionally regarded as birds of ill omen.
2577. About 40.
2578. The grayhen, whereas the male is known as the blackcock.
2579. Probably Bernier's Teal and the Madagascar Pochard, both of which are restricted to a few small lakes in Madagascar. Possibly no more than 20 pairs of the teal remain. The Campbell Island and Aucklands Island Flightless Teal also have very restricted ranges, but both have traditionally been regarded as races of the more widespread New Zealand Brown Teal. The most recent taxonomic revision combines the two insular teal into a separate species.
2580. Yes. The voice of a kiwi is among the loudest in the forest.
2581. The island was named after the overpowering stench ("funk") of the Great Auk feather industry. Hundreds of thousand of auks were rendered in boiling cauldrons, and the fires were fueled by the fat-laden birds themselves, and this was obviously odoriferous.
2582. Yes.
2583. The egret was introduced from Florida in 1958.
2584. The Black Skimmer, because its repetitive call sounds like a barking dog.
2585. The Trumpeter Swan.

Answers

2586. No, they mate ashore at the nest site.
2587. Between 7 and 20 percent.
2588. Two species inhabit the Juan Fernandez Islands, 500 miles (805 km) west of Chile.
2589. Not really, but many have strong voices.
2590. All are cavity nesters.
2591. Three: Little Blue (Fairy), Yellow-eyed and Fiordland Crested Penguin.
2592. The 9-inch-long (23 cm) Japanese Grosbeak.
2593. The Greater Roadrunner.
2594. Yes.
2595. Between 11 and 13, as well as a number of subspecies.
2596. The Black-footed Albatross.
2597. Probably the male.
2598. No.
2599. No.
2600. The waterfowl. Many ducks have a distinct row of lamellae that, together with the tongue, form an efficient straining apparatus. The tongue movement sucks water in through the tip of the beak and, on closing the bill, the water is forced out at the base of the beak, trapping food particles against the lamellae.
2601. Generally black, but some finches have red, and more rarely, yellow heads.
2602. Las Islas Diego Ramirez, Chile, 60 miles (97 km) southwest of Cape Horn.
2603. Up to 5.6 feet (1.7 m).
2604. The Lucifer Hummingbird.
2605. Only one. According to much of the literature, dippers are confined to the Americas and Eurasia, but the Eurasian Dipper is resident in the Atlas Mountains of extreme northwest Africa.
2606. The raptors drown their prey.
2607. In bygone times, birds being fattened for the pot often had their eyelids sewn shut, which supposedly quieted them down. The English term *hoodwinking* was derived from this practice.
2608. Yes, but cassowaries can also be inquisitive. Their close relative, the Emu, is notoriously curious.
2609. The African Saddle-billed Stork. The other storks lay 3-5 eggs.
2610. Gigantic hummingbirds.
2611. The Australian Musk Duck. Large males are 24-29 inches (61-74 cm) in length, and weigh up to 8.5 pounds (3.9 kg), but more typically weigh 4-5 pounds (1.8-2.3 kg).
2612. A widespread Eurasian finch.
2613. Yes.
2614. An Old World birding term referring to the general size and shape of a bird, as well as a combination of other characteristics that aid in identifying a bird in the field, such as its flight style, speed, behavior and feeding techniques.
2615. About 15 feet (4.6 m). When running at sprint speed, its stride may exceed 25 feet (7.6 m).
2616. No. Just why the endemic gull is so scarce is not known, but it is not considered endangered. The population probably never numbered more than 800 birds in historic times.
2617. Heroners.
2618. The parrots show "affection" to one another.

Answers

2619. Yes.
2620. Johann Friedrich Brandt (1802-1879), a German zoologist who first described the bird in 1837.
2621. The Scarlet Ibis and Rufous-vented Chachalaca.
2622. Clark's Crow.
2623. Warm-weather penguins must shed heat, and when the bare facial skin is flushed with hot blood from the central core, excess heat is dissipated.
2624. Most are noisy birds.
2625. *Audubon Field Notes*.
2626. No, the most notable American example is the Reddish Egret, but it also has a white morph.
2627. In current usage, tippet refers to the elongated facial feathers of some grebes.
2628. The female.
2629. Yes, it rarely ventures inland more than 1,000 meters.
2630. The scavenging Hall's (Northern) and Antarctic (Southern) Giant-Petrels.
2631. Males are not territorial because honeyguides are obligate brood parasitic nesters.
2632. Wasp larvae.
2633. No.
2634. The island of Bali (Indonesia). The birds are more commonly known as Bali Mynas.
2635. The Wood Thrush. Males weigh up to 1.9 ounces (64 gms).
2636. A kingfisher hole.
2637. No, most species are polygamous.
2638. Yes.
2639. No, but they generally kneel to drink.
2640. No, toucans are related to woodpeckers and are included the same order (Piciformes).
2641. Probably because the gamebird was introduced from France.
2642. No.
2643. Probably the frigatebirds. Post-fledging feeding often continues for a year, and even up to 18 months. Juveniles are generally, but not always, fed only by the female.
2644. About one-third.
2645. The eagles have blue eyes, an unusual raptor eye color.
2646. Yes.
2647. When it pulls up sharply from steep dives during courtship flights, the nighthawk's wing feathers produce a whirring roar.
2648. None.
2649. The wife of Admiral Sir James C. Ross, of Arctic and Antarctic exploration fame.
2650. The Canvasback.
2651. Seeds from the sealed-in female's food fall to the ground and germinate. By examining the seedlings at the foot of the tree, it is possible to judge not only the date of incubation by the age of the seedlings, but also the number of years that the nesting hollow has been used.
2652. Because its saffron-colored plumage resembles the color of the robes worn by holy men.
2653. If food is scarce, the number of nestlings may be reduced by the parents which consume the youngest, weakest or most lethargic chicks.
2654. The Eurasian Hooded Crow.
2655. The grouse seldom flies more than 100-200 meters. If flushed three or four times in rapid succession, an exhausted bird can reportedly be picked up by hand.

Answers

2656. According to National Turkey Federation figures, about 18 pounds (8.5 kg), up slightly from 12 pounds (5.4 kg) in 1986. The increase possibly reflects a shift away from red meat.
2657. Hummingbirds.
2658. Bright yellow or orange.
2659. The Common Raven.
2660. Yellow. The color is caused by a yellow, fat-soluble, carotenoid pigment called lutein.
2661. The Red-billed Tropicbird.
2662. Skuas fly by close to a cliff-side nest, forcing parent kittiwakes to flush, thus exposing the chicks. Adults are also caught in mid-air.
2663. None.
2664. Yes. In one instance, a cock copulated with no less than 21 females in a single morning.
2665. The Old World honeyguides are the only known wax eaters.
2666. None, all are restricted to Asia and Africa.
2667. No, but both races of the similar-appearing Royal Albatross have black cutting edges, as does the Amsterdam Albatross.
2668. The Bobwhite. While in such a formation, a predator can be spotted approaching from any direction.
2669. Only 10 birds, and perhaps as few as four.
2670. No.
2671. The raptor generally bites the snake's head off.
2672. No, the two African terrestrial ground-hornbills have relatively long legs.
2673. The adult Brown Booby.
2674. Vanity and luxury.
2675. No, it was a name for the smaller male Lanner Falcon.
2676. No, it was about the size of a small crow.
2677. About 20.
2678. The Wild Turkey.
2679. The Shag.
2680. Most are, but some, such as the Western and Horned Grebe, are somewhat colonial nesters.
2681. Yes.
2682. No. Most are, but the Comb and Muscovy Duck are sometimes polygamous, as are probably some stiff-tails and the obligate brood parasitic-nesting Black-headed Duck.
2683. Nearly 25 percent.
2684. Limited flight is generally possible within 3-4 days.
2685. The Atlantic, Mississippi, Central and Pacific flyways.
2686. No, the huge bird was herbivorous.
2687. Nothing. The single egg is laid directly on top of its volcano-shaped mud nest.
2688. The Palmchat. It was formerly included with the silky-flycatchers and waxwings.
2689. No, it is one of the least known, and was not even discovered until 1939, and the female was unknown to science until June of 1947. A close relative is the Eyrean Grasswren, that was rediscovered in September of 1961, after an absence of 86 years, having last been seen in December of 1875, when six specimens were collected.
2690. The European Pochard.
2691. The Broad-billed Prion.
2692. Pale blue in front and orange-red to the rear.

Answers

2693. White.
2694. Yes. Their bills can also be red or yellow, but 80-90 percent of all hummingbirds have black or dusky-colored bills.
2695. No, but some sea-ducks, such as scoters and eiders, may use their wings initially to assist them in submerging. Bottom-foraging guillemots use their feet to some extent.
2696. The Masked Booby.
2697. The Magpie Goose of Australia and the Hawaiian Goose (Nene).
2698. The female alone incubates and rears the young. Its ground nest is concealed, but an incubating or brooding flightless bird is defenseless against introduced terrestrial carnivores. The fact that nesting females are so vulnerable clearly must be one of the main reasons that far more male than female New Zealand Kakapos currently exist.
2699. The International Council for Bird Preservation (ICBP) was founded in 1922. Its name was changed in 1993 to Birdlife International.
2700. Probably the Northern Goshawk.
2701. Yes, but some Limpkins nest above ground as well.
2702. Its nest, sometimes festooned with bits of lichens, is shaped to resemble a lodgepole or spruce cone. It is positioned amid a cluster of real cones, generally beneath an overhanging branch that provides additional cover.
2703. Yes.
2704. The African tit creates a mock opening in its nest in the form of a sack-like cavity below the actual entrance. Upon leaving the nest, the actual entrance is squeezed shut. This may deceive a nest robber into abandoning its efforts upon finding the mock cavity empty.
2705. Yes, generally so, as do high winds.
2706. Stiff-tail ducks are primarily vegetarian, but the Musk Duck is essentially carnivorous.
2707. Yes.
2708. The Flammulated Owl.
2709. No, the gull breeds throughout the year.
2710. Yes, the finches have extremely powerful beaks.
2711. A relatively common western bird, the jay is absent from the rest of the continent except for a small relict population in the scrublands of central Florida.
2712. It is dominant over other vultures at a carcass.
2713. The Blue Petrel is larger and has a white-tipped tail, whereas the tails of all prions are black-tipped.
2714. The Jabiru, but it is not the stork of Latin America with the same name.
2715. Only two: Snowy (Greater) and Black-faced (Lesser) Sheathbill.
2716. Approximately 25-30 percent.
2717. The Bar-headed Goose. A migrating flock was observed at 30,758 feet (9,375 m) over Mount Everest. Even at such an extraordinary height, the birds were honking!
2718. Yes, but the eagle also feeds on reptiles, birds and mammals that it kills.
2719. It provides greater maneuverability in flight.
2720. Yes, many species do.
2721. Yes.
2722. Virtually all the birds were painted life-size.
2723. No, only the males, but females may hiss. In most cormorant species, males are more vocal than females.

Answers

2724. A small, crested Asian cuckoo-falcon with long wings. Unlike true falcons, it has two tooth-like projections on its upper mandible, rather than one.
2725. About 184.
2726. A modified unbranched feather shaft.
2727. Yes. Feathers are given to the chicks from day one, sometimes even before feeding them for the first time.
2728. The exact purpose is unknown, but the decoration of the tail and wingtip area possibly distracts a potential predator.
2729. The falcon stores food.
2730. An Old World warbler named for its repetitive song.
2731. The Isle of Man in the Irish Sea. Manx is the adjectival form of the island's name.
2732. They lie down with their necks extended out on the ground.
2733. The Black-necked Crane of China, in 1876.
2734. The storm-petrels.
2735. The Long-billed Dowitcher.
2736. *The Auk.*
2737. No.
2738. No, but the males does.
2739. The Superb Lyrebird.
2740. While highly unusual for rails, the flightless bird was white. It disappeared about 1830, and is known from only two existing skins.
2741. A crane.
2742. No, broadbill nests are large, conspicuous, pear-shaped bags with a long, drawn-out neck, and with a porched entrance in the lower half. Nesting material may hang below the nest like a ragged beard, and nests may be suspended by a slender string, often over water.
2743. The King Bird-of-paradise, that is about 9.8 inches (25 cm) long, weighing slightly less than 1.7 ounces (50 grams).
2744. Yes.
2745. The dingo, the wild dog of Australia.
2746. No, a chick positions its bill crosswise in the bill of a provisioning adult like an albatross chick.
2747. Birds of prey have few enemies, thus it is safe for them to migrate during the day. In addition, thermals, which the raptors use to gain altitude, are not a nocturnal phenomenon.
2748. Sea lettuce, not kelp as is commonly believed.
2749. Yes, but only on unattended eggs or small chicks.
2750. No, the eggs of a jacana are unusually glossy.
2751. None. The large shorebirds occur throughout the world, but are absent from New Zealand and North America.
2752. Six of the 16 species.
2753. Wipe it on the ground.
2754. No, the parrots are sometimes completely oblivious to the approach of an intruder and can almost be touched.
2755. Females don't care for the chicks at all, and possibly never see their own young. The male drives young off to fend for themselves at about nine months of age. After leaving their father, juveniles remain together for some time.
2756. The Australian White Pelican.

2757. Yes, but in recent years, the breeding population has not exceeded 5-10 pairs. Ironically, the little alcids nest on Eldey Island, where the last Great Auk was killed.
2758. Yes. When the vultures come down to feed on road-kills, they themselves fall prey to speeding vehicles.
2759. A raptor trapped as an adult and trained to hunt.
2760. Bright yellow.
2761. An average of only about seven pounds (3.2 kg), despite its great height.
2762. No, the petrel was named for Alexander Wilson (1766-1813), who is sometimes known as the *father of American ornithology*.
2763. Six: Kansas, Montana, Nebraska, North Dakota, Oregon and Wyoming.
2764. Because of similarities in the skull and syrinx. Even so, the recent shift remains controversial, and some authorities include them in a separate family of their own (Tityridae).
2765. Around sunset.
2766. The Turkey Vulture, that nests as far north as Canada, as far south as Tierra del Fuego, and even in the Falkland Islands.
2767. No, its speculum is white.
2768. The Sooty Tern. Population estimates range between 400 million and a billion birds.
2769. The tsetse fly. The extensive use of DDT obviously had a negative effect on many birds.
2770. None is truly endangered yet, but a number are quite rare and even threatened.
2771. The monals, but only males. All are highland residents of the Himalayas and western China.
2772. *Chattering.* The filtering action itself is sometimes termed *suzzling*.
2773. Yes, especially on foggy or misty nights. On March 15, 1908, at least 100 Tundra (Whistling) Swans were swept over the raging falls and dashed against the rocks or crushed by the ice below. The falls have been called the Niagara Falls Swan Trap.
2774. Yes, both are in the same order (Coraciiformes).
2775. Tree-nesting offers some protection from terrestrial predators.
2776. About 30 mph (48 km/hr). An intruder may be struck with considerable force, and on one occasion, a biologist was knocked out by a diving Brown Skua.
2777. The Van Dyke.
2778. None.
2779. Herons.
2780. Dark green when fresh, the egg turns almost black if kept in the dark, but fades otherwise.
2781. Yes.
2782. The Common Murre averages about 2.25 (1 kg). The Thick-billed Murre is slightly longer, but averages about two ounces (57 gms) less in weight.
2783. The Greenland Falcon.
2784. Yes.
2785. The Nene (Hawaiian Goose).
2786. Yes.
2787. No, gulls are conspicuously absent from the region, and only three species breed there.
2788. Yes, terete bills have a circular-shaped cross-section.
2789. No, their eggs are remarkably thick-shelled.
2790. Petrels flying in during the day are usually quickly eliminated by opportunistic Brown Skuas.

Answers

2791. Yes.
2792. Yes.
2793. The motmots.
2794. No, at least 18,000 species of fish are described.
2795. Five: waterfowl, cuckoos, honeyguides, icterids and weavers.
2796. The female European Northern Goshawk.
2797. About 10 days, and 7-10 days for the five species of echidnas.
2798. The striking area of orange-chestnut surrounded by pale buff-orange on its primary feathers was said to resemble a sunset.
2799. None. Woodpeckers are also absent from New Zealand, Madagascar and New Guinea.
2800. The New Zealand Kakapo (Owl Parrot). The endemic parrot is unique in that it is nocturnal and uses a *track and bowl* lek. Males clear and maintain distinct tracks to the bowl that is generally flanked by a natural reflecting surface that assists in projecting the deep, foghorn-like, booming calls more than 1.5 miles (2.4 km) across dark valleys.
2801. Its call is similar to the sound of a hammer on an anvil.
2802. The Hooded Pitohui, a common bird of New Guinea, but several other pitohuis may be poisonous as well. It has recently been demonstrated that the brilliant orange-and-black, jay-sized bird has feathers and skin laced with a potent toxin that is believed to deter predators. Many insects, fish, amphibians and reptiles have noxious compounds that make them distasteful, but chemical defense was previously unknown in birds. The neurotoxin, known as homobatrachotoxin, is seen nowhere else in nature except in the poison-dart frog of Latin America.
2803. The Whiskered Auklet.
2804. The Northern Fulmar. Some are pale, others are dark, while a number are intermediate in color.
2805. No, it had a tail of loose curly feathers.
2806. Yes.
2807. The Lapland Longspur.
2808. The California Quail.
2809. Yes, their long bills are slightly decurved.
2810. At least 50,000.
2811. Yes and no. The Black Vulture will nest again, but the Turkey Vulture apparently does not.
2812. The Whooper Swan. The entire base of its bill and the lores are a bright lemon-yellow.
2813. A tern, because of its long wings and buoyant flight.
2814. No, but Two-banded Plovers are relatively common in the Falklands. Three-banded Plovers are African birds.
2815. Insects and worms. Roulrouls are among the few galliform birds that depend primarily on an animal diet.
2816. The robin scorched its breast bright red while stealing fire from the sun.
2817. Yes, so far as is known, but the Highland Guan may prove to be an exception.
2818. Yes, when carrion is scarce, it may kill marmots, hares, tortoises, lizards and possibly lambs.
2819. Six.
2820. Yes.
2821. Jenny Lind (1820-1887).

Answers

2822. The hanging-parrots.
2823. No, it was apparently initially domesticated in Brazil.
2824. No, unlike their near relatives, the waterfowl.
2825. Bright red.
2826. Yes, but some nest sites may be relatively exposed.
2827. Flutes and whistles.
2828. No, the Canvasback is the largest. Drake Canvasbacks average 2.76 pounds (1.25 kg), but weigh up to 3.75 pounds (1.7 kg), whereas male Redheads weigh between 1.8 and 3 pounds (0.8-1.4 kg).
2829. Usually the outer primary.
2830. Vertical leaps of 3-4 feet (0.9-1.2 m) are common, but jumps up to six feet (1.8 m) have been recorded.
2831. The shorebird protects its eggs from mountain sheep and caribou by sitting tight until the last possible moment, and then flying up in the face of the animal to frighten it.
2832. Yes, but the large forest kingfisher was introduced from the Australian mainland.
2833. No. For 34 years the pitta had not been seen and was feared extinct, but was rediscovered in Thailand in 1986.
2834. For use as targets, and up to 50,000 were shot weekly at shooting clubs.
2835. No, males tend to be more numerous.
2836. Between 1959 and 1963, there were between 300 and 400 aircraft/albatross collisions. After finally recognizing that killing some 54,000 adult albatrosses was ineffective, the elevated sand dunes that produced updrafts surrounding the runway were leveled, leading to a significant reduction in strike frequency almost immediately.
2837. Guano exploitation and over-harvesting of penguin eggs. Guano mining causes disturbance and removes the substrate that nesting penguins burrow into. Between 1844 and 1930, over 400,000 eggs were collected annually from Dassen Island alone. Legal egg collecting ended in 1969. Currently, penguins must compete with commercial fishing fleets and are vulnerable to oil-spills.
2838. By mimicking the call of a local owl.
2839. Colorful neotropical parrots.
2840. Pinkish, but White-tailed Tropicbirds nesting on Christmas Island are tinged a strong peach, apricot or golden color and are locally known as Golden Bosunbirds.
2841. The Sedge (Short-billed Marsh) Wren, that nests from Canada south to Tierra del Fuego.
2842. No. The raptor was named after William Cooper by C. L. Bonaparte from a specimen (or specimens) collected by W. Cooper.
2843. Abbott's Booby.
2844. The Long-tailed Tit.
2845. The western form of the Spruce Grouse.
2846. No. It ranges from Mexico to Argentina, but the Greater Yellow-headed Vulture is exclusively South American.
2847. The 11 species of woodswallows.
2848. Funk Island. The population is estimated to consist of between 396,000 and 500,000 pairs.
2849. No, both species are Asian, and are replaced in Africa by the Marabou Stork.
2850. Yes, the petrels readily feed on floating carrion.
2851. Yes, but only as adults.
2852. Yes.

Answers

2853. On March 16, 1978, the super-tanker *Amoco Cadiz* went aground two miles (3.2 km) off Portsall, France, spilling at least 68 million gallons (250,740,720 liters) of crude oil. During the brief 1990/1991 Persian Gulf War, Saddam Hussein's invading Iraqi forces pumped more than 80 million gallons (300,283,200 liters) of crude oil into the Gulf, already one of the world's most polluted bodies of water, resulting in the largest oil spill in history. The Gulf is relatively shallow and years may be required for the water to flush through the narrow Strait of Hormuz.
2854. Yes.
2855. Pinkish.
2856. Nearly 90 percent.
2857. Northern Royal Albatrosses nest at Taiaroa Head, Otago Peninsula, South Island, New Zealand. The first authentic record of breeding occurred in 1920, but the egg was stolen. The first chick was hatched in 1935, but it was killed. It was not until 1938 that a chick finally fledged. The colony is very small, consisting only of about 17-20 breeding pairs.
2858. Not generally, although in Florida, Anhingas are occasionally observed on salt or brackish water.
2859. Up to 30 seconds.
2860. The Hispaniolan Woodpecker.
2861. Possibly 30 mph (48 km/hr), but 20 mph (32 km/hr) is probably more realistic.
2862. Yes.
2863. Traveling the shortest distance between two points.
2864. The New Guinea Pesquet's Parrot. Its diet consists entirely of fruit, and the unique parrot drinks with a suction-pump action of the tongue.
2865. Five: Blue Duck, Paradise Shelduck, Black Scaup (Teal), Brown Teal (with two races) and the Campbell Island Flightless Teal. The Aucklands Island Merganser disappeared shortly after the turn of the century.
2866. Sight, in most species.
2867. Possibly as an adaptation for digging and probing.
2868. No, all species have extremely short tails, and appear tailless.
2869. Probably the Eskimo Curlew. The last reliable sightings in Texas were in 1987, and two birds were reported in the Northwest Territories. Four curlews were reported in Mar Chiquita, Argentina in October of 1990. The Black Stilt, a New Zealand endemic, is another possibility. The population probably does not exceed 50 birds with only 15-20 pairs of pure-blood breeding birds remaining.
2870. The Indigo Bunting.
2871. The megapode inhabits Australian semiarid country in habitat locally known as mallee.
2872. No, Princess Stephanie (1864-1945) was from Belgium.
2873. The Streamertail. The hummingbird's image adorns coins, currency and stamps.
2874. Their tongues.
2875. Curiously, the chicken is not mentioned at all.
2876. No.
2877. The African Palm-nut Vulture, also known as the Vulturine Fish-eagle.
2978. The large Cactus Wren of southwestern U.S.
2879. Its powerful bill enables the bird to pry bark from trees to expose grubs. The bird wedges the projecting tip of the upper mandible under a piece of bark, and then pushes its whole body forward and rotates it around the beak pivot until the bark is dislodged.

Answers

2880. Nearly five feet (1.5 m).
2881. At least 700.
2882. The White-fronted Goose.
2883. The Rhinoceros Auklet.
2884. Yes, based on accounts in some of the earlier literature, but this is not confirmed, and many contemporary ornithologists doubt such speeds are attainable in still air. The Russians reportedly documented a speed of 105.5 mph (170 km/hr) for the White-throated Spinetail swift during courtship display flights.
2885. The two seriemas. These long-legged, semi-terrestrial birds prey on snakes, animals, insects and large amounts of fruit and vegetable matter. Seriemas are more closely related to rails and bustards than to raptors.
2886. Unlike typical plovers that lay a four-egg clutch, the Wrybill invariably lays only two eggs.
2887. Yes.
2888. Shivering, but shivering is not restricted to birds.
2889. Bill-clattering. Storks are essentially mute, but are not voiceless.
2890. Yes.
2891. The dowitchers, but so little data are available, it might be premature to assume this is typical. There is some evidence that both sexes of Short-billed Dowitchers incubate, but the female takes little part in brood care.
2892. The Antarctic (Southern) Giant-Petrel.
2893. The naturalist John Gould, in 1840.
2894. Its eyes are remarkably small for a nocturnal bird.
2895. Cotinga comes from an Indian name for the White Bellbird meaning *white-washed.*
2896. The Brown Pelican.
2897. Yes.
2898. No. They take off with a pattering run across the water, and when airborne, remain low, seldom flying very far.
2899. Yes. In 1942, Superb Lyrebirds were introduced from Australia and occupy a restricted range.
2900. The White-tailed Kite.
2901. No, fish are undoubtedly more numerous.
2902. A crane.
2903. No one knows because its bower has never been viewed by an ornithologist.
2904. As of the late 1970's, less than 700 birds, but by 1993, it possibly did not exceed 350, with less than 100 breeding pairs. In the wild, the ibis are confined to a narrow strip of desert coastline in Morocco in no more than a half-dozen colonies. Approximately 800 Waldrapp are housed in zoological collections where fortunately they breed well. Some captive-reared Waldrapp may be introduced into Spain, but not within the historical range.
2905. Yes.
2906. Col. Sanders of Kentucky Fried Chicken (KFC) fame.
2907. No, it is a three-toed bird. The only didactyl bird is the Ostrich.
2908. Prey is impaled on a spike. Australian butcherbirds also impale prey.
2909. In the 1940's, southern California cagebird dealers illegally shipped House Finches to New York for sale as "Hollywood finches." When the U.S. Fish and Wildlife Service cracked down, the dealers released finches on Long Island, and the birds subsequently became established in the northeast.

Answers

2910. *Bird Lore.*
2911. Reportedly, only birds close their eyes in death.
2912. None.
2913. Stones swallowed by the extinct Dodo to aid in crushing food. Bezoars were invariably found with the skeletons of both Dodos and solitaires, and the stones were once valued for medicinal purposes.
2914. The King Vulture.
2915. The Light-mantled Sooty Albatross.
2916. The Lammergeier (Bearded Vulture).
2917. Probably the Sanderling.
2918. At least eight: two shearwaters, five gadfly petrels, and the recently rediscovered Fiji Petrel.
2919. The American Flamingo. If flamingos are feeding in deep water, the ducks are often in attendance, taking advantage of food stirred up by the long-necked birds.
2920. The sunbird feeds extensively on insects, but also preys on small lizards.
2921. The cormorants and penguins.
2922. Yes, generally so.
2923. Not currently, but Horned Screamers formerly did.
2924. The 28-inch-long (71 cm) central tail feathers of the Helmeted Hornbill are twice the length of the other tail feathers. Including the tail, the hornbill spans 50 inches (127 cm).
2925. The name refers to the horny sheath covering the base of their short, stout bills.
2926. There are conflicting opinions on this. Some authorities believe that robins can, whereas other experts challenge this.
2927. About 20 percent.
2928. A flock of waterfowl in flight, particularly geese when in long, uneven lines.
2929. Although primarily insectivorous, the Australasian woodswallows have brush-tongues, enabling them to lap up nectar and pollen.
2930. *The Pigeon that took Rome.*
2931. Only 6-7, compared to nine in most passerines.
2932. Up to 12 feet (3.7 m).
2933. Birds not selective in their choice of habitats.
2934. The Black Vulture.
2935. No, the birds are probably related to crows.
2936. The duck is virtually extinct. The only one seen since 1970 was a male trapped in fishing gear in August, 1991, and moved to a botanical garden where it subsequently died. Its disappearance is a result of unrestricted over-hunting, habitat loss and the introduction of exotic fish that compete for food.
2937. Possibly to rid their feet of fish slime; the birds may dunk their heads underwater as well.
2938. The northern swans.
2939. Yes.
2940. About 18 months.
2941. The Pink Pigeon. As of mid-1993, the original wild population was estimated at only 16 to 22 pigeons. At least 150 are maintained in some 20 captive collections where the species breeds regularly. About 15 captive-reared birds have been released on Mauritius. The species is highly endangered due to the introduction of exotic predators, such as the crab-eating macaque, as well as cyclones, food shortages and inbreeding.

Answers

2942. Approximately 1,900 breeding pairs. In 1967, prior to extensive habitat destruction, at least 2,300 pairs existed. Only 60 percent of the population may breed within a given year.
2943. Greenbuls.
2944. The Ruddy Shelduck.
2945. In 1838, in South Africa, but Ostrich farming did not really get going until about 1860. Were it not for Ostrich farms during the height of the feather craze, it is possible that the Ostrich would have become extinct. Ostrich feathers have been used for adornment for at least 5,000 years, as evidenced by both Mesopotamia and Egyptian art.
2946. Only about 125.
2947. Probably to escape the hot tropical sun, and perhaps also biting insects. Some penguins breed on the mainland, and burrow nesting offers a degree of protection from terrestrial predators, such as the Patagonian gray fox.
2948. Ghillies.
2949. Yes.
2950. The ducks doze while paddling with one foot, propelling themselves in a tight circle.
2951. The behavior is possibly a relatively recent adaptation, and may have evolved to enable the birds to avoid intense solar radiation. Burrow nesting also favors a nidicolous species.
2952. Hall's (Northern) Giant-Petrel breeds on some southern islands of Tierra del Fuego.
2953. No. The pouch or wattle is not hollow and cannot be inflated with air, but it can be thickened considerably and distended at the base. It is the only stiff-tail with a wattle.
2954. An average of 180-200 times, with a maximum of 310 songs prior to pair formation, but falling to 30-160 songs an hour afterwards.
2955. Burdekin is the Australian name for the Radjah Shelduck.
2956. Only rarely. Oystercatchers feed primarily on bivalve mollusks. The muscle that holds the two shells closed is cut and the soft flesh extracted.
2957. Yes.
2958. The dippers.
2959. Yes.
2960. No, the cormorants nest on the ground in huge colonies.
2961. Yes, but only seasonally.
2962. The tuna.
2963. At least 708.
2964. No, most rollers are essentially blue.
2965. The Slaty-headed Parakeet, in eastern Afghanistan.
2966. Blackish.
2967. Three: Noisy, Red-bellied and Rainbow Pitta.
2968. The White-throated Rail of Aldabra Island.
2969. No, its neck is completely feathered. It seldom feeds on carcasses, so there is no advantage to having a featherless neck.
2970. The Imperial Parrot *(Sisserou),* an Amazon parrot attaining a length of 20 inches (51 cm). It is very much at risk and possibly only 60 remain in the wild. The Imperial Parrot is the national symbol of the West Indian island of Dominica.
2971. The Indian Peacock.
2972. Yes, during the breeding season.
2973. The Tufted Puffin.
2974. Vivid red and yellow.

Answers

2975. The Common Eider. Both ducks have especially long, sloping head profiles.
2976. On March 4, 1971, the Natural History Museum of Iceland paid £9,000 for an extinct Great Auk. The museum was reportedly prepared to go as high as £20,000 to obtain the valuable specimen.
2977. The high-altitude partridges resemble grouse.
2978. A ground-hornbill.
2979. Their toes are folded forward.
2980. About 518, of which 301 breed (or may have bred) in the state, and 276 are regular nesters.
2981. The false eye spots may deceive predators approaching from behind.
2982. The three catbirds: Green, Spotted and White-eared.
2983. Yes.
2984. The name refers to its migratory habits: a "passer-through," a wayfarer or traveler. Even its Latin name *Ectopistes migratorius* related to its migratory habits.
2985. The Snow Petrel.
2986. 1934.
2987. Yes.
2988. A penal settlement was established on Norfolk Island (New Zealand) in 1788. Convicts and guards alike were left to fend for themselves, and when hard-pressed for food, the "colonists" were greatly dependent on the birds, and the petrel colony was essentially exterminated within the first two years. The species is also known as Solander's or Providence Petrel.
2989. *My Little Chickadee.*
2990. The Laysan Albatross, with a population exceeding 1.5 million birds.
2991. Eight.
2992. The liver.
2993. None. Eggs are simply laid on the ground, in tree cavities, between rocks, or in cliff cavities or caves.
2994. The New Zealand rock wrens.
2995. No feathers surround their eyes. A toucan's eye is encircled by bare skin.
2996. The chachalacas.
2997. Yes. All participating females lay in a single nest where the male alone incubates. The number of eggs laid varies between 20 and 50.
2998. Native grave mounds.
2999. The Sand Martin.
3000. Not as a rule, but tropicbirds can be inquisitive and may approach a ship, circle a few times, and then disappear over the horizon. The White-tailed Tropicbird is the most inclined to visit a ship, and may briefly even land in the rigging of a vessel.
3001. The bitterns and tiger-herons, but most herons are colonial.
3002. Yes. In 1961, the Chestnut-bellied Sandgrouse was introduced onto the big island.
3003. The two-toed pygmy silky anteater, an elusive, nocturnal, completely arboreal mammal. In Trinidad, the local name *poor-me-one* is applied to both the potoo and the silky anteater.
3004. Cuba, with six species. The Cuban race of the Ivory-billed Woodpecker is essentially identical to the probably extinct mainland birds. The Ivory-bill is now considered extinct, since a team of Dutch and Cuban ornithologists spent three months in 1993 searching without success for any sign of the bird in the last remaining habitat in eastern Cuba.
3005. The Eared Grebe.

Answers

3006. Their bills turn bluish.
3007. No.
3008. A sounding lead.
3009. Over 200.
3010. Four: the widespread Galapagos Mockingbird, Espanola (Hood), Floreana (Charles) and San Cristobal (Chatham) Island Mockingbird.
3011. Nests are typically collected three times a season. The swiftlets rebuild each time and are allowed to rear their chicks the third time, and the nest is collected after the young fledge. Rights to the caves are jealously guarded, and nest removal has been carefully controlled for centuries. The birds are possibly declining and more stringent harvest controls may be required. Nest poachers are a problem in some areas because they are unconcerned about the chicks they kill, and favored sites are protected by armed guards.
3012. Not naturally, but Eurasian Blackbirds and Song Thrushes were introduced. There formerly was a New Zealand Thrush, but it was not related to the true thrushes. Last seen in 1930, it is most likely extinct.
3013. No, both species are crepuscular.
3014. A large hawk (sphinx) moth with a flight style strikingly similar to that of a hummingbird. The insects have a long proboscis used to draw nectar from flowers.
3015. Yes.
3016. Swallow-like mud nests built on the walls of bat caves. Nests may be located in the light at the cave entrance, but the aberrant passerines typically nest deep within the cave in almost total darkness.
3017. Varieties of domestic geese.
3018. Not as breeding birds, but Common Puffins occasionally winter off the coast of Morocco.
3019. Only six mounted skins and two incomplete skeletons, all dating from about 1840-1850.
3020. The Yellow-browed Toucanet of the Peruvian Andes.
3021. No, but both groups were named for their far-carrying, bell-like calls.
3022. No, most Clapper Rail races inhabit saltwater habitat.
3023. Yes.
3024. Maroon.
3025. Large insects.
3026. The narrow, round entrance hole on one side of the huge domed nest, invariably opens downward and is overhung by the upper structure, making the nest inaccessible from the rear.
3027. The beautiful clouded leopard. It has longer canines relative to its size than other cats, a reflection of its specialized diet.
3028. Fiji and Samoa, a one-way journey of nearly 2,000 miles (3,219 km) over water.
3029. No.
3030. Rictal bristles, which are modified feathers.
3031. The Swallow-Tanager. Its nest is excavated in a bank and may extend two feet (0.6 m) into the bank.
3032. The Emperor Penguin. Its single egg averages one pound (0.45 kg), only 1.6 percent of the weight of a 60-pound (27 kg) female.
3033. The rare Damara Tern. Its much reduced population may not exceed 3,000 birds.
3034. The Little Blue (Fairy) Penguin.
3035. No, they are generally drabber in color.

Answers

3036. The Indian Black (Pondicherry) Vulture, because it is dominant over other scavengers on a carcass.
3037. Pesquet's Parrot.
3038. No, the 25-day fledgling period is long for a small passerine.
3039. Spatulate-shaped. The sides of the upper mandible droop down to form large flaps.
3040. Possibly to confuse potential predators, but it may also be used to maintain flock cohesiveness.
3041. Its pinkish or horn-colored bill turns black or reddish.
3042. About 6,000 birds.
3043. Brandt's Cormorant of the North American west coast.
3044. Yes, but there are exceptions, such as the hornbills.
3045. Their eggs are nearly identical in size, color, and pattern to the eggs of the host species.
3046. Yes, as a rule.
3047. Pale green.
3048. With shotguns using number 10 or 11 shot, to minimize damage to the delicate plumage. South American Indians skillfully downed hummingbirds with tiny clay balls from blowguns.
3049. *Thunderbird* wine.
3050. Both sexes have white neck ruffs.
3051. The Bufflehead duck.
3052. The Iceland Gull.
3053. Yes.
3054. No, but waterfowl, even seemingly distantly related species, readily hybridize in captivity. Wood and Mandarin Ducks are commonly maintained together, but no Wood Duck/Mandarin hybrids have been documented. However, captive Wood Ducks have hybridized with at least 25 species of ducks, but none of the resultant hybrids was fertile.
3055. The jaegers.
3056. Strutting.
3057. Between 12 and 18.
3058. Yes. If a bird has a tuft of head feathers that can be raised like a "horn," it has corniplume feathers.
3059. Probably the Snowy Owl, which has been recorded in -80.5°F (-62.5°C).
3060. Yes, but just barely. Even so, all three screamers are surprisingly capable swimmers.
3061. Yes.
3062. Egrets. The kingfishers hover above the egrets, diving on fish disturbed by the wading birds.
3063. An individual who issues bad checks.
3064. The Lesser Frigatebird has a white flank-patch and "spur" across the auxiliaries, and the Christmas Island Frigatebird has a diagnostic white belly-patch.
3065. The Ostrich and Emu.
3066. No, both their necks and legs are extended.
3067. Yes.
3068. No, unlike other terns, noddies swallow fish that are later regurgitated for their chicks.
3069. Not really. The Dodo's flesh was said to be rather bitter and very tough, but this mattered little to sailors craving fresh meat. The closely-related solitaires were much more tasty.
3070. No, but swallows commonly do.

Answers

3071. *Thought* and *Memory*.
3072. The three phalaropes.
3073. The Anhinga and darters. If so, they are the only pelecaniform birds exhibiting this trait.
3074. The Yellow-headed Caracara. The raptor is sometimes called a tickbird because it feeds on cattle ectoparasites.
3075. Yes. In some geese, more than 12,000 skin (cutaneous) muscles are present for feather control alone.
3076. Yes.
3077. The Least Tern, a name alluding to its small size and habit of striking the water while diving for fish.
3078. No. Older males are whiter, and the Wandering Albatross is among the very few sexually dimorphic seabirds.
3079. The flying squirrel.
3080. The Ringed Turtle-Dove. It is not a native, but the dove is established in the Los Angeles area, and locally from Miami to Baltimore on the east coast. (The domesticated form *S. risoria,* is now known to be a derivative of *S. roseogrisea,* the African Collared-Dove.)
3081. The West Indian Whistling-Duck, that is greatly reduced throughout much of its West Indies range, possibly numbering less than 2,000 birds. It is well established in captivity, where it breeds readily.
3082. Yes.
3083. Yes, sometimes with fatal results.
3084. The bird is about to molt. The plumage of many pre-molt penguins becomes lackluster and brownish.
3085. New Guinea, with at least 45 species.
3086. The Winter Wren.
3087. The parrots.
3088. The Helmeted Guineafowl, so named for its call.
3089. The long wings act as balancing aids when the birds run. Running rheas often zigzag rapidly, sometimes turning sharply at right angles.
3090. The Akialoa, a bird about eight inches (20 cm) long, that was last recorded in 1967.
3091. The Yellow-throated Caracara.
3092. No, larks are generally rather somber in color, but there are exceptions, such as the Horned Lark.
3093. About 524, with an additional 13 introduced species.
3094. Yes.
3095. The White-throated Bee-eater of West Africa. During the winter, forest squirrels strip away the fibrous oily skin of oil-palm nuts and as the strips of skin are dropped, waiting bee-eaters catch them in the air as skillfully as they would a flying insect.
3096. The Little Curlew derives a degree of protection from predators that avoid the eagle's territory.
3097. Jerdon's Courser. The bird was blinded by the torch of a local hunter who immediately scooped it up with his hand. Its status is not known, but no doubt the courser is very rare.
3098. The cockatoos.
3099. No, but a juvenile Wandering Albatross does.
3100. Bennett's (Dwarf) Cassowary.
3101. Probably Alexander the Great.

3102. Yes.
3103. Tristan da Cunha in the South Atlantic, where 50,000 Great Shearwaters are harvested annually. Cory's and Little Shearwaters are harvested on the Cape Verde Islands and Madeira, and Audubon's Shearwaters in the Caribbean.
3104. The Comb Duck.
3105. The Neotropic Cormorant.
3106. The Torrent Duck of South America.
3107. Lammergeier translates to *lamb-killer,* a name dating back to the time when it was assumed the vulture preyed on lambs.
3108. *If one seeks to soar with the eagles, it is best not to hoot with the owls.*
3109. The tiny eggs are only 0.6-0.7 inches (16-18 mm) in length by 0.55-0.59 inches (14-15 mm) in diameter, or about the size of a House Sparrow egg, but slightly more spherical in shape.
3110. Reddish.
3111. *The Eagle and the Hawk.*
3112. A large neotropical parrot.
3113. Yes, but local movements may be undertaken.
3114. The Australian Wood Duck or Maned Goose.
3115. The Demoiselle Crane.
3116. No.
3117. Yes, the birds capitalize on stirred-up insects.
3118. About 6.5 hours. The general rule is 25 minutes per pound (0.45 kg), but only 20 minutes per pound for un-stuffed turkeys.
3119. A starling.
3120. The grebes.
3121. The all-time champion appears to be a male African Gray Parrot with a vocabulary of nearly 800 words. It won the Best Talking Parrot-like Bird title at the British National Cage and Aviary Bird Show for 12 consecutive years (1965-2976), and retired undefeated.
3122. 1889.
3123. The dippers and wagtails.
3124. The White-tailed Sea-Eagle.
3125. A pelican, so named because of its shape.
3126. The vireos.
3127. In 1984, between 30,000 and 50,000 birds, but more recent estimates indicate 70,000-100,000. The most reliable figure is probably about 33,000.
3128. No, Greenland is one of the few places where the pest has failed to gain a foothold.
3129. In most birds the right ovary is greatly reduced, but this reduction is far less extreme in raptors.
3130. The aberrant Magpie Goose. The long-legged birds can settle on surprisingly slim branches.
3131. No, not compared to the tiny size of the birds.
3132. The sound produced by the quivering train might serve a courtship function, but the main purpose may be to shift the iridescent feathers about so the colors catch the light from different angles.
3133. White, but its plumage is tinged with pale salmon.
3134. Yes.

3135. Probably some of the hawk-moths, that can attain speeds of 33.3 mph (53.6 km/hr). Possibly some small-bodied tropical butterflies are even faster.
3136. The Siberia-nesting gull fledges in only 16-17 days.
3137. No, but lyrebirds run swiftly.
3138. John James Audubon.
3139. Two: Canada and Egyptian Goose; the Swan Goose is sometimes called a Chinese Goose.
3140. Yes.
3141. A miniature forest of seedlings that sprout from regurgitated seeds.
3142. A small resident population is restricted to northern Morocco, but it has been reduced to probably less than 100 birds. Technically, this makes the Great Bustard the largest flying bird of Africa, heavier even than the endemic Kori Bustard.
3143. The honeyeaters (Meliphagidae).
3144. The Barnacle Goose may nest on precarious ledges, high on steep cliff faces, at times close to Gyrfalcon nest sites.
3145. No, tropicbirds typically do not fly lower than 100 feet (30 m).
3146. Its enormous nest is cemented into a rock crevice and may weigh nearly nine pounds (4 kg), whereas the little nuthatch only weighs about 1.4 ounces (40 gms).
3147. Toucans position their bills along the plumage of the back and fold their tails up over the bill. The sleeping birds are transformed into a featureless ball of feathers.
3148. New Zealand, where about 32,000 pairs nest.
3149. The Turkey Vulture, but apparently erroneously so.
3150. The Scarlet Tanager.
3151. Yes.
3152. Yes, but it is uncommon. In one study, less than one percent of females had beards, although beards are far more common on hen domestic turkeys.
3153. The nuthatches, but this is generally a habit of Old World species. The American Red-breasted Nuthatch may smear the edges of its nesting hole with pine pitch to discourage predators.
3154. The pigmentation may be caused by the consumption of large quantities of green sea urchins. Alaskan sea otters feed extensively on sea urchins and also have pink or purple-tinted bones.
3155. The large honeyeater generally does not build a nest. It typically makes use of a roosting nest constructed by a Gray-crowned Babbler, but also uses old nests of friarbirds, miners or Apostlebirds.
3156. The feathers are crinkle-edged and uniquely broadened in a peculiar undulating fashion, giving the appearance of a ruffled coat.
3157. Feather dusters.
3158. No, most turacos construct rather flimsy twig nests.
3159. The African Congo Peacock.
3160. No. A Royal Albatross chick fledges in about 240 days, whereas about 278 days are required for a Wandering Albatross.
3161. No, the Kakapo has surprisingly small eyes for a nocturnal species.
3162. A swanherd.
3163. The Tibetan (Hume's) Ground-Jay is entirely terrestrial, often digging a nest cavity in earth banks.
3164. The Wedge-rumped (Galapagos) Storm-Petrel.

Answers

3165. Dr. Elliot Coues.
3166. The Dunnock.
3167. No.
3168. No, it is lower-pitched.
3169. The Scarlet Tanager.
3170. The Hawaiian Goose (Nene).
3171. No, a Black-footed Albatross is more apt to follow ships.
3172. Yes.
3173. Six: Ashy Storm-Petrel, Elegant Tern, Xantus' Murrelet, California Thrasher, Yellow-billed Magpie, and California Condor.
3174. The 3.3 million square miles (8,547,000 sq km) of African desert contain an estimated 214 million birds.
3175. The Spectacled Eider, with a population estimated between 300,000 and 500,000 birds.
3176. It is not a corvid, but rather is in the bell-magpie family (Cracticidae).
3177. The Puerto Rican Parrot. Breeding thrashers usurp prime nesting holes and destroy parrot eggs. In 1975, the Amazons were reduced to only 13 in the wild. By 1987, the population was up to about 40, including captive birds, but the destructive 1989 *Hurricane Hugo* greatly depressed the wild population.
3178. The Black-headed Duck. South American natives call it *pato sapo* (toad duck), because courting drakes assume a "toad posture," where the head is held low, the throat and cheeks are inflated to twice normal size, and the head is pumped rapidly.
3179. Purple dorsally with a white abdomen.
3180. The New Zealand kiwi.
3181. Pesquet's Parrot.
3182. Yes.
3183. Yes, but a becalmed albatross has much difficulty becoming airborne and remaining aloft.
3184. The Australian Black Swan.
3185. No, sandpipers depend more on tactile clues.
3186. Yes.
3187. No, surprisingly few mammals are named for birds.
3188. No, the gull was named after Sir John Franklin, a noted English explorer who disappeared in the high Arctic in 1874.
3189. The monals.
3190. Probably the Silver-eye (Gray-backed White-eye). It apparently reached New Zealand on its own in 1856, from Australia.
3191. The juvenile Scarlet Ibis that is grayish-brown above with a white rump and underparts.
3192. Between 56 and 58 days.
3193. Teddy Roosevelt in 1909, in response to public indignation over the millinery trade. More than 300,000 birds were killed on Laysan Island alone in less than six months in 1909. At the same time, it is certain that over a million birds were killed on other Pacific islands, such as Lisianski, Midway and Marcus Islands. All Leeward Islands except Midway Atoll were included in the refuge.
3194. The lovebirds.
3195. On a palm trunk at the junction of the trunk and fronds. Nests can completely encircle a tree and may be occupied by up to 30 pairs. Although the birds are communal nesters, each pair has its own compartment.

Answers

3196. A type of thin, woolen cloth.
3197. Unlike most waterfowl, its clutch of 5-10 pure white or faintly yellow eggs were perfectly spherical, and were said to resemble unpolished billiard balls.
3198. *Robin* Williams.
3199. Yes. Some megapodes that inhabit Southwest Pacific islands even undertake roosting flights from larger islands to small offshore islands.
3200. A conspicuous eye-ring of black feathers.
3201. Sickle-shaped. It is often used to describe the shape of some scapular feathers, such as those of the drake Falcated Duck.
3202. No, eight weeks are required for fledging.
3203. Orange or reddish.
3204. Tit chicks hatch at almost exactly the same time that large numbers of green caterpillars suddenly emerge. The caterpillars constitute the major food source for the chicks.
3205. No, surprisingly few have been bred.
3206. Tanagers are the most colorful of New World birds, with those in the largest genus *Tangara* seeming to exhaust the color patterns possible on sparrow-sized birds. Hummingbirds also might be considered, but the tiny birds must be viewed under ideal conditions to be appreciated.
3207. Every four years.
3208. Probably not before their fifth or sixth year.
3209. The New Zealand race of the wide-ranging Purple Swamphen is locally known by the Maori name Pukeko.
3210. Bright orange-red to bright red at peak of breeding.
3211. Yes, many bathe and preen thoroughly after feeding.
3212. The Freckled Duck. The estimated population size was only 19,000 birds in 1983, and by 1990 it might have dropped to 10,000.
3213. The Cooper Ornithological Society, in 1899.
3214. No, all birds have very short large intestines, often no more than an inch (2.5 cm) in length.
3215. Capercaillies and Black Grouse hybridize frequently, and German hunters formerly called the hybrids *rackel fowl*.
3216. Not generally. A male tends not to display on dull days because the shimmering colors of its crown and gorget will not flash and produce the desired dazzling effect. On sunny days, a courting bird actually orients its dive so that its body is facing the sun, thus allowing the iridescent feathers to glitter and sparkle to best advantage.
3217. The tuatara, a primitive lizard-like reptile unique to New Zealand.
3218. *I'd Rather Be A Hammer Than A Nail.*
3219. Caribou, on rare occasions.
3220. Eleven, with a range of 7-18 eggs.
3221. Three: Atlanta Falcons, Philadelphia Eagles and Seattle Seahawks.
3222. The Dovekie (Little Auk).
3223. The Northern (Red-necked) and Red (Gray) Phalarope.
3224. The Black Vulture.
3225. The Wood Stork.
3226. A crane.
3227. Probably the Chimango Caracara.
3228. The Curl-crested Aracari.

Answers

3229. No, its bill is significantly shorter.
3230. The Kakapo (Owl Parrot). The cat-like whiskers at the base of its bill indicate the New Zealand parrot has a superior sense of touch, a very functional nocturnal adaptation.
3231. The woodpeckers, that repeatedly pound on wood or other substances. The rhythmic sounds of other animals are produced by structures, such as feathers, that are no less a part of the animal than their vocal capabilities.
3232. Yes.
3233. No, unlike many of the Old World cuckoos. Some, perhaps all, are polyandrous.
3234. A gadfly petrel, also known as the Magenta Petrel, until recently known only from the type specimen collected in 1867. Rediscovered in 1967 when several were seen at the Chatham Islands (New Zealand), its breeding grounds remain unknown.
3235. Yes, but some may form small flocks of 20-30 birds.
3236. The two races were readily distinguishable, with western birds generally paler, with the green of the back and rump bluish, rather than tending towards yellow, although the yellow of the wing was more prominent.
3237. James' Flamingo of the high Andes, currently known as the Puna Flamingo.
3238. Yes, puffins are often caught in mid-air and may be swallowed in flight.
3239. Earthworms, but the endemic bird also preys on insects and small reptiles.
3240. Brilliant red.
3241. No.
3242. The two giant-petrels. Northern Fulmars at shore whaling stations can almost be considered shore-feeders when the birds converge on whale carcasses being hauled ashore.
3243. *Parus major minor.*
3244. The Great Blue Heron.
3245. No, it is reluctant to fly, even from danger, and can occasionally be caught by hand in broad daylight.
3246. No.
3247. Two. The Peruvian Brown Pelican was formerly considered a race of the Brown Pelican, but most taxonomists now afford it full specific rank (*Pelecanus thagus*).
3248. The birds were killed in great numbers during the 19th century for their magnificent tail plumes. Lyrebirds are currently protected and can be surprisingly common in suitable habitat.
3249. The Ring-billed Duck. The neck ring of breeding drakes is indistinct and only evident at very close range, whereas the bill ring is quite conspicuous.
3250. A distinct oily odor.
3251. The Whiskered Auklet's population is estimated at 20,000-50,000 birds. The largest known colony is on Buldir Island where about 3,000 pairs breed.
3252. The Hazel Hen.
3253. The Coscoroba Swan, that seldom permits a close approach.
3254. The Eurasian Nutcracker.
3255. Yes.
3256. No, the long-legged, terrestrial raptor has short toes relative to its large size.
3257. The Brown Booby. It is also the most numerous sulid of the Southwest Pacific.
3258. The swan dive.
3259. The Fiordland Crested Penguin of southern New Zealand. Its specific name *pachyrhynchus* refers to the thickness of the bill.

Answers

3260. Yes.
3261. Small bustards.
3262. No, the pale morph is far more common.
3263. Male carpenter bees of some species are nearly an inch in length (2.5 cm), and when in an amorous mood, occasionally intercept and attempt to seduce anything that flies, including birds, in the mistaken belief that the birds are female bees.
3264. The name alludes to its thinness or scrawny appearance. Similarly, the Great Egret is called "long white."
3265. During the day, with females taking the nocturnal shifts.
3266. The California Quail. The initial introduction took place in 1864, followed by additional attempts in 1870. The quail is well established.
3267. The horn is not hard or rigid, but rather is muscular, and is more accurately described as a miniature trunk or proboscis.
3268. The Demoiselle Crane.
3269. The Tropical Kingbird.
3270. No, all four are ground nesters.
3271. The Snowy (Greater) Sheathbill.
3272. White, tipped with black.
3273. No, it is a lowland bird.
3274. Yes.
3275. Only one: Eurasian Golden-Oriole.
3276. The Kelp Gull.
3277. Gray.
3278. Abdim's Stork.
3279. Southern Louisiana.
3280. No, its nest is not much larger than those of smaller hummingbirds.
3281. The megapodes may move up to 300 tons a year, with as much as a cubic meter of litter added or removed at a time.
3282. Yes.
3283. No, rayaditos are neotropical ovenbirds.
3284. Yes.
3285. The grouse.
3286. The Cape Verde Islands race of the Soft-plumaged Petrel.
3287. White, but some species have yellow, pink or dark sooty to almost black down.
3288. Nowhere. It is an Old World breeder.
3289. He believed the Redstart, a common bird in Greece during the summer, turned into a robin. The European Robin does not breed in Greece, but is an abundant winter visitor.
3290. The Western Tanager ranges as far north as the Alaskan border in western Canada.
3291. At least four egrets for every ounce of aigrettes.
3292. No, but the Lesser Scaup is.
3293. The lowlands of the Demilitarized Zone (DMZ), a region known for a sparse human population. The DMZ has separated North and South Korea since the cease-fire of 1953, and consists of a 2.5-mile-wide (4.0 km) band of mostly mountainous terrain spanning the peninsula for 149 miles (240 km). One of the major crane wintering sites is only 20 miles (32 km) from Seoul.
3294. The aortic arch turns to the right in birds as opposed to the left in mammals.

Answers

3295. Yes.
3296. The Wood Stork, because its soaring flight rivals that of a Turkey Vulture.
3297. African Gray Kestrels, Barn Owls and Verreaux's Eagle-Owls when the nest is still in the platform stage. Sparrows and other small birds may utilize the nest while it is still active.
3298. The extinct Labrador Duck.
3299. On May 3, 1977, a white domestic goose named *Speckle* from Goshem, Ohio, laid an egg weighing an unbelievable 1.5 pounds (0.7 kg). A normal-sized goose egg averages 6-7 ounces (170-227 gms).
3300. Yes, at least partially diurnal.
3301. Yes.
3302. Yes, but the dull-colored juveniles have brownish heads.
3303. The slugs are skinned.
3304. Abbott's Booby.
3305. The eagle preys primarily on phalangers and pigeons, but also takes some fish.
3306. Probably the Sora rail that migrates nearly 3,000 miles (4,828 km) one way.
3307. The Magellanic (Upland) Goose was named for Ferdinand Magellan (1480-1521).
3308. The two African oxpeckers.
3309. The Silvery-fronted Tapaculo reaches Costa Rica, where it is the only tapaculo of the country.
3310. Yes.
3311. No, but the birds were legally hunted up until 1972.
3312. Yes.
3313. The endangered Hawaiian Crow (Alala).
3314. The Australian Pelican.
3315. White.
3316. No.
3317. The American Woodcock.
3318. Yes.
3319. The Great Bustard. The penalty for taking either the bird or its eggs was a substantial fine and one year imprisonment. The bustard was one of the first British birds afforded legal protection, but protection came to late, and the bird disappeared in Britain.
3320. The Bobolink.
3321. The myna.
3322. A single nest was discovered in early August, 1990, in British Columbia. Up until that time, only 13 nests had been recorded worldwide, and as of 1992, only tree nests of the murrelet had been found between southeastern Alaska and central California. Nests typically consisted of moss-covered platforms on large moss-covered limbs. In tree-less parts of Alaska, the murrelet nests on the ground or in rocky cavities, but in forested parts of their range, only in trees.
3323. Yes.
3324. No, sandgrouse lack salt glands.
3325. A fish-eagle.
3326. The Caspian Tern was named after the Caspian Sea.
3327. The bellbirds.
3328. No. Both sexes produce a variety of bubbling and booming noises, but only displaying females utter the loud, resonant, high-intensity boom. Males are more inclined to grunt.

Answers

3329. A drake Mallard.
3330. No, it is a solitary nester.
3331. Based on feather counts, nearly 3,000.
3332. A sonogram (audiospectrogram) is a visual reproduction of a bird's call produced electronically by a sound spectrograph.
3333. *To ruffle one's feathers*.
3334. No, but their long, bifid tongues roll into a "tube" for gathering nectar.
3335. Yes, probably at least partially by scent.
3336. Bright red.
3337. Seriemas are raised with chickens because the sharp-eyed birds act as guardians, giving advance warning of predators or intruders.
3338. No, the Yellow-crowned Night-Heron is more diurnal.
3339. No, unlike most terns, fish are swallowed by Sooty Terns.
3340. Yes.
3341. The Western, Hepatic, Summer and Scarlet Tanager.
3342. The dove's-foot.
3343. The Little Friarbird.
3344. The Australian Blue-winged Kookaburra.
3345. No, it is an Asian pheasant.
3346. In dense clusters of up to 50 birds.
3347. Yes.
3348. The Skylark.
3349. Yes, despite the fact it lacks webbed feet.
3350. About 300 million. A single large blue whale consumes up to eight tons of krill daily.
3351. Yes, but only slightly so: 7.6 vs. 6.5 inches (19.3 vs 16.5 cm).
3352. Nothing, so far as is known. The tongue doesn't appear to be functional.
3353. The frogmouths, so named because of their especially wide bills.
3354. The Auckland Islands Brown Teal.
3355. Yes.
3356. The cuckoo hops and darts around in trees with fluid, squirrel-like movements.
3357. No, but the two terrestrial African ground-hornbills are the only exceptions.
3358. No, seed or fruit-eating birds have longer small intestines.
3359. Five: Mute, Whistling (Tundra), Whooper, Trumpeter and Coscoroba Swan.
3360. Bright orange or reddish.
3361. The cormorants, penguins and formerly the alcids.
3362. No, despite the fact that turacos are commonly called plantain-eaters.
3363. The Little Spotted Kiwi. Males weigh 1.9-3.0 pounds (0.9-1.35 kg) and females normally weigh 2.2-3.1 pounds (1.0-1.4 kg), but weigh up to 4.3 pounds (1.95 kg) when gravid.
3364. Yes.
3365. The megapode kicks barrages of sandy stones and litter at competitors.
3366. According to U.S. Fish and Wildlife Service estimates, more than half of the original wetland habitat, or 117 million acres (46,945,200 hectares).
3367. The Lapland Bunting.
3368. A large, wingless cricket that pierces swiftlet's eggs and sucks out the contents. If the crickets cause chicks to fall from the nest, they are consumed by snakes, civets, rats or monitor lizards.

Answers

3369. About a quarter.
3370. Yes. The average size was 35.4 inches (90 cm) tall for the extinct forms versus 23.6 inches (60 cm) in modern penguins.
3371. Three: Fork-tailed, Ashy and Leach's Storm-Petrel. The Black Storm-Petrel nests just south of the Mexican border on the Coronado Islands, and *possibly* on some southern California islands.
3372. No, unlike some of the Old World white pelicans.
3373. Yes, many do.
3374. No.
3375. Anna's Hummingbird begins nesting in late winter and early spring.
3376. Yes, on those of a warthog.
3377. Yes, many are highly nomadic and some are even migratory.
3378. The Siberian Red-breasted Goose. A heavy gander weighs slightly less than three pounds (1.7 kg).
3379. Nocturnal fledging offers some protection against predators, such as large gulls.
3380. New Zealand and its subantarctic islands with at least 14: three kiwis (Brown, Little Spotted and Great Spotted), six penguins (Little Blue, Yellow-eyed, Fiordland, Snares Island, Erect-crested and Rockhopper), two ducks (Campbell Island and Aucklands Island Brown Teal), two rails (Weka and Takahe) and the Kakapo (Owl Parrot).
3381. The Horned Grebe.
3382. Pigeon milk averages about 35 percent fat content, whereas cow's milk averages only about five percent.
3383. Its huge arched bill is bright yellow.
3384. No, it is a bunting.
3385. The barbets.
3386. The nocturnal, flightless Kakapo (Owl Parrot) of New Zealand. The distinct owl-like facial discs are evidence that the parrot has superior hearing.
3387. The 14 frogmouths.
3388. No, but the Antarctic (Southern) Giant-Petrel has a white and a dark-color morph.
3389. No. The tip of the tail of the Japanese Waxwing is red, but the other two waxwings have yellow-tipped tails.
3390. None has webbed toes.
3391. No, its bill is relatively short.
3392. The eiders may defecate on the eggs, resulting in an extremely fowl-smelling nest, possibly deterring mammalian predators. Some other ducks do the same thing. Once the fecal matter dries, the smell disappears, but because ducks lack a strong olfactory sense, the smell is of little consequence.
3393. Probably the Red-breasted Merganser in 1898, in Denmark.
3394. Most cowbirds are obligate brood parasitic nesters, hence do not brood at all.
3395. The Sooty Albatross.
3396. The woodpeckers.
3397. No, but the Oriental White Stork formerly did. Storks were widely distributed throughout Japan 100 years ago, but the last documented breeding occurred in 1970. The last remaining colony was protected, although poaching continued. Mercury contamination finally doomed the stork as a breeding species. The species is endangered, and only 3,000 are estimated to remain in China and Russia.

Answers

3398. Some of the storm-petrels.
3399. Three: Cooper's, Harris' and Swainson's Hawk.
3400. Yes. Unlike their mute parents, chicks are extremely vocal, producing loud screaming and wailing sounds.
3401. Reportedly, in about a half-hour.
3402. Calling kiwis adopt an extreme upright posture, with neck and legs fully stretched and bill pointed skyward.
3403. About 16.5 pounds (7.5 kg), but up to 24 pounds (10.9 kg).
3404. The Peruvian Booby.
3405. The swallow.
3406. Nesting petrels. The birds have a variety of wailing, moaning, screaming, cackling and squalling calls. At many breeding stations, the petrels nest in dense vegetation and call loudly when underground in their burrows, and after dark this can be nerve-racking.
3407. Yes, especially during the winter.
3408. The Northern Lapwing.
3409. No, all three species have very short tails.
3410. Immortality.
3411. No, it is a South American cotinga.
3412. A resting Rock Dove may breathe only 29 times a minute, but its breathing rate increases to 180 times a minute when it is in flight.
3413. The males of the nocturnal Owl Parrots don't breed annually because when displaying, the birds boom continuously all night long for as long as 10 weeks. The booms are emitted in a regular monotonous rhythm of about one boom every two seconds, repeated for many hours. During such times, males feed very little and literally wear themselves out, and 2-3 years may be required for them to recover. Most of the males gather at the leks, thus many will not be active the following year. Lone males tend not to boom because the stimulus of other males is required.
3414. The five tragopans construct bulky tree nests. The African Congo Peacock is also a tree-nester.
3415. The pigeons and doves, and the neotropical tinamous. With feathers that detach easily, a predator often ends up with only a claw or mouthful of feathers, instead of the bird itself.
3416. Only 40 million. Approximately 20 million were taken by hunters, and another 40 million were lost because of natural causes, such as predators, malnutrition, etc.
3417. Yes.
3418. The Socotra Cormorant of the coast and islands in the Persian Gulf, but not Socotra Island itself, from where the type specimen was collected. Nests typically consist of simple scrapes on gravel or on stony ridges of low islands, or among boulders. Colonies usually number 50-100 pairs, but assemblages of as many as 10,000 pairs are known. In 1981, up to 250,000 birds were recorded nesting in the Hower Island group off Qatar, with a nest density of 4-6 nests per square meter. The cormorant has been impacted by the 1990/91 Gulf War and associated oil spills, and is currently regarded as near-threatened.
3419. Yes, but only on nestlings.
3420. Yes, but the neck of its near-relative, the Dodo, was very short.
3421. Yes.
3422. The grouse benefit from food that foraging caribou dig up.

Answers

3423. Dr. James G. Cooper named the warbler in 1861, after Lucy Hunter Baird, the 13-year old daughter of Spencer F. Baird. Baird later became secretary of the Smithsonian Institution.
3424. Two.
3425. The Louisiana (Tricolored) Heron.
3426. The Whooping Crane.
3427. Lead-gray.
3428. Yes.
3429. Stresemann's (Ethiopian) Bush-Crow. While not a colonial breeder, it is apparently normal for three birds to attend the nest, although based on egg patterns, only one female lays.
3430. Yes.
3431. The eagle.
3432. Nine.
3433. Yes, most species do.
3434. The unique Cloven-feathered Dove of the humid forests.
3435. Penguins.
3436. The Indian Peacock.
3437. When cock Black Grouse display at a lek, producing a bubbling sound, they are *rookooing*.
3438. No. The first-laid (alpha) egg of all six eudyptid penguins is always *significantly* smaller than the second-laid (beta) egg.
3439. Yes, its plumage is velvety-black.
3440. No, the rather delicate structure is strictly ornamental.
3441. The introduction of Arctic foxes by fox farmers to the goose's breeding islands in the Alaskan Aleutians.
3442. White.
3443. Yes, but only rarely.
3444. The tropical American Boat-billed Heron.
3445. The 17 Old World honeyguides.
3446. Yes, but it was introduced.
3447. Not until 1965, high inland on the very steep slopes of the seaward Kaikoura Mountains of New Zealand's South Island.
3448. The white, dark-tipped tail feathers of a juvenile Golden Eagle.
3449. No, a Eurasian Woodcock weighs about twice as much, and is up to 13.5 inches (34.3 cm) long.
3450. There is no difference; the booby lacks a juvenile plumage.
3451. The name comes from *tangara* in the language of the Tupi Indians of Brazil.
3452. No, it is a small chicken.
3453. No, cormorants typically forage in inshore waters.
3454. To *goose* someone.
3455. Unlike most mammals, virtually none.
3456. Yes.
3457. Typically pointed up at about a 45-degree angle.
3458. The Carolina Wren.
3459. The Count Raggiana Bird-of-paradise.
3460. No, their necks are very short, although a roller's head is relatively large.
3461. None. All tits nest in cavities.
3462. The duck is usually silent, suggesting it was dumb.

3463. The 2,000 copies were sold out within the first week.
3464. The penguins generally smack each other with their hard flippers, although some species are prone to bite as well.
3465. The tropicbirds, that are obligated to shuffle about on their bellies. Their legs are very short and set far back on the body, and both legs and feet are weak.
3466. The pelicans.
3467. Yes.
3468. The Sharp-tailed Grouse, although several other species are extremely rare.
3469. No, fish are transported cross-wise in the bill.
3470. Its tongue may serve a temperature-perceiving function. The bare skin of the head and neck might also be used.
3471. No, but all were until the recent taxonomic revision that subdivided them into three families. The neotropical barbets remain in the old family Capitonidae, the Asian species are currently in the family Megalaimidae, and the African barbets have been placed in the family Lybidae.
3472. Morillon is an obsolescent British fowler's name for the juvenile Common Goldeneye, once believed to be a different species.
3473. A shepherd's crook at the end of a long pole used to catch swans.
3474. Ten, but there are a few exceptions, such as the Marvelous Spatuletail.
3475. No, the Sooty and Light-mantled Sooty Albatross nest in solitude, or in very small groups.
3476. Trumpeters have a very distinct hunch-backed posture, so much so that in Suriname they are known as *kamikami* (camel back).
3477. Wilson's Phalarope.
3478. No, the calls of the pen (female) are slightly lower.
3479. The Common (Ring-necked) Pheasant.
3480. There is no difference; both names apply to the same bird.
3481. The Turkey Vulture.
3482. The Common (Atlantic) Puffin.
3483. Yes.
3484. The Labrador Duck.
3485. No, it is a small, New World tyrant-flycatcher.
3486. The large flightless birds were driven from the shore over spread sails to the ships. The auks were later killed and preserved. In 1578 alone, at least 400 French, English and Spanish transatlantic vessels called at Funk Island to slaughter auks. Some ships salted down as much as 4-5 tons, and 200 years later the birds were still numerous enough at the island for commercial exploitation.
3487. Long necks enable the vultures to penetrate deep into carcasses.
3488. The Jamaican Tody.
3489. No, the birds have heavy, almost conical, bills.
3490. Not black, but rather dark chocolate-brown.
3491. At least 883, of which 127 occur only as long-distance migrants that are not known to breed in the Republic.
3492. The Red-breasted Merganser is a ground nester, and dark down in the nest provides some concealment. The cavity-nesting Common Merganser has little need to conceal its eggs.
3493. No, its feet are noticeably weak.
3494. The Egyptian Vulture.

Answers

3495. The Western Meadowlark. Its voice has been described as having the flute-like quality of the Wood Thrush with the rich melody of the Baltimore Oriole.
3496. The three waxwings.
3497. *The pen is mightier than the sword.*
3498. Discovered in 1918 at Cape Sable, Florida, the endangered sparrow was once believed to be a distinct species, and was the first new bird discovered in North America in the 20th century. The bird was later determined to be a race of the Seaside Sparrow.
3499. Yes, occasionally in Alaska, northern Canada and Greenland.
3500. Pale yellow.
3501. Three: Hawaiian Hawk, California Condor and Mississippi Kite.
3502. The Eurasian Dotterel, at 4,265 feet (1,300 m).
3503. Between 90 and 125 days.
3504. It was not a bird, but rather a flying archosaurian reptile.
3505. The northern swans.
3506. Yes.
3507. The Papua-Australian Cicadabird, that was named after its cicada-like call.
3508. Yes, although not exclusively.
3509. An owlet.
3510. The Black (White-capped) Noddy.
3511. Between 10,000 and 30,000: mostly Common Puffins, Common Murres and Razorbills. Over 119,000 tons of oil were spilled when the ill-fated tanker went on the rocks.
3512. The Green Woodpecker.
3513. Such positioning probably makes flight easier by reducing wind resistance.
3514. Yes, penguins are far more dependent on their thick, waterproof plumage.
3515. *Counting one's chickens before they hatch.*
3516. 1964. Breeding commenced in 1970, along the south shore of the Salton Sea.
3517. Yes, the vultures make surprisingly good pets, following their owner about like a dog.
3518. The Golden-crowned Sparrow.
3519. So far as is known, only the Red-legged Honeycreeper and possibly the Blue-black Grassquit.
3520. Possibly, but dark morph is essentially restricted to Lesser Snow Goose. A few large Blue Geese have been seen, and these *might* be Greater Snow Geese.
3521. The Short-tailed Albatross, with yellow on its head and neck.
3522. Yes.
3523. Approximately 300,000.
3524. About 25 percent.
3525. The Bay-winged Cowbird.
3526. While their bills are tiny, their gapes are extraordinarily wide, so swifts have no real need for rictal bristles to aid them in flycatching.
3527. *A dead bird does not help the appearance of an ugly woman, and a pretty woman needs no such adornment.*
3528. Yes. As a foraging spoonbill sweeps its bill through the water, the bill instantly snaps shut when live prey is encountered.
3529. Between 2-3 mph (3.2-4.8 km/hr), but the birds can swim much faster if pursued.
3530. The Brown-crested Flycatcher (formerly known as Wied's Crested Flycatcher) has been observed catching and consuming hummingbirds in Arizona.

Answers

3531. Its range expansion was natural.
3532. Their long thin bills are held angled upward.
3533. The ridge formed by the junction of the two rami of the lower mandible near the tip of a bird's bill.
3534. Yes.
3535. No.
3536. The four Marx brothers: Groucho, Harpo, Chico and Zeppo.
3537. Yes, across the hindneck.
3538. Apparently from the old Norse word *alka*, reflecting the vocalizations of some species.
3539. A specific term for the prominent nail on the tip of the upper mandible of some birds.
3540. The peacock flounder.
3541. Both repeatedly toss food into the air, catching it with the tips of their long bills. It may be the birds are crushing food prior to swallowing it, but I have observed such behavior countless times, and this does not appear to be the case.
3542. At least two were killed by hailstones in 1938.
3543. About 140 times, on the average.
3544. About 3.2 wingbeats per second.
3545. Yes.
3546. The shrikes.
3547. Red.
3548. A domestic duck that was probably developed from a strain of Indian runner duck, but the show breed apparently no longer exists. It had very short legs placed far back on the body, forcing the duck to stand in a very erect posture, like a penguin.
3549. The 13-inch-long (33 cm) California Thrasher.
3550. The Rough-legged Hawk.
3551. *Yardbird* or *jailbird* is American slang for an individual serving time in prison.
3552. A few robins were transferred from tiny Mangere Island to predator-free South East Island. Its clutch size is two eggs, but only a single chick is reared, and it is dependent on the parents for some time after fledging. Two broods a season are not known, but the birds re-lay if their clutch is lost early. Consequently, eggs were removed to stimulate the laying of a second clutch, and were placed under incubating Chatham Islands Warblers. The Black Robin builds an open cup nest and its chicks excrete their droppings over the side of the nest, whereas the warbler constructs an enclosed nest and its chicks excrete their droppings in fecal sacs. Therefore, the fostered robin chicks became wet and messy in the domed nest. The problem was solved by supplying the fostering warblers with a freshly built warbler nest every five days. Ultimately, the robins were fostered under Chatham Tomtits on South East Island, a species that utilizes an open nest.
3553. No, the birds have "toothed" upper mandibles.
3554. No.
3555. White.
3556. Coulman Island in the Ross Sea where 27,900 chicks were counted in late 1990. The nearby Cape Washington colony is nearly as large with about 24,000 chicks in 1990.
3557. They become instantly alert and bolt.
3558. The inquisitive birds are not in the least bit wary.
3559. Probably the Roseate Tern. In 1977, only 630 pairs were counted in six main colonies. By 1985, the entire European breeding population may not have exceeded 550 pairs.

3560. Rachel Carson, in her 1962 classic book, *Silent Spring*.

3561. No, Megapodes do not occur in Africa. However, it is estimated that as many as one million megapode eggs are harvested by Papua New Guinea natives alone annually.

3562. The Short-eared Owl.

3563. A hawker.

3564. A single egg.

3565. The Red-billed Quelea.

3566. The Harlequin Duck.

3567. The rare endemic New Zealand Stitchbird.

3568. The Rainbow Lorikeet was pictured in Peter Brown's *New Illustrations of Zoology,* published in 1774.

3569. Some species of caracaras.

3570. Cuba, in 1931. The Caribbean Flamingos were not pinioned and flew off. In 1937, another flock was established, but this time the birds were pinioned and breeding commenced in 1942.

3571. An owl.

3572. Yes, but not unless their snow-white heads and necks are stained orange-yellow by iron deposits in the water. Staining also occurs in Snow Geese, and Trumpeter and Tundra (Whistling) Swans.

3573. No, males have distinctly longer upper mandibles.

3574. Panama.

3575. Yes.

3576. A Red-breasted Merganser pursued by an airplane in Alaska attained a top air speed of 100 mph (161 km/hr). The previous record was for a Canvasback flying at 72 mph (116 km/hr). The Oldsquaw Duck is also among the fastest flying of waterfowl and has been reliably clocked at 71 mph (114 km/hr) in level flight.

3577. Yes.

3578. Yes.

3579. Yes.

3580. Pink.

3581. A stylized crowned-crane.

3582. If an individual with an aching back, on hearing the call of a Whip-poor-will, turned somersaults timed with the bird's calls, the backache would disappear.

3583. No. A Turkey Vulture tilts and teeters in flight, whereas a Black Vulture has a much more stable flight.

3584. Geese.

3585. Siberia.

3586. Yes.

3587. The Scarlet, Hadada and Sacred Ibis.

3588. No, in some species, such as the hermits, both sexes are dull colored.

3589. A goose wing.

3590. Yes, but the birds generally perch in the shade.

3591. Yes, if caught on the wing by the broad-winged birds.

3592. The glittering chitinous parts of insects regurgitated by the nesting pair. By the time the chicks fledge at about three weeks of age, the nest hole is befouled with excrement and numerous regurgitated insect hard parts.

Answers

3593. Yes.
3594. After his brother, Sir Edward Sabine, an eminent British astronomer and physicist.
3595. No, it is a widespread Old World thrush.
3596. None, but the *Western* Meadowlark is the state bird of six states.
3597. The name refers to the half-moon shaped white spot behind and below the eye of a drake in breeding plumage.
3598. Yes.
3599. No, one chick generally perishes, usually as a result of sibling rivalry.
3600. No, but for many years, biologists believed that the birds did. Nectar is probably obtained as a result of capillary action.
3601. Yes.
3602. The Australian Black Swan.
3603. The Barn Owl.
3604. Yes, the aquatic passerines have a musty odor.
3605. Reportedly up to 55 mph (87 km/hr).
3606. Parrot fever or psittacosis.
3607. Birdos.
3608. A male Indian Peacock served up in a lordly dish, garnished with its head and train.
3609. No one knows, but the introduction of cats to Guadalupe Island (Mexico) undoubtedly played a major role.
3610. The gallinaceous birds, presumably because of the lyrebird's superficial resemblance to neotropical chachalacas.
3611. Yes, as a rule.
3612. No, their feet are very weak.
3613. The three umbrellabirds of Central and South America.
3614. Seven: Common Murre, Pigeon Guillemot, Xantus' Murrelet, Marbled Murrelet, Cassin's Auklet, Rhinoceros Auklet and Tufted Puffin.
3615. Probably the Dalmatian Pelican at about 85 days, but some young are not independent until they are 100-105 days old.
3616. A cuckoo-shrike.
3617. Over 1,000 years ago, Chinese traders came to Borneo to trade for the swiftlet nests, and the trade may have commenced as early as the T'ang Dynasty (618-907 A.D.).
3618. No. All passerines, raptors and pigeons, to name only a few, have a single brood-patch, but some birds have no brood-patches at all, whereas others have two, or even three.
3619. The Masked Duck, in which both sexes have white speculums.
3620. Wrynecks have cryptic plumage, similar to nightjars.
3621. The name apparently comes from the Old Norse or Icelandic *lonr*, Swedish *lom,* or the Old Dutch *loen,* all of which loosely translate to "lame," "clumsy," "slow" or "stupid fellow," probably referring to a loon's clumsiness ashore.
3622. No, it is rather vocal.
3623. Grunts, snorts and hisses.
3624. An estimated 6,000 tons.
3625. No, it flies slowly and low, often in chevrons, with its head supported on its shoulders.
3626. Gannets have a mostly feathered gular pouch, but a booby's is bare.
3627. Yes. Unlike Turkey Vultures that fly with their wings held in dihedral fashion (V-shaped angle), Black Vultures soar with wings held level.

Answers

3628. No. Linnaeus was a famous professor of medicine and botany at the University of Uppsala, Sweden, from 1741 until his death in 1778.
3629. The endangered Red-cockaded Woodpecker.
3630. Only two: New Zealand Quail (about 1868) and Himalayan Quail (about 1870).
3631. Five: Wilson's Storm-Petrel, Wilson's Phalarope, Wilson's Snipe, Wilson's Plover and Wilson's Warbler. The Wilson Ornithological Society is also named after him, as is the Society's quarterly journal, *The Wilson Bulletin.*
3632. Yes, if trees are not available, the raptors may nest in bushes or on rocks.
3633. Yes, some species occasionally consume mice.
3634. A bittern.
3635. The King Vulture.
3636. The reed roofs of old huts. More typically, nests are in dense thickets or in burrows with tunnel-like entrances.
3637. Probably the Old World pittas. Their olfactory system is apparently the most advanced of the passerines.
3638. The Turkey Trot.
3639. The hummingbirds.
3640. Their triangular-shaped bills are goose-like and designed for feeding on water-lily seeds.
3641. The cheek.
3642. The Northern Fulmar.
3643. Yes.
3644. Males have an unusual looped trachea unknown in other songbirds.
3645. No, the eastern birds are slightly brighter in color.
3646. The Crane Hawk.
3647. No, but some of the 19 species of Australian and New Guinea bandicoots have an extremely short gestation period of only 12 days, the shortest of any mammal. The American opossum reportedly has a gestation period as short as eight days, but 12-13 days is more typical.
3648. Bright red, but yellowish at the base.
3649. Anvils.
3650. Probably as an adaptation to the dim light of the forest.
3651. Kirtland's Warbler. The 4-foot-high (1.2 m) stone monument is on the lawn of the county courthouse in Mio, Michigan, and was dedicated on July 27, 1963.
3652. Cruel captains would be condemned forever to flutter over the sea as storm-petrels. Like some other petrels, storm-petrels are also known as "soul birds." A general belief among seafaring men is that the touching of a storm-petrel is a sure sign of bad luck, possibly bringing death to the man that dared touch it, or to his family.
3653. The name refers to its nesting habitat on rocky places.
3654. The Wood Thrush.
3655. Chapman's survey of 700 hats disclosed that 542 were decorated with feathers or whole birds, of at least 40 species.
3656. The Brown Thrasher.
3657. Yes, generally so.
3658. The Barn Swallow. Up to 67 percent of its 7.5-inch-long (19 cm) body is accounted for by the two long tail streamers.
3659. Colombia, with at least 143 species.

Answers

3660. Yes.
3661. Both sexes have orange crests.
3662. Rockfowl have no head feathers. Both species are bald, and the birds are commonly called bald- crows.
3663. No, but oil and stomach contents may be spit (projectile vomiting) at other intruders.
3664. The Egyptian Vulture, that picks up eggs and hurls them against a rock to break them open.
3665. All six.
3666. No.
3667. The Pine Grosbeak.
3668. No, a bird-of-paradise is featured on the crest.
3669. It feeds at the surf line when on the coast, unmindful of flying spray.
3670. The Purple Martin.
3671. The aracaris.
3672. Its clattering, cackling calls resemble the sound of old-fashioned clappers.
3673. Harris' Hawk.
3674. In the southwest, but just barely.
3675. The goslings are polymorphic, ranging in color from pure yellow to gray.
3676. Gigantic sandpipers.
3677. No, the egg tapers at both ends.
3678. The recovery stroke.
3679. Treeswifts perch, and instead of pursuing insects on the wing, they forage rather like flycatchers, darting out at intervals from lookout perches to snatch passing insects.
3680. Not generally.
3681. No, its legs are fully feathered.
3682. The Australian Black Swan.
3683. The Eurasian Wryneck, because of the simultaneous appearance in the spring of both the wryneck and the cuckoo.
3684. No, although it was occasionally sold in the market places of India, selling for about 15 rupees.
3685. Pigeons and doves have a softer mouth condition, enabling them to swallow successive gulps of water without lifting their heads.
3686. The Red-throated Loon.
3687. Yes, but there are exceptions, especially in woodpeckers and swiftlets.
3688. All North American herons occur in the West Indies, and all but the American Bittern breed in the islands.
3689. No. Some birds, such as gannets, frigatebirds, Anhingas, darters and cormorants lack external nares and must breathe through their mouths. In gannets and cormorants (and possibly some other birds as well), a permanently open slit is present at the gape enabling the birds to breathe without opening their mouth. The slit is protected by a horny lid that closes when they dive.
3690. The adult African Black Heron, but some reef-egrets have dark, almost black, color morphs.
3691. Yes.
3692. Probably, but the Northern Pintail *may* be more numerous in some years. In North America, the Mallard population is augmented by large numbers of captive-reared birds.

Answers

3693. Its long legs enable the specialized neotropical raptor to reach down into holes to extract prey.
3694. The Gentoo Penguin, especially in the Falkland Islands.
3695. A term describing the territorial or patrolling flight of male birds. In current usage, it is usually applied only to the characteristic twilight display flights of the Eurasian Woodcock.
3696. Yes, but it appears reluctant to fly and tends to rely on thermals.
3697. Yes, kiwis can run surprisingly rapidly with long, loping strides. They run about as fast as a human.
3698. No, its bill is slightly upturned.
3699. Chicks lay in a depression they have dug, kicking sand into the air that settles on their backs, making them difficult to detect.
3700. The erection of electric power poles. The lack of suitable nesting sites precluded their nesting in the Transvaal prior to this.
3701. Red.
3702. The erect head feathers extending over the base of their bills, and the rounded shape of the head, resembles a helmet.
3703. John James Audubon.
3704. Adult males have red throat pouches, whereas females have varying amounts of blue on the featherless red throat pouch.
3705. A goose neck (goose-necked lamp).
3706. There are numerous reasons, but one of the primary advantages of flocking is that it provides a degree of protection from predators. A predator has more difficulty concentrating on a single individual in a flock than on one bird flying alone.
3707. Three: Black Rail, Yellow Rail and Purple Gallinule.
3708. No, it had a rather long tail.
3709. Gould's Petrel.
3710. The Hawk-headed (Red-fan) Parrot.
3711. To cool a bird, most of which have a rapid hot metabolism.
3712. Five: White-tailed, Mississippi, Swallow-tailed, Everglade (Snail) and Hook-billed Kite (a very rare south Texas breeder).
3713. A quarter-pound (0.1 kg): 1.25 pounds (0.6 kg) versus 1.0 pounds (0.45 kg).
3714. The somewhat moth-eaten specimen was deposited in a museum in Oxford, England, and the decision was made to destroy it in 1775. An assistant fortunately pulled the head and a foot from the fire at the last moment. All that remains today of the Dodo are two heads, two feet and some skeletons.
3715. The cotingas, manakins and some birds-of-paradise.
3716. Spreading animal fat or oil over the water often attracts petrels.
3717. The Lesser Snow Goose.
3718. No, both parents feed chicks.
3719. For the first few days after hatching, chicks are said to spit oil at their parents. Once their parents are recognized as "friends," the chicks cease spitting oil at them.
3720. Habitat destruction and trophy hunting.
3721. Sir Peter Scott, the son of Robert Falcon Scott, of British Antarctic exploration fame.
3722. Ivory from the solid casque of a Helmeted Hornbill.
3723. Reportedly in the unbelievably brief time of 45 days.
3724. No, the Imperial Heron is second largest, exceeded in size only by the Goliath Heron.

Answers

3725. Colombia.
3726. No. The Black-billed Magpie is larger: 17.5-22.0 inches (44.5-55.9 cm) versus 16-18 inches (40.6-45.7 cm) long.
3727. The bird has very small or rudimentary wings.
3728. The Anhinga and darters.
3729. No. The Red-crested, Southern and Rosy-billed Pochards are not superior divers, unlike the other pochards.
3730. Lake Nakuru in Kenya is one of the greatest of ornithological spectacles. Only 9,884 acres (4,000 ha) in extent, the lake supports 1.5-2.0 million birds, or about 375-500 birds per 2.5 acres (1.0 ha).
3731. No, it is one of the most threatened birds of continental Africa. Absent throughout most of its former range, it probably only survives in viable numbers in limited areas of Liberia and the Ivory Coast.
3732. The crossbills.
3733. The neotropical Boat-billed Heron.
3734. The Sarus Crane, but the Siberian White Crane and Demoiselle Crane winter in India.
3735. Occasionally in the burrow of a pika (mouse-hare).
3736. Captain James Cook (1728-1779), the famous explorer and world circumnavigator.
3737. The cotton-top.
3738. The South American Inca Tern.
3739. The Siberian White Crane.
3740. The name was given by Audubon because the bird was "overlooked" by members of the Lewis and Clark expedition.
3741. No, it is usually solitary or lives in pairs.
3742. Yes, the Laughing Kookaburra often takes lengthy snakes that it drops from a height to stun them.
3743. The Princess Stephanie, Queen Carola, King of Saxony and Emperor of Germany Bird-of-paradise. The Count Raggiana Bird-of-paradise was also named by the Germans, but a count is not a member of a royal family.
3744. Apparently from the Icelandic word *sanderla*, suggesting its sandy habitat.
3745. Nothing, but the quail are highly poisonous to dogs that eat them.
3746. Red.
3747. At least 24 species, 12 of which definitely became established, with three others probable.
3748. The Poo-uli, a tiny insect-eating and snail-eating bird, went straight from discovery in 1973 to the endangered species list, and may already be extinct.
3749. The domestic goose has curled feathers that do not lie flat, with long streaming tertiary feathers. The feathers generally appear to be growing in the wrong direction.
3750. The nearly extinct Crested Ibis.
3751. Arthur Bernard Singer.
3752. Yes.
3753. Yes.
3754. Audubon's Shearwater.
3755. The Hoatzin.
3756. The Sage Thrasher. Its bill is unlike the long, curved bills of the more typical thrashers.
3757. Yes.
3758. Yes, but the first sighting did not occur until 1987.

Answers

3759. In most birds, young migrating for the first time are on their own, but in northern geese, swans and cranes, the parents must show the young the way.
3760. No. Field Martin is an alternate name for the kingbird.
3761. The *Blackhawk.*
3762. The birds are greatly, sometimes wholly, dependent on the actions of predatory fish, especially tuna, that drive large schools of fish to the surface. Some species, such as the Sooty and White (Fairy) Terns, are collectively called *tuna birds.*
3763. The Crow-billed Drongo.
3764. *Duck.*
3765. A large, neotropical, grosbeak-like fringillid.
3766. Yes, all three South American screamers have very long toes well suited for this purpose.
3767. No, the eastern birds are regarded as superior singers.
3768. The upper surface of an Andean Condor's wing is white, whereas in the California Condor, the under surface of the wing is white.
3769. Yes.
3770. The White-winged Potoo. One was sighted several years earlier in October of 1985.
3771. New Zealand.
3772. No, a constant body temperature is required during incubation.
3773. The cottonmouth.
3774. Blue-violet, but its head is orange-yellow to yellow.
3775. Probably the Sooty Shearwater, with an overall population numbering in the tens of millions.
3776. Yes, the Stanley Crane of Africa, also known as the Paradise or Blue Crane.
3777. Very carefully. Change-over at the nest is accomplished by meticulous, sideways-shuffling steps.
3778. The dippers.
3779. No, its name is from the Latin *lucifer* that translates to "light bearer." The reference is to the colorful plumage rather than Satan.
3780. Trumpeter Swans construct huge nests in the water, but Tundra Swans nest ashore, often some distance from water.
3781. Bright rose-red.
3782. The gray-and-white kite with black shoulders has a distinct black line on the underwings shaped like the letter *W* when the wings are extended. The partially nocturnal raptor hunts mice by moonlight, but is often seen by day at inland prairie fires, capitalizing on fleeing animals.
3783. The Gray-headed Albatross.
3784. *Half-webbed goose-duck.*
3785. No, they are extremely pugnacious, attacking birds as large as crows or harriers.
3786. The last Cuban Macaw was shot at La Vega (Zapata Swamp) in 1864, and the Hispaniolan Macaw disappeared around 1830.
3787. No.
3788. Yes, including even the male Resplendent Quetzal.
3789. Islas Diego Ramirez (56°30'S), about 60 miles (97 km) southwest of Cape Horn, Chile. A number of species breed there, including the Thorn-tailed Rayadito, Austral Thrush and Gray-flanked Cinclodes.
3790. Bright yellow.

Answers

3791. Yes.
3792. Wrangle Island, Siberia, where in productive years, 30,000-40,000 geese bred, but in recent years the number of Russian-breeding Snow Geese has fallen dramatically. All winter in North America.
3793. McKay drowned in 1883 when a small boat in which he was riding capsized. Foul play was suspected, but never proved. He was a U.S. Signal Corps observer in Alaska who collected birds for the U.S. National Museum.
3794. The name is from a French word meaning *bearded,* referring to the prominent rictal bristles surrounding a barbet's bill.
3795. The Bearded Tit (Bearded Reedling). It has recently been reclassified as a parrotbill.
3796. Goose-stepping.
3797. The Black Crowned-Crane.
3798. Mud cups plastered to a vertical rock wall, and their nests are generally placed near a cave entrance or rocky outcrop in a forested ravine.
3799. New Zealand. The finch is not native, having been introduced from Britain during the 1860's.
3800. Yes.
3801. Yes, as a rule. Nesting in a cavity provides concealment and protection, hence chicks need not fledge as rapidly.
3802. Approximately 50 times as large, accounting for 13-30 percent of the bird's weight.
3803. No, it is an icterid (Icteridae).
3804. The flamingos.
3805. The Cuban Sandhill Crane.
3806. None, all are restricted to the Western Hemisphere.
3807. Three thousand or more nests.
3808. Yes, some species are among the most beautifully-plumaged birds.
3809. Yes.
3810. The Spanish Imperial Eagle. Restricted to Spain and Portugal, possibly less than 100 pairs remain, a number of which inhabit the Coto Doñana Nature Reserve in southern Spain.
3811. The Barn Swallow.
3812. The sheathbills, and possibly the oystercatchers. The aberrant South American shorebird was recently elevated to full family rank.
3813. Eight.
3814. Yes.
3815. Some do, most notably the African Pitta.
3816. Yes.
3817. When incubating birds are off the nest for a prolonged period, even days at a time, it is called egg neglect. Such behavior is not unusual for a number of seabirds, especially petrels and some alcids, and also swifts.
3818. The White (Fairy) Tern.
3819. The flamingos.
3820. The Northern Fulmar.
3821. Roman crested geese, a breed of domestic sacred geese, alerted the Romans. To commemorate the event, a golden goose was carried in procession to the capitol every year, and the dogs whipped as punishment for their silence.
3822. No.

Answers

3823. The South American Maguari Stork.
3824. The tiny Pygmy Tit of Java is barely three inches (7.6 cm) long.
3825. Many curassows have shiny, often brightly-colored ceres.
3826. Probably no more than 80 skins have been preserved, the oldest of which is dated 1825. The paucity of museum skins of this unique and much sought species may be taken as a measure of its rarity, even in the last century, when no considerations of conscience or conservation would have restrained a collector's zeal.
3827. The fused tail bone (pygostyle).
3828. The Snowy Owl.
3829. Yes, but only Cassin's Auklet, and only at the Farallon Islands, California.
3830. Yes, some are. One famous female known as *Old Blue* was still breeding at 13 years of age. Between 1979 and 1983, *Old Blue* was the *only* breeding female and all 50 robins alive in 1987 descended from her. No other Black Robin has come close to achieving such age and had she not survived so long, it is likely that the species would have become extinct.
3831. Yes.
3832. *Dove-eyed.*
3833. Yes.
3834. Over five million copies in the half-century since the field guides were first published.
3735. The birds are said to be *ploughing*.
3836. No, but avian blood contains more red blood corpuscles per ounce than the blood of any other animal.
3837. The Common Nighthawk and Mississippi Kite.
3838. June, 1972. Within a few years, some avian species negatively impacted by DDT began to recover.
3839. John James Audubon, who obtained 60 birds in Kentucky.
3840. The cassowaries, but not regularly. Their normal diet consists of fallen rain forest fruit, but cassowaries have been recorded eating Little Tern nestlings and eggs, domestic fowl, as well as fish, frogs, snails, rats, mice and even carrion.
3841. The bird often eats thistle seeds, and thistledown is used as nesting material.
3842. The disease was first described in Manx Shearwaters *(Puffinus puffinus)*.
3843. The large Mexican parrot feeds on pine seeds.
3844. The tyrant-flycatchers.
3845. No, but he was a falconer.
3846. The Whooping Crane, but only during the winter in south coastal Texas.
3847. The Ivory Gull.
3848. At least 1,195.
3849. The Song Sparrow.
3850. The smaller Red-throated Loon has a greater ability to take off from a more confined area.
3851. The Brahminy Kite.
3852. No.
3853. At dusk, with males typically calling first.
3854. Ross' Goose. At about three weeks of age, goslings from different broods may combine to form groups of up to 200 young. This early dissolution of the family unit deviates from typical northern goose behavior. Brood merging may also occur in some Canada, Snow and a few Greylag Geese.

Answers

3855. Yes.
3856. Yes. Some biologists suggest the Eurasian Snipe does not, and instead engages in complex lekking behavior. If so, this might indicate that the Old Wold race of the Common Snipe is a separate species.
3857. The American Bittern. When in the freeze position, it stands perfectly still with its bill pointing to the sky.
3858. Yes.
3859. No, it tends to be very short, sometimes only 10 days.
3860. The Willow Ptarmigan.
3861. No, fish are brought to the surface to be swallowed.
3862. No. The nasal tubes are joined externally, but the tubes are not divided by a septum, thus the nostrils are not separate.
3863. The New Zealand kiwis.
3864. Yes, there are a few instances of defensive Mute Swans killing children who were tormenting them.
3865. The goose is from Australia. Until 1847, colonial Australia was known as New Holland.
3866. No, peacocks fly with regular, rather slow wingbeats, without the glides typical of most large galliform birds.
3867. No. In fact, one of the vernacular names for Kittlitz's Murrelet is "short-billed murrelet," as opposed to "long-billed murrelet" for the Marbled Murrelet.
3868. Its flight is undulating, like a woodpecker.
3869. Yes.
3870. Until the recent taxonomic revision, three: Smooth-billed, Groove-billed and Greater Ani. The South American Guira Cuckoo is now regarded as an ani, probably more because of behavior than appearance. To be technically correct, four anis are currently recognized.
3871. Yes.
3872. One who plunders or ravens.
3873. The tiger-parrots.
3874. Yes, males may fight so fiercely with one another that blood flows.
3875. Not at all: potoos are solitary.
3876. No, a *nest* site may be as high as 100 feet (30 m).
3877. The Emperor Goose, because of its littoral foraging habits.
3878. Twenty-four: *Four and twenty blackbirds baked in a pie.* The blackbirds are actually European thrushes.
3879. Nesting trees are often slowly killed as a result of the profuse amounts of toxic guano.
3880. The neotropical jacamars.
3881. Frogs and tadpoles.
3882. The colony was visited a day or so prior to the actual egging and all eggs were smashed. Thus, when eggs were collected, all were fresh.
3883. A murmuration of starlings.
3884. Sometime prior to the 12th century, in Great Britain.
3885. The New Zealand Fiordland Crested Penguin nests in dense virgin rainforests along the southwest coast of the South Island.
3886. Yes.
3887. 1971.
3888. No, virtually all have been females or juveniles.

Answers

3889. No.
3890. The shorebirds just reach Venezuela as a breeding species at Islas Los Roques.
3891. No, although the Acorn Woodpecker is an exception.
3892. The Least Auklet.
3893. Yes.
3894. The waterfowl. Many banded ducks and geese are shot by American hunters, greatly enhancing the probability of recovering banded birds.
3895. No, its eggs are smaller.
3896. 1901.
3897. Mainly as fuel for lamps.
3898. No, the birds only molt once annually.
3899. The Long-tailed Jaeger.
3900. Yes, in most instances.
3901. The storks do not locate food by sight. Rather, they wade in murky 8 to 10-inch-deep (20-25 cm) water, groping with their bills partially open for minnows, frogs and small alligators. When the lower mandible encounters a moving object, the bill immediately snaps shut. The storks may move a foot around to stir things up.
3902. No, but the American wood-warblers are.
3903. Yes, but sometimes this can be fatal. Six young Common Terns were found dead in 1962, having choked to death on small inflated blowfish lodged midway down their throats.
3904. No, cassowary chicks are striped yellowish-buff and brown or black. At 3-6 months of age, the chicks acquire the uniform brown of first-year plumage birds.
3905. The Stilt Sandpiper. The Asian Curlew Sandpiper is a casual visitor to the East Coast, but is not a regular breeder.
3906. A wattlebird.
3907. This old English expression was applied to lawyers, who often carried a pen quill stuck behind their ear.
3908. No, the raptors are neither cuckoos nor falcons.
3909. Yes, with cheerful, twittering notes.
3910. The Indian Hill Myna.
3911. About a dozen, all of which were killed and brought in by the lighthouse keeper's cat.
3912. Yes, its bill is raptor-like in shape.
3913. Dark red.
3914. Probably the Horned Lark. It is often called the "prairie bird."
3915. Yes, cygnets are vulnerable to leeches.
3916. The gnatcatchers: both are in the family Polioptilidae.
3917. Yes.
3918. No. The vireo was named by Audubon in 1844 for his friend John Graham Bell, a taxidermist who accompanied him on his 1843 Missouri River expedition.
3919. Their shape supposedly resembled chickens.
3920. Stamp their feet or charge and kick viciously.
3921. The extinct Great Auk.
3922. The male Ribbon-tailed Astrapia, a bird-of-paradise, with a tail 36 inches (0.9 m) long. Exclusive of the tail, the bird itself is only about 12 inches (30 cm) long.
3923. Many of the eagles, hawks and falcons.
3924. The Pied-billed Grebe.

Answers

3925. A large, domed structure often exceeding a foot (0.3 m) in diameter. The nest typically consists of green and yellow mosses and fine grasses, with a side entrance.
3926. Yes.
3927. A *duck*.
3928. 1934. In 1990, the WBSJ had more than 30,000 members.
3929. The Tody Motmot and Blue-throated Motmot.
3930. Feign death or regurgitate foul-smelling food.
3931. Yes, at times.
3932. No.
3933. Deep scarlet or orange.
3934. The Western Grebe, that weighs as much as four pounds (1.8 kg), with a wingspan of up to 40 inches (102 cm).
3935. The Whooping Crane, probably because of its dependence on animal food.
3936. Wilson's and Collared Plover.
3937. Limberneck.
3938. No, it has a black cap.
3939. Yes.
3940. River *Phoenix*, who collapsed on the sidewalk outside a Sunset Strip nightclub, and died at the age of 23. He starred in the hit movie *Stand by Me,* and gained fame playing young Indiana in *Indiana Jones and the Last Crusade*.
3941. The Ash-throated Antwren (*Herpsilochmus parkeri*).
3942. Ice, and severe icing can be fatal.
3943. The African and Asian weavers.
3944. No.
3945. No. While some authors indicate that Sandhill Cranes weigh 8-14 pounds (3.6-6.4 kg), compared to an average weight of 8-10 pounds (3.6-4.5 kg) for Whooping Cranes, a Whooping Crane male may weigh as much as 17 pounds (7.7 kg).
3946. The shoes of a Whip-poor-will.
3947. Yes. In addition to feeding like typical herons, the bird forages with its bill partially submerged, thrusting forward and scooping with each step, using the wide bill to capture sedentary prey in the mud and leaf litter.
3948. *The Curlew.*
3949. Both sexes have black, fleshy knobs.
3950. The New Zealand Kea does not nest in trees. It is among the very few ground-nesting parrots.
3951. Yes, for short distances.
3952. Yes.
3953. Bonaparte's Gull. Herring Gulls also nest in trees, but not nearly to the same extent.
3954. At least 160 feet (49 m).
3955. Because the birds might gain the advantage of early-morning light and warmth of the rising sun.
3956. The Magpie Tanager.
3957. The aberrant shorebird is about 13 inches (33 cm) long. The strongly curved bill alone accounts for about three inches (7.6 cm) of its overall length.
3958. The Great Blue Heron.
3959. Around 1500 B.C.

Answers

3960. The name refers to their thick or swollen "knee," which is actually the heel joint.
3961. John Xantus (1825-1894), a very energetic bird collector, originally from Hungary.
3962. Eggshells are usually consumed or carried off and dropped some distance away, possibly reducing the risk of betraying the nest location to a predator. This may also eliminate injury to nestlings from the shell's sharp edges.
3963. 1886.
3964. In excess of 200 million metric tons, most of which consists of penguins.
3965. Yes.
3966. The neotropical King Vulture.
3967. A *raven stone*.
3968. Yes.
3969. Probably because wing-loading is high. If a sequential molt occurred, energetically inefficient flight would result.
3970. Its fine song, like the lark, is uttered in flight.
3971. The South American Coscoroba Swan.
3972. Two: Gambel's and Montezuma (Mearns') Quail.
3973. Honeyguides are obligate brood parasitic nesters and their chicks have a pair of sharp mandibular and maxillary hooks. The interlocking "teeth" are used to kill the host's chicks.
3974. The membrane is white or silvery. Most other birds have essentially transparent membranes.
3975. A small, oropendola-like icterid with a conspicuous frontal shield or casque, like an oropendola.
3976. Some Sooty Terns were still breeding at 30 years of age.
3977. The Water Thick-knee (Dikkop). The birds are not molested by the reptiles, and may serve as sentinels, giving early warning of approaching danger. The birds benefit because the crocodiles function as effective predator deterrents.
3978. Yes.
3979. Sea snakes.
3980. Males of these Old World cuckoos are smaller than the polyandrous females.
3981. No, the Bronzed Cowbird does not occur in Canada.
3982. The Arizona or Brown-backed Woodpecker.
3983. Leadbeater's or the Pink Cockatoo.
3984. About 800 species on some 500 varieties of birds.
3985. Two: Great White Pelican and the smaller Pink-backed Pelican.
3986. Geese. For centuries, Russian fighting geese were selectively bred for male aggression, and fights were staged in goose-pits. Combatants pecked and beat each other with their wings until blood was drawn or one was killed. Unlike cock-fighting that takes place year-round, the "sport" was restricted to the start of the breeding season. Goose-fighting was banned over a century ago.
3987. Yes, very much so.
3988. Black.
3989. The Yuma Clapper Rail.
3990. The White-fronted Goose, also known by the popular name of "laughing goose."
3991. The striking courser not only is not a plover, it no longer inhabits Egypt. Formerly abundant along the Nile in southern Sudan and Egypt, it was extirpated there early in the 20th century.

3992. Alistair Maclain. In 1969, the book was made into a movie starring Clint Eastwood and Richard Burton.
3993. Incubating males reportedly do not drink, eat, or even defecate.
3994. No, the name is derived from the conspicuous white or silvery ring encircling their eyes.
3995. The two Australian lyrebirds. The single egg weighs slightly over two ounces (57 gms).
3996. It lays variable-colored eggs.
3997. Some of the megapodes, such as the Orange-footed Scrubfowl.
3998. The albatross is reluctant to follow ships, thus is "shy."
3999. The Old World kestrels and the elanine kites, but the harriers probably use flapping flight more continuously.
4000. Daniel Boone.
4001. Twelve, the same as mammals.
4002. No.
4003. Yes.
4004. The California Least Bell's Vireo.
4005. No, but the buffalo (bison) is.
4006. No, all are freshwater birds.
4007. The Samoan Tooth-billed Pigeon.
4008. The three South American screamers. The uncinate process strengthens the ribs by means of a cross-strut.
4009. Bounty Islands and Antipodes Islands (New Zealand).
4010. Not really. It is primarily a shorebird of the high plains and semi-desert regions of the American west.
4011. Yes.
4012. No, avian echo-locating ability is inferior. Bats detect and intercept small flying insects via echo-location, but birds only require it to negotiate their way through presumably familiar caves, and to avoid collisions with other birds.
4013. Not generally, but Australian Welcome Swallows have been seen catching moths attracted to lights at night.
4014. Howard *Hawks*.
4015. Yes, but such behavior is not typical.
4016. The Rodrigues Solitaire, a turkey-sized, Dodo-like bird.
4017. The name is a corruption of buffalo head, in reference to the puffiness of its head.
4018. The African Black Heron. It stands motionless in the shallows holding its wings forward in an extended position so that the wings form a canopy over its head. The resultant patch of shade may constitute a false refuge into which fish are lured, and the shade may also cut down on the surface reflection or glare. The Tricolored Heron and Reddish Egret also canopy-feed, but not nearly as consistently as the Black Heron.
4019. Hong Kong is the major market where in excess of 60 percent of the nests are consumed. This amounts to about 100 tons valued at $25 million. A single bowl of the Edible-nest Swiftlet bird's-nest soup may cost as much as U.S. $50. The Chinese communities in North America are second only to Hong Kong in consumption, accounting for approximately 30 tons of nests annually.
4020. A flushing bar. It forces ground-nesting birds, mainly Ring-necked (Common) Pheasants, to flush off their nest before the mowing machine grinds them up.
4021. Generally the seventh and eighth.

Answers

4022. The Surf Scoter. American hunters formerly referred to all three scoters as sea coots.
4023. Only the Tuamotu Sandpiper. The White-winged Sandpiper formerly inhabited Tahiti and Moorea, but is now extinct, and only a single museum specimen exists.
4024. The Rose-breasted Cockatoo.
4025. When geese rapidly descend and alternately side-slip on set wings, first in one direction and then in the other, they are wiffling.
4026. The Mew Gull and the two kittiwakes.
4027. The drongos, and they may also take small birds.
4028. It has a long, straight crest resembling a spike, as opposed to the Indian Peafowl's crest of 24 feathers that have naked shafts, but are webbed at the tips.
4029. The Blood Pheasant.
4030. The Black-faced Cuckoo-shrike.
4031. The parrots, but hookbill is essentially an aviculturist name.
4032. A hide.
4033. Brailing renders a bird incapable of flight by binding the manus to the forearm so the treated wing cannot be unfolded.
4034. A densely concentrated flock of birds on the water.
4035. No, unlike most African antelope that do seek shade.
4036. The birds have very short, much compressed, and sharply curved, parrot-like bills.
4037. Yes.
4038. Yes.
4039. The Eskimo Curlew, that weighs up to a pound (0.45 kg).
4040. Yes, it is among the few birds that actually spears fish.
4041. The three kiwis.
4042. Yes.
4043. Yes, the birds tend to feed more actively at night.
4044. The aberrant South American Magellanic Plover.
4045. The two openbill storks.
4046. No.
4047. The endangered Yellow-eyed Penguin of New Zealand.
4048. The sapsuckers.
4049. *Ballooning.*
4050. No, the Lesser Sheathbill's sheath is far more prominent and upturned.
4051. No, the unique New Zealand plover is surprisingly tame.
4052. Its short, mellow whistle is typically repeated 6-7 times.
4053. Yes, some of the smaller bitterns are dimorphic, but dimorphism is not typical in most herons.
4054. No, but their feeding behavior is shrike-like. No true shrikes inhabit Australia.
4055. The Wrentit, because of its loud, easily recognizable song.
4056. Large pouch lice.
4057. Yes.
4058. Ichabod *Crane*.
4059. Yes.
4060. Kiwi droppings are strong smelling. The bird itself exudes an earthy, sweetly musty, acrid, mammal-like odor that may linger for days, if not weeks, making kiwis very vulnerable to dogs.

Answers

4061. No, rollers hardly ever walk, and resort to hopping about clumsily.

4062. Yes.

4063. The drastic decline of the specialized alcid is directly related to a 95 percent reduction of the original coastal old-growth conifer forests, which have been logged in the past century. Some of the older trees are 1,500 years old, and trees where murrelet nests have been found were 300 to 600 years old. An estimated 5,000, 2,000 and 2,000 murrelets remain in Washington, Oregon and California respectively, compared to 30,000 to 50,000 in British Columbia and 50,000 to 250,000 in Alaska. Many were formerly drowned in gill nets in California, and the birds are vulnerable to oil pollution.

4064. The great owlet moth, that ranges from southern U.S. to southern Brazil. A specimen collected in Panama had an 11.97-inch (30.4 cm) wingspan.

4065. The Greater Roadrunner.

4066. The Hooded Vulture.

4067. A flicking rate of about 13 times a second.

4068. The Northern Pintail, Green-winged and Blue-winged Teal.

4069. Yes. Even though their tails are used for support, the feathers are not stiffened like those of woodpeckers.

4070. Pablo Picasso.

4071. The Gyrfalcon, because in those times falconry was very much a class sport. Peregrine Falcons could be owned by high noblemen, Merlins by high noblewomen, and short-winged hawks, such as Northern Goshawks and European Sparrowhawks, by land owners and members of the clergy.

4072. A clutch.

4073. The Mountain Avocetbill and Fiery-tailed Awlbill.

4074. A spring of teal.

4075. Tit originated from the old Icelandic word *titr*, meaning something small, and mouse is a corruption of the old Anglo-Saxon word *mase*, which translates to a kind of bird.

4076. Yes, unlike typical swifts, treeswifts perch on the topmost branches of trees.

4077. Yes.

4078. No, pale-morph birds tend to migrate farther south.

4079. A headache.

4080. At least a half-mile (0.8 km), but sometimes more than 2.5 miles (4.0 km), depending on atmospheric conditions.

4081. The Loggerhead Shrike.

4082. The Avicultural Society, a British organization founded in 1894. Its journal, *Avicultural Magazine,* has been published continuously since that time.

4083. None.

4084. It was named by Linnaeus after the blind son of Venus because the little wings (feathers) on the grouse's neck were likened to Cupid's wings.

4085. The raptor feeds largely on young birds, and when attacking weaver broods, it often hangs head downwards with flapping wings. The double-jointedness of its legs undoubtedly enables the bird to more easily extract nestlings from tree holes and other enclosed nests.

4086. None.

4087. Aristotle.

4088. Yes, but it occurs inland as well.

4089. Swinhoe's and the Mikado Pheasant.

Answers

4090. Yes.
4091. *No one gets justice when the jury or judge is biased.*
4092. McKay's Bunting.
4093. Both countries have an eagle pictured on the flag.
4094. General Winfield Scott, an American army officer who served in the Mexican War.
4095. No, the raptors roost in groups of two or three, but up to 30 have been reported.
4096. Females line their nests with downy feathers plucked from their backs and thighs, a very unusual passerine trait.
4097. Chile and New Zealand.
4098. Woodcock chicks have two egg teeth, the second of which is located on the tip of the lower mandible.
4099. Yes, as a rule.
4100. The Crested Caracara.
4101. The last *documented* sighting occurred near Lake Okeechobee, Florida on April 2, 1904, when Frank M. Chapman observed two small flocks, for a total of 13 birds. Sightings were reported until 1938, none of which was regarded as valid.
4102. A dabbling duck, such as a Mallard or Northern Pintail.
4103. The large bitterns.
4104. An average of 13 days, with a range of 12-15 days.
4105. The Mountain Black-eye.
4106. The chicken has a feathered tarsus.
4107. Three have wispy, upstanding crests: Northern and Southern Lapwing and Black-headed Plover.
4108. St Louis, Missouri, in 1870.
4109. Probably the kingfishers or rollers.
4110. The Blue-footed Booby.
4111. The White (Fairy) Tern. Long claws are no doubt beneficial to chicks hatching on a tree branch.
4112. The nostrils of the endangered New Caledonia bird are covered by a flange that prevents dirt from entering when a Kagu is digging with its bill.
4113. No, the predatory birds use their bills.
4114. The bustards, especially the Great, Indian and Kori Bustard, when their gular sacs are greatly inflated.
4115. Approximately 850 species, of which 600 are permanent residents.
4116. A penguin.
4117. The Tufted Puffin. Common and Horned Puffin chicks have white ventral down.
4118. Lucy's Warbler. The 4-inch-long (10.5 cm) bird weighs only 0.3 ounces (5.67 gms).
4119. Tucked in among the feathers of their rump or back.
4120. Probably near Oaxaca, Mexico.
4121. The beaver or muskrat. Their lodges may be used as nest sites.
4122. A small colorful tanager.
4123. No, the primaries of only one wing should be cut.
4124. Insects.
4125. The cuckoos.
4126. Ross' Goose. Wavy is a corruption of the Indian word *wa-wa,* meaning wild goose.
4127. Gobbler spurs attain a length of 1.25 inches (3.2 cm).

Answers

4128. The Giant Canada Goose, with ganders attaining weights exceeding 20 pounds (9 kg).
4129. His cow would produce bloody milk.
4130. Stropping.
4131. Aldo Leopold.
4132. As quarry for falconry, a royal sport.
4133. A male. Females have dark eyes.
4134. No, their wings are well developed.
4135. No, it is about the size of a sparrow.
4136. Franklin's Gull.
4137. Yes.
4138. The Cuban race of the Sandhill Crane.
4139. No, they typically jerk their tails from side to side.
4140. All have selected the American Robin as the state bird.
4141. Red.
4142. The falcons may bob their heads.
4143. The function is unknown, but is probably associated with courtship.
4144. Florence *Nightingale* (1820-1910), the English pioneer of modern nursing.
4145. The Sooty Tern. It typically does not breed until 6-8 years of age, with some terns not breeding until their 10th year.
4146. The serrations or "teeth" reflect an adaptation to grazing lifestyles.
4147. Yes.
4148. The larvae of a moth devours the nests, weakening the structures, sometimes causing nests to fall to the ground.
4149. The golden-plover.
4150. Between 4.0 and 4.75 inches (10-12 cm).
4151. *The Journal of the Bombay Natural History Society,* published continuously since 1886.
4152. The broad black chest marking recalls the letter V.
4153. The Reddish Egret.
4154. Yes.
4155. Their 7th, 8th and 9th cervical vertebrae have highly modified hinges between them, facilitating a swift strike.
4156. Yes, from sea level up to 10,000 feet (3,048 m).
4157. 1884, Vienna, Austria.
4158. The Tawny Frogmouth and its close relatives build flimsy stick nests. Other species construct pad-like structures consisting of their own down with an external covering of spider-web and lichens.
4159. The alternation of bird songs which may occur when two or more territory holders are responding to one another.
4160. The Parasitic Jaeger.
4161. The name is believed to be of North American Iroquoian Indian origin.
4162. Thrush or sour crop.
4163. No.
4164. Yes, the birds often perch on the backs of cattle and other animals to feed on ectoparasites.
4165. The dull-colored female Riflebirds have longer bills than the males.
4166. Yes, as well as to India, Java and Japan.
4167. Their heads are often thrown back over the shoulders, even when they are calling in flight.

Answers

4168. Yes.
4169. Yes, but with difficulty.
4170. The Cactus Wren.
4171. No, but some domestic breeds, such as the Toulouse and African Goose, have dewlaps.
4172. The adult cassowary.
4173. The Black Sicklebill, that attains a length of about 43 inches (110 cm), and weighs just over 11 ounces (320 gms).
4174. Pigeon-toed.
4175. No, two drongos are gray to whitish.
4176. The Great Egret. It occurs throughout the New World, Europe, Asia, Africa and Australasia, but in recent years the Cattle Egret may have become even more widespread.
4177. Many red flowers lack fragrance, and these are preferred by hummingbirds, possibly because non-fragrant flowers are less likely to attract bees competing for nectar.
4178. Yes, in most instances.
4179. Retracted, so that the head is held back on its shoulders, with the long bill resting on the folded neck and chest. This position brings the bill closer to the center of gravity.
4180. The Least Grebe.
4181. Six: one in the New World, one Eurasian, and four others restricted to islands, mainly in the East Indies.
4182. Yes.
4183. Yes, but near the end of the rearing period the chick is so large that it may poke its head through the domed roof of the nest to be fed.
4184. No, bright colors are confined to the legs, feet, bill and lining of the mouth.
4185. The Bobwhite.
4186. Probably because the beak enables birds to remove ticks from other parts of the body.
4187. No.
4188. The beard may have a tactile function, possibly preventing the Bearded Vulture from pushing its rather long beak too far into the hollow of a large bone in quest of marrow.
4189. The grebe lays a two-egg clutch, but is unique in that it takes only a single chick away with it from the nest, abandoning the other egg.
4190. A knitting yarn spun from the wool of merino sheep.
4191. The neotropical Boat-billed Heron.
4192. The Siberian White Crane.
4193. Sisquac Condor Sanctuary, in 1937. Consisting of about 1,200 acres (486 hectares), it included a great cliff and waterfall where condors roosted and came to drink and bathe. The 35,000-acre (14,164 ha) Sespe Condor Sanctuary was established 10 years after Sisquoc, in 1947.
4194. No.
4195. The aquatic passerines have very dense, soft plumage.
4196. The range was never determined. The duck disappeared before its breeding grounds were discovered, but was so named because the type specimen was collected in Labrador.
4197. Blakiston's Fish-Owl.
4198. The glamorous sirens were creatures variously depicted as birds with women's heads, breasts and arms. They used their charms to lure mariners to their destruction. Modern sirens include the manatees, dugongs and sea cows, and these ungainly creatures are believed to be the mythical mermaids.

Answers

4199. The Cattle Egret.
4200. The Southern Lapwing.
4201. Yes, but the long necks are not evident because of their long, loose feathers.
4202. Yes. Food is offered from the parent's bill, but most shorebird chicks feed themselves.
4203. Yes.
4204. The Jackal Buzzard and Fox Kestrel.
4205. In some species, four to seven years, a remarkably long period of time for a passerine.
4206. The Greater Roadrunner. Its calls include dove-like coos, a whine, a whirr and even a growl.
4207. With his rebellious crew on the verge of mutiny, Columbus took new hope at the sight of land birds, and actually changed course to follow the birds to shore.
4208. White.
4209. The eggs or chicks may be moved.
4210. No, a swift's wing bones are more massive.
4211. The Tambourine Dove.
4212. The Gull-billed Tern.
4213. No, it is a bird of mountains and high plateaus. In China, it breeds as high as 15,000 feet (4,572 m). During the non-breeding season, it may occur at lower altitudes.
4214. The Cattle Egret and the Black-headed and Whistling Heron. Even these terrestrial herons depend on rainy season flushes of insects and vertebrates to obtain sufficient food for successful nesting.
4215. The Common Wood-Pigeon.
4216. No, it is among the most diurnal.
4217. All winter in the southern U.S. and Mexico.
4218. In kiwis and megapodes, the yolk accounts for an incredible 61 percent of overall egg volume. Upon hatching, a third of many kiwi chicks still consist of unabsorbed yolk. For the first several days after hatching, the plump chicks cannot stand because their legs are spread so wide by the big yolk sac enclosed in their belly.
4219. Only ten.
4220. No.
4221. It has a black face mask.
4222. No.
4223. The darker shades predominate by about two to one.
4224. Yes.
4225. The Dark-winged Trumpeter. The other two, the Gray-winged and Pale-winged Trumpeter, are widespread throughout the Amazon basin.
4226. Probably because it is white like a Snowy Owl.
4227. Not really. The interior of its huge bill consists of a network of cellular structures filled with air.
4228. Lewis and Clark, during their Rocky Mountain and Pacific expedition.
4229. Seventy-three days after separation from the tom.
4230. The pelican breeds at any time of the year if sufficient rains fall or water is available.
4231. Probably the Pink-footed Goose.
4232. About 269 species, of which 95 are regular breeders.
4233. No, all build domed roof nests with side entrances.
4234. Budgerigar is a corruption of the Australian aborigine name *betcheryygah*.

Answers

4235. The Purple Martin.
4236. The Canada Goose. Although normally a ground breeder, some races may use old Osprey, hawk, owl, Common Raven or heron nests.
4237. The Florida Keys.
4238. The name may have come from the color of nankeen, a cotton cloth from China.
4239. The British Storm-Petrel.
4240. Cave swiftlets may feed at night with the assistance of artificial illumination. The Waterfall Swift (Giant Swiftlet), a species with relatively large eyes, is also sometimes a crepuscular feeder.
4241. The bird is marked with two or more colors in distinct blotches, generally black and white, such as a magpie.
4242. The White-crowned Sparrow.
4243. Yes, I have observed such behavior on numerous occasions.
4244. No, it is endangered and is on the verge of extinction. As of mid-1993, possibly no more than 30 remained, of which only 12 were in the wild. In 1993, eight eggs were removed from three nesting pairs, seven of which were fertile, and six hatched.
4245. 1849.
4246. No, a saltatorial bird is a jumping species.
4247. 1916, when the Migratory Bird Treaty Act was passed.
4248. To keep warm, the birds may bunch together, often hanging vertically with their tails pointing straight down. At times, they sleep hanging upside down.
4249. Grace Darling Coues, the 18-year-old sister of Dr. Elliott Coues.
4250. Death.
4251. The great Khan maintained 200 Gyrfalcons, as well as 300 other raptors. Armies of 10,000 beaters were mobilized for his hawking expeditions.
4252. The Long-tailed Jaeger.
4253. The Goldcrest and Firecrest, with the 3.5-inch-long (8.9 cm) Goldcrest constructing the smallest nest of British birds.
4254. None. All species apparently dig burrows in earthen banks, slopes and even flat ground.
4255. At least 96 species; 32 definitely are established with 12 others probable.
4256. A kite or buzzard.
4257. No.
4258. Three: Wilson's Plover, Wilson's Phalarope and Baird's Sandpiper. Temminck's Stint is a rare Alaskan straggler.
4259. No, but the endemic raptor is close to it. Considered one of the rarest birds of prey in the world, it was until recently known only from 12 specimens collected between 1874 and 1930. In 1988 a documented sight record occurred, and in February of 1990 the decomposed remains of a serpent-eagle were discovered.
4260. They flutter their gular pouches.
4261. Harlequin means a variegated pattern, and both the drake Harlequin Duck and the beetle have such a pattern. *Harlequin* was a fictional character in an Italian comedy, a likeable buffoon who wore parti-colored pants.
4262. Blue, but Oilbird eyes reflect red when struck by a beam of light in the dark.
4263. Metallic purple.
4264. Yes.
4265. No, their nests are quite flimsy.

4266. Yes, but not naturally. Ostriches were introduced from southern Africa in the 1870's for feather-farming, and a small feral population still exists in South Australia.
4267. The 10-inch-long (25 cm) Altamira (Lichtenstein's) Oriole.
4268. The conspicuous holes are tell-tale signs of foraging kiwis.
4269. No, they are New World flycatchers.
4270. Tar and feathers.
4271. No, despite its 1 to 3-egg clutch, seldom is more than a single chick reared.
4272. The horn-like covering of the bill.
4273. No.
4274. No fewer than 36 species, with most of the remaining birds migrating to southern Greenland.
4275. The Common Poor-will (*Phalaenoptilus nuttallii*) was named after Thomas Nuttall in 1844, a distinguished 19th-century botanist and ornithologist.
4276. The Northern Flicker.
4277. Yes.
4278. An elite organization open only to birders who have seen 600 or more of the approximately 650 breeding species of North America in the contiguous 48 U.S. states.
4279. Not generally, but the birds may glide.
4280. Yes, as a rule.
4281. Bass Rock at the mouth of Firth of Forth, Scotland. Even the Northern Gannet's scientific name reflects the site: *Sula bassanus*.
4282. Albert's Lyrebird.
4283. The clapping noise results when its outstretched wings strike each other above the body.
4284. *Larus* is derived from the Greek for *ravenous seabird.*
4285. Yes, when feeding on some types of over-ripe, fermenting fruit, birds can become intoxicated, sometimes falling to the ground in a stupor.
4286. The Black Kite.
4287. Linnets.
4288. Yes.
4289. Yes, many species do.
4290. Yes, if Baja California is considered North America.
4291. A partridge from Africa and Asia.
4292. No, the hummingbird may be the most abundant and widespread Jamaican bird.
4293. No, but the Elegant Tern does.
4294. Crowtits, but the birds are neither crows nor tits.
4295. Yes, but there are numerous exceptions, especially in some galliform birds.
4296. The Australian Wedge-tailed Eagle.
4297. No, not so far as is known.
4298. The Galapagos tortoise. Mockingbirds and ground-finches also perch on the backs of land iguanas and more infrequently, on marine iguanas.
4299. The Stork-billed Kingfisher.
4300. No, the flycatchers are typically associated with water.
4301. Such behavior is typical of some raptors, especially during the breeding season.
4302. Yes.
4303. Yes, but only with difficulty. Their feet are not webbed and must be moved very rapidly because they present only slight resistance to the water.

Answers

4304. The tropical raptor is a cavity nester that favors hollow trees.
4305. Yes, its tongue is large and paper-like, but is probably not functional.
4306. John Wayne.
4307. Yes.
4308. The Hawaiian Goose (Nene) and Florida Duck.
4309. Their pompous, measured gait reminded British colonial troops of those seldom-loved officers. British colonials referred to the Greater Adjutant as the "beefsteak bird" because of the red color of its head, neck and throat pouch during the breeding season.
4310. The Pygmy Tit of mountainous western and central Java.
4311. Sandgrouse never perch in trees.
4312. Yes.
4313. The Black-and-white Warbler. It creeps over trunks and limbs of trees like a Brown Creeper or nuthatch.
4314. The African Akun Eagle-Owl *may* be entirely insectivorous. It has been observed hawking insects in forest clearings, and the stomach contents of eight specimens contained only insect remains.
4315. Frank M. Chapman of the American Museum of Natural History, in 1900.
4316. Unlike cuckoos, whydah chicks do not oust the host eggs or chicks, but rather are reared with the nestlings of the host.
4317. The Brown Booby and the dark-color morph Red-footed Booby.
4318. Yes, as a rule.
4319. No, they are thrushes.
4320. When polished, the wood has many eye-like markings.
4321. The Common Loon.
4322. Considered a delicacy, the birds are broiled to a crisp on skewers, and eaten bones and all.
4323. Eating the uncooked heart of a kingbird.
4324. The South American Oilbird.
4325. The Bufflehead duck.
4326. Apparently to conserve heat that would be lost through the bare skin of the bird.
4327. The egret was named after the sacred Egyptian bird, the Sacred Ibis.
4328. It is an inland, desert-breeding species that flies at night in order to reach the distant sea.
4329. The birds assume a nearly vertical body position.
4330. No, it was named after the male's persistent, loud, repeated *zit-zit* call.
4331. Yes.
4332. Up to 25 percent.
4333. Three: Long-billed, Short-billed and Asian Dowitcher.
4334. The Rose-breasted Grosbeak.
4335. Not more than 5,000 falcons, and probably less.
4336. Yes, both its upper and lower jaws contained vertical, peg-like, enamel-tipped teeth.
4337. To keep crows away because they fly off with golf balls.
4338. No, not considering its size and weight. By comparison, an Emu egg weighs only about 50 percent more, but its shell weighs four times as much.
4339. The large bone consisting of the last thoracic vertebra, fused with the lumbars, sacrels and anterior caudal vertebra. The synsacrum functions as the hip bone or pelvis.
4340. In 1975, on Dent Island, a tiny offshore, predator-free island adjacent to Campbell Island. The population is small, probably consisting of no more than 30-50 birds in the wild.

Answers

4341. No, unlike sandpipers, plovers depend more on visual clues.
4342. Kittlitz's Murrelets, particularly in the northern part of their range, may nest adjacent to streams used to transport their chicks to the coast.
4343. Yes.
4344. The Least Grebe has a 14-inch (36 cm) wingspan and weighs a maximum of five ounces (142 gms).
4345. Gauchos pursue rheas on horseback, and ensnare them with bolas. Bolas are three-pronged ropes with weights at the ends that entangle the legs and necks of running birds, causing them to fall.
4346. Swiftness of foot is apparently an adaptation for rushing ahead of incoming waves and then back-tracking for food items left stranded or exposed by the receding water.
4347. A jacana nest is typically rather flimsy, and incubating males push their wings beneath the eggs, separating them from the damp nest.
4348. No.
4349. Yes, its bill is slightly hooked.
4350. Yes.
4351. Yes, at the tip.
4352. Between 15 and 25 mph (24-40 km/hr).
4353. The Gray Jay. The glands presumably enable the bird to probe into bark crevices and cones to extract food items on a mucous-coated tongue, or to form compact food balls for storage in crevices in conifer trees.
4354. On the surface of the open sea.
4355. The Herring Gull.
4356. The toucans, because their long bills resemble bananas.
4357. No. Their nests are typically frail structures, and eggs are usually visible through the bottom of the nest.
4358. Up to 90 percent.
4359. The Killdeer.
4360. To protect introduced California Quail which the slow-flying raptors could not catch anyway. Over 200,000 harriers were sacrificed as a result.
4361. The Pacific White-fronted Geese.
4362. No.
4363. No. At about 100°F (37.7 ^{o}C), the kiwi has one of the lowest bird temperatures.
4364. At the time of the cormorant's discovery, Steller was marooned on the Komandorskiye Islands with Vitus Bering.
4365. Yes.
4366. Yes.
4367. Hibernated by burying themselves in the mud at the bottom of a stream.
4368. The giant Argus of Greek mythology, who had a hundred eyes and was ever-watchful. The name alludes to the chain of beautiful eye-spots on the long and wide secondary feathers. In common usage, argus-eyed means sharp-eyed.
4369. Probably the Black-throated Sparrow.
4370. No.
4371. No nest has ever been seen by an ornithologist.
4372. An Asian estrildid finch.
4373. The male Painted Bunting.

Answers

4374. Brush-tongues, used for collecting nectar and pollen.
4375. Courtship feeding.
4376. The American Kestrel.
4377. The Yellow-billed Loon.
4378. Yes.
4379. Yes.
4380. The Marabou Stork. South African natives refer to it as the king of the vultures, but the stork is not really dominant at a carcass, it only appears to be. Their bills are not very efficient at cutting meat, and they depend on morsels dropped by vultures, hyenas or jackals. Lumps of meat weighing up to 2.2 pounds (1 kg) can be swallowed whole.
4381. About 35.
4382. No, the gamebirds are notoriously difficult to keep because they are susceptible to a number of diseases. However, in recent years, an increasing number of aviculturists are maintaining and propagating grouse.
4383. No, both rockfowl (*Picathartes*) are bizarre passerines.
4384. The megapodes (incubator birds), because the birds test the temperature of their nesting mounds.
4385. It attacks with a rapid surprise approach when birds are feeding on the ground.
4386. Between 35 and 40 nests.
4387. The availability of suitable nesting cavities. The erection of bird houses has encouraged the Tree Swallow's steady increase, and it has become one of the commonest bird-box inhabitants in North America.
4388. Yes.
4389. 1898.
4390. Probably the Red-throated Loon.
4391. Six: Eastern, Western, Cassin's, Tropical, Gray and Thick-billed Kingbird.
4392. The doves.
4393. An egret.
4394. The aberrant Crab Plover.
4395. No. The warbler breeds across palmless southern Canada and winters from Florida southward, not in palms, but rather in open fields and scrublands.
4396. No, their diet consists almost entirely of animal matter.
4397. By panting.
4398. The kink is caused by the lengthened sixth cervical vertebra. It articulates at an angle with the fifth cervical, providing an increased surface for muscle attachments.
4399. Yes.
4400. Their claws.
4401. No, the birds are seldom observed in flight.
4402. The whistling-ducks, especially the White-faced Whistling-Duck.
4403. Dark sooty-brown.
4404. Yes, but only the entrance hole and the single inner nesting chamber.
4405. Gray, but females are brown.
4406. Yes.
4407. The birds prey on insects and other animals flushed by the advancing fire.
4408. Yes.
4409. Yes, but there are numerous exceptions.

Answers

4410. No. Unlike their near relatives, the falcons, caracaras have long, but broad, rounded wings.
4411. The sapsuckers.
4412. White or whitish.
4413. In October, 1968, a Wilson's Phalarope carcass was found on Alexander Island (71°S), Antarctica.
4414. The insular sparrow is threatened. The population fluctuates between 50 and 400 birds, but perhaps has never been more numerous. The vulnerable birds have specialized nesting habitat requirements, and introduced goats and the presence of the U.S. Navy on the island are potential threats.
4415. The shorebirds had dangerous levels of DDT contamination.
4416. Yes, slightly earlier.
4417. Yes, even open mailboxes are used.
4418. No one knows. Presumably its bill would be just as functional if it curved to the left.
4419. Peregrine Falcons.
4420. Yes.
4421. Yes, Sunbitterns lack a juvenile plumage stage.
4422. Owl pellets may last for years.
4423. Nearly 20, some with regularity.
4424. Just over three feet (0.9 m).
4425. Vireos and shrike-vireos. Peppershrikes have much heavier, more strongly-hooked bills than those of vireos, but both groups are in the same family (Vireonidae).
4426. The mousebirds.
4427. No, their eggs are elongate.
4428. Yes.
4429. The American Ruddy Duck. The small feral population resulted from escaped birds.
4430. Sixteen: Ostrich, Emu, Boat-billed Heron, Shoebill, Hamerkop, Osprey, Secretary-bird, Hoatzin, Plains-wanderer, Limpkin, Kagu, Sunbittern, Crab Plover, Oilbird, Cuckoo-Roller and Common Hoopoe. The most recent revision also contains 16 families, but the Boat-billed Heron has been included with the herons, the Magellanic Plover and the Ibisbill have been elevated to family status, and the Common Hoopoe has been split into two species.
4431. No, but loons do.
4432. A drumming Ruffed Grouse.
4433. The Ornate Hawk-Eagle.
4434. About 15 mph (24 km/hr).
4435. As of 1993, the Mallard, Canvasback and Northern Pintail have been depicted five times.
4436. A dove.
4437. Blood-red.
4438. The Black-legged Kittiwake.
4439. In the few species examined, males maintained a relatively constant temperature, averaging about 91-92°F (32.7-33.3°C).
4440. No. Although often stated, such behavior has never been documented and has a basis only in aborigine legend.
4441. No, chickens were probably introduced into South America in pre-Columbian times by Polynesian seafarers.

Answers

4442. Loosely translated, raptor means *snatcher*.
4443. No. More than 70 species of parrots occur in Brazil, but despite being called the continent of parrots, only 52 inhabit Australia.
4444. Despite its common name, the tern was not reported from the Aleutian Islands until 1962, well after its discovery on Kodiak Island in 1868.
4445. The acorns are shelled.
4446. Yes, it is very tame and can be closely approached.
4447. The murrelets. Newly-hatched chicks are among the heaviest alcids relative to adult weight.
4448. No, sunbirds are monogamous.
4449. Yes.
4450. Not really. Its flesh was fishy and shot birds often rotted before they could be sold.
4451. No. The toes of Wilson's Phalarope are evenly fringed, not lobed at each joint as in the other two phalaropes.
4452. No. Along with the waxbill finches, goldfinches are among the very few higher passerines that do not remove nestling fecal matter from the nest.
4453. Powdered bittern bills supposedly induced sleep.
4454. Philadelphia, where the type specimen was collected *(Larus philadelphia).*
4455. A type of coal with iridescent colors.
4456. Pirouetting.
4457. The barks, whines and yips of a young puppy.
4458. The goose-beaked whale.
4459. The Goliath Heron.
4460. A plump of waterfowl.
4461. The Eskimo Curlew.
4462. The woodpeckers.
4463. Small bats that potoos catch on sallies from a lookout perch, in the manner of flycatchers.
4464. Adelie, Chinstrap and Gentoo Penguins all have very long tails that sweep behind them like a brush as they walk.
4465. The Lesser Goldfinch.
4466. The goshawk hunts by walking, rather like a miniature Secretary-bird.
4467. The bittern flies with dangling legs.
4468. Yes.
4469. Yes.
4470. Their long tails are distinctly wedge-shaped, consisting of 12-16 pointed feathers.
4471. The advantages of a laterally-deflected bill are not fully understood, but the bill is apparently perfectly designed for extracting prey, such as mayfly larvae and fish eggs, clinging to the underside of stones and rocks. Such foraging sites may not be accessible to other river-bed plovers that have thicker-tipped, straight bills.
4472. No, toucans are among the noisiest of jungle birds.
4473. No, nests of these shy, elusive birds are exceedingly difficult to find.
4474. No.
4475. A haystack.
4476. Generally between 18 and 25 days of age. The partially down-covered chicks are accompanied by the male parent and are not able to fly until 39-46 days of age.
4477. Yes.

Answers

4478. The incubation period is unknown.
4479. No, Lesser Yellowlegs are far more gregarious.
4480. The name is a corruption of the word "nuthack." Several species of Old World nuthatches hew acorns and hazelnuts open by wedging them into crevices and hacking them with the bill.
4481. No, but Parasitic and Pomarine Jaegers do.
4482. Yes, the fat layer may equal up to 50 percent of its overall weight.
4483. Roosts were dynamited, slaughtering thousands of crows.
4484. The Black-crowned Night-Heron, because of its call, often uttered in flight.
4485. Yes, nests may be placed on top of muskrat or beaver lodges.
4486. Aphids.
4487. The Roseate Tern.
4488. Generally not until 2-5 years of age, and possibly as long as 6-7 years.
4489. Ants.
4490. Grouse.
4491. The Calliope Hummingbird may make use of its old nest by either refurbishing it with a fresh lining or building a completely new structure on top.
4492. No, lapwings are less dependent on water.
4493. Eggs are carried in the mouth, whereas chicks may be carried between the wing and thigh.
4494. No, they are simply called penguin chicks.
4495. The two Australian scrub-birds.
4496. Yes, some of the smaller species do.
4497. Yes.
4498. Africa, where over half of the 72 species occur.
4499. The tropical seedeaters perch on a tuft of grass or weed, vault into the air and deliver a weak ecstatic song or call, and then flutter down to the perch again. A whole field of singing grassquits has been likened to popcorn in a popper.
4500. Yes, if the roofs are flat.
4501. Thick-knees do not have a hind toe.
4502. No, their tiny nests of bark and feathers are so minuscule that the single egg virtually fills the nest.
4503. Yes.
4504. A rabbit burrow.
4505. Yes.
4506. The bittern.
4507. Becard is derived from the French *bec*, meaning "beak," and *becarde* is French for "shrike," alluding to the shrike-like beak of the becards.
4508. The naturalist John Gould.
4509. The King Penguin. Its bill length may exceed 5.5 inches (14 cm).
4510. The stiff tail is used as a brace. The parrots feed on fungi gleaned from forest tree trunk and limb bark by creeping about in nuthatch fashion.
4511. No, it is only able to flutter a few hundred feet (61 m).
4512. The Cattle Egret.
4513. Probably from birds escaped from falconers.
4514. Alexander Wilson collected the type specimen in a magnolia tree in Mississippi.
4515. No, their eggs are practically round.

Answers

4516. The screech-owl.
4517. No, the right lobe is larger.
4518. No, prominent bristles shield the nostrils.
4519. Yes.
4520. About 66 percent.
4521. Near Minneapolis, Minnesota, in 1895.
4522. No, they are primarily frugivorous.
4523. *Four Feathers.*
4524. Nearly 11,000 feet (3,353 m).
4525. No, but swallows are.
4526. No.
4527. No, the dumpy, neotropical puffbirds are rather inactive.
4528. The Red-eyed Vireo, that ranges from Canada to Argentina.
4529. Yes.
4530. Small downy young may be trampled to death.
4531. Aristotle.
4532. No.
4533. The Ladder-backed Woodpecker.
4534. The Red-throated Loon.
4535. Yes, but only slightly so.
4536. Yes.
4537. No, both species fly weakly, and apparently reluctantly.
4538. Acorns are sawed open by means of a sharp keel on the roof of the grackle's mouth.
4539. Yes, the International Crane Foundation. The last species, the Wattled Crane, was hatched in 1993.
4540. Yes, courting males perform a distinctive *sky dance*.
4541. *Turkey day*.
4542. The Ring-billed Gull.
4543. It consists of a single egg, a most unusual clutch size for a passerine.
4544. Yes, its throat usually shows flecks of red, often forming a patch of color.
4545. A person with very sharp vision.
4546. Some often-used nests weigh over 1,000 pounds (454 kg).
4547. Probably never. Its single egg is generally laid atop a broken stump.
4548. Yes, especially the female Great Spotted Kiwi.
4549. Yes, but not efficiently.
4550. About 183 species, of which 39 just barely reach the Arctic. Definitions of the Arctic differ, but it is generally considered a region of permafrost where the mean temperature of the warmest month (July) does not exceed 50°F (10°C).
4551. No.
4552. North Island Kokakos have brilliant ultramarine-blue wattles, but the probably-extinct South Island birds had rich orange, and sometimes almost vermilion, wattles. Nestling Kokakos have pink wattles.
4553. A bullfinch. The name probably came from bull fence.
4554. The Falcon.
4555. When Blue-eyed (Imperial) Shags were sighted, Shackleton was aware that his long-sought destination was close at hand.

Answers

4556. Only one. The Sharpbill is in a monotypic family and very little is known about it.
4557. The Kaka.
4558. From their habit of harrying prey.
4559. Swan skins were fashioned into powder puffs.
4560. Craveri's Murrelet. No accurate counts have been made, but possibly only 6,000-10,000 exist.
4561. The African Marabou Stork and the two Asian adjutant storks.
4562. No, the little African birds have a direct, steady flight that is as straight as an arrow.
4563. The Apostlebird. The jay-sized bird is dark gray, and it jumps in long leaps over the ground and from branch to branch.
4564. The grebes.
4565. Unlike other snake-eagles, it repeatedly hovers at some height.
4566. No. A number of cotingas are quite small, ranging in size from 3.5 inches (8.9 cm) up to 18 inches (45.7 cm) in length.
4567. The hummingbirds.
4568. No, they have relatively weak bills and depend on other species for nest cavities.
4569. No, even during migration, bustards seldom fly higher than 200-300 feet (61-91 m).
4570. Three: Lear's, Spix's and Buffon's (Great Green) Macaw.
4571. The owl.
4572. Yes, generally so. This possibly prevents chicks from drowning in the often very wet nest cup.
4573. Guineafowl tend to run rather than fly, making them undesirable as gamebirds.
4574. Yes.
4575. The Brent goose. The down in its well-lined nest nearly equals the quality of Common Eider down.
4576. Yes. Ground-nesting Ospreys are particularly common in Baja California, Mexico.
4577. None.
4578. Yes. Cranes glide, and even soar, especially during migration when the birds ride thermals upward and then glide down in the desired direction.
4579. The Limpkin, because the bird was very good eating.
4580. No, it is more numerous north of the tree line.
4581. No. The New Zealand shorebirds circle in a clock-wise direction, presumably because their bill curves to the right.
4582. Usually, but not always. Chicks come to the entrance to be fed, and at times may move to a nearby site.
4583. The Laysan and Black-footed Albatross.
4584. The White-naped Swift of Mexico weighs as much as 7.5 ounces (213 gms).
4585. Red.
4586. The Black-legged and Red-legged Kittiwake.
4587. No, but when they do fly, the birds may glide for considerable distances.
4588. Between 24 and 25 days, an unusually long period of time for a small passerine.
4589. Mousebird chicks hatch naked, thus lack down.
4590. A herd, a term also applied to a group of swans or cranes.
4591. The Rusty Blackbird nests across North America as far north as the tree line.
4592. No, its claw is nearly straight.
4593. Both ducks feed more extensively on soft plant foods, thus are less likely to ingest lead

shot picked up from the bottom of a marsh.

4594. Probably from the Swedish word *leka*, meaning to "gather" or "assemble:" a gathering.
4595. Yes, in some instances.
4596. A bittern.
4597. Yes.
4598. The 10 species of spiderhunters, Southeast Asian sunbirds devoid of brilliant colors. Spiderhunters also differ from typical sunbirds in being larger and coarser in build, with long, heavy, curved bills.
4599. *Campephilus* is Greek for "caterpillar-loving," but the woodpecker favors wood-boring insects.
4600. The darters and the Anhinga
4601. Yes.
4602. Most are, but the Old World Corn Bunting is a notable exception.
4603. No. They most generally nest on the ground or near to it, but some pittas nest as high as 30 feet (9 m) above the ground.
4604. Yes.
4605. The Hawaiian Audubon Society.
4606. Only one, Buller's Albatross. The Short-tailed Albatross is sometimes called Steller's Albatross. A race of the White-capped Albatross is commonly known as Salvin's Albatross, and an alternate name for the Yellow-nosed Albatross is Carter's Albatross.
4607. No, barbet wings are short and rounded, and their weak flight is seldom sustained.
4608. Observers formerly believed that its preferred food was the fruit of the hawthorn.
4609. The Western Grebe, and probably Clark's Grebe.
4610. Not quite. An immature Bald Eagle had 7,182 feathers, compared to 13,913 for a coot. The more densely-feathered coot reflects the greater need for insulation that is typical of aquatic birds.
4611. One thousand years.
4612. Bee-keepers believe that kingbirds consume too many bees.
4613. No. Stilt nests are often bulky affairs, whereas avocet nests more typically consist of rough scrapes in the ground, occasionally lined with grass or pebbles. If the water rises, avocets may add material, and their nests may reach a foot (30 cm) or more in height.
4614. The pochards, especially the Greater and Lesser Scaup.
4615. No, but the vast majority do. Most cotingas hatch chicks heavily covered with down, but the chicks are still helpless.
4616. The Summer Tanager.
4617. The dogs were formerly used in hunting woodcocks.
4618. Yes, for its diminutive size.
4619. Fourteen, but the Old World broadbills have 15 neck vertebrae.
4620. No.
4621. The coursers and pratincoles.
4622. Yes, although flocking is not typical of most northern flycatchers.
4623. The Eurasian Smew.
4624. Both upstroke and downstroke provide power.
4625. Yes, a few species are surprisingly capable swimmers.
4626. Its feet are not used because of the danger of being bitten. A shrike harasses a mouse until an opening is seen, and then seizes the rodent by the neck with its bill.

4627. Captain Cook in 1778 released three goats so his men would have fresh meat on their return voyage.
4628. Bright red.
4629. None.
4630. Prince Albert, Queen Victoria's consort.
4631. Yes, the birds apparently never gather in flocks.
4632. No, its flight is fast, but its wingbeats are slow.
4633. No, pygmy-parrots are confined to New Guinea and adjacent islands.
4634. Yes.
4635. Yes, if dust-bathing can be considered bathing.
4636. The Bananaquit, but only casually in southeastern Florida.
4637. Yes. Wild Turkeys were introduced as early as 1815, with birds imported from Chile.
4638. Head twisting is an aggressive behavior intended to warn intruders to stay away. Many people believe this reflects nearsightedness, but this assumption is incorrect.
4639. Yes.
4640. Yes.
4641. Yes, the birds feed primarily in trees, seldom venturing to the ground.
4642. No. Their pleasant, whistled trill is such that many people consider it to be among the loveliest of nocturnal sounds.
4643. It is in the parrotbill family, a group of mostly Asian passerines.
4644. The Dickcissel, because it has a yellow breast with a black bib, and it often associates with meadowlarks. The Lark Sparrow is also known as a "little meadowlark."
4645. Yes, especially if approached on foot. The terrestrial birds are not wary of automobiles or people on horseback, a trait that has served them poorly. Bustards are excellent eating, and have been subjected to intense hunting pressure in some areas.
4646. The Golden Eagle.
4647. Yes.
4648. All have a long, straight hind claw.
4649. Black. Good light is required for the iridescent plumage to sparkle.
4650. No, but Black-legged Kittiwakes commonly do.
4651. Three: Red-cockaded, Red-headed and White-headed Woodpecker. Technically, the Ivory-billed Woodpecker should also be included because its bill is part of the head.
4652. Yes.
4653. Despite their tiny size, hummingbirds are surprisingly pugnacious and ferocious, and if only one feeder is available, it may be defended by a single hummingbird who will chase away any other hummers that appear. Hummingbirds have been known to engage hawks and eagles in aerial combat, allegedly attacking the eyes.
4654. *Aquila* is Latin for eagle, thus an aquiline nose is an eagle-like nose. People with very high-bridged noses with an abrupt downward curve are said to have aquiline noses.
4655. Yes.
4656. Yes, Johann Reinhold Forster (1729-1798) accompanied Cook aboard the *H.M.S. Resolution* on his second voyage around the world. He was the first to describe the Eskimo Curlew, and the Emperor Penguin is named after him (*Aptenodytes forsteri*).
4657. Native trappers ate captured females because they believed the non-colorful birds could not be sold. As soon as trappers learned the brown females were just as valuable as the brightly-colored males, females became available.

4658. No.
4659. In the 1760's, a number of well-to-do Englishmen formed an organization known as the Macaroni Club. Its members were described as exquisite fops, indolent, and fond of gambling, drinking and dueling, and were noted for a distinctive hair style known as a macaroni coiffure. When English sailors first saw the Macaroni Penguin, they were instantly reminded of the dandies and their peculiar hair style.
4660. At least 667.
4661. The nutcrackers.
4662. Between 16 and 20 million.
4663. The Marbled and Kittlitz's Murrelet.
4664. Yes, bats may be taken by American Kestrels.
4665. Yes.
4666. Yellowish.
4667. About 47 days, a remarkably long time for a passerine.
4668. The cockatoos, but in other parrots, the female alone incubates.
4669. The middle toe is not longer. All four toes are of equal length.
4670. Goldfinches require thistledown for their nests, as well as new thistledown seeds to feed their chicks, and this is not available earlier in the year.
4671. Darwin's Rhea.
4672. Yes, at times, especially Adelie and Chinstrap Penguins breeding along the Antarctic Peninsula.
4673. The booby prize.
4674. Yes, as well as storm-petrels and ducks, especially when fish are in short supply.
4675. Two skins were presented to the King of Spain following Magellan's circumnavigation voyage in 1522.
4676. The Crested Tern.
4677. The todies and the Palmchat.
4678. The grouse is unique in migrating *up* to high altitudes during the winter and *down* to lower elevations in the spring.
4679. By carrier pigeon.
4680. It captures insects on the wing like a swallow.
4681. Its large globular nest of thorny twigs is so impregnable that concealment is unnecessary.
4682. A falconer's goshawk was not entitled to wear royal-purple jesses until it had killed what was considered to be the most noble of all game, the Red-crowned (Japanese) Crane.
4683. Four: White-breasted, Red-breasted, Brown-headed and Pygmy Nuthatch.
4684. In the beam of a light, their eyes gleam like red-hot coals. Many nightjars sit on roads at night where their glowing eyes are illuminated by car headlights.
4685. Only the two African crowned-cranes. Their blue eggs are unmarked, but the eggs of other cranes are dull white to brown, and spotted with darker shades of brown.
4686. Florida, Texas and Arizona.
4687. The Greater Scaup.
4688. The owl is only two feet (60 cm) tall, but its eyes are nearly as large as the eyes of a full grown human male.
4689. No, bustard nests are shallow, crude, poorly-lined scrapes in the ground.
4690. Calliope was the Greek goddess of eloquence, a name meaning *beautiful voice.* It is not clear why a bird lacking a melodious voice was dedicated to this particular mythical muse.

Answers

4691. The most recent taxonomic revision places loons between the alcids and sandgrouse.
4692. The three endemic mesites (roatelos) of Madagascar with five pairs of powder-down patches.
4693. Yes.
4694. The two oxpeckers, but they copulate on the ground as well.
4695. The male Summer Tanager.
4696. Yes, some do.
4697. No one knows. The loss of suitable habitat because of drainage and the cutting of river-bottom forests is possibly partially responsible.
4698. While not typical, the storks reportedly occasionally feed on floating carrion.
4699. In most gallinaceous birds, a pair of outpockets of the digestive tract is located at the junction of the small and large intestines known as the caeca. The caeca possibly provides a place for bacterial breakdown of cellulose and similar fibrous materials that cannot be handled by the bird's digestive enzymes.
4700. Yes.
4701. The Bateleur.
4702. Yes, but only if the fish are fresh.
4703. Guineafowl, hence the origin of the family name Numididae.
4704. A Least Bittern.
4705. The Acorn Woodpecker.
4706. Up to 34 percent.
4707. Yes and no. The duck-billed platypus incubates its eggs externally, but the five echidnas (spiny anteaters) incubate their eggs in a pouch.
4708. Not particularly.
4709. Yes.
4710. The Labrador Duck. Its bill was relatively soft around the edges, perhaps indicating a diet of soft-shelled invertebrates, as well as a specialized foraging technique. The duck was possibly ultimately a victim of over-specialization, because very specialized foraging techniques and food preferences might have made it unable to cope with changes in the molluscan fauna that could have resulted from an increased coastal human population and associated pollution.
4711. No. It was named in honor of Mrs. Virginia Anderson, the wife of Dr. W. W. Anderson (1834-1901), an assistant army surgeon who discovered the warbler in New Mexico in 1858.
4712. The Great Horned Owl, because it readily preys on captive birds.
4713. The Ruffed Grouse. In 1970, about 3,700,000 were shot.
4714. No, it has a white rump.
4715. The Tupi Guarani Indians of Brazil's Amazon region. Tanager, aracari, toucan, jacamar, jabiru, ani, jacana, anhinga, macaw, hoatzin, tinamou, seriema and cotinga are all names that originated with the Tupis.
4716. Yes.
4717. No, it is limited to the forested highlands.
4718. No, little hornbills are quite bold and are often found near human dwellings.
4719. Yes, but this is unusual in raptors.
4720. Yes, their calls are loud, clear, shrill, far-carrying, repeated whistles.
4721. The family name loosely translates to *caterpillar-eater*.

Answers

4722. The white beluga whale of the Arctic, because of its high-pitched, bird-like vocalizations.
4723. To compensate for its long incubation period, kiwi eggshells are less porous to reduce water vapor loss, preventing the embryo from withering through dehydration.
4724. Yes.
4725. Yes.
4726. These are the only species in which neither sex has any red in the plumage.
4727. Yes, and juveniles may participate as well.
4728. No.
4729. No. Even when only a few days old, chicks leave the nest and creep about the nearby branches, crawling with their wings, beak and feet. They return to the nest to be brooded at night.
4730. A remedy for eye diseases.
4731. Adults have black-capped heads. The misleading name originated with the white-headed juveniles.
4732. The Gadwall.
4733. Yes, the birds are very tasty and succulent.
4734. On the left hand for right-handed falconers.
4735. Yes, their bills are generally significantly shorter than those of typical rails.
4736. Until recently, yes. However, the extent of rainforest destruction is such that species are being lost before they have been discovered. Thus South America or Southeast Asia perhaps have worse records.
4737. No, it is yellow.
4738. Yes.
4739. The growing feathers retain their waxy sheaths until just before fledging, giving the chicks a strange bristly appearance.
4740. Carl Linnaeus, who described and named 133 American birds.
4741. With upraised wings.
4742. The name originated from their shrieking, shrill calls.
4743. The Sandhill Crane, with a population exceeding a half-million birds.
4744. Aside from its small size, the petite goose has a *prominent* white neck-band.
4745. The Lesser Yellowlegs.
4746. Millions were shipped to Europe and the U.S. to be fashioned into pins, brooches and adornments for ladies' hats. As many as 400,000 hummingbird skins were imported in a single year by one London dealer alone.
4747. *To feather one's nest.*
4748. No, its name reflects the bird's alarm call that sounds like two stones being knocked together.
4749. Yes. An Alaskan Peregrine Falcon was observed wading into a cold river where it caught a fish. Such behavior is atypical, and 95 percent of a Peregrine's prey consists of birds. A Peregrine was also seen to catch and eat a crab off Toroshima Island.
4750. The megapodes (incubator birds) that depend on natural fermentation generating heat, and the obligate brood parasitic-nesting birds that depend on the heat of others for incubation and brooding.
4751. The draining of marshes. This destroyed the necessary habitat for the apple snail, the kite's only food. In former times, many were also shot for "sport."
4752. None.

Answers

4753. No, condors were regarded as predators and shot on sight.
4754. Yes. Although primarily nocturnal, nighthawks are also surprisingly active during the day.
4755. Yes.
4756. The Hooded Merganser, presumably because it preys on tadpoles.
4757. Land is near and a large seabird rookery is located nearby. The ammonia odor is from the copious amounts of seabird guano.
4758. In 1809, Lewis died under mysterious circumstances while traveling to Washington. Some historians believe his death was a suicide, whereas others suggest he was murdered by the two servants who accompanied him.
4759. Only seven, all of which are parakeets.
4760. Yes, the birds have a shrill, long-drawn, eerie scream.
4761. About 1,450.
4762. Four: Chuck-will's-widow, Whip-poor-will, Common Poorwill and Pauraque.
4763. Deer mice. Western Gulls also take chicks, and Barn Owls are major predators of adults.
4764. Yes.
4765. Also known as the Bare-eyed Pigeon or Red-eyed Squatter, it has an extensive area of bare scarlet or chrome-yellow skin around and in front of its white eye.
4766. Reportedly up to four pounds (1.8 kg), and possibly more.
4767. Yes.
4768. The greatly enlarged fleshy tongue acts as a piston during filter-feeding, pumping at the rate of 5-6 times a second. In Lesser Flamingos, the pumping rate is about 20 times a second, but lesser amounts of water are taken in.
4769. As harbingers of winds, mists and fog, and the birds were also considered to be the spirits of seamen swept overboard in gales. The killing of an albatross is believed to bring bad luck.
4770. Spotted Sandpipers are polyandrous.
4771. Inflate their necks.
4772. No, the birds feed on the ground.
4773. Yes, very much so.
4774. None.
4775. Yes, but such behavior is not typical.
4776. All sandgrouse have legs feathered to the feet, but at least two Asiatic species (Pallas' and Tibetan) have feathered toes as well.
4777. A doormat of animal dung.
4778. The three plantcutters. Until recently, the birds were in a separate family, but they are currently included with the cotingas.
4779. Pallas' Cormorant. It was named after Peter S. Pallas, a Russian biologist and zoogeographer of Siberia, who described the species.
4780. Males squat down, waving their extended wings, while swinging their head and neck from side to side.
4781. The 1880's.
4782. Yes.
4783. Six to eight feet (1.8-2.4 m).
4784. The beautiful pheasants were named for Pan, the Greek god of pastures and woods. Pan is usually represented with the horns and legs of a goat, and the name tragopan is a reference to the two fleshy, erectile horns on the heads of adult cocks.

Answers

4785. Yes, unfledged chicks probably went to sea at about 2-3 weeks of age.
4786. Probably the Romans. The pheasants were raised in game preserves by numerous rulers as early as the 11th century.
4787. She was frozen in a 300-pound (136 kg) block of ice and shipped from Cincinnati to Washington D.C., where she was mounted. *Martha* is still on exhibit at the Smithsonian Institution.
4788. M. T. Brunnich (1737-1827), a Danish zoologist and director of the Copenhagen Natural History Museum.
4789. Argentina and Chile.
4790. Yes and no. The normal skua clutch size is two eggs, and two chicks often hatch and both may be reared, but if food is scarce, only a single chick is reared, or none at all.
4791. Apparently from the Danish word *sidsken*, or the Swedish word *siska* meaning "chirper."
4792. Queleas breed 3-4 times a year, often in regions inaccessible to control units, and their clutch of three eggs hatches in only 11-12 days.
4793. The Rock Ptarmigan.
4794. Only three: White, Black and Abdim's Stork.
4795. Two: Blue Grouse and White-tailed Ptarmigan.
4796. Yes, and many are surprisingly swift.
4797. No, but Green Sandpiper eggs have been found in old squirrel nests.
4798. Yes.
4799. No, the birds are highly vocal, chattering and whistling almost incessantly.
4800. White.
4801. Yes.
4802. Benson's Quail.
4803. The cuckoo fish.
4804. Rarely.
4805. Generally not. While the Red-legged Seriema occasionally nests on the ground, more typically it nests in trees up to 10 feet (3 m) above the ground. The Black-legged Seriema generally nests in low bushes or trees.
4806. No, their bills are unusually soft basally, but are hard and strongly hooked at the tip.
4807. Texas.
4808. Yes, but their nests are located in a tree cavity.
4809. Yes.
4810. The huge, powerful birds are temperamental and unpredictable and apparently tire easily. When tired, they simply squat down and quit. However, wild Ostriches are said to be able to run up to 30 mph (48 km/hr) for 15-20 minutes without exhibiting signs of fatigue.
4811. The South American Guira Cuckoo and the three typical black anis.
4812. 1940.
4813. Up to 3.3 feet (1.0 m) of loose sand.
4814. Yes.
4815. No, it has a flat flight profile.
4816. Yes.
4817. Yes, some jacanas are black, rather than cinnamon-red.
4818. Females lack a white band across the tail and have buffy, rather than white, throats.
4819. Up to 33 percent.
4820. Yes, a long, loose crest of stiff-shafted feathers.

Answers

4821. About 15 inches (0.4 m), compared to 5.5 feet (1.5 m) in a Snowy Owl.
4822. The flushed gamebirds may explode up and collide with branches, tree-trunks or other obstacles, possibly because they are ineffective at steering.
4823. Yes, Cuban Flamingos were introduced into Kauai in 1929, but the introduction failed and the three birds survived only for about a year.
4824. A crane.
4825. Blue.
4826. The Black-and-yellow Silky-flycatcher.
4827. No, their wings are positioned well to the rear of the center of their body.
4828. Yes, but just barely in southwestern Canada.
4829. The vireos. *Vireo* is from the Latin, meaning "to be green."
4830. The White-eyed River-Martin, but it has not been reliably recorded since 1980, although one was reportedly trapped by a native in 1986.
4831. The pipits.
4832. Yes. A Black Oystercatcher weighs up to 30 ounces (0.86 kg), whereas an American Oystercatcher weighs 22 ounces (0.6 kg).
4833. Hornbill ivory, a substance the ancient Chinese valued more than jade or ivory.
4834. Yes.
4835. A squirrel nest.
4836. Yes, during the breeding season the warbler is almost invariably found in with pines.
4837. Egret plumes. Ospreys have no plumes, and the raptors were not sought by plume-hunters.
4838. Retort-shaped. It resembles a bottle lying on its side.
4839. No, they roost in bushes or low trees, but the Bobwhite of the east typically roosts on the ground.
4840. Its hard, dry droppings cannot be squirted.
4841. None.
4842. King Solomon, in about 1,000 B.C.
4843. No.
4844. The Northern Shrike.
4845. *Immutabilis* is Latin for unchanging, implying the bird lacks a juvenile plumage.
4846. No, its head sheen is purplish, but that of the Greater Scaup male is greenish.
4847. At least 750.
4848. The single chick is often positioned between the flank and wing of the brooding parent. Its head is frequently visible and the chick may be fed while being brooded.
4849. Yes and no. In 1933, 18 Cattle Egrets from India were liberated near Darby, where it was hoped the birds would control noxious insects in cattle country, but the egrets ultimately disappeared. In the 1940's, egret migrants from Asia commenced to colonize the country.
4850. No, the little birds are heard far more often than seen.
4851. The openbill storks, because of their snail-eating habits.
4852. Yes.
4853. A type of coal that burns with a crackling sound.
4854. The Western and Clark's Grebe.
4855. The duck's off-white, flesh-colored bill flushes bright pink.
4856. None.
4857. The large, flightless bird was good eating. It was hunted not only by natives, but also by Arctic explorers, as well as sealers, whalers and sea otter hunters.

4858. No, the Common Potoo has bright yellow eyes.
4859. Yes, and this results in a very loud noise.
4860. Alabama, in 1937.
4861. Yes.
4862. The Orchard Oriole.
4863. The Kokako. Its song is unlike that of any other New Zealand bird, consisting of long, rich, musical, organ-like notes that are audible from great distances.
4864. The Yellow-breasted Chat is 7.5 inches (19 cm) long, and weighs up to an ounce (28 gms).
4865. Guineafowl lack leg spurs.
4866. The Great Cormorant.
4867. Yes.
4868. The Green-tailed Towhee.
4869. None.
4870. The Black-necked Stork (Australian Jabiru).
4871. The name is Latin for *anteater*, an allusion to one of its foods.
4872. Yes.
4873. The Egyptian Vulture.
4874. No, in the extreme southern portion of its range, the ducks occupy suitable habitat at nearly sea level.
4875. Probably the cosmopolitan Barn Owl.
4876. Buzzard's Bay.
4877. In 1972, the estimate was 134 million birds.
4878. In flight, the birds appear headless.
4879. Yes, but such encounters can be fatal for the owls.
4880. Yes.
4881. For trout-fly feathers.
4882. The average is just over 1.25 inches (3.2 cm).
4883. Yellow.
4884. Yes, very much so.
4885. *Just ducky*.
4886. The Golden Eagle, and its image surmounted the staffs borne in front of every Roman legion.
4887. Yes.
4888. Water must be drained from its pouch.
4889. Probably in the Northwest Territories of Canada (northwestern Mackenzie), and possibly the Chukotka Peninsula, Russia.
4890. The fresh green leaves that male bowerbirds use to decorate their rudimentary bowers or stages are secured using the specially modified bill. As many as a hundred leaves may decorate the bower, and these are gathered daily.
4891. The Procellariiformes: (albatrosses and petrels). The Wandering Albatross is one of the largest of birds, whereas the Least Storm-Petrel is among the smallest.
4892. Yes.
4893. Generally at dusk. Chicks are prompted to jump by males calling on the water below. Once in the water, the chick and the male of the pair swim out to sea, presumably to be far enough from land by daylight to escape attacking gulls.
4894. Yes.

4895. The deflated pouch fades to orange, but that of a courting male frigatebird is bright red.
4896. No, most parrots have rather smallish eyes.
4897. Yes.
4898. A roasted Ostrich.
4899. Yes, but only the egg-laying monotremes (duck-billed platypus and the echidnas).
4900. A *roller*.
4901. The tinamous are in the first-listed taxonomic order.
4902. No. Egg covering does not occur until after the entire clutch is laid and incubation has commenced. Covering the nest may be more to keep the eggs warm than for concealment.
4903. No, Limpkins float high on the water.
4904. Bright blue.
4905. The Mexican Jay. Progeny from previous years remain with the family long after they have reached sexual maturity to help the breeders rear the current crop of young.
4906. The crane. Shortly after the Norman conquest of England there was a general interest in genealogy, and the branching form of a family tree was referred to as a "crane's foot" (*pied de grue)*, from which is derived the word pedigree.
4907. The Old World coucals.
4908. The birds are shy and keep to themselves.
4909. The Ruffed Grouse.
4910. Both sexes incubate.
4911. Probably the pigeons and doves, with more than 50 species.
4912. The Rosy Finch.
4913. The Mangrove Cuckoo of Florida.
4914. The South Island Pied Oystercatcher.
4915. Dippers. In some areas the aquatic birds readily prey on fingerlings and trout fry.
4916. About 98.
4917. A bundle of 13 arrows is clutched in the left foot and an olive branch in the right foot. The olive branch signifies America's peaceful ideals, and the arrows signify the willingness to defend those ideals.
4918. Fourteen.
4919. Yes.
4920. The Yellow Warbler.
4921. *Bye Bye Birdie*.
4922. The Asian shorebird is generally restricted to high mountain torrents and pebbly streams.
4923. No, none of the six woodhoopoes is crested.
4924. Alcatraz originally came from the Arabic word *al-quadus*, meaning a "bucket to hold water" or "water carrier," and it may also be related to the Greek *kados*, meaning a "water pot." The Spanish *alcatraz* is an allusion to the pelican's reputed use of its gular pouch as a bucket.
4925. No, Dutch explorers discovered the swan in 1697.
4926. The twenty-dollar gold piece.
4927. Based on sketches on the walls of the Tomb of Ti, it appears that the stork inhabited Egypt about 5,000 years ago.
4928. No.
4929. The name refers to the long plumes grown during the breeding season, known as "aigrettes."

Answers

4930. Only as far south as the highlands of Guatemala.
4931. The Osprey.
4932. The society was founded to combat the plumage trade.
4933. *Flamingo Road.*
4934. Not really. Insects are taken in small quantities, but the neotropical manakins feed primarily on small fruits generally plucked on the wing.
4935. No, not compared to other types of owls.
4936. No. The neck of the Australian Jabiru (Black-necked Stork) is completely feathered, but that of the neotropical Jabiru is naked.
4937. No.
4938. Yes.
4939. Yes. However, some, such as the Chukar and Hungarian Partridge, were successfully introduced into North America.
4940. No, but most nightjars have rictal bristles.
4941. No, it utters high-pitched whistling notes that sound ridiculous compared to the large size of the bird.
4942. *As scarce as hen's teeth.*
4943. Pinkish. The color results from the copious amount of penguin guano, consisting primarily of krill remains.
4944. No. Their burrows are characteristically short, at times allowing daylight to penetrate the incubating chamber. Some burrows may be 3-4 feet (0.9-1.2 m) deep, sometimes with bends, and may be dug months, and even years, before use.
4945. No, Wilson was an accomplished poet.
4946. A snake. A wryneck's scaly plumage pattern, combined with the undulating movements of its extended neck and its hissing, might confuse a potential predator.
4947. Duckpins.
4948. A swift.
4949. No, at times the storks nest on rocky ledges.
4950. Short toes allow the eagle to completely surround the body of a snake with its feet, enabling it to grip and crush more efficiently.
4951. No, Dr. C. J. Meller (c.1836-1869) was superintendent of the Mauritius Botanical Gardens in 1865.
4952. Yes, often using sticks and leaves, but most petrels do not line their nests.
4953. No, the birds fly weakly and only for short distances.
4954. The Sunbittern.
4955. The Lark Bunting.
4956. No.
4957. White to pearl-gray.
4958. Yes, but the down disappears once the young fledge.
4959. The uncontrolled human growth rate. Most of Earth's serious problems, including widespread famine and the alarming loss of habitat, are directly related to the population of a dominant species out of control. In excess of 3.5 billion humans blanket the globe, and the population could triple in the 21st century. However, because of religious and moral considerations, surprisingly few conservation organizations are willing to seriously campaign for birth control and abortion. The population is currently far outstripping available natural resources, much to the detriment of a huge percentage of life forms, and

the situation will only worsen in the future. Many hundreds of species of birds alone are probably doomed to extinction in the next few decades, if not sooner. Possibly a million plant and animal species are currently threatened globally, of which 1,000 are birds threatened with extinction, with another 5,000 declining.

4960. In 1916, on the Isle of Skye, but the eggs were taken by an English vicar.
4961. Yes, but generally only as chicks.
4962. Yes.
4963. A company of wigeon.
4964. No, loons crash-land into the water chest first.
4965. Playing tapes of their calls usually elicits a response from any rails within hearing.
4966. Mainly yellow with some red.
4967. No.
4968. The breeding-plumaged drake Pacific Eider (a race of Common Eider) and the drake King Eider. Males of the Northern Eider subspecies sometimes also have a black V.
4969. The Derbyan Parakeet.
4970. Yes.
4971. No, Ruff leks are typically located on open hillsides.
4972. The kiwi and the Kakapo (Owl Parrot). Beautiful warm and fragrant feather capes were fashioned from the sweet-smelling moss-green skins of the flightless Kakapo, and these were more valuable than mink or sable ever became.
4973. Poisonous sea snakes.
4974. Females were more brightly colored than males.
4975. The Blue Duck. Not only are pairs intolerant of their own kind, but they also vigorously defend their stretch of a mountain stream against other waterfowl and aquatic birds.
4976. Nesting miners burrow deep into sandy banks, like gold or coal miners.
4977. Only as far south as Belize (formerly British Honduras).
4978. No, wintering birds generally move only far enough north to reach open water.
4979. No, sluggish, lowland streams and ponds with forested banks are favored.
4980. No, not even gall bladders.
4981. The Arctic Loon.
4982. The owls.
4983. Yes, but some southern storm-petrels have only one coat of nestling down.
4984. No, the South American gamebirds are quite plump.
4985. Isaac Sprague, an artist and botanical illustrator.
4986. The Seychelles Fody, an endemic weaver locally known as toq toq because of its call.
4987. Skimmers fly very close to the water, so their wingbeats must be very shallow to prevent the tips from striking the water.
4988. A single black spot.
4989. The Snowy Owl. Males are mostly white, but females are barred.
4990. In the lee of waves, hence the descriptive name "storm-petrel."
4991. The Merlin.
4992. None.
4993. The color apparently comes from the uropygial gland during the breeding season, and is spread when the birds preen.
4994. Yes, although there are a number of notable exceptions.

Answers

4995. The South American Maguari Stork, but in Venezuela it may use bushes or small trees.
4996. Yes.
4997. The pads prevent abrasion of their heel joints by the rough floor of the nesting cavity. The pads drop off prior to fledging.
4998. South America, where in excess of 2,500 species of birds breed.
4999. There is no difference.
5000. The Gull-billed Tern.
5001. Three: Humboldt, Magellanic and Adelie Penguin.
5002. Yes, the population is estimated at about a half-million, but has declined in recent years.
5003. A spoonbill.
5004. The Red-faced Cormorant.
5005. Yes, in many species.
5006. Yes, in some species.
5007. The black plumage is attained at about three years of age, but the casque and wattles keep growing and the bare parts on the head and neck start to gain the lurid colors.
5008. No, its 1-3 pale bluish-green white eggs are relatively small.
5009. With Dayak headhunters, an ear-plug fashioned from hornbill ivory signified the taking of a head.
5010. Yes, it preys chiefly on lizards.
5011. Hornbills, because of their large bill, long tail and habit of calling when flying high over the treetops.
5012. An Old World babbler.
5013. Yes.
5014. The Brown Booby. It may force a Blue-footed Booby to cough up its fish by grabbing it by the tail or wing, sometimes causing both birds to crash into the sea.
5015. Yes, except when Ruddy Turnstones are migrating.
5016. No.
5017. The Laughing Gull, that may actually perch on the head of a pelican in an attempt to steal its fish.
5018. Yes.
5019. A male.
5020. Allen's Hummingbird. In the field, females of the two species are very difficult to reliably separate.
5021. Yes.
5022. Coffee beans.
5023. Most kingfishers dig nesting burrows, but kookaburras commonly nest in tree hollows.
5024. Yes.
5025. The tiny Crowned Slaty Flycatcher, *Griseotyrannus aurantioatrocristatus.*
5026. Guillemots may use their feet, enabling them to remain underwater in a somewhat stationary position. Other alcids only use their wings.
5027. A maximum of 200 meters, but rarely that far.
5028. When their chicks are threatened.
5029. No, but a bony orbital shield is typical of many raptors.
5030. The King Penguin.
5031. Dive-bombing terns may repeatedly defecate in the face of an intruder.
5032. Barn Owls are occasionally taken.
5033. They have a distinct juvenile plumage.

5034. The Resplendent Quetzal.
5035. Young are fed not only by their parents, but also by helper birds. Upon emerging from the burrow at about 30 days of age, juveniles fly up from a distance to snatch food from an attendant's bill as they sweep past. This presumably provides practice for the nunbird's habit of foraging on the wing.
5036. Apparently so.
5037. No, a flying Kea can be very vocal.
5038. The worldwide drift-net fisheries set more than 22,000 miles (35,400 km) of net daily. This resulted in an annual incidental kill of possibly a million seabirds, as well as 100,000 marine mammals and millions of other non-targeted species, such as turtles, sharks, rays and bony fish.
5039. The recently extinct flightless grebe of Lake Atitlan.
5040. Yes, some bowers have existed at the same site for 50 years.
5041. Yes, and the birds are regarded as a distinct subspecies.
5042. Black.
5043. Histoplasmosis, a respiratory disease also known as pigeon fever. The fungus causing the disease develops when pigeon droppings react with spores in the soil. It is not known how common the disease is because most human sufferers believe they merely have a cold.
5044. Yes, but Magellanic Penguins breeding at the southern extreme of their range probably do migrate, and a percentage of northern east coast breeders move as far north as Brazil.
5045. The large fatty knob on top of the bill. It may be bitten off and consumed immediately after an eider is captured.
5046. No.
5047. No, it has a blue or pink eye-ring, but the closely related Royal Albatross has a black eye-ring.
5048. By knocking two stones together.
5049. Steller's Eider.
5050. Fowl cholera practically eliminated some Antarctic Peninsula colonies in recent years.
5051. An early departure by females may relieve pressure on the food supply, to the benefit of those remaining behind. In phalaropes, males incubate and rear the young.
5052. A variety of curly-haired poodle.
5053. A displaying Indian Peacock.
5054. No, their plumage is not waterproof at all. During heavy rains, soaked birds may die of hypothermia.
5055. The White-throated Needletail, some of which breed in Siberia and winter as far south as Tasmania.
5056. *A sitting duck.*
5057. The Gray Catbird.
5058. Based on some studies, a pair rearing chicks requires about 600 pounds (272 kg) of food annually, of which 20-30 percent is lost or wasted. On the average, the breakdown consists of about 110 pounds (50 kg) of carrion, 200 pounds (91 kg) of mammals, and 160 pounds (73 kg) of birds. A single eagle requires about 200 pounds (91 kg) annually.
5059. Yes, the birds may circle to great heights, but soaring is not typical of most terns.
5060. At least 40,000, in about 1,500 different locations.
5061. Yes, some do.
5062. Seven. The only two exceptions are Fischer's and Lilian's Lovebird.
5063. Eagle Scout.

Answers

5064. The endangered flightless rail lays a two-egg clutch.
5065. Pigeons. Their calls consist of delightful coos.
5066. Yes.
5067. Yes, some did.
5068. Yes.
5069. Yes, occasionally off northwestern Baja California.
5070. No, the collective name for the seven neotropical potoos is Creole in origin.
5071. Brownish, but the ring is only discernible in good light.
5072. The Philippine Cockatoo.
5073. Yes, the raptors may use old bird nests in trees, but cliff nesting or in small caves is more typical.
5074. A cast of hawks. In falconry jargon, a cast refers to two hawks, not necessarily a pair.
5075. No, Pseudonestor is an alternate substantive name for the Maui Parrotbill, one of the Hawaiian honeycreepers.
5076. Larry *Bird.*
5077. King and Emperor Penguins, but some crested penguins may incubate in a semi-upright position.
5078. H.R.H. The Duke of Edinburgh.
5079. Lidth's Jay, that is restricted to Amani and Tokunoshima Islands.
5080. The loons, grebes and alcids.
5081. The Great Horned Owl.
5082. The name came from the Dutch *manneken*, meaning "a small man" or "dwarf." Manakins are tiny neotropical birds, about the size of a titmouse.
5083. For their feathers.
5084. Yes.
5085. Patuxent Wildlife Research Center (Laurel, Maryland), in 1967.
5086. Armadillos may take eggs, and in South America, eggs as large as rhea eggs may be preyed on.
5087. Although among the most numerous of high Arctic ducks, the birds are delicate and very difficult to breed in captivity. Oldsquaw are particularly prone to heat stress, especially during the molt. Most ultimately contract aspergillosis, a fungal disease that is nearly always fatal.
5088. The male Lesser Frigatebird.
5089. Probably hares.
5090. Nevada, in 1947.
5091. Goose bumps.
5092. No, the terrestrial birds have flattened toenails.
5093. The Barn Owl.
5094. The snipe nest in burrows excavated by other birds.
5095. The European Robin.
5096. No, the chick is doomed.
5097. Five: Cuban, Hispaniolian, St. Lucia, St. Vincent and Puerto Rican Amazon. A race of the Yellow-headed Amazon is called the Panama Parrot.
5098. Yes, in some instances by as much as 25 percent.
5099. No. John Reeves (1774-1856) was an English naturalist and collector who worked in China from 1812 to 1831.
5100. Red.

Answers

5101. The gorgeous Painted Bunting.
5102. A simpleton or a fool. The name is an allusion to the ease with which the woodcock was formerly captured in the spring.
5103. Yes.
5104. The Greater Roadrunner.
5105. The vultures don't often fly into power lines, but many have been electrocuted when perching on power pylons.
5106. Yes.
5107. Yes, some do, such as the Whimbrel.
5108. Yes, its small bill is surprisingly strong.
5109. Yes, generally so.
5110. No, shallow water is preferred.
5111. Yes.
5112. About 285 pounds (129 kg).
5113. On the backs of larger birds, especially cranes.
5114. Iceland, from the word *skufr*.
5115. Lucy.
5116. The Bristle-thighed Curlew. Its Polynesian name is *kivi*, but in the Maori language, the *v* sound has given way to the *w*. The curlew, while smaller than a kiwi, has mottled brown plumage and a long decurved bill. It probes for its food as does the kiwi and is very aggressive, so there are a number of similarities. Even so, some authorities suggest that the kiwi was named for its call.
5117. Yes, on occasion.
5118. Yes, although pittas are primarily terrestrial.
5119. Bright chestnut.
5120. Plovers locate their prey by sight and fewer birds no doubt reduce the physical disturbance to the substratum, which might otherwise make prey unavailable. When disturbed, many prey species either become immobile or move to depths beyond the range of a plover's bill. Even during the non-breeding season, some plovers defend feeding territories.
5121. The introduced Gray (Hungarian) Partridge.
5122. Baffin Island, Canada, in 1929.
5123. The Southern Ground-Hornbill.
5124. No, their straight bills are comparatively long.
5125. Most of the endemic Bermuda Petrels were consumed by British settlers, and the winter famine of 1614-1615 was such that the Cahow was essentially eliminated. Introduced black rats and pigs no doubt hastened the decline.
5126. Herons, which led to the protection of some rookeries.
5127. No, but male Willow Ptarmigans do, and cocks are very defensive. Gulls are attacked without hesitation, as are humans. In one instance, an aggressive cock even attacked a grizzly bear that stumbled over its mate's nest.
5128. The Razorbill.
5129. Chicken pox.
5130. Texas, followed by New Mexico.
5131. Based on calculations, approximately 252,000 wingbeats.
5132. Yes.
5133. Yes.
5134. The Bornean White-fronted Falconet favors dragonflies.

Answers

5135. Pest control measures include the use of flame-throwers, dynamite bombs and poisonous aerial sprays.
5136. Yes.
5137. A delightful musical whistling of its wings.
5138. No, they are more insectivorous.
5139. No, the cuckoo chick pushes host chicks out of the nest.
5140. 1938.
5141. Stones may aid in crushing and grinding up hard-shelled beetles.
5142. No, the cormorant nests on most islands in the Chatham Islands (New Zealand). The population may be less than 1,000 birds, but does not appear to be under any threat.
5143. No, not in the true sense of the word. The avian bill corresponds roughly to mammalian lips, although it is hard and dry.
5144. No, the name comes from the distinctive humming sound of their rapidly whirring wings.
5145. No, he only saw the warbler once. While rare during Audubon's time, the bird has probably benefitted more than any other eastern warbler by the changing land uses brought about by the development and subsequent decline of agriculture.
5146. No, it has a square tail.
5147. Adults weighed up to 12-14 pounds (5.5-6.4 kg).
5148. The female is said to bite the chosen male, but Reeves may breed with more than one male.
5149. A swift's food-ball may consist of up to 600 insects.
5150. The Glaucous, Iceland, Ivory, Ross' and Glaucous-winged Gull.
5151. An owl train.
5152. The New Zealand race of the Little Pied Cormorant and the Stewart Island Shag.
5153. In some instances, in excess of 150 miles (241 km).
5154. The Marabou Stork. Not only do the storks prey on chicks, but they capture adult flamingos by stabbing them in the back with their huge, powerful bills; the victims are then drowned and consumed.
5155. No, but the owl inhabits all other continents except Antarctica.
5156. Three: Blue-throated, White-eared and Berylline Hummingbird.
5157. A flock of geese on the ground.
5158. Reportedly, the birds were used as lamps. Their high oil content is such that a lighted cotton wick inserted into the throat of a dead storm-petrel would burn for a considerable length of time.
5159. No, their eggs are a dark off-white, speckled with gray and brown, an unusual trait for a cavity-nesting bird.
5160. Yes.
5161. The Crested Caracara.
5162. Yes, especially if suitable nesting habitat is scarce. Otherwise, pairs may nest alone.
5163. Bateleurs remain in immature plumage for about five years. Several more years are then required to completely pass into adult plumage.
5164. No, although some authorities suggest it was. Despite its comparatively small wings, it apparently could fly reasonably well.
5165. No, all five species have flat bills.
5166. About 70 per square inch (11 per sq cm). This is rather dense covering for a bird, but is minuscule compared to mammals. The sea otter has the densest fur of any mammal, with up to 806,250 hairs per square inch (125,000 per sq cm).
5167. Yes.

Answers

5168. The Weebill, a tiny 3.5-inch-long (9 cm) bird weighing a fifth of an ounce (5.7 gms).
5169. The California Least Tern.
5170. It has a tendency to roost in reed beds. It is currently known as the Long-tailed Cormorant.
5171. The Keel-billed Toucan.
5172. To crane one's neck.
5173. The distal half and the base are red, but the bill is black in the middle. Atop the bill is the conspicuous yellow saddle from which the stork takes its name.
5174. No, the Beardless-Tyrannulet is smaller.
5175. Twenty-one days.
5176. No, and it was believed extinct for many years. Up until 1961, when rediscovered, only about 20 museum specimens existed, the last of which was taken in 1889. During the 1960's, the number of breeding pairs was only 40-50, but had increased to about 138 pairs by 1983. The only other scrub-bird, the Rufous Scrub-bird, is not much more numerous, with a population probably not exceeding several hundred breeding pairs.
5177. The esophagus.
5178. Leach (1790-1836) was a world authority on the crustacea.
5179. More than 50 percent.
5180. The 50-day incubation period is undoubtedly a result of the female being off the nest for long periods of time. Throughout incubation, the single egg is deserted for 3-6 hours each morning. A lyrebird egg reportedly hatched in 28 days when incubated continuously under a domestic hen.
5181. Three: Spruce Grouse, Sage Grouse and Willow Ptarmigan.
5182. About $300 million, in 1990.
5183. Eagles might be used to attack slow-flying, canvas-covered German war planes. Fortunately for the raptors (and presumably the German pilots as well), the idea never got off the ground.
5184. The Sora rail.
5185. October 23rd.
5186. The Hooded Merganser.
5187. Yes, but palm trees are definitely favored. In the mountains where palm trees do not grow, other trees may be used, but the communal nest is generally much smaller and is seldom occupied by more than two pairs.
5188. Bright red or orange.
5189. Approximately 750 tons.
5190. No, domestic turkeys have white-tipped tails, but the tails of wild birds are brown-tipped.
5191. The Great Horned Owl.
5192. Probably the Common Murre. In 1970, approximately 577,000 pairs bred in Britain, of which 80 percent bred in Scotland.
5193. A concretionary nodule or geode about the size of a walnut. The Ancients believed that eagles took eaglestones to their nest to facilitate egg laying. Pliny the Elder (23-79 A.D.) indicated that eagles placed the stones in their nests to effect many cures and if a bird did not have a stone, it could not reproduce. Pliny advised pregnant women to keep an eaglestone with them to prevent miscarriages.
5194. Both are named after barnacles. The Brent's specific name *bernicla* is either from the Latin name for barnacle or was derived from the Norwegian word for barnacle, a reference to the legend that the geese hatched from barnacles clinging to driftwood. In old Roman Catholic Ireland, the Brent goose was called the barnacle.

Answers

5195. No, humans can often approach to within a few feet (0.5m).
5196. Generally at dawn. In some areas, the raptor serves as an alarm clock for natives during the summer.
5197. No, its eggs are rounder.
5198. No, the warbler nests on the ground.
5199. Florida, no doubt because of the number of Bald Eagles that nested there.
5200. A cocker.
5201. Yes, it breeds as far south as southwestern England.
5202. A toucan.
5203. Yes.
5204. Its call is a loud explosive *kut*. From a distance, the calls sound rather like wood being struck with an ax.
5205. Yes.
5206. Probably because the host species are usually much smaller birds than the cuckoos.
5207. Yes.
5208. Yes. In California orchards, considerable crop damage may be done to peaches, apricots, nectarines, plums and cherries.
5209. Its head is completely feathered.
5210. The birds commonly land on ships.
5211. No, Swainson was a brilliant, versatile British ornithologist of the early 19th century.
5212. Generally after the eggs are laid, but pelicans tending large chicks may retain the horn.
5213. The Peregrine Falcon and American Kestrel.
5214. Its short bill is feathered for at least half its length, an adaptation to sub-freezing conditions.
5215. The male Standard-winged Nightjar. During the breeding season, the ninth primary is elongated with the broad webs confined to the tips like rackets. In flight, the 9-inch-long (23 cm) bare shafts of the two primaries are not visible, thus the 7-inch-long (17.8 cm), broad web rackets appear to be two small following birds.
5216. No, the eggs are left unattended for several hours or more a night while the male is out feeding.
5217. The Philadelphia Eagles.
5218. With noose-traps set out at night along stretches of water where the birds roost. Birds are also caught with bolas, or driven ashore when flightless during the molt. Flamingos are eaten and the feathers used as dance ornaments.
5219. Only females of the typical sunbirds incubate, but both sexes of the 10 spiderhunters incubate.
5220. Northern Goshawk females weigh more than three pounds (1.4 kg).
5221. Yes and no. The hollow bones of flying birds lack marrow, but in flightless birds, such as ratites and penguins, the solid bones contain marrow.
5222. No, they resemble shorebird chicks.
5223. Yes, most do.
5224. Duck cloth, a material used in making small sails, bed ticking and certain types of clothing.
5225. Gambel, an early California field ornithologist, died of typhoid fever in 1849.
5226. Yes, the parrots feed primarily on dry seeds.
5227. The Black-headed Grosbeak is well known for its courtship flight song, and is one of the very few members of the entire family that sings on the wing.
5228. The Pied-billed Grebe.

Answers

5229. The Golden-cheeked Warbler. The warbler was added to the ever-expanding endangered species list in 1990.
5230. The Bare-throated and White-crested Tiger-Heron.
5231. Yes, the crest is erected when the grouse are alarmed.
5232. The base of its slightly upturned bill is often blue.
5233. Snipers.
5234. No, Pigeon Guillemots are slightly heavier.
5235. The cranes stood on one leg grasping a stone in the raised foot of the opposite leg. If sleep overtook the bird, it would drop the stone, awakening itself and other sleeping cranes.
5236. No, but the large, conspicuous upper tail coverts are white.
5237. The Sandwich Tern.
5238. Red.
5239. The Bufflehead.
5240. The cave-dwelling Oilbird.
5241. A movable flap or operculum covers the nostril completely, perhaps offering some protection against sand or dust.
5242. No, a single-egg clutch is more typical.
5243. Males defend bee nests and when females visit the nests to feed on wax, the males copulate with them. Males with the best food supplies obtain the most copulations, so it can be said that, in effect, males trade food for sex. This unorthodox mating system has been dubbed *resource-based non-harem polyyamy.*
5244. White, whereas that of the Red-winged Blackbird is yellow.
5245. An owl fly.
5246. No.
5247. From gander, the term for a male goose, probably alluding to its large size.
5248. Between 104-110°F (40-43.3^{O}C), with the norm approximately 105°F (40.5^{O}C).
5249. No, a group of penguins is simply a known as a group, flock, gathering, etc. I propose the name *kingdom:* a kingdom of penguins.
5250. About 44 skins and a few separate skeletal items.
5251. No, right angle bends are common.
5252. No, it is one of the noisiest.
5253. No, the birds have rudimentary tongues.
5254. Yes, but just barely, whereas the legs of a Yellow-crowned Night-Heron extend well beyond its tail.
5255. A murder of crows.
5256. No, Asian broadbills tend to be more brightly colored.
5257. Pallas' Fish-Eagle. House Crows also prey on the eggs.
5258. The Black and the Surf Scoter.
5259. A bird of prey trap, consisting of a framework of foot snares surrounding the outside of a small cage containing a live pigeon or other bait. When the raptor strikes the cage to capture the prey, it becomes ensnared.
5260. King Penguins inhabit the subantarctic, a region lacking fast-ice as is typical of the high Antarctic. With open water close at hand, the penguins can forage near the colony and relieve one another during incubation. The travel time over the ice to reach open water, sometimes 100 miles (160 km) or more away, precludes partner exchanges in the Emperor Penguin.
5261. Not until their fourth or fifth year.

Answers

5262. Yes, although carrion is not a typical food source.
5263. The Montezuma Quail.
5264. The Imperial (Blue-eyed) Shag.
5265. No, most honeyguides are rather drab.
5266. The Blue-headed Pitta of the lowland forests, and the Blue-banded Pitta of the submontane forests.
5267. Not normally, but wayward Ospreys have been recorded.
5268. Glossy black.
5269. The name might be in reference to the long central tail feathers of some African whydahs that recall a widow's train.
5270. Yes, immatures are predominately brown, as opposed to the gray of adult males.
5271. Yes, several specimens were collected in Cuba during the last century.
5272. Yes, Montezuma (about 1480-1520) was killed during the Aztec uprising against the Spanish conquerors.
5273. The Snowy Sheathbill. The short carpal spurs are used for fighting among themselves.
5274. Crayfish.
5375. Yes, especially moonlit nights.
5276. The Great Tinamou. The indians of the forest tribes of Brazil and Colombia believe that the jaguar imitates the call of the tinamou, in order to lure it into range for a kill.
5277. The parrots with nine species and the pigeons with six.
5278. Only about 13 percent.
5279. The Eurasian and African Hoopoe. Until recently, the two constituted a single species: the Common Hoopoe. The Madagascar form may be a separate species also.
5280. No, sibias are in the babbler complex.
5281. January, 1957, at Laguna Colorada, Bolivia.
5282. In no other U.S. state.
5283. No, it is one of the largest, measuring more than a foot (0.3 m) in length.
5284. All material, including prey, is carried in the bill, because the feet are not suitable for grasping. Both sexes carry nesting material, and both construct the large platform structure that in often-used nests may be 6.6-8.0 feet (2.0-2.5 m) across.
5285. The flycatcher was named after Thomas Say (1787-1834), a well-known American entomologist.
5286. No, the Ring-billed Gull was named after Delaware *(Larus delawarensis)*, but the California Gull is the only gull with the name of a state in its common name.
5287. Yes.
5288. Charles Haskins Townsend. Townsend's Shearwater was named in his honor.
5289. No, the birds migrate by day.
5290. No, the Eared Grebe is more gregarious.
5291. The falconer is usually on horseback, taking the weight of the bird on a prop resting on his knee or saddle.
5292. Males had hollows in their backs where eggs were deposited in flight, and females incubated while the male remained airborne. The skins, prepared by New Guinea natives, lacked feet, because the natives made certain ornaments out of them. The Greater Bird of Paradise's Latin name *apoda* (from the Greek meaning lacking feet) refers to this. Because of this apparent lack of feet, Europeans thought the birds never landed.
5293. No, it is perhaps the least known, no doubt due to the remoteness and inaccessibility of its habitat in the Amazon basin.

5294. A distraction display of certain tundra-nesting birds, especially the Purple Sandpiper. The bird jumps into view when a potential predator is a few feet (0.6 m) from the nest, then zigzags away, crouched low and hunched over. Its wings are dropped and appear like hind legs, the feathers are fluffed out and look like fur, and the bird utters mouse-like squeaks. This rodent-like behavior generally lures a fox or weasel away.
5295. No, it also occurs in the Comoro Islands.
5296. The African Skimmer.
5297. Snakes.
5298. No, pygmy-owls have long tails, but an Elf Owl's tail is short.
5299. The Long-bill Curlew, that weighs up to two pounds (0.9 kg).
5300. The Cattle Egret.
5301. A frothy substance produced on plants by the nymphs of certain insects to envelop their larvae.
5302. No, while much threatening and bluffing occurs, serious fights are rare.
5303. Not generally, but lizards have occasionally been recorded in its diet. The parrot typically feeds on a wide variety of plants ranging from small ground plants, tussocks, sedges and herbs to large forest trees. Its diet changes seasonally.
5304.The single, pale greenish-yellow to greenish-buff spotted egg is reportedly incubated for 30 days.
5305. Twice as much. Anna's Hummingbirds weigh between 0.12 and 0.2 ounces (3.4-5.8 gms).
5306. Yes, the birds may skim the sand near the nest with their bills, but such behavior has only been recorded in American birds. Latin Americans call the bird *rayador* (one who draws lines).
5307. The Common Quail.
5308. The Tawny Owl.
5309. Yes.
5310. The Double-crested Cormorant, and its kinked neck is distinctive.
5311. No, the birds can often be closely approached.
5312. A distinctive parrot with a number of readily recognizable forms of semiarid inland southern Australia.
5313. The Indian mongoose.
5314. No, it is relatively large for a warbler.
5315. The White-headed Stilt, an irregular visitor to New Britain from Australia.
5316. None.
5317. Snap their bills, which in larger owls can result in a loud noise.
5318. Primarily willow buds and twigs.
5319. Yes.
5320. A muster of peacocks.
5321. The Congo Peacock and Stresemann's (Ethiopian) Bush-Crow.
5322. The Common Merganser.
5323. Twelve.
5324. Condors are dependent on thermal uplifts for soaring, and thermals do not develop until well after the sun rises.
5325. Yes, it has a bony helmet, as well as wattles that hang from the gape of the bill.
5326. Phoenix, Arizona.
5327. No, the large neotropical gamebirds generally don't commence breeding until at least two years of age.

Answers

5328. No. H. B. James (1846-1892) was a naturalist and businessman who lived in Chile.
5329. The Orchard Oriole.
5330. The Western Gull.
5331. The explosive noise so disconcerts a predator that it may miss its prey, or a hunter may miss his shot.
5332. Yes.
5333. Guineafowl.
5334. Yes.
5335. To pry open bivalves.
5336. Ants. Mortality in some regions can be extensive, accounting for up to 15 percent of nest failures.
5337. Dried horse dung or cow chips.
5338. Four: Ash-throated, Willow and Alder Flycatcher, and the Rose-throated Becard. The becard was formerly considered a cotinga, but is currently included with the tyrant-flycatchers.
5339. Up to 15 inches (38 cm) long, and the tail is forked for more than half its length.
5340. Lucy's Warbler.
5341. The internal shell or "bone" of a cuttlefish, a squid relative. Dried cuttlebone is used as a bird food and mineral supplement and, when powdered, as a polishing agent.
5342. Probably as an adaptation for underwater swimming. The nostrils of other petrels point forward.
5343. No.
5344. The Smooth-billed Ani.
5345. Displaying males apparently use the mounds as stages.
5346. Chicks are boiled for their fat, that is sold as a remedy for tuberculosis.
5347. No, males have lower voices.
5348. No, only males have white throats. Females have buff-colored throats.
5349. No, the wasp-eating raptor is gregarious.
5350. Yes.
5351. No.
5352. In 1910 or 1911, Falkland Islands Magellanic (Upland) Geese were introduced to provide fresh meat for sealers and whalers, but by 1950 all had been extirpated. Another introduction was attempted in 1959, but the birds disappeared shortly thereafter.
5353. The Sage Grouse. The birds are primarily dependent on large sagebrush for food from October to May, and for cover throughout the year.
5354. Yes.
5355. It lacks the rictal bristles typical of most flycatchers.
5356. At least 150 pounds (68 kg).
5357. The Waved Albatross.
5358. A small farmer.
5359. The male Anna's Hummingbird.
5360. The Brown-headed Cowbird.
5361. No, it is exclusively a ground nester.
5362. The Black-bellied Plover.
5363. The Rockhopper Penguin.
5364. Koloa is the local name for the endemic Hawaiian Duck.
5365. The Waldrapp (Hermit or Bald Ibis).

Answers

5366. The tiny island of Buldir in the Aleutian Islands (Alaska) with 12: Pigeon Guillemot; Ancient Murrelet; Parakeet, Crested, Rhinoceros, Cassin's, Least and Whiskered Auklets; Horned and Tufted Puffins; Common and Thick-billed Murres.
5367. The Canada Jay.
5368. The Lesser Adjutant stork.
5369. Yes.
5370. Shake quelea chicks out of nests and devour them whole.
5371. No, the bird is almost foolishly tame.
5372. Construct a new nest over the old one and lay another clutch of eggs.
5373. Yes, although their bills are not as colorful as the swollen bills of the drakes.
5374. They are smaller with white, rather than black, throats. Juveniles also lack the distinctive white eye-ring, and have bluish, as opposed to black, dorsal plumage.
5375. New Zealand, with 30 percent.
5376. The birds are attracted to salt spread on highways to melt snow and ice.
5377. The Emerald Cuckoo.
5378. Despite their close relationship to waterfowl, all three screamers have sharp, fowl-like bills.
5379. No, it is primarily a Canadian breeder.
5380. Some burrows exceed five feet (1.5 m) in length.
5381. Yellow, but males are red.
5382. Yes, in some areas considerable crop damage can result.
5383. The Australian Pelican.
5384. Despite their reputation for being almost exclusively vegetarian, mousebirds occasionally prey on young birds. The mobbing birds are more aware of this than are most biologists.
5385. Paisley, Scotland. Wilson emigrated to the U.S. in 1794, at the age of 28.
5386. Probably to protect the birds from stinging bees.
5387. It usually spreads its wings and points its bill in the direction it intends to fly.
5388. The phalaropes.
5389. No, both feet are paddled in unison, but the feet are paddled alternatively when swimming on the surface.
5390. Yes, but only the giant-petrels that kill weakened seal pups, as well as introduced rats and rabbits. Birds are preyed on as well, and even at sea, giant-petrels have been known to attack and drown small albatrosses. On numerous occasions I have seen them attack and kill recently-fledged penguins at sea, even those as large as King Penguins.
5391. A path of small stones.
5392. Yes, the raptors may kill new-born pigs.
5393. No, some of the smaller African species, such as Red-billed and Yellow-billed Hornbills, lack casques.
5394. The Antarctic (Southern) Giant-Petrel.
5395. Yes, some catbirds build stage-type bowers.
5396. Townsend fell ill, apparently from the effects of powdered arsenic with which he had prepared so many bird skins, and died on February 6, 1851.
5397. The Andean and Guianan Cocks-of-the-rock.
5398. Yes, the grouse have narrow, pointed tails.
5399. No.
5400. Catch raindrops in their expanded gular pouches.
5401. *In fine feather.*

Answers

5402. Neither. They tend to squat down and conceal themselves.
5403. Because not all races have a neck-ring.
5404. The duck has not been seen since 1902. All that remains of this austral, insular merganser are 25-26 skins (including downy young), three virtually complete skeletons, some additional skeletal parts and an alcohol-preserved carcass.
5405. Dive and swim underwater.
5406. Possibly because a large segment of the parakeet population inhabits the Argentine pampas where large trees are not numerous, and nesting hollows are scarce. The birds capitalize on the few suitable trees present.
5407. No, they were named for their feline-like calls.
5408. Possibly 50 years, and some researchers speculate that 70 years might be achievable.
5409. Yes, the birds often call continuously on the wing.
5410. No, colonies may be located as far as three miles (4.8 km) inland, but generally with a view of the sea.
5411. About nine pounds (4 kg).
5412. Yes.
5413. To its right.
5414. Yes.
5415. No, it is most abundant in the southern Great Plains.
5416. Its cylindrical casque is open in the front.
5417. No, its preferred nesting habitat is along river beds.
5418. The Cinnamon Teal.
5419. The Elegant Trogon.
5420. The tern was named for Sandwich, Kent, England, where the type specimen was collected.
5421. No, the shorebirds zigzag rapidly away.
5422. The Hawaiian Goose or Nene.
5423. Yes, in wooded areas the alcids may land in trees and flutter down to walk the remaining distance to their burrows.
5424. The crop serves no digestive function, but is enlarged for use as a resonance chamber, a situation that is unique in the entire order.
5425. Florida.
5426. This may enable them to keep an eye on other vultures, and when they locate food, King Vultures follow them down.
5427. Adult females.
5428. The Crows that lived in the upper basins of the Yellowstone and Big Horn Rivers of western U.S.
5429. None, all have crests.
5430. The Mountain Quail, that inhabits northern Baja California. All other northern quail range well south of the border.
5431. Yes, but this is apparently typical of many shorebirds.
5432. When the trees reach about 18-20 feet (5.5-6.1 m) in height, the habitat is no longer suitable. This is why fires are important to the survival of this endangered warbler.
5433. The neotropical cotingas.
5434. Traffic-killed jackrabbits.
5435. Lady birds.
5436. Not until 1896.
5437. No, the birds migrate during the day.

Answers

5438. Pinkish.
5439. Audubon died in 1851 at the age of 66.
5440. Yes and no. As a rule, a crane will run, but it may also leap into the air from a standing position.
5441. The Egyptian Vulture.
5442. The American Black Duck.
5443. Bright blue with a black tip. At other times of the year, its bill is brownish-yellow or orange.
5444. Yes, even though their long toes are not webbed.
5445. Apparently because the storks are mistaken for raptors.
5446. From the phonetics of the bird's incessant calls on the breeding grounds.
5447. Yes, at times.
5448. Hans Christian Andersen.
5449. No, during the breeding season the Eastern Great White Pelican is cosmetically tinged rosy-pink with secretions from the uropygial gland. As a result, it is sometimes known as the Rosy Pelican.
5450. The Greater Flamingo.
5451. Some of the coucals. The larger female may take two or more mates, each with a different nest.
5452. Probably the Ring-necked Pheasant, but it has been introduced throughout most of its current range.
5453. The Bobolink, because of its rice-eating habit.
5454. The Sage Thrasher.
5455. The Black Swift.
5456. The western Indian Ocean coral island of Assumption, but wholesale exploitation of the island's guano deposits doomed the booby as a breeding species here. It currently breeds only on Christmas Island (eastern Indian Ocean), an island administered by the Australian government. The booby was named after W. L. Abbott (1860-1936), who collected the type specimen on Assumption Island in September of 1892.
5457. The Common Ground-Dove, that is only 6-7 inches (15-18 cm) long, or about the size of a sparrow.
5458. No, all are drab, nondescript birds.
5459. No, the Lesser (Darwin's) Rhea occurs as high as 12,000 feet (3,658 m) or higher.
5460. No.
5461. Bantamweight and featherweight divisions. A bantamweight boxer weighs 113-118 pounds (51.0-53.5 kg), and a featherweight division boxer weighs 118-126 pounds (53.5-57.0 kg).
5462. The little-known New Guinea Forest Bittern and White-crested Tiger-Heron.
5463. The rare Northern Spotted Owl, that is dependent on old-growth forest, generally more than 100 years old. Such forests are highly valued by logging companies. The tree-nesting Marbled Murrelet is also becomming controversial.
5464. The eagles partially ingest a snake leaving a short length hanging freely from their bills. Chicks pull at this protruding section of tail and retrieve the entire snake.
5465. Honey is an ideal medium for fungal growth that can cause injury to a hummingbird's tongue. Super-saturated sugar water should always be used instead of honey.
5466. No, their winter movements are rather limited.
5467. Yes.

Answers

5468. The last European record of a Great Auk is somewhat confusing. According to one account, an auk captured alive in Waterford Harbor (southeast Ireland) in 1834 was subsequently beaten to death by the local people, who believed it to be a witch. Another account indicates that the huge alcid was promptly collected for a museum specimen. The confusion may be the result of yet another account relating to an auk captured on St. Kilda (once an important breeding site) in 1840 by an islander who kept it alive for three days, but suspecting it was a witch responsible for a recent storm, he subsequently killed it.
5469. Yes, their tiny wings are beaten continuously and the webs of their feet are spread to slow them down. Ducklings of other arboreal-nesting waterfowl do the same thing.
5470. The Gray Noddy.
5471. Yes.
5472. Yes.
5473. A Pileated Woodpecker always calls with a series of notes, whereas an Ivory-billed Woodpecker typically calls with a single high-pitched note.
5474. Sea-snakes.
5475. The female Green-winged Teal with an average weight of 0.68 pounds (0.3 kg).
5476. Yes.
5477. The Gila Woodpecker.
5478. Yes, but only in southern Texas.
5479. The Spotted Sandpiper. Both birds are spotted and walk with a distinct bobbing motion.
5480. A nye of pheasants.
5481. Columbus.
5482. Yes.
5483. Purple.
5484. The Parakeet Auklet, but the Crested Auklet is nearly as large: 10 versus 11 ounces (286 vs. 312 gms). The much heavier Rhinoceros Auklet also breeds in Alaska, but it is not a true auklet.
5485. Large, downy King Penguin chicks. Oakum was hemp fiber formerly used for caulking sailing ship seams. King Penguin chicks are covered with long, shaggy down, reminding sailors of the fluff-covered lads who shredded hemp rope to mix with tar to form the caulking material, hence the name *oakum boys.*
5486. No.
5487. Yes.
5488. The pendent wattles are completely feathered.
5489. The kiwi.
5490. Both sexes construct a roofed nest with a side entrance, usually tucked into a cranny.
5491. Yes, in addition to insects and a variety of invertebrates, many of which are poisonous, and potentially dangerous prey, such as scorpions and small snakes, as well as shrews, small rodents, toads, lizards and even young birds.
5492. No, the African hornbill generally feeds on the ground.
5493. He was a Philadelphia medical doctor and eminent entomologist, and Le Conte's Thrasher is named in his honor.
5494. No, it favors large rivers and lakes. Black Skimmers do frequent the coast and, less frequently, African Skimmers.
5495. Yes, which is peculiar because elephants are intolerant of oxpeckers.
5496. Southern Patagonia.
5497. Blue.

Answers

5498. No, the eagles also occur in Baja California, Mexico.
5499. Yes.
5500. Yes.
5501. The Great Curassow, Crested Guan and Chestnut-winged Chachalaca. All were introduced in 1928, but none became established. Over the years, numerous other gamebird exotics were also introduced, including the Lesser Prairie-chicken, Montezuma (Mearns') Quail, Crested Partridge (Roulroul) and Copper Pheasant. All these failed as well.
5502. *A feather in one's cap.*
5503. The Red-tailed Tropicbird.
5504. Yes.
5505. No, sunbirds very rarely hover at flowers, although they occasionally do. Typically, they perch while feeding. Sunbirds with short bills often pierce the base of long flowers to obtain otherwise inaccessible nectar.
5506. The rabbit.
5507. Collisions with electric power lines, which may also be the greatest single cause of Whooping Crane mortality.
5508. Most birds-of-paradise display from perches, often at a great height above the ground, thus the ventral colors are more visible to females. Bowerbirds display on the ground and are viewed by females from above.
5509. The Lammergeier (Bearded Vulture).
5510. The Shama thrush (Magpie-robin), that is well known for its fine song.
5511. A few breed in the mountains of western Alaska.
5512. Their nesting islands are devoid of trees.
5513. Africa.
5514. The Acorn Woodpecker. It often nests communally and groups may mutually share incubation duties and chick rearing.
5515. The goshawk is not migratory.
5516. *The Thornbirds.*
5517. Yes, on Buldir Island in the Aleutian Islands, auklet remains have been found in eagle nests.
5518. The Green Kingfisher.
5519. Yes, on occasion.
5520. The extensive surface area of their huge secondary feathers enables the cranes to become airborne.
5521. The Thick-billed Murre.
5522. At least 540 feet (165 m).
5523. Yes.
5524. The striking Gouldian Finch.
5525. The Everglade (Snail) Kite. Males are overall dark blue-black except for their white rumps, whereas females and juveniles are rusty-black above with heavily-streaked buffy underparts.
5526. 1924.
5527. Falcons squirt their droppings straight down.
5528. No, hybridization is infrequent, no doubt because the ranges of the two quail barely overlap.
5529. No, the female has a conspicuous bright yellow eye, whereas the eye of the male is brown.

Answers

5530. Yes, apparently so.
5531. *Hawkeye.*
5532. This may be a necessity for chicks hatching underground.
5533. The Australian lyrebirds. Juveniles may be fed by the female for up to eight months after fledging.
5534. The name comes from one of their common call notes.
5535. The American Coot.
5536. Pink with dark tips, but adults have black bills.
5537. Yes, eggs may be laid in January when snow is still on the ground.
5538. The flightless bird was buried in hot coals, feathers and all, encased in a large lump of clay.
5539. *A bird's-eye view.*
5540. The nostril openings are elliptical and directed slightly upward. Those of a Royal Albatross are more prominent with the round openings directed forward.
5541. Their tails may be flicked.
5542. The Pygmy Nuthatch.
5543. In some parts of the country, thousands of wintering coots converge on golf courses. Coot droppings on putting greens adds some challenges to the game.
5544. The Rook, also known as the castle.
5545. According to Greek legend, Alcyone's husband drowned in a storm and so great was her grief, the gods permitted him to live again as a kingfisher. Alcyone also was metamorphosed into a kingfisher. The gods provided that every year there should be from 7-14 days when the sea lies still and calm so that Alcyone might tend her floating nest. This calm period has become known as the halcyon days, the days of peace and tranquillity. Many kingfishers are in the genus *Halcyon.*
5546. Usually within two weeks or less.
5547. Yes, there is a single record from the Chukotski Peninsula from the end of the last century.
5548. The wild dogs that occurred there (from the Latin *canis*).
5549. From its call of *wonga* or *wonka,* audible from a half-mile (0.8 km) away.
5550. Yes. The pigeons did not breed there, but wandered to Cuba, and even Bermuda.
5551. Drakes of the European race have a horizontal white stripe along the side of their body, as opposed to the conspicuous white vertical stripe down the side of American birds, except for those of the Aleutian Islands.
5552. Not generally.
5553. No, they have rather short legs.
5554. Yes.
5555. The Neotropic Cormorant, as well as Bigua, Black, Brazilian, Common and Sonora Cormorant.
5556. Large females have a wingspan of about 6.5 feet (1.9 m), but long wings are not an advantage to a predator pursuing prey through the jungle. Forest eagles that hunt beneath the canopy have no real need to soar. The slightly smaller Philippine Monkey-eating Eagle has a greater wingspan of about seven feet (2.1 m).
5557. The Pileated Woodpecker.
5558. Yes, but not commonly.
5559. The Red-billed Tropicbird.
5560. No, their nests are relatively small.
5561. *A feather merchant.*
5562. The Long-tailed Jaeger.

Answers

5563. Yes, many do.
5564. No, it has a very small gape.
5565. Land crabs, hence its alternate name of Crab Hawk.
5566. King Edward IV, in 1482.
5567. The expansion of the railroad system and the invention of the telegraph. The expanded railroad system allowed commercial hunters to reach even the most distant colonies and ship birds to urban markets hundreds of miles away. The telegraph enabled hunters everywhere to learn quickly of the discovery of new colonies. In the second half of the 19th century, more than 5,000 commercial pigeon hunters and trappers were engaged in the trade. Trappers often used alcohol-soaked grain as bait because drunken birds were easily collected. Pigeons in countless barrels on the sidewalks of large cities often rotted for lack of buyers.
5568. An out-dated American initiation rite where participants ventured into the boondocks with a naive tenderfoot. Left out in the woods all night alone, the greenhorn calls out or whistles while holding a gunny sack open in an attempt to catch the elusive creatures. There are numerous variations to the theme, but the mythical snipe are never caught.
5569. Feeding on birds is atypical, but a few instances of pelicans consuming birds are documented. In one case, a young Brown Pelican fed on not only chicks, but also on the eggs of several species of herons. An Australian White Pelican swallowed a Gray Teal along with her duckling.
5570. Clark's Grebe.
5571. The Imperial (Blue-eyed) Shag. On January 28, 1989, the Argentine naval supply vessel *Bahia Paraiso* struck the rocks in front of the U.S. Palmer research station of and sunk, spilling approximately 170,000 gallons (643,518 liters) of diesel and jet fuel. Some birds were killed outright, but the overall direct kill was *relatively* low because diesel and jet fuel dissipates rather quickly.
5572. On the backs of cranes.
5573. The Black Guillemot does not migrate.
5574. No. The parrots *may* have formerly bred north of the border, but no nests were discovered.
5575. No, juveniles are much darker.
5576. The White-winged Scoter weighs up to four pounds (1.8 kg), with a wingspan of 41.5 inches (105 cm).
5577. At least 18.
5578. Killed birds.
5579. At least 1,250.
5580. No, but such behavior is typical of loons.
5581. No, a Downy Woodpecker is less wary.
5582. Collectors noted that green birds were always males, and red birds were invariably females, and the connection was finally made.
5583. A pochard.
5584. Yes.
5585. The Red-faced Warbler.
5586. A small Australian passerine started a 1959 brush fire by carrying a lighted cigarette back to its nest.
5587. The Black-legged Kittiwake.
5588. Almost entirely black both above and below, with no trace of white on the underparts. Facial markings are indistinct, and juveniles lack crests.

Answers

5589. Yes.
5590. A covert of coots.
5591. Moist deciduous woodlands.
5592. The 4.5-5.0-inch-long (11.4-12.7 cm) Bridled Titmouse.
5593. The Great Blue Heron and Great Egret. The Great White Heron is a color morph of the Great Blue Heron.
5594. A parliament of owls. The term generally applies to a group of owls roosting in a tree.
5595. The Bristle-thighed Curlew molts during the winter on mid-Pacific islands.
5596. Yes, especially if owlets are present.
5597. Ethiopia.
5598. The White-tailed Ptarmigan.
5599. The Sooty Shearwater.
5600. Swan Lager and Emu Beer.
5601. The Marbled Murrelet, because its low cries as it flew through the dense, foggy coastal coniferous forests resembled those of a lark.
5602. Egypt, hence it is also known as the Egyptian or Nile Tern.
5603. Because visual signals are usually ineffective.
5604. No, the cave-dwelling caprimulgid is chestnut-colored with black bars and white spotting.
5605. Severe hunting pressure at the end of the 18th century. The population formerly numbered about 25,000 birds. Hawaiians hunted them in volcanic craters during the molt when the geese were flightless, but the natives did not take a significant toll. European and American shooters failed for many years to recognize that the Nene nested in for what was them the middle of the traditional hunting season. Mule loads of Nenes were brought down from the highlands and were often sent by clipper ship to California to feed the forty-niners, or were salted down for use on whaling ships. Hunting was finally prohibited in 1911, but by then the geese were virtually gone. Introduced predators also took a toll. Fortunately, a large captive population currently exists and hundreds of captive-reared birds have been reintroduced, but the introduction program has not been as successful as hoped. At least 2,000 geese have been released since 1960, but the wild population numbers only about 350-500.
5606. No, but the tiny birds have long primaries.
5607. The Ribbon-tailed Astrapia, in 1938.
5608. Like ratites, its sternum lacks a keel.
5609. A harrier.
5610. The Evening Grosbeak. The Pine Grosbeak is not a true grosbeak.
5611. Germany. Turkeys were originally introduced in 1571, with subsequent introductions prior to 1939, followed by the release of 127 additional birds between 1959 and 1966.
5612. Pale yellow.
5613. The Old World falcon is a social hunter, with up to 50 birds hunting together.
5614. May.
5615. The Australian Freckled Duck.
5616. Wild turkeys.
5617. Yes, coastal habitat up to the snow line is utilized.
5618. White.
5619. The rabbit. Its specific name *lagopus* is from the Latin, meaning "hare's foot" or "hare-footed," in reference to the feathers on its legs and feet.
5620. Large males exceed ten pounds (4.3 kg).

Answers

5621. Conifer seeds.
5622. No, she was a free Black (Creole) from Hispaniola.
5623. The Glaucous Gull.
5624. No, each range is separate, as opposed to sympatric or overlapping geographic ranges.
5625. Yes, but they are more inclined to feed on fruit.
5626. No, their bills are used.
5627. The loss of appropriate habitat combined with Mallard hybridization, resulting in a loss of genetically pure birds.
5628. No, their nostrils are partially concealed by bristles.
5629. No, frigatebirds are non-migratory.
5630. Yes, but only in males. Females have reddish-brown tails.
5631. Twilight.
5632. The Arctic fox.
5633. The Purple Martin.
5634. The little alcids are stuffed into seal-skin sacks that are buried in the rocks. Seal blubber and the Dovekies combine to form *kiviak*, a great delicacy eaten in winter when food is scarce.
5635. Fantails are Old World flycatchers that take most of their prey on the wing, and the habit of fanning their tails may be a means of startling insects into flight. Their long, rounded tails are constantly moved from side to side and up and down, spreading them in a fan and closing them again.
5636. No, they rummage through leaves and litter with their bills.
5637. No, not as a rule.
5638. No, it is an especially solitary member of the family. Because of its heavy bill, it is also known as the Thick-billed Plover.
5639. Yes.
5640. Yes, Common Murres are actually dark chocolate-brown.
5641. The cotingas. The largest species is about 80 times as heavy as the smallest.
5642. The ground squirrel.
5643. The Ruddy Shelduck.
5644. Yes. Nests may start out high and dry, but if the water level rises, the stilts may quickly construct a more elaborate structure resulting in a well-built floating platform.
5645. Yes. New Guinea natives hunt the birds, but hunting is limited to providing material for traditional native adornment and an extensive barter trade.
5646. About nine inches (23 cm).
5647. No, juveniles have paler-colored eyes.
5648. The Blackburnian Warbler.
5649. In one study, about 100 pounds (45 kg).
5650. The Galapagos Dove.
5651. Red, but the remainder of the bill is yellow.
5652. They have been caught in nets set at 180 feet (55 m), and the Bank Cormorant has been known to dive to 300 feet (70 m).
5653. No, the region is devoid of corvids.
5654. A crow.
5655. The Common Merganser.
5656. Yes, but not as breeding birds. Patagonian Burrowing Parrots and Austral Parakeets are rare visitors.

Answers

5657. The Congo Peacock.
5658. Yes, the gamebirds remain surprisingly abundant.
5659. Yes.
5660. No, it has a purplish glossed head, but the head of a Common Goldeneye male has a greenish sheen.
5661. Probably from the Gaelic *tarmachan* and the Irish *tarmochan*, but exactly why is unknown. The name is wrongly spelled with a *P* because of the supposed Greek origin. The name was apparently first applied in 1599.
5662. Yes, but the bird was introduced.
5663. The Hermit Warbler.
5664. Males have a distinct non-breeding plumage, a rather unusual situation in tropical finches.
5665. The Osprey.
5666. North Dakota.
5667. No, the stork also occurs in southeastern Asia, as far east as the Philippines.
5668. The Pomarine Jaeger.
5669. The cliff.
5670. Yes, both are identical to males.
5671. The Common Starling and Crested Myna, both of which were introduced.
5672. The military borrowed Peterson's technique of highlighting specific field marks of birds. Charts were drawn up pointing out the "field marks" of both friendly and enemy aircraft so that the planes could be properly identified, presumably to prevent gunners from shooting down allied planes.
5673. All construct mud nests, and were collectively known as the mudnest-builders.
5674. Yes.
5675. He is a swindler or a cheat, especially in gambling. The name came from the Rook's thievishness.
5676. Flat-topped acacia thorn trees. In East Africa, trees that have been pruned to a dense thicket by browsing giraffes, making them impenetrable from below, are favored.
5677. Yes.
5678. No.
5679. The name probably originated from the locality of Ouidah of the West African coast, from where many of the birds were shipped for the cagebird trade.
5680. It runs toward an adversary with flapping wings and then jumps into the air, kicking at the opponent.
5681. The widespread Great Horned Owl.
5682. The Garganey Teal.
5683. The Ruff. Its leg color ranges from yellow to red, with variable bill color.
5684. The African Fish-Eagle. It is far more vociferous than most other eagles, a fact reflected in its specific name *vocifer*.
5685. The Pink-eared Duck. Zebra duck is a far more descriptive name because the "pink ears" are barely visible, except at very close range.
5686. Green; hence it is commonly called the Green Pheasant.
5687. Spain.
5688. No, but the Sandhill Crane does.
5689. Four to six years.
5690. The Morepork (Boobook Owl), found throughout Australia and Tasmania.
5691. The Black Swift. The 7.5-inch-long bird (19 cm) weighs up to 1.68 ounces (48 gms).

Answers

5692. The Puna (James') Flamingo.
5693. The adult female Double-wattled (Southern) Cassowary.
5694. Yes.
5695. Rudolph Valentino, who played a Russian cossack.
5696. The Lanner Falcon.
5697. No, they often sit motionless for long periods of time.
5698. The Ringed Kingfisher.
5699. The lakes generally do not freeze because of naturally heated hot springs. During especially severe winters the water may freeze, resulting in swan mortality.
5700. Lewis' Woodpecker.
5701. Breeding and molting. Molting penguins are not waterproof and the fasting birds cannot enter the water. Most molt ashore, but Emperor and Adelie Penguins molt on pack-ice, although Adelies also molt ashore.
5702. The Tropicbirds. Fights can last for days, resulting in deep gashes and scars on their heads and necks. Fights occur when nest sites are scarce, and the availability of suitable nesting sites is the prime factor in limiting breeding.
5703. The birds bite one end, mince the caterpillar back and forth in the bill, and shake it vigorously at one end until the gut and its toxic contents come free. Caterpillars are beaten against a branch, and the hairs are consumed.
5704. Siberia.
5705. Approximately 100,000.
5706. The Red-billed Quelea and Wattled Starling.
5707. No, it is a small Old World accipiter.
5708. It nests in trees, but the true peacocks nest on the ground.
5709. A feather.
5710. 1938.
5711. *Fregata minor* is inconsistent with its common name. Linnaeus originally placed the frigatebird in the same genus as the pelicans, and because it was smaller, it was named *minor*.
5712. Bradley was shot dead in 1905 by the captain of a schooner being loaded with egrets in Florida. The murderer got off scot-free. The murder was in many ways responsible for the 1906 law making the use of wild bird feathers illegal in the millinery trade, and within eight years of Bradley's death, the traffic in wild bird plumages went wholly underground. In 1900, the first of the Audubon wardens was hired (even before the official founding of the national society), and by 1904, 34 wardens were employed in 10 states.
5713. Green, and leaves are most often used.
5714. The Arctic Tern.
5715. September 19, 1945, near Napier, New Zealand (North Island Brown Kiwi).
5716. Charles Barney Cory (1857-1921). He was an American ornithologist who was a noted authority on West Indian birds, and one of the founders of the American Ornithologists' Union.
5717. The Old World Marsh-Harrier. Its introduction into Tahiti, Bora Bora and Moorea has been suggested as the cause for the extirpation of the lory from those islands. The introduction of the mosquito, an avian malaria vector, may have caused the disappearance of the lories from Niau. Introduced rats also prey on eggs and chicks.
5718. No, the white morph is more numerous.
5719. No, the little perching ducks are ill at ease out of the water, spending little time ashore.

Answers

5720. The Wood Thrush.
5721. No, juveniles have reddish-brown bills, but adults have bright red bills.
5722. Newfoundland.
5723. No.
5724. The Red-breasted Merganser.
5725. The California Quail.
5726. Yes, its internal temperature can reach 113 F° (45°C), incredibly near the lethal limit in birds.
5727. Yes.
5728. The umbrellabirds.
5729. The name is European and means *goose hawk.*
5730. Yes.
5731. The Black Duck, and dusky duck is a far more descriptive name.
5732. No, but in August of 1936, nine King Penguins were released in an attempt to establish them in Norway, followed by additional penguins in 1938. Sighted off and on for years, the penguins disappeared by 1950.
5733. The Bulo Burti Boubou, in 1992. Feathers and a blood sample enabled scientists to conduct DNA tests, and photographs were taken of the new bushshrike. Even so, the failure to take a specimen has resulted in criticism by some ornithologists, who insist on a type specimen. Description of a new species without a type specimen, or a piece of one, is almost without precedent.
5734. The Rockhopper Penguin.
5735. An Osprey.
5736. Buller (1838-1906) was a New Zealand lawyer and leading authority on the birds of the region. Buller's Shearwater and Buller's Albatross are named in his honor.
5737. The Long-billed Curlew.
5738. Bright red. *Phoinix* is Greek for crimson and the name was given in allusion to the flames from which the bird arose.
5739. The Australian Darter.
5740. Yes.
5741. Hooting, and one of its local names is "hooter."
5742. The Parakeet Auklet.
5743. The Sulphur-bellied Flycatcher.
5744. The Sedge Wren.
5745. Yes, the alcids gather on the water near the colony for a mutual display known as a *water dance.*
5746. The Pine Grosbeak, that is 10 inches (25 cm) long, with a wingspan of 15 inches (38 cm).
5747. From the Latin *corvus* and *marinus,* meaning "sea crow," which evolved into the French *cormoran*. The generic name *Phalacrocorax* is Greek in origin, meaning "bald raven," in reference to the bare facial skin.
5748. Eurasian carp. The fish have destroyed much of the aquatic vegetation the ducks require.
5749. The rather prolonged nestling period is 42-60 days.
5750. The nest may be enlarged.
5751. Yes.
5752. None, but under the breast feathers there are a number of subcutaneous blood vessels that dilate during incubation.
5753. It doesn't nest in trees, but prefers the ground, generally near the base of a jack pine.

Answers

5754. A *crow*bar.
5755. The Torrent Duck.
5756. No, it was named after John Henslow, a prominent 19th-century English botanist.
5757. Yellow or reddish.
5758. Yes.
5759. Yes.
5760. It was named after the red robes worn by Roman Catholic cardinals.
5761. No.
5762. The Sacred Ibis preys on Jackass Penguin eggs and chicks, as well as cormorant chicks, in South Africa.
5763. No, and in Texas, the gamebirds are even fed by tourists.
5764. Feign death.
5765. Sulfur.
5766. Yes, Anhingas are rarely reported from eastern Oklahoma.
5767. The Spur-winged and Egyptian Goose. The African Pygmy-goose also commonly perches in trees, but is really a perching duck.
5768. Yes.
5769. The Old World orioles. Figbirds differ from the orioles in that they have slightly decurved bills, and their lores and eye region are devoid of feathers.
5770. Yes.
5771. The central portion of each eye is blue.
5772. The Snowy Sheathbill.
5773. The feathers are twisted and blunt at the tips.
5774. No, but the ranges overlap south of the border.
5775. Turkeys were considered stupid and cowardly, and the Indians feared that these characteristics would be acquired if the birds were consumed.
5776. Yes, but not commonly, although it may wade into water deep enough to cover its back.
5777. The Downy Woodpecker.
5778. Its tail attains a length of 27 inches (69 cm).
5779. Some months were based on, or named after, birds or bird migrations prevalent at that time of year.
5780. Brandt's Cormorant. The name refers to the specific name *penicillatus,* which is Latin for "a painter's brush" or "pencil of hairs," referring to the elongated nuptial plumes.
5781. Yes.
5782. Nothing. The birds can be surprisingly tame and may allow themselves to be touched, or even lifted off the nest.
5783. The Surf Scoter.
5784. The Snow Petrel.
5785. No, both incubation and rearing periods are longer in euphonias.
5786. The owls roost near the trunk of a tree and elongate themselves by compressing their feathers. Hence, they resemble part of the trunk itself.
5787. Yes, I have often seen skuas capture introduced rabbits in the Auckland and Kerguelen Islands. Introduced rats are no doubt taken as well, and a weakened new-born seal pup might be killed.
5788. The gamebirds were successfully introduced into Louisiana, but introductions into other mainland states failed, although a few may still exist in Florida. The species is well established in Hawaii.

Answers

5789. To avoid the wholesale conscriptions of Napoleon because of the war then developing between England and France.
5790. All have spotted breasts.
5791. The Topknot Pigeon.
5792. Their preferred food is available during the winter. Crossbills commence nesting as early as January, although the birds breed in every month of the year.
5793. Yes.
5794. The Gray Jay, because of its confiding nature and general coloration.
5795. Their chicks are fed primarily on birds migrating through the Mediterranean at that time of year.
5796. No, males have white wing-patches, whereas those of females are gray.
5797. Wrens; from WRNS.
5798. The Red Grouse.
5799. The Spectacled Guillemot.
5800. Yes, on rodents.
5801. The Limpkin.
5802. About 30,000, in 1990.
5803. Seven, but the Greater Antillean Nightjar nests only in the Florida Keys.
5804. Assume a concealment posture in which its ear-tufts are raised and its eye slits are arranged so as to distort its usual outline and pattern. This may be done to avoid mobbing.
5805. Prior to the El Niño event of 1982/83, the population consisted of 16,000-20,000 penguins, most of which were in Chile. Subsequent to the big El Niño, the population was reduced by as much as 70 percent, but by 1990 had recovered, only to be negatively impacted again by the El Niño of 1991/92.
5806. Yes.
5807. Only one, the endangered Fairy Pitta.
5808. Spain. In 1977, the population was estimated at 9,000 pairs and increasing, but the population is declining throughout the remainder of its range.
5809. Grayish, with a faint pinkish wash.
5810. The Pyrrhuloxia.
5811. The Herring Gull. It was subsequently considered a separate species, but is currently considered a race of the Iceland Gull.
5812. The Red-bellied Woodpecker.
5813. Brewer's Blackbird, named after Dr. Thomas M. Brewer, a Boston physician and ornithologist.
5814. The South Polar Skua.
5815. Yes.
5816. Yes. Smaller hornbills lay up to six eggs, whereas the larger species typically lay two eggs.
5817. Audouin's Gull. The gull was formerly exploited by Spanish fishermen who sought its eggs for making a traditional dish. In 1983, the population consisted of about 3,500 breeding pairs. The gull is unevenly distributed on some Mediterranean islands, and may be increasing in the western Mediterranean.
5818. The hummingbirds, which loosely translates to *flower pickers.*
5819. The Ivory Gull.
5820. *Bird on a Wire.*
5821. The Northern Fulmar.

Answers

5822. The Tui.
5823. St. Matthew and Hall Islands (Bering Sea).
5824. No, they typically leap in feet first, but not always.
5825. The distinctive cotinga moos like a cow.
5826. No, it is soft at the base.
5827. Yes, especially in the southern part of their range in Tierra del Fuego.
5828. Bennett's Cassowary.
5829. Lawrence's Goldfinch.
5830. Bright blue and pink, whereas the head of the White-necked Rockfowl is bright yellow.
5831. Yes, especially ptarmigan during the fall and winter.
5832. The almost mythical Golden-fronted Bowerbird. Females had never been seen prior to 1981. It now appears that the bowerbird is locally common, but is known only from the Foya Mountains in New Guinea.
5833. No. A Laysan Albatross averages 5-7 pounds (2.3-3.2 kg), whereas a Black-footed Albatross weighs 7-8 pounds (3.2-3.5 kg).
5834. An aberrant, ground-dwelling warbler.
5835. Yes, the little rodents puncture Least Auklet eggs, kill chicks, and occasionally even wound adults.
5836. The insular fish-duck was probably never numerous, and shipwrecked mariners and sealers took many. Introduced predators could have also been partially responsible. In 1901, a relatively large number were shot by museum collectors, and this could have been the ultimate reason for extinction.
5837. Yes.
5838. The Great Blue Turaco attains a length of 30 inches (76 cm) and weighs more than 2.5 pounds (1.1 kg).
5839. No, the 27-28 day incubation period is long for a passerine.
5840. The Boran Cisticola.
5841. Some of the smaller bitterns, such as the Little and Least Bittern. In most herons, the male brings nesting material and the female arranges it in the nest.
5842. The bandi*coot*.
5843. It doesn't; there is no sexual dimorphism.
5844. The Moluccan Cockatoo.
5845. A fall of woodcock.
5846. Yes.
5847. No, females are typically more numerous, and pair bonds are not really formed.
5848. The Black-throated Diver.
5849. Leach's Storm-Petrel and Fork-tailed Storm-Petrel.
5850. No, but its legs are feathered to the feet.
5851. For the most part, the species is probably Chinese (northwestern Manchuria), but also inhabits extreme southeastern Russia, and possibly the border regions of North Korea. This little-known, rare fish-duck is commonly known as the Chinese Merganser.
5852. Possibly. The male definitely shares in the brooding of the young.
5853. Barbara Streisand and George Segal.
5854. No, cliff rocks may be used, but trees are favored.
5855. Between March 27 and April 30.
5856. No, the ducks typically elevate their long tails at an acute angle, except when preparing to dive.

Answers

5857. The seven species of neotropical potoos.
5858. Chicks remain in the nest for about 20 days and then climb about the bushes or move to the ground where they are fed by the adults.
5859. Brown.
5860. A pochard duck.
5861. From its preference of perching and roosting in trees.
5862. The two meadowlarks and the Bobolink.
5863. The Pomarine Jaeger.
5864. The vireo may hang upside-down while foraging in twigs.
5865. Kamchatka Peninsula and Sakhalin Island.
5866. Possibly because when squeezing down the narrow hole, a wet kiwi might be able to divest itself of excess water. A small hole also attracts less attention.
5867. The Ancient Murrelet.
5868. The American Coot, but the duck seldom mixes with other birds.
5869. Pennsylvania, 1934.
5870. Yes.
5871. Traill's Flycatcher. Both birds were formerly believed to be the same species.
5872. No, hummingbirds migrate in solitude.
5873. No, it is about the size of a sparrow.
5874. Yes.
5875. Ivory-colored.
5876. The Summer Tanager.
5877. The Lesser White-fronted Goose has a bright orange or yellow eye-ring.
5878. The Anhinga and darters.
5879. None.
5880. It retains its strikingly-patterned plumage year-round.
5881. The White-breasted Nuthatch.
5882. Iridescent black, but in the California Brown Pelican, the pouch is bright red.
5883. Bright yellow or yellow-orange that greatly contrast with its black legs. Except for its feet and bright yellow eyes, the egret is totally black.
5884. At least 45.
5885. The Yellow-billed Cuckoo, that was named for its call.
5886. As far as 50 miles (80.5 km).
5887. Bright coral-red.
5888. Their flight is weak, and if a river is too wide, trumpeters flying across it may tire and plunge into the water.
5889. No, the Greater Prairie-chicken is much more common and widespread.
5890. People believed crane fat would improve hearing, and it was also considered a treatment for deafness.
5891. About 2.5 months.
5892. The agave (century plant), a species of the arid mesas and foothills of western Texas.
5893. No, but a *subspecies* of the Bahama (White-cheeked) Pintail is restricted to the Enchanted Isles, where it is known as the Galapagos Pintail.
5894. Jamaica, where the predatory mongoose had previously been introduced.
5895. The trogons. The name refers to their shining plumage and possibly also to their fluttering flight.
5896. A feather.

Answers

5897. Yes, and they do not return until they are mature.
5898. The trachea is said to loop outside the breastbone. Some authorities challenge this, and the loop has not been seen in the many hundreds of ducks examined in recent years.
5899. Yes, but only slightly so: 1.5-3.0 ounces (41.9-90.0 gms) versus 1.5-2.5 ounces (47-72 gms).
5900. No.
5901. No, most have rather un-melodious voices.
5902. Pale yellow.
5903. The *Velociraptor;* typically called a raptor.
5904. Despite a fast exceeding three months, the emaciated males are still capable of producing a protein and lipid-rich crop secretion to feed their chicks for up 15 days. If their mates have not returned by that time, the males are forced to abandon the chicks.
5905. The zoologist Baron Lionel Walter Rothschild (1868-1937), founder of the Tring Museum. The Tring Museum in Hertfordshire is a branch of the British Natural History Museum specializing in ornithology.
5906. The Great Crested Grebe.
5907. Florida.
5908. Yes.
5909. Two, but in California and Gambel's Quail, the crest consists of 6-9 feathers.
5910. The Bufflehead.
5911. Yes.
5912. The Southern Screamer. The name comes from its loud call, emitted frequently in flight.
5913. Lizards, and they are known as lizard-cuckoos.
5914. The megapodes (mound-builders) and the cracids (curassows, guans and chachalacas).
5915. The Sparrow.
5916. Yes.
5917. Yes: *Himantopus mexicanus*.
5918. It pierced a reed and blew through it.
5919. *Snowflakes.*
5920. To its partially-webbed toes. In reality, the toes are only slightly lobed at the base, but this is sufficient to enable the shorebird to walk on soft mud without sinking.
5921. Each 250 feet (76 m) of altitude corresponds roughly to one degree of latitude, or about 69 miles (110 km) of northward progression.
5922. The Dolphin Gull.
5923. No one knows, but the parrot is certainly among the rarest of birds. A single individual was seen and photographed in 1990 in Brazil. The lone bird was apparently paired with a Blue-winged Macaw, leading to suspicion it was the last of its kind in the wild. At least 20 captive birds are known, but there may be other unreported ones in private collections.
5924. Yes, the cotinga lives in groups, all members of which may attend a single nest.
5925. Probably the Bearded Tit and Corn Bunting, that may fledge in the brief period of only nine days.
5926. The Common Raven.
5927. Yes, many do.
5928. Both sexes incubate and care for the chicks.
5929. No, most eudyptid penguins have red eyes.
5930. No, and less than two dozen of the specialized South American ducks have ever been maintained in collections.

Answers

5931. The Black-throated Sparrow.
5932. No one knows. Some biologists speculate that the bright blue or red bare faces of several species appear to imitate the eyes of forest cats, and possibly serve to frighten predators or competitors.
5933. Probably not.
5934. The Pomarine Jaeger.
5935. Yes, both occur in Texas and Louisiana where the two icterids apparently do not hybridize.
5936. About 14 times as much energy.
5937. No, the underwing lining of Xantus' Murrelet is white, as opposed to dark in Craveri's Murrelet.
5938. Juveniles resemble adults, but some have grayish-yellow, rather than bright yellow bills.
5939. Chicks generally leave the nest the same day as hatching.
5940. No.
5941. No, their eyes are typically squeezed tightly shut.
5942. By voice. Breeding males sing constantly all day, and are relatively easy to locate in dense cover.
5943. Yes, but the double crest is not always apparent.
5944. No, their necks are held erect, but the neck of a Mute Swan curves gracefully.
5945. No, only the American Avocet assumes a winter plumage.
5946. A lyrebird nest weighs about 30 pounds (13.6 km).
5947. No, but extensive winter ice encourages northward movement.
5948. No, the scavengers do not occur in Baja, but Turkey Vultures are very numerous.
5949. White-tailed Sea-Eagles, Snowy Owls and Peregrine Falcons.
5950. No, but only the pigeons, doves and Common Snipe do not.
5951. No, its nest is well hidden in a bush, and although of rough construction, it has a softly lined cup.
5952. The Procellariiformes (albatross and petrels). At least 72 species have been recorded in New Zealand waters.
5953. Yes, approximately 24,000 pairs bred there in 1985.
5954. Yes, in some areas.
5955. Approximately 1.3 million, in 1989.
5956. The Common Wood-Pigeon.
5957. *Vulture's Row.*
5958. The domestic turkey. The dim-witted birds sometimes appear reluctant to come in out of a driving rain, and may freeze to death on cold nights, even if warm shelter is only several meters away. They are also prone to panic, and on one occasion over 13,000 turkeys trampled themselves to death on a California ranch when a low-flying jet roared over.
5959. Only two: Steller's Jay and Clark's Nutcracker.
5960. An average of 43 days. When combined with the 23-day incubation period, this results in a protracted period for egg/chick care, especially for such a small bird. In the larger Black-nest Swiftlet of Borneo, the incubation/fledgling period may be as long as 90 days.
5961. Possibly because the pheasants nest in trees, and primaries aid the chicks in reaching the ground.
5962. Provisioning adults thrusts their bills deep into the throat of their gaping chick to disgorge food.
5963. A muskrat lodge.
5964. No, the incubating bird slips quietly away from its nest.

Answers

5965. Around 10,000.
5966. Yes.
5967. The Freckled Duck of southern Australia.
5968. Minnesota, with possibly as many as 5,000 pairs, and the population is fairly stable. Maine is probably second with about 1,100 pairs, but this figure may be conservative.
5969. Yes.
5970. The Connecticut, Prothonotary and Tennessee Warbler.
5971. The Black Guillemot.
5972. No, the larger, more aggressive Brown Skua preys more extensively on penguins, whereas the South Polar Skua is far more dependent on marine resources.
5973. Mourning Doves.
5974. They have white spots on their dark backs, and a race confined to Florida has white, rather than red eyes.
5975. Feather-footed, fin-footed or wing-footed.
5976. Waxbill finches.
5977. Yes, both are in the same order (Coraciiformes).
5978. The first egg ever seen was collected on an ice-floe between 1839 and 1843, by a member of a French expedition under the command of Dumont d'Urville. Nothing was known of the penguin's breeding habits at the time, and the significance of the egg was not recognized. The first eggs collected for science were the three acquired by Dr. Edward Wilson, Birdie Bowers and Apsley Cherry-Gerrard at Cape Crozier (Ross Island) during the winter of 1911.
5979. No, their bills are tiny.
5980. The Great Cormorant.
5981. No, their nests are domed.
5982. Twelve of the 20 species.
5983. About 20 foot-stomps per second.
5984. This is often stated, but at least two gallinaceous birds and a jacana apparently have green pigment as well. In turacos, the pigment is known as turacoverin.
5985. The Chinstrap Penguin.
5986. The crow's nest.
5987. The extinct Great Auk.
5988. Yes, but only slightly.
5989. Yes.
5990. For their oil and flesh.
5991. The Brown Skua.
5992. Australia.
5993. A heron.
5994. None.
5995. The Hooded Grebe, discovered in 1974 by the Argentine ornithologist Mauricio Rumboll, in Laguna Las Escarchadas (Patagonia) at an elevation of 2,438 feet (743 m).
5996. At the nest site.
5997. The name refers to the whitish or grayish feathers on the heads of breeding adults, implying old or ancient.
5998. Yes, it is a viral disease confined to birds.
5999. No, its flight is silent, which is no doubt of great value to the eagle as it swoops down on its prey.

Answers

6000. Yes, sacred since the early days of Hinduism, it has been strictly protected. The cranes have lost much of their fear of humans and prosper over much of northern India. However, some populations are declining, and in parts of India the cranes have disappeared altogether.
6001. A bazaar of murres.
6002. The Scarlet Macaw.
6003. To crake or craking.
6004. No, but the Short-eared Owl is.
6005. The Veery.
6006. A large cuckoo.
6007. No.
6008. No more than 50, and possibly as few as 30.
6009. No, although some are certainly nomadic.
6010. No, they have short, finch-shaped bills.
6011. Yes. Magellanic and Falkland Flightless Steamerducks sometimes nest in abandoned Magellanic Penguin burrows, but the ducks more typically nest on the surface.
6012. The Antarctic (Southern) Giant-Petrel.
6013. Jeff *Jaeger*.
6014. Yes.
6015. The Boat-tailed Grackle.
6016. The Gentoo Penguin, whose tail feathers may exceed 5.75 inches (19.7 cm) in length.
6017. No, for such a large owl, its voice is relatively weak, carrying only about 490 meters.
6018. The two African ground-hornbills.
6019. Yes.
6020. The name comes from the River Phasis in Asia, where the birds were said to be numerous.
6021. To control mosquitos.
6022. Bright red.
6023. No.
6024. The coot.
6025. A rookery.
6026. The Southeast Asian Black Magpie.
6027. Yes.
6028. A snipe.
6029. Known locally as *l'arimanon*, meaning "coconut-bird," the name refers to the little parrot's habit of nesting in cavities or hollows of coconut palms or in rotting coconuts.
6030. No, its specific name *jamaicensis* implies Jamaica.
6031. Three. Only the Whooping and Sandhill Cranes breed in the Western Hemisphere, but the Common Crane is a rare visitor.
6032. Yes. The mandibles are rapidly vibrated, and courting males hold their vibrating bills against their inflated throat pouches, which function as sound boxes.
6033. Yes, but only as casual visitors.
6034. Yes.
6035. Both feet and flippers are used, but the feet provide most of the power.
6036. Yes.
6037. Yes.
6038. The Blue-headed Vireo.
6039. The Rose-breasted Cockatoo.

Answers

6040. No, but vultures do gather at village dumps in Asia and Africa.
6041. Yes.
6042. The male Guianan Cock-of-the-rock. Its broad, truncate inner secondaries have outer webs lengthened into long, springy filaments.
6043. The Bank Swallow, because it often nests under eves.
6044. Only about nine ounces (255 gms).
6045. The original name was craneberry because the berry recalled the shape of a crane's head.
6046. No, the bunting is actually whiter during the summer.
6047. No, most do, but a few species have four toes.
6048. Yes.
6049. An estimated 130 to 150 pairs, but this figure is probably conservative.
6050. The Double-crested Cormorant.
6051. Between three and six times more efficient.
6052. Ten: five grouse, three ptarmigan and two prairie-chickens.
6053. Bright red.
6054. The Common Snipe, because its spring appearance coincides with the appearance of spawning shad.
6055. Yes, as long as both partners are alive.
6056. The Boat-tailed Grackle.
6057. John Stevens Henslow, after whom Henslow's Sparrow is named.
6058. A lame duck.
6059. Yes.
6060. Yes, at least three species.
6061. A puffinry.
6062. The crabs are stabbed.
6063. Yes, but in general, hummingbirds lack melodious songs. The Wedge-tailed Sabrewing has a very pleasant song and is also known as the nightingale or singing hummingbird. The soft musical songs of the Vervain and Wine-throated Hummingbirds are also quite lovely.
6064. The gier eagle.
6065. To protect military carrier pigeons, because hunting Peregrines readily take pigeons.
6066. No, the birds are large, neotropical cotingas.
6067. No, it is one of the few non-colonial breeding herons.
6068. The male Australian Gang-gang Cockatoo.
6069. The Red Phalarope.
6070. The Waved Albatross has been observed attacking boobies as they take off after a successful dive, forcing them to regurgitate their recent catch.
6071. A miniature murre.
6072. No. Adults of this gigantic worm reportedly attain a length of 23 feet (7 m), with a body diameter of about an inch (2.5 cm). Such an extreme length is probably unlikely, especially considering that earthworms are easily stretched. The average length is only 4.0-5.5 feet (1.2-1.7 m.
6073. Yes.
6074. Nest-prospecting Manx Shearwaters break puffin eggs or toss them out.
6075. Yes, and this is one means of separating the two birds in flight.
6076. No. Their powerful claws act as crampons, and their short hind toe may be an asset for descending steep slopes.

Answers

6077. Yes, some do.
6078. Colorful Australian parrots.
6079. The Galapagos Mockingbird.
6080. Guillemots have a whistling call.
6081. Ostriches lack a uropygial gland, thus their feathers are not waterproof.
6082. The Andean Condor.
6083. Possibly because the relatively delicate process of shell formation is best accomplished during the night, while the female is immobile.
6084. No, they have round white spots, but the white spot is crescent-shaped in male Barrow's Goldeneyes.
6085. As rapidly as 20 minutes.
6086. No.
6087. The Vikings.
6088. No, the little flycatchers were named for their calls.
6089. None.
6090. The Common Crane.
6091. A single egg, unlike most terns.
6092. Limpkins often flick their tails.
6093. Green, hence it is also known as the Green Peacock.
6094. Yes, but only the four southern diving-petrels.
6095. The Red-breasted Merganser.
6096. Termites.
6097. The Parasitic Jaeger.
6098. Probably the Little Green Bee-eater. In 1977, several nests were discovered at an altitude of 1,302 feet (396 m) below sea level at the Dead Sea. Several burrows were 5.2 feet (160 cm) in length, so the nesting birds were possibly 1,307 feet (nearly 400 m) below sea level.
6099. No.
6100. No, in the U.S., the jay only occurs in Texas.
6101. Tinamous are practically tail-less, and tails function as rudders in flight.
6102. The Purple Martin. Its Mayan name is *cozumel.*
6103. No, the mean clutch size is slightly over 11 eggs.
6104. The thick-knees (stone-curlews).
6105. No.
6106. No.
6107. Yes, some owls even inhabit London.
6108. A partridge.
6109. It was taken as a sign of approaching war.
6110. The Short-eared Owl. The Barn Owl is a rare vagrant and may have bred in at least one locality up to 1951.
6111. No, flamingo chicks are down-covered.
6112. An Ostrich feather was symbolic of justice because of its perfect symmetry.
6113. The Jackdaw.
6114. Carmine or crimson.
6115. The Bearded Bellbird. The cotinga was named for the numerous black, string-like wattles hanging from the male's throat.
6116. The Noisy Pitta.
6117. At least 141.

Answers

6118. The Gentoo Penguin.
6119. Only 8-9 species inhabit lowland New Guinea rainforests. All others are highland birds.
6120. No.
6121. The Pukeko (Purple Swamphen).
6122. A pelican, and there is convincing evidence that the Shoebill might be related to pelicans.
6123. Woodpecker and barbet holes.
6124. No.
6125. No, it is a noisy cotinga.
6126. A parrot that had been trained to greet him.
6127. No, the birds generally nest in trees or bushes.
6128. Dark blue.
6129. Occasionally in Hawaii.
6130. No.
6131. The Snowy Owl.
6132. Audubon noted it ate myrtle berries, especially during the winter.
6133. The Black-browed Albatross, with a population that may exceed 600,000 pairs.
6134. Yes, but the birds also sleep in a prone position.
6135. No, it has a wedge-shaped tail.
6136. The Wood Duck. Such a nest site can result in mortality if a female becomes trapped and cannot escape. Prospecting goldeneye drakes may also go down a chimney and subsequently perish.
6137. Up to 86 days.
6138. The name of the endemic honeyeater refers to the two white tufts of feathers on each side of its throat, recalling the two white *labels* worn by clergymen.
6139. About 4,500 birds, most of which are in Florida and Louisiana.
6140. Yellow.
6141. Pufflings.
6142. Yes, but coarse, hair-like feathers are widely scattered about its head and neck.
6143. Yes.
6144. The Limpkin, because of its distinctive limping gait.
6145. The Sandhill Crane.
6146. Yes, on Emperor Goose and King Penguins as well. The birds are usually taken at sea, but the predatory seals may lunge up onto the ice to grab a penguin.
6147. The Canada Goose, American Ruddy Duck, Egyptian Goose and Mandarin Duck.
6148. The shorebirds were named for their calls.
6149. Yes.
6150. About six feet square (0.56 sq m).
6151. The Great Frigatebird.
6152. Abert's Towhee, named after Major James W. Abert, who collected the first specimen known to science in New Mexico around 1852.
6153. The Ruby-throated and Rufous Hummingbird.
6154. No, *The Condor* is a quarterly publication.
6155. Following its near-extinction in the mid-1960's, from a low of 25 or even fewer birds, the kite has recovered to some extent with a high count of 668 in 1983. In some years the population has fallen to 250 birds. The Florida drought of the late 1980's severely impacted the raptors because of their almost exclusive dependence on the apple snail, a mollusk that perishes in dry conditions.

Answers

6156. Formerly considered a distinct species, it is currently a race of the Savannah Sparrow.

6157. The Italians deliberately set August 18 for the start of the shooting season so that the greatest toll of migrating passerines can be taken. The 1,700,000 gunners shoot an estimated 200-400 million migrants annually, scattering 52,000 tons of lead shot over the countryside.

6158. Children's kites were named after the sudden twisting, diving, or rising in the wind flight of the raptors.

6159. The adult male Resplendent Quetzal of Central America.

6160. Heermann, a medical man attached to the Pacific Railway Surveys, was a field collector of birds. In 1865, at the age of 38, while collecting birds near San Antonio, Texas, he stumbled and was killed by his own gun.

6161. No, their long, slender claws are well suited for capturing voles and other small fast-moving creatures on the ground.

6162. The Chipping Sparrow, because of its habit of using horse hair as nesting material.

6163. Probably the Hawaiian race of the Hawaiian (Dark-rumped) Petrel. It nests in the walls of the Haleakala Crater of Maui at the 7,000 to 8,000-foot-level (2,134-2,438 m).

6164. The extinct New Zealand Huia. Females had a long, thin, decurved bill, whereas that of the male was short and straight.

6165. Hummingbirds.

6166. Traditional spots where Lammergeiers drop bones to break them open. Some ossuaries have been used for centuries.

6167. The Common or American Egret.

6168. The bee-eaters.

6169. A vestal virgin who broke her vows and became, by Mars, the mother of the twins Romulus and Remus. According to legend, Romulus was the founder and first king of Rome.

6170. The number of birds during northward migration reminded people of the tremendous flights of Passenger Pigeons.

6171. The Great Spotted Woodpecker feeds extensively on pine seeds, and pine trees are numerous and widespread.

6172. No, but in recent years a few Chinstrap Penguins have been observed showing some Gentoo Penguin markings, but whether or not the birds truly are hybrids is not known. In captivity, if the closely-related species in the genus *Spheniscus* are maintained together, hybrids result, such as Humboldt/Magellanic or Humboldt/Black-foot Penguin crosses.

6173. The Ring-necked Duck.

6174. Immatures frequent bower sites of adult males, often shifting leaves and twigs about in the owner's absence as they mimic his building activities.

6175. Common and Roseate Terns, because of their association with schools of mackerel.

6176. The Crested Auklet.

6177. No.

6178. The female is more brightly colored than the male.

6179. The pale-color morph of the female Pomarine Jaeger.

6180. Vultures feed on the ground, thus are vulnerable to terrestrial predators. The birds feed rapidly, and a large crop enables them to store a large amount of food that can be leisurely digested later when they are safely roosting in a tree or on a cliff face.

6181. Probably the Eared Grebe.

6182. Yes.

Answers

6183. Probably the Brown-headed Cowbird, that lays up to 40 eggs a season during a six-week period. They generally are egg producers for only two seasons, thus females could conceivably lay up to 80 eggs in their lifetime. More typically, the cowbirds average 12-15 eggs per season.
6184. No, the owls are more inclined to bark, especially during the breeding season.
6185. Three: Gould's (New Zealand) Harrier, New Zealand Falcon and Nankeen Kestrel.
6186. The Light-mantled Sooty Albatross.
6187. Yes, at least in those species studied, with honey-bees predominating.
6188. Townsend's Shearwater.
6189. No, it is a Madagascar endemic related to the vangas.
6190. The Striated Caracara nests as far south as Islas Diego Ramirez, southwest of Cape Horn, Chile.
6191. Gray.
6192. The Wood Thrush (*H. mustelina*). The Latin name means *weasel-like,* alluding to its tawny color.
6193. Yes.
6194. The penguin is fasting.
6195. Yes.
6196. Ginger Rogers.
6197. Yes.
6198. Yes.
6199. About 84 percent.
6200. Unrestricted hunting and the introduction of rabbits to Laysan Island in the mid-Pacific Ocean. Rabbits essentially denuded the entire island, removing the cover required by the indigenous ducks and other avian endemics.
6201. The Hooded Oriole, because it nests in native Washington fan palms.
6202. Yes, courting pairs call out with far-reaching duets of high, trilling songs.
6203. Yes, a number of hybrids have been recorded in Iceland, even though the King Eider is not a common Iceland breeder.
6204. The Neotropic (Olivaceous) Cormorant.
6205. It has an extremely large mouth and long head.
6206. The Osprey.
6207. More than 18 million. Vast numbers were frozen and exported to the U.S., where the tinamous were sold as South American partridges.
6208. No, although as many as four broods may be raised.
6209. A miniature heron. The sandpipers stealthily approach wary insects with retracted necks and crouched legs, and then rapidly thrust at the prey.
6210. The White Ibis, but Scarlet Ibis juveniles are somewhat darker overall.
6211. Henry P. Attwater, a pioneer conservationist in Texas. The grouse is a race of the Greater Prairie-chicken.
6212. No, the raptors feed primarily on insects and mice.
6213. Traditionally, they have been closely linked with the corvids, but it may be that birds-of-paradise are more closely related to starlings.
6214. About 10 percent. The same is true for bee-eaters and rollers. Young are starved for a day or so to encourage them to leave.
6215. Yellow.
6216. An eagle.

Answers

6217. It has a preference for aquatic habitats and easily catches small fish on the surface in flight with its feet, like a fish-eagle. It is also known as the Vulturine Fish-Eagle.
6218. The three puffins, especially the Horned Puffin.
6219. The mockingbirds and thrashers.
6220. The adult neotropical King Vulture.
6221. No, but males do.
6222. No, but there are exceptions. Some queen termites may live longer than 50 years, although few survive past 15 years,. The North American 17-year locust lives almost two decades.
6223. Because fish must surface more often and are more easily captured.
6224. No, but the five species in the genus *Tinamus* may roost in trees. Most tinamous do not roost, as the birds are notoriously inept at maneuvering, and landing on a branch is difficult. The few that do roost in trees select a rather thick horizontal branch where they can rest comfortably on their folded legs, because they do not perch. The tarsus of tree-roosting species is rougher than other tinamous, providing a better grip.
6225. Yes, a habit shared with bee-eaters.
6226. There is no difference.
6227. None.
6228. No, but Parasitic Jaegers sometimes do.
6229. Yes, the black includes all the primaries and half the secondaries.
6230. The terrestrial birds commonly follow large, arboreal, fruit-eating and nut-eating animals to take advantage of dropped food.
6231. Ross' Goose is often cited as the smallest, but the Brent is actually smaller. Males weigh only 3.4 pounds (1.5 kg), as opposed to an average of 4.0 pounds (1.8 kg) for a Ross' gander. The Cackling Goose, a race of the Canada Goose, is the smallest of all, with females weighing only 2.8 pounds (1.3 kg), scarcely larger than a Mallard.
6232. The Tovi Parakeet.
6233. Yes, non-breeding flocks may consist of hundreds of birds.
6234. At least 70.
6235. The adult female Double-wattled (Southern) Cassowary. Its huge spike-like claw is up to five inches (12.7 cm) long and more than an inch (2.5 cm) in diameter at the base.
6236. Chicken soup.
6237. About 300 beats per minute.
6238. The King Eider.
6239. Only three: Lava and Swallow-tailed Gulls of the Galapagos, and the Silver Gull of New Caledonia.
6240. Generally not before their fourth or fifth year.
6241. The United States.
6242. The Scissor-tailed Flycatcher.
6243. The Forest Wagtail moves its tail from side to side, but typical wagtails bob their tails up and down.
6244. Yes, some Asian hill and tree-partridges are arboreal, a trait reflected in their generic name *Arborophila.*
6245. No.
6246. The Great Blue Heron.
6247. About four times larger than the predicted size.
6248. No, it is generally encountered in oak canyons between the 5,000 and 8,000-foot-level (1,524-2,238 m).

Answers

6249. Yes, although the storks are generally encountered alone or in pairs. Up to 50 may gather at favored roost sites.
6250. The Australian Owlet-Nightjar.
6251. A marine animal similar to a jellyfish. It has a transparent, gooseberry-shaped body with a pair of long, trailing tentacles.
6252. The Willow Ptarmigan.
6253. A parrot.
6254. The Everglades (Snail) Kite. Two-thirds of its eggs generally fail to hatch.
6255. Primarily a subantarctic nester, Gentoo Penguins also breed along the northwest sector of the Antarctic Peninsula. The southernmost rookery is considered to be Petermann Island (65° 10'S), but in January of 1988, I located several pairs nesting among Adelie Penguins in the Yalour Islands, some five miles (8 km) to the south (65° 14'S).
6256. No, the gamebird runs at only about two mph (3.2 km/hr).
6257. Yes, but they more typically feed from the surface.
6258. The harrier.
6259. The Falkland Islands, where approximately three-quarters of the population breeds (400,000 pairs).
6260. Yes, both incubate for about 30 days.
6261. No, adults have yellow bills, but juveniles have black bills.
6262. The Purple-backed Thornbill has a bill only 5/16th's of an inch (7.9 mm) in length.
6263. The Peruvian Booby.
6264. Yes.
6265. Yes, as a rule.
6266. The Long-tailed Jaeger.
6267. The Ancient Murrelet.
6268. No, the specialized storks are nocturnal feeders.
6269. No.
6270. The Andean Condor.
6271. Lincoln's Sparrow.
6272. The Northern Hawk-Owl.
6273. No.
6274. Following the savage land battle for Saipan Island in the Marianas group in 1944, the Japanese Imperial Navy responded by sending its aircraft carriers to Saipan waters. A major carrier duel resulted in which American naval aviators shot down many of Japan's front-line pilots, while sustaining relatively few losses themselves. The battle was a huge disaster for Japan, not only because of the loss of so many aircraft and pilots, but primarily because an air base was gained within bomber range of Tokyo.
6275. Vaux's Swift, that only weighs 17 grams (0.6 oz).
6276. No, abdominal organs are typically taken first.
6277. The Wood Duck.
6278. Alexander von Humboldt in 1799, in Venezuela.
6279. It is divided into two or more subspecies.
6280. Members of the expedition exterminated the last of the rabbits that were introduced in 1903 or 1904, by Captain Max Schlemmer. The rabbits negatively impacted the native avifauna by denuding the island. Schlemmer intended to start a coconut plantation, but introduced rabbits to add variety to his diet. It is also said he intended to start a rabbit-canning business.

Answers

6281. Yes.
6282. An average of 140 days.
6283. No, the chunky alcid weights more than 2.5 times as much as a robin: 7.8-8.5 versus 2.5-5.2 ounces (220-240 versus 70-90 gms).
6284. The raptors watch where the turtles lay their eggs.
6285. The White (Fairy) Tern can carry as many as 12 fish.
6286. No, it is more numerous.
6287. The Hawaiian Goose (Nene) is restricted to the lava flows on the big island of Hawaii and Maui.
6288. None.
6289. Yes.
6290. The extinct Dodo.
6291. The brain.
6292. Approximately 350,000, in 1989.
6293. No, a trumpeting Emperor Penguin points downward. The closely-related King Penguin points to the sky when calling.
6294. The Crab Plover.
6295. Black.
6296. Feral pigs that take eggs and chicks.
6297. Yes. The world's largest surviving lizard at nearly 10 feet (3 m) in length and weighing up to 365 pounds (166 kg), consumes ground-dwelling birds. More typically, it feeds on wild deer, hogs and carrion.
6298. Bright blue. Monarchs are Old World flycatchers.
6299. Rhea.
6300. No, but some may re-lay if a clutch is lost early.
6301. Yellow and green.
6302. The bright scarlet bird disappeared in 1923, apparently the victims of a violent gale that wiped out the last three millerbirds on the islands, and reduced the Laysan Rail population to two birds.
6303. The Marabou Stork and the aberrant Shoebill, although the Hamerkop partially retracts its neck. The Marabou may extend its neck during short flights.
6304. Someone in the house was destined to die.
6305. The Jamaican Least Pauraque was last recorded in 1859, 13 years before the introduction of the Indian mongoose. It is possible the elusive nightjar still exists in the semiarid and little known Hellshire Hills.
6306. The giant-petrels, fulmars and broad-billed prions.
6307. The Bateleur eagle.
6308. The New York Historical Society.
6309. Yes, but not always.
6310. The frigatebirds. Their huge wing area and light weight is such that they have the lowest wingloading of any bird measured, and they can remain airborne for days.
6311. The Pelagic Cormorant, that breeds as far north as Wrangle Island, Russia.
6312. Presumably this makes them more inconspicuous, because the predominantly green-plumaged birds blend well with the green foliage, and resemble hanging leaves.
6313. No, only a single hybrid is documented, which is surprising because their ranges overlap. Hybridization may be more frequent than the record suggests because hybrids might not be recognized as such.

6314. *Sula* is the Icelandic word for gannet.
6315. A neotropical guan.
6316. Yes.
6317. The Bobolink breeds as far north as Canada, and winters in southern Brazil and northern Argentina.
6318. No, two are recognized. The Lesser Roadrunner (western Mexico to Nicaragua) is similar to the Greater Roadrunner (southwestern U.S. to southern Mexico), but is smaller and its foreneck and chest are unstreaked.
6319. The drake Tufted Duck and Red-crested Pochard.
6320. Yes.
6321. No, both sexes of the Hispaniolan birds are identical.
6322. The bird has proven to be notoriously difficult to keep for any length of time because of the problem of duplicating its highly specialized diet of leaves and foliage. In 1989, two males and four females were collected in Venezuela, and subsequently exhibited in the New York Zoological Park.
6323. The Slaty-tailed Trogon.
6324. No, its tail is distinctly wedge-shaped.
6325. No, most winter along the Pacific coast.
6326. Yes and no. The raptors often succeed 90 percent of the time, but in some areas the success rate falls to possibly only 25-33 percent.
6327. At least 38.
6328. None.
6329. Yes, some wrens are regarded as superior singers, and their melodious voices are unusually loud.
6330. The Lovely Cotinga. Males are glossy blue with a triangular patch of rich purple on the abdomen and another on the throat, separated by a blue breast-band.
6331. Black.
6332. Yes.
6333. Bermuda was settled by the British in 1609, and the winter famine of 1614/15 was so severe that the starving settlers nearly exterminated the birds. This prompted the governor, who was alarmed at the scale of the massacre, to issue a proclamation protecting the petrels during the breeding season. This became law in 1621, one of the first bird-protection laws enacted.
6334. Its black crest has a very pronounced forward curl.
6335. Pine needles.
6336. The shags flew out to meet incoming whaling ships and would fly alongside the crow's nest, where they were easily knocked down on deck. The birds were much sought after as food.
6337. The House (English) Sparrow.
6338. Male Montezuma Oropendolas attain a length of 20 inches (50.8 cm).
6339. No.
6340. As of 1985, 3.2 million metric tons, of which 13 percent originated from tanker wrecks or groundings.
6341. The crow.
6342. Not as a rule, although some do so occasionally.
6343. Probably because of the puffiness of the puffin's round belly, as if it were swelling and puffing out.

Answers

6344. Yes.
6345. The Arctic Warbler, in Alaska.
6346. Yes, but huge winter flocks cause major crop damage.
6347. No, but males assist with chick-rearing.
6348. Yes.
6349. In 1963, an aircraft carrying a large consignment of birds from Cuba to Europe made an emergency landing at Nassau. Some birds died during the delay, but many others were released, and the grassquit subsequently became established.
6350. *The Emu.*
6351. The Eared and Pied-billed Grebe.
6352. Lovebirds.
6353. No, the insular duck is a surprisingly capable flier.
6354. The Peregrine Falcon.
6355. Yes.
6356. Only 10 days, but adult size is not attained until 6.0-7.5 weeks of age.
6357. The Pied-billed Grebe.
6358. Possibly because their smaller size is an advantage when pursuing faster prey. A larger size might also be an advantage for females protecting the nest.
6359. Cassowaries have no tail quills.
6360. No, despite their name, Common Loons are not abundant anywhere.
6361. Ostriches are often preyed on by lions, and some rheas are no doubt taken by pumas. Feral cats may pose a threat to kiwis.
6362. Black.
6363. The Tawny Eagle.
6364. No, but the females of many species do.
6365. No, it flies with slow, powerful wingbeats, alternated with glides.
6366. The swallow initially appeared in the late 1950's, and since then its spread has been explosive.
6367. Mudskippers.
6368. No.
6369. No comprehensive census work has been conducted due to inaccessible breeding sites and dense vegetation. Crude estimates indicate 5,000-10,000 pairs, with some estimates as low as 1,000 pairs.
6370. The four South American seedsnipes.
6371. The Andean Condor.
6372. The Red and Red-necked Phalarope.
6373. The adult male Painted Bunting.
6374. Based on one study, the raptors were only successful in killing prey once in 15 attacks, or about 6.6 percent of the time. Some misses were possibly mock attacks.
6375. No, it is one of the most nocturnal owls.
6376. Kansas.
6377. The Red-eyed Cowbird.
6378. Yes.
6379. No, the little seabirds typically forage close to shore.
6380. Yes.
6381. The Gray Jay, because of its habit of perching on a moose.
6382. A total of 347, of which 98 have bred on the Mediterranean island.

Answers

6383. About 6.5 pounds (2.9 kg), but domestic drakes weigh up to 11 pounds (5 kg).
6384. Nuthatches.
6385. The Ibisbill of the Himalayas and the Tibetan Plateau.
6386. The Boreal Owl.
6387. Yes. Unlike shoveler ducklings, Pink-eared Duck young have the highly modified spatulate bill at hatching, evidence of a specialized lifestyle from the very beginning.
6388. Yes.
6389. Four: Lapland, Chestnut-collared, McCown's and Smith's Longspur.
6390. No, the Lesser Prairie-chicken is more inclined to occupy drier habitat.
6391. The 12-inch-long (30 cm) parakeet weighed about 10 ounces (284 gms).
6392. The nuthatches, and it is in the same family (Sittidae).
6393. The three-toed sloth.
6394. Red.
6395. The Ivory Gull.
6396. No, their nests are large and bulky.
6397. Yes.
6398. T. H. Huxley. The phrase alludes to the fact that birds may have evolved from reptiles.
6399. The Pukeko (Purple Swamphen).
6400. The Galapagos Hawk.
6401. Its eyes protrude from its head. This affords the bird a 360 degree field of vision, enabling it to keep a potential source of danger in view while facing in the opposite direction.
6402. Sea otter pups. When females dive for food, pups are left on the surface where their bleating cries may attract an eagle.
6403. A pelican's foot.
6404. Yes, in some species flocks up to several thousand swim lined up abreast and cooperate in driving shoals.
6405. In a bizarre breeding strategy, *either* parent of a kite pair may desert the family, leaving the other to feed the young while he or she wings off to form a new pair bond and rear another brood.
6406. The Ringed Turtle-Dove, but only sterile males resulted.
6407. The White-tailed Sea-Eagle, Rough-legged Hawk, Gyrfalcon, Peregrine Falcon and Merlin.
6408. No, their undulating flight is interspersed with gliding.
6409. Foraging birds dig and kick about much forest litter in quest of food.
6410. Generally mammal fur.
6411. Based on calculations, each wing would be 32 feet (9.8 m) long.
6412. Grebe fur. The feathers were in great demand for adorning women's hats and shoulder capes.
6413. Snow Buntings.
6414. The Gray Catbird.
6415. No, the falcons hunt in pairs.
6416. The birds were released by Vikings at sea to determine if land was near. If the birds failed to return, it was assumed land was close at hand.
6417. Five, all of which had feather impressions. A separate isolated feather impression has also been discovered.
6418. The shorebirds.
6419. Yes.

Answers

6420. The legendary Mary Pickford.
6421. Considering the enormous size of the egg, the female would probably not be able to lay it if it were round.
6422. The first record dates from 1200 B.C., when four birds were released in opposite directions to carry the message of Pharaoh Ramses III's assumption of the throne.
6423. Suet.
6424. Yes, in 1921, a Lammergeier was observed at 25,000 feet (7,620 m).
6425. Up to three pounds (1.4 kg).
6426. Only in Australia.
6427. Yes, the fur is used to line their nests.
6428. The Banded Stilt.
6429. Yes.
6430. *Island of birds.*
6431. Yes.
6432. The Yellow-eyed Penguin of New Zealand and several of its subantarctic islands. Pre-molt males weigh as much as 18.7 pounds (8.5 kg), but northern Gentoo Penguins are nearly as large.
6433. Probably the corvids.
6434. Lost lead weights or sinkers are ingested by foraging swans, resulting in fatal lead poisoning, leading to serious local declines.
6435. The megapodes (incubator birds) and the obligate brood parasitic nesters, such as honeyguides and some cuckoos.
6436. Yes.
6437. No, females are essentially voiceless.
6438. Approximately 290, or about 80 percent of its diet.
6439. Thrasher is probably a variant of a dialect form of thrush: *thrusher.*
6440. Between 5,000 and 10,000.
6441. The Great Tinamou.
6442. Dogs scare off or divert the attack of male rheas. A charging rhea can prompt a horse to shy or bolt.
6443. Yes. About 93 percent of seabirds breed in colonies, whereas only about one terrestrial bird in seven is a colonial nester.
6444. Probably because white birds are able to approach fish more closely than dark ones before being detected.
6445. No, mating take place on the water.
6446. Vultures feed on carrion, and a strong, grasping foot is not required to restrain or capture prey.
6447. The Gray Heron.
6448. No, the little birds are solitary.
6449. On cliffs.
6450. No, the name came from the Malay word *luri* for the colorful parrots.
6451. The Blood Pheasant.
6452. The nitrogenous heron guano adds nutrients to the water, thus fish thrive.
6453. Yes.
6454. No, the pelican's heart-rate is about 150 beats per minute, whereas that of the turkey is about 93 a minute.
6455. Some cuckoos.

Answers

6456. No, a kiwi's bill is flexible.
6457. The oil may be used in suntan oil.
6458. No, aracaries have longer tails.
6459. The tiger-herons.
6460. The lyrebird.
6461. No, females typically have longer bills.
6462. Foretell the future.
6463. From its distinctive call.
6464. Africa, with at least 30.
6465. Yes.
6466. Nothing. Newly-hatched chicks lose weight until they leave the nest at 2-3 days of age and make their way to the sea.
6467. Cooper's Hawk.
6468. The Black Skimmer.
6469. Yes.
6470. The Bufflehead, but this ability is not typical of most diving-ducks.
6471. Its clutch consists of two eggs, unlike most pheasants that lay more eggs.
6472. A babbler from the highland forests of eastern Madagascar.
6473. Wintering American Wigeons.
6474. A host of sparrows.
6475. The red represents the blood of slaughtered Indians.
6476. Yes, it nests on the Pacific coast islands of Clarión and San Benedictio, and off northern Yucatan.
6477. No. The huge parrot often forages on the ground in cattle corals, sits on fences, drinks out of cement water-troughs, and may nest in trees next to ranch houses.
6478. The flightless Weka rail.
6479. Poland.
6480. Probably because chicks taste like very greasy mutton.
6481. The Crowned Hawk-Eagle, but the Martial Eagle is larger.
6482. The Pittsburgh Penguins.
6483. The Bare-eyed Cockatoo.
6484. Yes.
6485. No, with closed eyes.
6486. Yes, at times.
6487. No, the sea-eagle will trigger flight, but the Wedge-tailed Eagle is generally ignored by the cranes.
6488. The parrots bite savagely.
6489. Sabine's Gull.
6490. Yes.
6491. The Kea.
6492. No.
6493. Yellow.
6494. No, red-color morphs predominate, but the situation is reversed in the Rocky Mountains.
6495. The droppings, enclosed in a fecal sac, are carried from the nest and deposited in a stream, or may be swallowed by the female.
6496. Yes.
6497. Yes, the platypus is said to be capable of swimming at twice the speed of most waterfowl.

Answers

6498. The pipits and wagtails.
6499. The Sandhill Crane, because of its trumpet-like calls.
6500. Yes. In the southern portion of their breeding range, two broods may be reared, an unusual situation in northern ducks.
6501. Yes, the beard is clearly visible from nearly 1,000 feet (300 m) with the naked eye.
6502. The male Bobolink in spring plumage.
6503. Grouse do not have leg spurs.
6504. The thousands of petrel nesting burrows allegedly undermined the heathland, reputedly making it dangerous for horses, and leading to soil erosion. Cats were introduced onto New Island (one of the prion's breeding strongholds) to reduce the population.
6505. Snipe-nosed.
6506. New Delhi, India. Within 30 square miles (77.7 sq km), 715 breeding territories were counted with 679 occupied nests, an average of 19.3 pairs per 0.39 square mile (1 sq km). Black Kites accounted for 83 percent of the total, followed by the White-backed and Egyptian Vulture.
6507. Sharp-tailed Grouse and probably prairie-chickens.
6508. No, flamingos use only their bills.
6509. Yes.
6510. No, their featherless faces are bright orange.
6511. Keratin. Finger nails, hair and reptilian scales also consist of keratin.
6512. The Fairy (White) Tern.
6513. The avian ankle joint.
6514. Chicks may gather in concentrated crèches.
6515. The Trumpeter Swan.
6516. The single egg is held between the legs on top of the outstretched toe membranes.
6517. Probably the grebes, and possibly the jacanas.
6518. No, it is barred.
6519. The Siberian White Crane.
6520. Yes, some species do.
6521. Chameleons, so far as is known.
6522. No, they are very confiding birds.
6523. The name is from the Latin: *caper* meaning "goat" and *mulgeo* meaning "to milk" or "to suck." People formerly believed that goatsuckers sucked milk from goats in the night.
6524. Probably the White-crowned Sparrow, but the Song Sparrow is a close second.
6525. The male.
6526. The Rock Ptarmigan.
6527. Both sexes of the Magpie Goose.
6528. Yes, generally so.
6529. Only about 5-6 percent.
6530. *Ancient wing.*
6531. Yes, very much so.
6532. The soft skin forming the corner of the mouth.
6533. Yes. On at least two occasions, Short-tailed Shearwater eggs were incubated, and stones, golf balls and even teacups have also been incubated. In one instance, a Fairy Penguin even brooded a litter of rabbits.
6534. The high and dry habitat the jays require is prime real estate in high demand by developers. In 1991, the jay population consisted of less than 5,000 pairs.

Answers

6535. The toenail functions as a brace, preventing the birds from being blown over backwards in strong winds.
6536. The Eastern Meadowlark.
6537. Kittlitz's Murrelet. The estimated world population is only 18,300, of which all but the estimated 550 Russian birds inhabit Alaska. While the number of birds killed in the spill was relatively low (somewhat in excess of 500 birds), even this small number represents about three percent of the world population.
6538. Yes, as a rule.
6539. The Sparrowhawk, currently known as the American Kestrel.
6540. The Puerto Rican Parrot. Half of its range was damaged, with some reports indicating that only six of the wild population of 46 Amazons survived the storm.
6541. The Harlequin Duck, but such a feat is unlikely.
6542. Yes.
6543. Their eggs are more elongate.
6544. Yes, and the name was also applied to the extinct Passenger Pigeon.
6545. Cory's Shearwater.
6546. One hundred pounds (45.4 kg).
6547. Yes.
6548. Yes, the birds sing night and day.
6549. Stephan *Crane*.
6550. No. Except for the two cocks-of-the-rock, in which nestling down is long, but sparse, cotinga nestlings are thickly covered with down. This is unusual in most passerines.
6551. Pinkish flesh, but its legs are black.
6552. Probably Australia.
6553. Yes.
6554. The Pied Avocet.
6555. Rhea.
6556. The Salton Sea (southern California).
6557. The Caribbean Flamingo.
6558. Norway. When nesting commenced in 1983, the eggs were the first in Scotland in more than 70 years. The first sea-eagle hatched in 1985.
6559. Yes.
6560. Yes. The Komodo dragon, for example, has an incubation period of about eight months.
6561. No, the glands are small relative to the bird's size.
6562. A *lorry*, but pronounced the same as the parrot (lory).
6563. Yes.
6564. Nearly 200 pounds (90 kg) of pressure.
6565. Yes.
6566. Presumably either sex is equally fast.
6567. The 5-inch-long (12.7 cm) cockatoo-like crest feathers.
6568. Migration.
6569. The Bateleur eagle.
6570. With the aid of a magic ring.
6571. 1840.
6572. The Spot-billed, Pink-backed and Dalmatian Pelican.
6573. The Great Gray Owl, because of the fanciful notion that it is a phantom or other object of superstitious terror or dread.

Answers

6574. Yes.
6575. The ovenbirds often nest on utility poles, resulting in short-circuits.
6576. No, prey is generally eaten on the ground.
6577. The flockmaster (master cock).
6578. The shortness of their legs does not permit a full down-stroke of their wings, forcing swans to run over the surface, striking the water alternately with each foot.
6579. Six square miles (15.5 sq km) of virgin swamp timber. The same area could support 36 pairs of Pileated Woodpeckers or 126 pairs of Red-bellied Woodpeckers. The Mexican Imperial Woodpecker requires as much as 6-7 square miles (15.5-18.0 sq km).
6580. The Andean Condor.
6581. Yes, generally so.
6582. Yes, flying insects appeared about 220 million years ago.
6583. No, the rare gull generally nests in swampy lowlands near lakes.
6584. The Rufous Hornero (*el hornero*).
6585. Yes.
6586. No. The sexes of both the Brazilian Merganser and the extinct Auckland Islands Merganser are (or were) virtually identical.
6587. About 75 times as thick.
6588. Tiger sharks. About one albatross fledgling in ten is devoured by a shark before it makes it past the reef.
6589. The rabbit.
6590. The quail.
6591. The American Wigeon.
6592. A siege of herons or bitterns.
6593. The Common Murre.
6594. Grouse drumming is of such a low frequency (40 cycles a second) that probably the sounds are of too low a frequency for the owls to hear. The owl's hearing ranges from 60 to 7,000 cycles a second.
6595. No, the Bronzed Cowbird's bill is significantly longer.
6596. The conversion of the coast prairies of Louisiana and Texas into agriculture lands for rice crops. In 1990, 494 grouse were counted, but this number may be conservative.
6597. Yes, some almost exclusively so.
6598. The Kaka.
6599. Because of the general's punctiliousness in dress and decorum.
6600. Larks are especially well represented in Africa, with at least 47 species.
6601. He has been cheated, duped or tricked. The cheat himself was known as a guller.
6602. The Double-eyed Fig-Parrot.
6603. No, all three species are quite tame and fearless.
6604. No, males tend to live longer.
6605. Yes.
6606. Primarily insectivorous, they *may* consume small chicks, especially sparrows and weavers, but more likely, while inspecting a nest for insects or eggs, small chicks are simply tossed out. As a result, woodhoopoes are often mobbed when they approach nests.
6607. Peru and Chile.
6608. Yes.
6609. Yes.
6610. No, the grouse is a prairie inhabitant.

Answers

6611. A juvenile Black Swan.
6612. Probably the ordinary house cat. In Wisconsin alone, cats may account for millions of birds annually.
6613. The Lapped-faced Vulture.
6614. Yes.
6615. A brood of chickens.
6616. Only two: Red-headed and Prong-billed Barbet.
6617. The mo*hawk.*
6618. Up to five pairs.
6619. Yes, as a rule.
6620. Probably the Budgerigar. It is by far the most numerous pet psittacine worldwide, with a population estimated at 500 million birds.
6621. Yes.
6622. Yes.
6623. No, kiwi ear openings are rather prominent.
6624. Yes, their tails are stiff, with the tips consisting of strongly stiffened bare shafts curved abruptly inwards.
6625. White.
6626. No, but adult males have colorful lappets.
6627. The mosquito.
6628. Russia.
6629. No, their nests are especially difficult to locate.
6630. No, but the Bohemian Waxwing is.
6631. The obligate brood parasitic-nesting Bronzed Cowbird.
6632. Probably the Brent goose.
6633. The owl is virtually fearless, and its ease of capture was believed to be a result of its bad daytime vision. To northern natives, the bird was known as the *blind one.*
6634. No, thrashers do not range south of Mexico.
6635. Yes.
6636. Two eggs, but rarely three are laid.
6637. Cotinga nests are often reduced to an absolute minimum for inconspicuousness. As a result, generally only one or two young are reared, the most the nest can accommodate.
6638. Frounce. The disease is usually contracted when raptors consume infected doves or pigeons.
6639. Yes.
6640. No, the warbler typically nests on the ground.
6641. No, but flickers apparently dust-bathe often.
6642. Yes.
6643. The frigatebirds, but frigatebirds everywhere have low breeding success, often with a 75 percent failure rate.
6644. Australia in 1974, and New Zealand in 1990.
6645. No, Little Auks typically copulate ashore.
6646. A ternery.
6647. The birds have very dense ventral plumage and the trapped cushion of air forces the little phalaropes to float high.
6648. Yes.
6649. The Hermit Thrush.

Answers

6650. Thrushes.
6651. Between 2,000 and 5,000 of the alpine parrots.
6652. The Boat-tailed Grackle.
6653. It becomes water-logged.
6654. No, it is an inland Asian resident.
6655. Traditionally Japan, but Korea may be equally, and possibly even more important.
6656. No.
6657. Crayfish, frogs and various types of fish.
6658. The Canvasback, presumably because of the shape of its head.
6659. The pelican was named after the Dalmatian coast of the Adriatic Sea, once an important wintering area.
6660. *Grus* comes from the Latin and translates to "a kind of bird," especially a crane. The name might refer to their hollow, guttural or grunting voices.
6661. No, the red throat is not attained until the first molt.
6662. No.
6663. Rhea fat was believed to relieve the effects of a venomous snake bite, and was used to treat rheumatism.
6664. Approximately 252.
6665. Yes, but the high Arctic-breeding eagles also nest on cliff ledges, or more rarely on the ground on small islands.
6666. The Marabou Stork, African Fish-Eagle, Black Kite, Kelp Gull and Fan-tailed Raven.
6667. Yes, but there are exceptions.
6668. Both species of South American rheas and, to a lesser extent, Emus.
6669. The Yucatan peninsula.
6670. Yes.
6671. The hyenas. Like Lammergeiers, they specialize on bones.
6672. No, the endemic corvid is dark sooty-brown.
6673. An advanced case of syphilis that affected his central nervous system, and he could barely walk unassisted.
6674. Yes.
6675. The *snake dance.*
6676. The rhea (probably the Lesser Rhea), but it is unlikely that any still exist.
6677. The Shy (White-capped) Albatross. Males weigh 9.8 pounds (4.43 kg)
6678. Terrestrial termite mounds. The birds are locally called antbed or anthill parrots.
6679. No, the flimsy nest structures generally collapse.
6680. No, goose referred to the female.
6681. First a honeyeater, then an oriole, then a bird-of-paradise, and finally a bowerbird. It is the most brilliantly plumaged of the Australian bowerbirds.
6682. Between 10 and 30 percent.
6683. No, but grebe chicks do.
6684. It is in the same order as the cuckoos, but was formerly regarded as an aberrant galliform.
6685. Yes, on Short-tailed Albatross chicks, but such prey is not typical.
6686. No, the nestlings reseal the entrance using their own droppings. Hornbill droppings are viscous and harden into a hard clay-like substance.
6687. No, the auk tends to crouch low over the egg.
6688. The Violet-crowned, Buff-bellied, Broad-billed, White-eared and Berylline Hummingbirds have either red or partially red bills.

Answers

6689. The Greater Roadrunner.
6690. No, the sea-geese tend to fly in long wavy lines or bunches.
6691. Spiders taken in a direct flight through the web.
6692. Only about 20 percent, with most mortality occurring during the first winter.
6693. A curlew.
6694. Yes, in some years up to 1.5 tons.
6695. *Dacelo novaeguineae* implies the kingfisher inhabits New Guinea, where it does not occur. The first specimen was incorrectly said to be taken from there.
6696. Yes, some species do, especially the six crested penguins.
6697. The grouse plunge into snow banks to elude predators.
6698. The Cape (South African) Gannet.
6699. Scarlet Ibis eggs were placed under incubating White Ibis. The resultant pure-fostered offspring disappeared over the years, but hybridization resulted. Some biologists indicate that escaped birds were responsible for the introduction. In Venezuela and Colombia, where the ranges of the two ibises overlap, hybridization is fairly frequent, leading some authorities to consider them conspecific.
6700. The fish-eating bats.
6701. An imaginary animal with the body and hind legs of a lion and the head and wings of an eagle.
6702. Yellowish, but most diurnal raptors except falcons lay eggs with a greenish inner shell.
6703. Reddish-buff.
6704. A swannery.
6705. *Nighthawks.*
6706. A building of Rooks.
6707. Yes, they attack birds as large as hawks and Ospreys.
6708. The Thick-billed Murre, whereas the Common Murre is better at pursuing pelagic fish.
6709. No.
6710. Yes, but if flushed birds run, rather than fly, they almost always run uphill.
6711. The Gray Heron.
6712. The Red-footed Booby.
6713. Yes, among non-passerines, some jacamars are outstanding for their elaborate songs.
6714. The Sandwich Tern.
6715. Yes.
6716. Yes, although it is often stated that only males have the wattle.
6717. The Northern Saw-whet Owl.
6718. Yes.
6719. The Egyptian Goose.
6720. A slow, undulating, butterfly-like flight.
6721. No, the hind toe is much reduced in size.
6722. Yes, cassowaries have bare heads and necks that are brightly colored in varying shades of red, blue, purple, yellow and white, depending on the species or subspecies.
6723. Probably to prevent the roundish eggs from rolling off the cliff edge.
6724. Yes, and afterwards the birds may lie down with their wings spread to dry.
6725. All have white heads.
6726. The two African oxpeckers.
6727. Chicks are fed from the bills of their parents, whereas most shorebird chicks must gather food themselves.

Answers

6728. Primarily birds, especially frogmouths.
6729. The Wood Duck.
6730. No, an Ostrich has cushioned soles that helps prevent it from sinking in loose sand.
6731. The cerebellum, the control center for balance and coordination. These functions are far less important for a flightless bird.
6732. No, the underside of the leaf generally faces up.
6733. Yes, but juveniles are washed with rust, especially on the head and neck.
6734. The largest colony in Wisconsin, in 1871, consisted of an area 100 miles (161 km) long by 3-10 miles (4.8-16 km) wide, or about 750 square miles (1,942 sq km). The colony was estimated to contain 136 million birds.
6735. A single egg.
6736. No, their nests consist of loose, rather messy, cup-shaped structures.
6737. No, more motmots occur in Central America.
6738. Yes.
6739. The 1,850 African birds represent 19 percent of the world's species.
6740. Behind a peeling slab of bark or in a hollow branch.
6741. Townsend's Solitaire, because it catches insects on the wing.
6742. Yes.
6743. No, harriers lay abnormally large clutches, regularly 4-6 eggs, but as many as 10.
6744. Although dull-colored, the bristlebird's plumage has a lacquered appearance. This possibly sheds water in rain-soaked, dense undergrowth.
6745. Brownish, but adults are basically black.
6746. The Smooth-billed Ani.
6747. The Tawny Owl.
6748. Yes.
6749. Yes.
6750. Hunting birds lie in wait near water and seize swifts in the air as they swoop down to drink.
6751. Seven: Ecuador (Galapagos Islands), Peru, Chile, Argentina, South Africa, New Zealand and Australia. A few pairs of Humboldt Penguins are rumored to breed in Colombia.
6752. The African Fish-Eagle.
6753. Yes, but its colonies are small, consisting only of up to a dozen or so pairs.
6754. The Boat-tailed Grackle.
6755. The long tail feathers of many species produce a humming sound in flight.
6756. The Tufted Puffin.
6757. Its small, beady eyes are black.
6758. Yes.
6759. The Kagu.
6760. The raptors frequently pause in flight to shake water from their feathers.
6761. Yes, generally so.
6762. Supposedly to protect salmon stocks. Between 1915 and 1951 bounties were paid out on more than 100,000 eagles. Even when the raptors were afforded special protection in 1940, Alaska was excluded, and the bounty system there continued until 1953.
6763. About 1,635.
6764. Prairie grasslands.
6765. Chameleons.
6766. John James Audubon. Both Bachman daughters died in 1840, at the ages of 22 and 23.

Answers

6767. No, even mixed species flocks may form.
6768. Yes, but it is perfectly capable of catching flying birds.
6769. It rears up to three broods annually.
6770. The Tennessee Ornithological Society.
6771. The Eurasian Green-winged Teal.
6772. The Red-necked Grebe.
6773. Not until 1972.
6774. Acorns.
6775. Coursers are swift runners.
6776. Probably because the birds feed extensively on hard-shelled crustaceans, such as crayfish.
6777. At least 22.
6778. No. While it is often stated that a Purple Martin consumes up to 2,000 mosquitos a day, mosquitos make up only between zero and three percent of its diet.
6779. Franklin's Gull.
6780. No, their oval eggs are rather pitted.
6781. It robs food from other birds, thus it is a parasitic feeder.
6782. The Crested Auklet. The length and depth of the male's bill is significantly greater and is hooked. On this basis, the auklets can be accurately sexed about 90 percent of the time.
6783. Males sweep low over the nest dropping food near the female that she offers to the nestlings.
6784. The auks were used as bait.
6785. The Carolina Parakeet. As many as 30 parakeets were recorded sleeping in the hollow of a single large sycamore tree.
6786. It readily nests in bird houses, and human houses as well. The little wren nests just about anywhere, including mailboxes, drainpipes, old hornet nests, flowerpots, baskets, tin cans, boots, gourds, toolboxes, even coat pockets, or in any cranny in an open shed or abandoned auto. Audubon painted a pair nesting in an old hat.
6787. Red.
6788. The vulture can hit a 10 by 10 meter rock with a dropped bone from 150 meters. It generally approaches the drop zone from downwind, perhaps to increase speed, and may dip or dive, possibly to increase accuracy of aim.
6789. Lord Love a Duck.
6790. It is twisted slightly to the right, enabling the bird to reach into the aperture of a snail and around the bend of its shell to cut the columellar muscle.
6791. The Rough-legged Hawk.
6792. Yes, the Crested Oropendola, for example.
6793. Yes, but drumming activity is most intense in early spring, and may take place during any hour of the day or night.
6794. Yes, as a rule.
6795. Their eyes are generally yellow, orange or red.
6796. Yes.
6797. No.
6798. Yes, a second brood may be reared.
6799. No. Chicks may be fed for as long as a year or more after fledging by the female, thus females can only breed in alternate years. Males are free to take a new mate each season and may breed annually, but with a new partner.
6800. No.

Answers

6801. Sometime prior to 1953, the Guanay Cormorant was introduced onto Isla San Geronimo, probably because of its guano-producing capabilities, but it failed to become established.
6802. A swan, from the myth that swans call out with a melodious song just before death. Singers are also called nightingales.
6803. In the I'On swamp near Charleston, South Carolina.
6804. The Great Gray Owl.
6805. No, the male does.
6806. In excess of 100 pounds (45.5 kg).
6807. Throw the foreign egg out, or bury it by building a new nest over the original one.
6808. Yes.
6809. The honeyeaters. At least 69 species occur in Australia, most of which are endemic.
6810. Yellow.
6811. Yes, the nests are repaired and re-used.
6812. Yes.
6813. Yes, the raptors are often flown at House Sparrows, but birds are not their normal prey.
6814. The Pied-billed Grebe.
6815. Named by Prince Jerome Bonaparte, a nephew of Napoleon Bonaparte, the name apparently was a political side-swipe reflecting his republican ideals. Bonaparte is better remembered among ornithologists for having cheated by describing and naming the bird when it had been discovered by others.
6816. Yes.
6817. Yes. The raptors are highly gregarious and even breed socially, with half a dozen or so nests located within several feet (0.6 m) of one another.
6818. No, they are wary.
6819. The Military Macaw.
6820. None has feathered legs.
6821. The Shore Lark.
6822. The Gray (Canada) Jay.
6823. The primates may descend to the ground to escape the huge forest eagles.
6824. To prevent the gourds from overheating.
6825. The huge extinct Haast's Eagle.
6826. Yellow, but adults have bright orange ear-patches.
6827. Approximately 65,000 tons.
6828. Yes.
6829. The South American cocks-of-the-rock.
6830. Ammonia.
6831. No, it has declined drastically in recent years. In Florida, there has been an 89 percent decline between 1961 and 1988, and a 73 percent decline in Michigan during the same period. Conversely, the population is increasing in Oklahoma and Wisconsin.
6832. Yes.
6833. The Yellow-headed Blackbird.
6834. Its head is essentially bald.
6835. The broad, flat nest consists of sticks, but is lined with bones, mammal skin, grass, moss, hair, and even old clothing and dry dung. Lammergeier nests typically stink.
6836. Bright reddish-orange.
6837. All are black-tipped, even those of juveniles.
6838. No, juveniles lack the glossy sheen of adults.

6839. It travels about in small flocks of 6-8 birds, hence the name *seven sisters.* When two or three flocks join up, the group is called a *sisterhood.*
6840. No, the edges are extremely sharp.
6841. No, the large pigeon is mainly terrestrial.
6842. No. The adult Japanese spider crab is the largest crustacean, with an outstretched claw-span of over 10 feet (3 m). The largest documented specimen had a claw-span of just over 12 feet (3.7 m) and weighed 41 pounds (18.6 kg). In 1921, an enormous crab was caught in a net that reportedly had a claw-span of 19 feet (5.8 m).
6843. Yes, a flock of Common Cranes, was observed flying over the English Channel at an altitude of 13,000 feet (3,962 m).
6844. No, it is one of the more northerly-nesting warblers.
6845. Yes, but only rarely.
6846. Between 100 and 115 days.
6847. Altricial comes from the Latin *altrix* meaning "nurse" or "wet nurse."
6848. The Goosander, Red-breasted Merganser and Smew.
6849. No, but many species of parrots do.
6850. Generally cocked at about a 45-degree angle.
6851. A goosery.
6852. The coots.
6853. A miniature mockingbird, especially the Blue-gray Gnatcatcher.
6854. The chicks are entirely precocial, departing the nest within days of hatching and heading out to sea.
6855. Bering Island (Kommandorskiye Islands).
6856. No.
6857. Two color phases are known, the more common being gray, a color typical of the savanna woodland birds. The red morph no doubt helps camouflage owlet-nightjars that roost among the rocks in arid zones.
6858. The Barking Owl. Its high-pitched stream has earned it the descriptive local name of "screaming-woman bird."
6859. About 12 percent.
6860. A species that turns up irregularly in small numbers in areas outside its normal range.
6861. In Australia, gibbers are rounded stones, and in some regions, millions form huge stony deserts. The Gibber (Desert) Chat is an endemic of such desolate arid habitat.
6862. About 327.
6863. The mosquito.
6864. The name probably is derived from old Anglo-Saxon or German words meaning burnt, referring to the black-brown color that gives the geese the appearance of being charred. Some authorities suggest the name originated from their calls.
6865. The Eurasian Spoonbill.
6866. In 1885, the Chilean Tinamou was introduced onto Easter Island where it subsequently became established.
6867. It was said that *If the birds be silent, expect thunder.*
6868. Yellow, but the remainder of the plumage is white.
6869. Yes.
6870. The penguin is about to molt.
6871. No, they are among the most solitary.
6872. Carrion.

Answers

6873. No, but there are numerous exceptions.
6874. A frog, but only occasionally.
6875. The Apostlebird, a reference to the Country Women's Association, because it is often encountered in groups. A group is known as a *happy family*.
6876. The clutch size may consist of up to three eggs, but typically only a single chick is reared.
6877. The Hawaiian Goose (Nene).
6878. No, the little petrels occasionally dive after planktonic organisms, but bob to the surface within a few seconds.
6879. Ibises in flight.
6880. The 16 large, well-developed white primaries probably serve to make the cock's courtship more spectacular.
6881. Yes.
6882. Chalky blue-green.
6883. The gannet.
6884. Black.
6885. The remains of all but two species have been uncovered at archaeological sites.
6886. Loon neck skins were used to make carpets.
6887. The Wandering Albatross. Older males are whiter than their mates.
6888. Yes, due to feather structure, its plumage, like that of he Ostrich, is smooth and soft.
6889. The Spur-winged Goose is no longer considered a perching-duck, thus it is the largest African goose. The Egyptian Goose is second, closely followed by the Ethiopian Blue-winged Goose.
6890. No, the owl is more numerous in Siberia.
6891. Birch seeds.
6892. Yes. All hatch in about 28-29 days, from the tiny 2.3-ounce (91 gms) Pygmy Falcon to the Gyrfalcon weighing 30 times as much.
6893. The Ruffed Grouse.
6894. The American White Pelican, but amphibians are not normal prey species.
6895. Yes.
6896. Chrome-yellow.
6897. No, the sexes are essentially identical.
6898. Yes.
6899. No, Ross's Goose juveniles are whiter.
6900. No, despite the fact that Rome is only 30 miles (48 km) from the sea, poultry was more popular.
6901. No, a single egg is laid, but puffins have two brood-patches.
6902. The Cattle Egret, because of the risk of a bird-strike with aircraft.
6903. The Diamond Dove.
6904. A chattering of choughs.
6905. The sturdy nests are used as purses.
6906. No.
6907. No, the owl is absent from the West Indies.
6908. No, males generally do.
6909. Yes.
6910. The Common Wood-Pigeon is not uncommon in London.
6911. Flamingos. All flight feathers are molted simultaneously, resulting in temporary flightlessness.

Answers

6912. Yes.
6913. No, potoos have long tails.
6914. Yes.
6915. Yes.
6916. As a water-loving species, the kite is typically associated with water.
6917. No, chicks do not dive well until several weeks old.
6918. Buff.
6919. No, the eagles are apt to mate on any day of the year.
6920. No, they are only slightly larger.
6921. All five.
6922. The bee-eater de-venoms the bee by rubbing the sting-end against a branch. The birds can apparently distinguish between the stinging worker and the non-stinging drone.
6923. Yes.
6924. A rabbit hole. When the chalk dunes and sandy heaths of southern England were reclaimed for agriculture in the second half of this century, rabbit holes disappeared, and breeding wheatears also disappeared in such areas.
6925. A smaller version of the Glaucous Gull.
6926. The grebes. When in such a position, the head is pointed downwards, with the closed bill mostly in the feathers on the side of the neck, and one or both feet tucked up under the wings, which are covered by the flank feathers.
6927. Yes.
6928. About six seconds.
6929. The Merlin.
6930. Twelve, with the central pair greatly elongated.
6931. White, and the bird is included in a group collectively called the white-winged cotingas.
6932. The Greater Prairie-chicken.
6933. Alcids and gulls.
6934. Yes, from Ospreys.
6935. Dive under the water.
6936. The Sandwich Tern.
6937. The Greater Roadrunner.
6938. Yes.
6939. Vivid ruby or magenta.
6940. No, a hoopoe walks or runs with a characteristic bobbing of its head.
6941. No, most are weak fliers.
6942. Dr. Elliott Coues. The name is a diminutive of murre.
6943. Yes.
6944. Both sexes have waxy, carmine-red bills with bluish tips.
6945. A swan in British Columbia had 451 shotgun pellets in its stomach.
6946. Brush-tongues divided at the tip into four parts, each frayed with the basal part curled into a long groove at each side.
6947. Yes, the owls live around New York City.
6948. Yes.
6949. Generally the head.
6950. In 1873, one of the famed clipper ships delivered a cargo of nine Pekin ducks to Long Island. From this small founding stock have descended all the many millions of ducks of the Long Island duckling variety subsequently produced.

Answers

6951. No, the unique shorebird winters in Southeast Asia.
6952. The King Vulture.
6953. The Chihuahuan Raven.
6954. Anna's Hummingbird.
6955. Yes, in addition to insects and snails.
6956. The Rockhopper, Fiordland and Snares Islands Crested Penguin, but only a single chick is ever reared. The other three eudyptid penguins lay two eggs, but the smaller alpha egg is generally broken.
6957. The Red-capped Dotterel.
6958. Yes *(Falco)*.
6959. Yes.
6960. Yes, some African Crowned Hawk-Eagles feed juveniles for up to 11.5 months after the first flight.
6961. Yes. In 1992 a male Whooping Crane paired with a female Sandhill Crane, and a chick resulted. The family wintered at Bosque del Apache Refuge in New Mexico. The hybrid has been called a "whoop-bill."
6962. The White Ibis.
6963. Probably monitor lizards.
6964. The Spotted Bamboowren is a small neotropical tapaculo.
6965. The Budgerigar. However, veterinarians examine more budgies than any other bird, so it may be that other species examined less frequently are just as prone to tumors.
6966. The South American Hoatzin, one of the few obligate folivorous (leaf-eating) birds. Hoatzins are among the smallest endotherms with this form of digestion, which can be compared to that of a cow. The food may remain in the gut for 20 hours or more. The New Zealand Pigeon may also utilize a form of foregut fermentation.
6967. Yes.
6968. *Josephine.* She died on September 13, 1965 by battering herself to death against the cage after being badly frightened by a passing helicopter.
6969. Yes.
6970. About 33 days.
6971. Howard *Hawks*.
6972. Yes.
6973. Duck or goose decoys are deployed, and waterfowl are called in using duck and goose calls.
6974. Both sexes incubated for 12-14 days.
6975. No.
6976. Yes, the floor and sides of the burrow are lined with vegetation, generally by the male, prior to laying.
6977. The raptor is often followed by more maneuverable crows and ravens that steal scraps of bone or marrow.
6978. The flamingos.
6979. Usually 9-11 hours.
6980. The Anhinga and darters.
6981. The Northern Bobwhite and Red-shouldered Hawk.
6982. Sunflower seeds.
6983. The Wisconsin Society for Ornithology.
6984. No, they often gather in large flocks.

6985. Yes, the Short-eared Owl, for example.
6986. Only the Eurasian Thick-knee.
6987. Yes, most sunbathe regularly with outstretched wings.
6988. The Western and Clark's Grebe.
6989. Yes.
6990. No, Red-tailed Hawks weigh 10-20 percent more.
6991. No.
6992. Yes.
6993. The loons, a condition that essentially precludes walking.
6994. The monotonous repetition of its phrases when singing reminds one of a preacher.
6995. Of the 115 species of Australian rainforest birds, 49 are virtually confined to the rainforest.
6996. The pronghorn antelope.
6997. Yes, the birds can generally be closely approached.
6998. No, alcids do not have a hind toe.
6999. Pinkish.
7000. The tanager (*Tangara chilensis*) does not occur in Chile.
7001. Its name is in reference to thorny brambles where the bird is occasionally found, but its preferred nesting habitat is in trees.
7002. Lewis' Woodpecker.
7003. When Bermuda was settled by the British, most of the island was ultimately occupied, and the few surviving Cahows were forced to nest on rocky islets where there is no soil into which to burrow. The petrels commence nesting in January, and resort to using the rock crevices favored by tropicbirds. The diurnal tropicbirds arrive in March, and when the petrels are out to sea during the day, Cahow chicks are evicted by the larger, more aggressive birds, and perish.
7004. Feathering.
7005. Pipers.
7006. The endemic honeycreeper is far better known by its Maori name Tui.
7007. The Kori Bustard, because of its fondness for the rubbery resin of acacia trees.
7008. The Boat-tailed Grackle. Enormous numbers may nest in heron colonies, and the birds sometimes even use heron nests.
7009. The Northern Harrier.
7010. Probably only several years.
7011. Yes, and the feat is accomplished with apparent ease.
7012. Yes.
7013. South America, with nine. Originally there were 10, but the Colombian Grebe disappeared about 1980.
7014. Precocial bird eggs have yolks constituting 25-50 percent of the overall egg weight, but the yolk of an altricial bird egg accounts for only 15-25 percent of the weight.
7015. On the ground.
7016. The birds feed on small seeds, mainly of native grasses, procured on or near the ground.
7017. In excess of 120 species.
7018. *Mousing* is a relatively recent term coined by Spotted Owl researchers. To *mouse* an owl is to feed it a live mouse once it has been called up. The technique provides information on the reproductive status of the owls. A single owl takes the bait and consumes it immediately, whereas if paired, it will take the first several mice back to its mate. If the pair has young, it is more likely to be the male that is attracted to the bait.

Answers

7019. No, the large Asian shorebird has loud, melodious calls.
7020. No, it typically faces the wind.
7021. Scarlet to ruby-red.
7022. These large cotingas live in groups and have a strange nesting strategy, and an uncommon one in tropical forest birds. The group protects the nest by mobbing an approaching potential predator, including birds as large as jays, toucans and hawks. Unlike most cotingas, the nest is neither small nor inconspicuous.
7023. The Red-crowned (Manchurian or Japanese) Crane.
7024. The diurnal raptors.
7025. Completely black.
7026. The birds were transformed into bullfrogs.
7027. No, Coleridge never saw an albatross. Some people believe the bird was actually a giant-petrel.
7028. Yes, as a rule.
7029. Yes, although this is not always the case in southern hemisphere species.
7030. Probably the Horned Grebe.
7031. The Hadada Ibis. Its loud, raucous call is an often repeated *Haa! Ha-de-daa.*
7032. Yes.
7033. Ruby-red.
7034. In the bower of the male.
7035. Wilson's Phalarope.
7036. The sea otter.
7037. Occasionally to Costa Rica.
7038. The Greater Roadrunner.
7039. The slain warriors were miraculously transformed into glittering hummingbirds.
7040. It may cling to a tree trunk and peck like a woodpecker, hence the sobriquet "woodpecker crow."
7041. The three kiwis, although some authorities suggest cassowaries are monogamous. However, polyandrous female cassowaries generally lay 2-3 clutches, each in the nest of a different male.
7042. Yes, many seabirds watch each other and fly long distances when they observe birds drop down to the water.
7043. No.
7044. An American alligator.
7045. Yes.
7046. No, they resemble crane chicks.
7047. Yes, the beard grows continuously.
7048. Yes.
7049. The Dovekie may be the most social alcid.
7050. *Empidonax* is Greek for "king of the gnats" or "of small insects." Some authorities suggest the name came from the Greek word for mosquito and was apparently given in the belief that the flycatchers fed extensively on mosquitos.
7051. Ptarmigan feathers.
7052. The Yellow Warbler.
7053. Swan's-down.
7054. Yes, well behind their tails.

Answers

7055. The German emperor Frederick II, who wrote a splendid treatise on the art of hunting with birds (*De Arte Venandi cum Avibus*). In actuality, this was the first real book on birds and was written about 1240, and illustrated with carefully executed illustrations of hawking scenes and bird life, representing the first known serious bird paintings. The work existed in manuscript form only, until published in 1596. This volume surpassed anything written on birds until Francis Willugby's book appeared more than 400 years later.
7056. Yes, their plumage was brilliant.
7057. Yes, on tadpoles.
7058. About eighty.
7059. The Eurasian Eagle-Owl.
7060. No, just the opposite.
7061. Redpolls have a high rate of energy intake owing in part to a special storage pouch in the esophagus. This is filled with food just before darkness so as to digest the food overnight. The birds also select high-calorie foods, such as birch seeds.
7062. The eagle is a juvenile, and its tail has a broad white band with a terminal band of black.
7063. Apparently because the shininess of these objects might be comparable to that of shed snake skins the birds commonly use to line their nests.
7064. The Ruddy Duck, with about 100.
7065. Hairy and Downy Woodpeckers. Once a Pileated Woodpecker scales bark off dead trees, the smaller birds move in seeking insects the larger bird may have missed or ignored.
7066. The Yellow-headed Blackbird.
7067. All are cavity or crevice nesters.
7068. The cormorants.
7069. About four billion.
7070. No, females migrate earlier in the fall.
7071. White, and the primaries are very conspicuous in flight.
7072. Yes, the females of many species do.
7073. No, it was named for Bernard Rogan Ross, chief trader for the Hudson's Bay Company in the mid-19th century.
7074. The Cattle Tyrant, because of the flycatcher's habit of riding on the back of a large quadruped.
7075. The White-throated Swift.
7076. The Aleutian Green-winged Teal is a separate race, and breeding-plumaged drakes closely resemble males of the Eurasian race. Rather than a broad vertical white stripe down the side, the white stripe is horizontal.
7077. No, it is a distinct species, as is the endemic screech-owl of Puerto Rico.
7078. No.
7079. The Kinglet Calyptura, that resembles a brightly-colored, thick-billed, and very short-tailed Goldcrest. It has one of the smallest ranges of any cotinga and is one of the least known. It has never been observed by a living ornithologist and has not been seen in this century.
7080. *Fred.*
7081. No, the cardinals generally abandon the nest.
7082. The Brown Creeper.
7083. The icterids (Icteridae).
7084. The snake-eagle, because it feeds predominately on snakes in that region.
7085. Dr. S. Dillon Ripley, the eighth Secretary of the Smithsonian Institution.
7086. No, there is no evidence of a pair bond at all.

Answers

7087. No. Craveri (1815-1890) was an Italian chemist and meteorologist, but he collected the first specimen of the little Mexican murrelet.
7088. Yes.
7089. Up to a pound (0.45 kg).
7090. The critic Brooks Atkinson (1894-1984).
7091. Yes.
7092. The Cock of the Plains.
7093. Rennell Island, where the Sacred Ibis and Eurasian Spoonbill occur.
7094. The abandoned nest of a Cliff Swallow.
7095. About 95 percent of the world's population of both species nest on tiny Isla Rasa (1,000 x 700 m) in Mexico's Sea of Cortez. About 120,000 pairs of the gulls breed there, along with 15,000 pairs of Elegant Terns.
7096. Tipping their arrows.
7097. The Groove-billed Ani.
7098. About 181, of which 59 are permanent breeding residents.
7099. Yes, chicks gulp fish down with tails protruding from their mouths while the heads are being digested in the stomach.
7100. Those of a Black Guillemot are white, whereas the underwing linings of a Pigeon Guillemot are dusky-colored.
7101. Ground-hornbills have 15 cervical vertebrae, whereas arboreal species have 14.
7102. Its stomach is so acid that a cow's vertebra reportedly entirely dissolves within one or two days.
7103. The guineafowl.
7104. Chicken-fried steak.
7105. The Aplomado Falcon.
7106. Yes, most do.
7107. Yes, but in nocturnal birds, it is more likely that the upper lid will be lowered.
7108. Great Britain.
7109. Yes, and the offspring are intermediate in appearance.
7110. With his tail.
7111. The Jackdaw.
7112. Up to 365 eggs a year.
7113. Pinkish-yellow in males, whereas that of a female is orange-red.
7114. No, reptile eggs tend to be white, or nearly so.
7115. No, their heads are small and thin.
7116. No. Although it is one of the finest of songsters, the thrush is quite shy.
7117. No, it is generally more distinct.
7118. Approximately 12 weeks of age.
7119. Fourteen, of which six definitely became established, with one other probable.
7120. The Brent goose.
7121. Bright orange-yellow.
7122. No, their voices are poorly developed, and their thin metallic calls do not carry far.
7123. The Reddish Egret (Florida and Texas). The Great White Heron has an even more restricted range (south Florida and the Keys), but it is a color morph of the Great Blue Heron.
7124. The Summer Tanager.
7125. The Skylark.

Answers

7126. Yes.
7127. The Greater Adjutant stork.
7128. The Red-throated Loon. The other loons generally feed in their breeding territories and swim to their young.
7129. The number of eggs on a grave specifies the number of enemies killed by the deceased man in his lifetime.
7130. Generally between 75 and 77 days of age.
7131. Between 17 and 20.
7132. The smell has been described as similar to that of fresh cow manure. Because of its unique foregut fermentation, the resulting fatty acids lend the bird its distinct odor.
7133. Several million.
7134. The Mountain Bluebird.
7135. An average of 2-15 square miles (5-38 sq km).
7136. The seriemas.
7137. The Pied-billed Grebe.
7138. Yes, apparently so.
7139. Swainson's Thrush.
7140. No, but both the plumes and crest are less conspicuous.
7141. Close up their ranks (ball up) and quickly change direction. This prevents the raptor from separating a bird from the flock, and because of its tremendous speed, the falcon will not dive through the flock as it might injure itself.
7142. No.
7143. The name apparently owes its origin to the Old English phrase *god wiht*, meaning "good creature," perhaps referring to the fact that the bird was good eating.
7144. None.
7145. Yes.
7146. Yes.
7147. Probably because the stinging ants dwelling within the hollow bull-horns discourage potential predators.
7148. As hand warmers during cold weather.
7149. Both bill and feet are used. Dirt is gouged away with the bill and pushed out with the feet.
7150. Penguins can sprint at speeds up to 15 mph (24 km/hr). Sailfish are said to be able to swim as fast as 68 mph (109 km/hr), and marlin at 40-50 mph (64.0-80.5 km/hr) in short bursts.
7151. Rufous.
7152. To adorn the ceremonial cloaks of the king Montezuma.
7153. About 100 pounds (45 kg), but by a year of age, it weighs about 187 pounds (85 kg).
7154. The name alludes to one of its preferred habitats, the woods.
7155. The American Wood Stork.
7156. No.
7157. Biologists have postulated that the seeds of the endemic tambalacoque tree needed to pass through a Dodo's gastrointestinal tract in order to germinate, and that the 300-year absence of the tree germinating naturally was a direct consequence of the Dodo's extinction. However, trees less than 300 years exist.
7158. Kirtland's Warbler. In the early 1960's, more than 70 percent of warbler nests were parasitized by cowbirds. Attempts to remove cowbirds from the endangered warbler's breeding grounds have only been partially successful.
7159. St. Patrick's day (March 17).

Answers

7160. All five species have yellowish bellies.
7161. Yes, many burrows do.
7162. The Pacific White-fronted Goose and Emperor Goose.
7163. No, it is also a disease of northern fur seals.
7164. The population of the aberrant African stork was estimated at 11,000 birds in 1986.
7165. No, but the endemic birds have an enthusiastic repertoire of squeaks and buzzes.
7166. The 14 tail feathers were used to fashion feather fans.
7167. Skimmers have been known to nest on gravel roof tops, but ground nesting is far more typical.
7168. The Northern Pintail, followed by the Mallard, American Wigeon and Northern Shoveler.
7169. No.
7170. An average of 5.5 ounces (160 gms) of earthworms, an amount nearly equivalent to the weight of the bird.
7171. No, it has exceptionally long wings.
7172. The Monk (Quaker) Parakeet.
7173. Yes, males are usually larger and more brightly colored.
7174. No, the terminal band is dark.
7175. About 1,300 feet (400 m), as a rule.
7176. About 25 times more.
7177. No, their bills are used.
7178. The European hare. It was imported for the commercial market and then released in numbers when the market declined.
7179. The Song Sparrow.
7180. Yes.
7181. The name came from Diomedes, the king of Argos. Next to Achilles, he was the bravest hero in the Greek army during the Trojan War. The gods exiled Diomedes to an island off southern Italy in the Adriatic Sea, and metamorphosed his companions into birds resembling swans.
7182. Yes, but such accidents are rare.
7183. Not generally.
7184. No, to poison a bird, lead shot must be ingested where it functions as grit in the gizzard. It is gradually worn down by the abrasive action of sand, gravel and other debris.
7185. While in Liverpool, he arranged for the exhibitions of Audubon's work, scheduled lectures and organized important introductions. Traill was a Scottish professor and zoologist.
7186. Its tail is dark, but the outer feathers are white.
7187. The name alludes to the golden nape of adults.
7188. The Watercock (Kora Rail) is used as a fighting bird, and wagers are placed, as in cock-fighting.
7189. The jays often bury acorns, some of which germinate.
7190. Yes, but only occasionally.
7191. Never; only males incubate.
7192. About 72 hours.
7193. Their large feet are well-suited for clambering about among reeds.
7194. No, but rocks weighing twice as much can be removed.
7195. The tropicbird family.
7196. Tinamou is a corruption of *inanbú,* a local Argentina name for the tinamou, that was probably a transliteration of a tinamou call.

Answers

7197. No, they are virtually indistinguishable.
7198. The Greater Adjutant stork, in the streets of Calcutta, India.
7199. None.
7200. The Antarctic Tern.
7201. The Eurasian Pygmy-Owl.
7202. Approximately 28 million (in 1989).
7203. The Noisy Scrub-bird. Upon rediscovery of the species in 1961 at Two Peoples Bay, a planned townsite was canceled and a nature preserve established instead.
7204. The hummingbirds.
7205. Cow dung or mud.
7206. No, the Red-legged Kittiwake has a darker back.
7207. Nowhere else, and even there it is restricted to loose colonies on Boatswainbird Islet, a flat-topped, steep-sided rock of only 7.4 acres (3 hectares). The frigatebird was extirpated from Ascension Island itself as a breeding species by feral cats in the early 19th century.
7208. The South American cotinga has bright orange down.
7209. The House Crow.
7210. No. When hovering, its wings are beating at the rate of 55 times a second, 61 times a second when backing up, and 75 times a second when flying straightaway.
7211. The Laughing Gull.
7212. No, rollers are cavity nesters.
7213. Yes.
7214. No. Some are bright orange, and are often considered a separate race known as the Equatorial or Apricot Cock-of-the-rock, as opposed to the Scarlet Cock-of-the-rock.
7215. No, the Puna Tinamou occurs in the high Andes, up to 17,388 feet (5,300 m), just below the permanent snow line.
7216. The potoos snap their bills.
7217. Ten years.
7218. Yes, in honor of Dr. David Livingstone (1813-1873).
7219. No, three to four years are required.
7220. The Red-billed Tropicbird.
7221. Some of the birds-of-paradise.
7222. Hummingbirds of various species were often shipped in the same container and the smaller hummingbirds used the long, straight bills of the large hummingbirds as perches, ultimately exhausting the birds. Once Sword-billed Hummingbirds were shipped in separate compartments, extensive mortality ceased.
7223. From the Greek: *aichme* (a spear) and *phoros* (bearing), a "spear-bearer," in reference to its sharp, pointed bill.
7224. Alaska, and probably Colorado.
7225. No, they are usually seen in pairs.
7226. The Green Heron.
7227. Yes.
7228. The Rock Sandpiper.
7229. Probably as an adaptation for running over soft ooze.
7230. Yes, their nostrils are not divided by a septum.
7231. The Cape buffalo. The weavers were named by early South Africa settlers because of their habit of following hoofed animal herds, presumably to take advantage of insects stirred up.
7232. A giant snake.

Answers

7233. Yes, some have distinctly notched bills.
7234. Sheep vertebrae, and in extreme cases, bowers may be decorated with more than 1,000 bones.
7235. Each adult builds a separate, tightly-woven globular nest in which it roosts alone throughout the year, excluding even its mate.
7236. Yes. The Peninsular Yellowthroat is a sedentary resident of the marshes of southern Baja California, but Baja is not generally considered a part of North America.
7237. Only four, with two other species probable.
7238. No, a Black-footed Albatross is easily annoyed, aggressively snapping at intruders, and it may resort to projectile vomiting.
7239. Yellow.
7240. Yes, Hooded Warblers have been recovered from the stomachs of neotropical green frogs in Mexico.
7141. Up to 60 revolutions a minute, and the birds pick at the water with the bill at each turn.
7242. Yes.
7243. Cuckoos are insectivorous.
7244. The Caspian Sea *(Podiceps caspicus*).
7245. Tinamous must diligently move aside the dense plumage surrounding the vent to avoid soiling it.
7246. The Mottled and Mexican Duck.
7247. None. The Barn Owl is the most familiar member of the genus.
7248. Approximately 20.
7249. The name is derived from the Latin *musca* "fly" and *capere* "to catch.
7250. The Ring-necked Parakeet.
7251. Fresh leaves are richer in protein and easier to digest. Up to 85 percent of a Hoatzin's diet consists of fresh green leaves.
7252. The American Coot.
7253. Yes, the kiwis, but the hind toe is small. Other ratites have either two or three toes.
7254. The Spotted Owl.
7255. No one knows, but the incubation period is probably similar to that of the Arctic Loon (28-30 days).
7256. The Caribbean race of the Brown Pelican (*Pelecanus o. occidentalis*). Males from Suriname weigh only 4.7-6.6 pounds (2.16-3.0 kg).
7257. About nine times as much.
7258. Because the happiness of a wedded couple would fly away sooner.
7259. Yes, as a rule.
7260. The Snowy Egret.
7261. Yes.
7262. J. N. "Ding" Darling, a conservationist and noted political cartoonist.
7263. Yes.
7264. The Least Auklet.
7265. The Eskimo Curlew.
7266. The name was probably derived from the Latin *ossifraga*, meaning "bone-breaker," and was apparently picked up by transference from the Lammergeier. This was obviously a mistake because the Osprey is a fish eater.
7267. Only two: the pheasants, quail, grouse and turkeys, and the cracids (Plain Chachalaca).
7268. Yes, on small ones.

Answers

7269. Small nuggets of gold were found in the gizzards of wild ducks shot there. The largest nuggets were worth only fifty cents, but the discovery nevertheless set the entire community to prospecting.
7270. The White-faced Ibis.
7271. No, its tail is black-tipped.
7272. No, it has a yellow rump.
7273. The Great Gray Shrike.
7274. Red.
7275. In theory, a minimum flight speed of 100 mph (161 km/hr). As a general rule, the heavier a bird, the faster it must fly to stay aloft.
7276. Fiddler crabs.
7277. The origin of the name is obscure, but it was perhaps named after the Muscovite Company that traded with South America. It was also possibly named after the Mosquito Coast of Nicaragua, where it was known as the *Musco Duck*.
7278. The Sharp-tailed Grouse.
7279. Bright blue.
7280. No, it rarely frequents such habitat, but rather is a bird of rocky shores and islands.
7281. The cormorants, with 38 species.
7282. The White-winged Scoter breeds as far south as extreme northern Washington and North Dakota.
7283. Yes.
7284. No, its tail is wedge-shaped.
7285. The Red-tailed Tropicbird.
7286. The Marbled Murrelet.
7287. Pale yellow or whitish.
7288. The brown feathers have a metallic green sheen.
7289. As of 1992, penguins, waterfowl, albatrosses, bustards, raptors and petrels.
7290. No, adult plumage is assumed at two years of age.
7291. Three: cranes, Limpkin and rails.
7292. Bonaparte's Gull weighs only 7.25 ounces (206 gms). The smaller Little Gull of Europe bred in North American for the first time in 1975, but it is not a native species.
7293. Yes.
7294. Xantus' Hummingbird and Xantus' Murrelet of Baja California.
7295. The two African ground-hornbills weigh nearly nine pounds (4 kg).
7296. At least 59, or about one-fifth of all raptors worldwide. All but six of the 59 breed in China.
7297. About 82 times as much. The average dry weight of its mud nest is 9.0 pounds (4.1 kg), whereas the bird itself weighs only 1.75 ounces (49 gms).
7298. Yes, but the birds are more vocal during the breeding season.
7299. No, their burrows often curve, and may attain a length of 14 feet (4.3 m).
7300. Copulation may take place either in the water or ashore.
7301. The skin of both sexes becomes yellow-orange.
7302. In the evening.
7303. Orange-yellow to red.
7304. An Osprey occasionally locks talons in a salmon or sturgeon too large to bring to the surface and drowning results.
7305. Yes, some species do.

Answers

7306. Apparently not. In the wild, drinking has only been observed on several occasions.
7307. None.
7308. The Carolina Bird Club (North Carolina).
7309. The swinging of a hanging gourd discourages European Starlings, a major nest site competitor.
7310. Young are basically white with a faint pinkish tint.
7311. No, females sit very tightly and one must nearly step on them before they flush.
7312. The Tricolored (Louisiana) Heron.
7313. Yes.
7314. Yes.
7315. No, the Northern Waterthrush is more numerous.
7316. Yes.
7317. No, nest construction is carried out by females alone. Males construct bowers.
7318. Females and juveniles of the Painted Bunting.
7319. No other state.
7320. The ovenbirds, most of which build covered nests, or nest in cavities. The nests of a number of species are quite elaborate, and in some instances, huge. The neotropical ovenbird family is one of the most diverse of all avian families.
7321. None. Corvids are worldwide in distribution except for New Zealand, Antarctica and some oceanic islands.
7322. Yes, although large numbers were discovered only recently.
7323. The badger.
7324. No. The top speed of a greyhound is about 40 mph (64 km/hr), and a panicked Ostrich can exceed 44 mph (70 km/hr), and possibly faster.
7325. Yes.
7326. Baja California.
7327. Between 1906 and 1959 the owl was considered extinct.
7328. The Yellow-shafted Flicker. About 45 percent of its diet consists of ants.
7329. The Convention on International Trade in Endangered Species of Wild Fauna and Flora lists endangered or threatened species. The CITES treaty was ratified on July 1, 1975.
7330. No, nest down is generally rare or lacking. However, their closest relatives, the waterfowl, are noted for extensive amounts of down in the nest.
7331. The Yellow Warbler.
7332. No, it also occurs in New Mexico.
7333. Yes, except for Antarctica that has no land birds.
7334. Most of the eagle's prey is arboreal, and at the moment of capture is usually clinging tightly to branches, and in the case of a sloth, the grip is extremely strong. The huge forest eagles snatch their prey in flight, almost without a break in flight. To perform this efficiently requires not only the high momentum resulting from the bird's great weight and speed, but also a secure first-time grip. The risks of injury are probably high in this type of attack, thus a great thickness of the tarsi may be required to prevent bone breakage.
7335. White.
7336. Only one, the European Nightjar.
7337. No, the entire feathered portion is black, but the nearly bare crown and lores are red.
7338. Yes, particularly in rip tides.
7339. A desert of lapwings.
7340. The South African (Cape) Gannet.

Answers

7341. Yes, their clear, melodious calls are such that Australians consider butcherbirds to be among their finest songsters.
7342. Yes, except around colonies.
7343. Bright coral-red. It is also known as the Red-billed Tree-Duck.
7344. No.
7345. Yes.
7346. About 40 percent.
7347. No, the birds are generally encountered as pairs.
7348. The Burrowing Owl.
7349. Yes, the name is an imitation of its cry of grating croaks.
7350. No, generally not until their second year.
7351. A clan.
7352. Henshaw could skin, make up and label a bird skin in less than three minutes.
7353. Yes.
7354. Yes, when the chicks are small, but it is not possible with larger chicks.
7355. The Macaroni and Royal Penguin.
7356. Epaulets.
7357. The Montezuma Quail, because it tends not to flee from danger.
7358. No, the owls are shy and wild.
7359. Yes, two species reach northern Argentina.
7360. The Cinnamon Teal.
7361. None.
7362. When in flight.
7363. The storks became torpid.
7364. The Bananaquit.
7365. Yes, but nest construction generally takes up to a week.
7366. Fish bones.
7367. Its feathered forehead is bright red.
7368. The population of the Galapagos race is not large, possibly not exceeding 100 pairs.
7369. Yes.
7370. Yes, some wrens pierce eggs to drink the contents.
7371. Three; two in front of the eye and one behind the eye.
7372. None.
7373. Arizona.
7374. No.
7375. No, the birds generally become instantly motionless.
7376. Not since the disappearance in the 1800's of the Flightless Night-Heron of Rodrigues Island.
7377. In Suriname (formerly Dutch Guiana), between 1877 and 1882.
7378. No, the bill of an Eared Grebe is longer and thinner.
7379. Tahiti.
7380. As much as 10 inches (25.4 cm), but 3-6 inches (7. 6-15 cm) is more typical.
7381. No, bustard secondaries are almost as long as the primaries.
7382. Pigeon pie.
7383. No, Sharpbills have been observed feeding on invertebrates, so they are clearly not exclusive fruit eaters, although this is stated in some of the literature.
7384. Five years.

Answers

7385. The Razo Skylark is endemic to Razo Island in the Cape Verde Islands off west Africa.
7386. Five to 30, and up to 7,500 pellets may fall into marshlands for every duck dropped.
7387. Not generally, but Brown and Peruvian Pelicans do feed on anchovies and sardines.
7388. Red, surrounded by bare blue skin.
7389. The Barn Owl.
7390. No, it is a cloud-forest cracid, seldom venturing below 7,500 feet (2,285 m).
7391. The thrasher is neither a mountain resident nor a mockingbird.
7392. No. The Bohemian Waxwing is 7.5-8.75 inches (19.0-22.2 cm) long, whereas the Cedar Waxwing is 6.5-8.0 inches (16.5-20.3 cm) long.
7393. No, their livers are larger.
7394. Yes, loons have only 14-15 cervical vertebrae, whereas grebes have 17-21.
7395. The Marbled Murrelt, but this has only been documented once. A murrelet egg was found in an old twig nest, probably that of a Band-tailed Pigeon. Normally, the single egg is merely laid on a large, horizontal tree branch, often in a moss-lined depression.
7396. Bird or feather lice. The parasites spend the whole of their life-cycle on the host where they live among the feathers, feeding on blood and other tissue fluids, and feathers. The lice of one genus are exceptional in that they live in the throat pouches of pelicans and cormorants.
7397. Yes, but just barely.
7398. Six Laysan Albatross nests containing chicks were discovered on Guadalupe Island in May of 1988, the first nesting record east of Hawaii.
7399. To line the bottom of bird cages.
7400. The U.S. Great Plains. Krider's Hawk is a pale-color phase of the Red-tailed Hawk.
7401. Yes.
7402. Birdlets.
7403. The Sarus Crane.
7404. Yellow, but the juvenile is brown-gray overall.
7405. George Sanders.
7406. Yes.
7407. No.
7408. Yes, but the birds may dip several times before lifting their heads. This method of drinking is known as suck-and-tilt where the bird sucks up water into the buccal cavity before raising its head to swallow the mouthful in a single gulp.
7409. The Himalayan Monal.
7410. About 19-20 inches (48-51 cm).
7411. The 4.5-inch-long (11.4 cm) Least Sandpiper, that weighs a maximum of 1.33 ounces (37.7 gms).
7412. He collected the first Bendire's Thrasher in Tucson, Arizona.
7413. Yes.
7414. Up to four months prior to egg-laying, a large pit is dug and filled with dead leaves. Following a heavy rain, the bird covers the wet leaves with sand, forming a smooth mound. Four months later the leaves have started to decay and give off heat, like a compost heap. Only then does laying commence.
7415. Foraging at refuse dumps and sewage treatment plants has resulted in decreased mortality because of an almost unlimited supply of food. The gull population has increased to nuisance levels near rubbish dumps, airports and tern colonies.
7416. The striking Three-wattled Bellbird of the humid montane forests of Central America.

Answers

7417. Yes.
7418. No, males precede the females.
7419. No, the bones are essentially solid.
7420. Generally between 4 and 12 percent.
7421. An average of five in the past 15 years.
7422. *Vociferus* implies that the bird is noisy, referring to its loud, repeated calls.
7423. An average of about 15 days.
7424. Yes.
7425. At least 100 have been recorded in a single cavity.
7426. Forty-five, as of 1990.
7427. Yes.
7428. No, its piercing, yelping cries or screams, are audible from a distance of a least one mile (1.6 km). Some of its calls are rather like the barking of a small dog.
7429. The Bald Eagle, but such unusual behavior has only been recorded at Amchitka Island, Alaska.
7430. Red, yellow, or orange.
7431. No, chicks have a single chest-band, but adults have two.
7432. 1972.
7433. Yes, as opposed to concave or grooved beneath.
7434. The Emperor Penguin. Its bill and flippers are 25 percent smaller relative to its size than any other penguin, no doubt increasing its resistance to extreme cold.
7435. Yes, and the jays even nest in loose colonies.
7436. The European White Stork. The name refers to the bill-clattering of breeding pairs.
7437. Yes.
7438. The Old World thrush has a powerful odor that distracts gamekeeper's dogs.
7439. The goose was regarded as the sacred bird of the god Geb, and while its flesh was widely eaten by Egyptians in bygone days, its eggs, which had symbolic power, were never consumed.
7440. Yes.
7441. The Peruvian Booby.
7442. Baraboo, Wisconsin.
7443. An estimated 71 tons.
7444. Yes.
7445. Conspicuous white wings may aid in luring a predator away from the nest.
7446. No, the woodpeckers are typically silent.
7447. The Northern Cardinal.
7448. No, brown pelican chicks are more vociferous.
7449. Because of nest site competition with introduced European Starlings.
7450. The *Gossamer Albatross*.
7451. No, the owl occurs as far north as Costa Rica.
7452. No, it is rare in Alaska.
7453. The superstitious fishermen use the livers when practicing *bruha,* or black magic.
7454. Yes.
7455. Yes.
7456. *Haematopus* is from the Greek *haimatopus,* meaning "bloody foot," referring to their brightly-colored feet.

Answers

7457. The Greater Roadrunner, as is the Lesser Roadrunner.
7458. Yes.
7459. The Australian Pelican.
7460. The underwing coverts.
7461. The name is of unknown origin.
7462. No, but typical swifts do.
7463. Yes, but only rarely. It is a Eurasian species.
7464. No, their tails are distinctively longer.
7465. Yes.
7466. The Willow Ptarmigan. In winter, up to 94 percent of its diet consists of willow twigs and buds, of which 80 percent may be of a single species (Alaska willow).
7467. Mostly dull sooty-black except for its conspicuous white spectacles.
7468. The frigatebirds.
7469. No, most wrens vocalize with their head held high and the tail tucked down and under.
7470. Yes, as least three species are known to plunge into water after fish on rare occasions.
7471. Crude estimates suggest about 50,000 birds.
7472. In general, yes.
7473. A small, dark albatross.
7474. Yes.
7475. Yes, on occasion.
7476. The Southern Screamer may gather in flocks numbering in the thousands during the non-breeding season. The Horned and Northern Screamer are far less gregarious.
7477. Yes, but actual fights are rare.
7478. Yes.
7479. Yes.
7480. The two African ground-hornbills.
7481. The gamebirds range in size from the tiny Asian Blue Quail only five inches (12.7 cm) long, to male peacocks that are 92 inches (234 cm) long.
7482. Yes, but this is an unusual shorebird trait. Chicks take food directly from their parent's bill during the first week after hatching, or pick food from the ground where it is dropped by an adult. Within 10 days chicks feed on their own.
7483. Yes, but it was primarily a seed eater.
7484. The Lappet-faced Vulture.
7485. A flash of white in the primaries, that is prominently displayed by aggressive birds, especially skuas.
7486. At least 50 percent. Eighteen species and subspecies have been lost in the past 150 years. At least 60 New Zealand birds are currently rare or threatened with extinction.
7487. Central Texas.
7488. Yes.
7489. Yes, as a rule.
7490. The first egg was laid by a duck raped by a water rat.
7491. Yes.
7492. No, but 33 species may breed in deserts.
7493. Yes, but the Scarlet Tanager is only a casual stray.
7494. A rattlesnake.
7495. Winston Churchill. The agent planted in England was code-named *Starling*.
7496. Yes, but the Lesser Yellowlegs is less of a fish eater.

Answers

7497. Cow dung.
7498. No, but the central tail feathers are elongated.
7499. Yes, but only rarely.
7500. Charles Lindbergh.
7501. The six crested penguins in the genus *Eudyptes*.
7502. Florida.
7503. The Red-cockaded Woodpecker. The 135-mph-winds (217 km/hr) destroyed nearly all the mature pines that served as nesting trees for 500 breeding pairs of the woodpeckers.
7504. The eyebrow. The term applies to a marking in some plumage patterns above the eye.
7505. Yes. Irruptive species move southward in some years in large or small numbers and for great or small distances. These irregular movements are generally, but not specifically, predictable.
7506. Yes.
7507. Idaho.
7508. While normally a scavenger, the stork is also an active predator.
7509. The destruction of mature riverine habitat.
7510. Yes.
7511. Chicks reach into an adult's open mouth to extract food from the crop.
7512. Yes, the bird has a thick, strongly curved, parrot-like bill
7513. The Cape of Good Hope (South Africa).
7514. Audubon's father was a French sea captain who was sometimes a slave trader.
7515. No.
7516. Princess Zenaide Charlotte Julie Bonaparte, the cousin and later wife of Charles Lucien Bonaparte. *Zenaida* is perhaps better known as the Mourning Dove's generic name.
7517. No, the birds have relatively small, pink-rimmed eyes, sometimes with small fleshy wattles.
7518. No, and on occasion they will feed from the hand.
7519. No, males do most of the nest construction.
7520. The White-tailed Tropicbird may breed at intervals of 6-9 months.
7521. Yes, apparently so.
7522. Yes, on rare occasions, hummingbirds are taken.
7523. No, the secretive tropical birds are generally encountered alone or in pairs.
7524. No, many of the whistling-ducks feed at night.
7525. Palmer returned to his native Australia where he was murdered in the gold fields. He was a bird collector best known for his work in Hawaii, where between 1890 and 1893, he collected all but seven of the then-known birds of Hawaii, including 15 species never before described.
7526. Barn and Cliff Swallows.
7527. The New Zealand Falcon.
7528. The Tufted Puffin is most apt to take flight, whereas the other two alcids tend to dive.
7529. *Morus* is Latin for "foolish" or "silly."
7530. The bluebirds may hover over potential prey..
7531. Frigate ships. Frigates were medium-sized ships of the 18th and early 19th centuries that depended on speed and maneuverability to elude pursuers. Frigate ships were often used by pirates to attack merchant vessels to relieve them of their cargoes, just as the kleptoparasitic frigatebirds relieve other birds of their catch.
7532. Only about two feet (0.6m), and only six feet (2 m) at night.

Answers

7533. The Red-legged Kittiwake.
7534. The flicker.
7535. The Willow Warbler.
7536. Yes. There are about 20 records from Baja, with the last confirmed sighting in 1937, although there were unconfirmed sightings in the early 1970's.
7537. Yes.
7538. No, it is one of the most frequent hosts.
7539. Trogon is Greek in origin and means *gnawer*, alluding to the hooked, dentate bill used to gnaw at fruit.
7540. Up to 10 huddling penguins.
7541. The Glaucous Gull of the far north.
7542. They frequent dark forest recesses, rather like hermits.
7543. The Red-whiskered Bulbul. Escaped birds founded thriving feral populations in some localities, and the bulbul is currently common around Sydney.
7544. Yes, it is more commonly known as the House Finch.
7545. The Marbled Godwit. Excluding the Long-billed Curlew, it is the largest North American shorebird. The 15-inch-long (38 cm) bird weighs up to 18 ounces.
7546. Guanacos, llamas, alpacas and vicuñas.
7547. The nests may be used by females and recently-hatched young as dormitories at night.
7548. Sometimes known as exotic Newcastle disease, Velogenic Viscerotropic Newcastle Disease (VVND) is a highly contagious, nearly always fatal, avian viral disease. When it initially broke out in Southern California in the early 1970's, it cost over $56 million to contain. The outbreak resulted in an avian quarantine of Southern California, a total embargo on bird importation into the U.S., and ultimately, a USDA-controlled quarantine for all imported birds.
7549. No, but one of the local names for the Australian Pratincole is roadrunner.
7550. The Inca Dove.
7551. Dr. Edmund *Jaeger*.
7552. The Golden Eagle.
7553. Yes, nests have been found as early as February.
7554. The Greater Pewee.
7555. The Cheer Pheasant.
7556. Forest destruction.
7557. No, body size tends to be larger in cold climates.
7558. A cheater or swindler, because he has the preying or grasping nature of a hawk.
7559. Yes. The little birds dislike clearings so intensely that some even refuse to cross firebreaks bulldozed through their territories.
7560. No, but gobblers do.
7561. The pigeons have a splash of red on their lower necks and upper chests resembling a bleeding wound.
7562. No, the cranes often forage in flocks.
7563. Yes.
7564. The bill is typically immersed for about two-thirds of its length.
7565. Cooper's Hawk.
7566. By voice. The Fish Crow has a nasal, falsetto call, as opposed to the harsh cawing of the American Crow. Fish Crows are slightly smaller, more slender, with more pointed wings and thinner bills.

Answers

7567. The Barn Owl.
7568. No, the black plumage of a drongo reflects bright metallic sheens of green to purple.
7569. The Gray Catbird.
7570. The Bald Eagle.
7571. Yes, except in the northwest.
7572. An Old World warbler.
7573. No, the journal is published quarterly by the Saskatchewan Natural History Society. The Virginia Society of Ornithology publishes *The Raven*.
7574. The four tropical and temperate penguins in the genus *Spheniscus* range from the Galapagos Islands astride the equator south to Islas Diego Ramirez, southwest of Cape Horn.
7575. The Bicolored Blackbird (a race of the Red-winged Blackbird) has solid-red epaulets lacking the yellow border.
7576. A feather aftershaft.
7577. The Montezuma Quail.
7578. Yes, but the young are paler in color.
7579. Yes.
7580. Greenish.
7581. Six.
7582. Yes.
7583. From the bill of their parents.
7584. South Carolina, in the I'On Swamp.
7585. The House (English) Sparrow.
7586. Southern Florida, California and the northeast.
7587. His library and papers were destroyed during the Civil War in a fire set by, in the words of Bachman, "Sherman's vandal army."
7588. The name is a reference to their frequent calling.
7589. A pigeon hole.
7590. The White-headed Woodpecker.
7591. No, the birds nest quite high in trees, even up to 70 feet (21 m) above the ground.
7592. The raptor enters caves to prey on bats or swiftlets, and it also takes swiftlet eggs and nestlings.
7593. No, it is a honeyeater.
7594. The Red-eyed Vireo, although such may no longer be the case. The vireo has declined significantly in recent years.
7595. Bright blue.
7596. The little flycatchers seldom drink, even if water is available, because their insect diet presumably provides ample fluids.
7597. No, the birds are gregarious year-round.
7598. Yes.
7599. No, most species are solitary.
7600. No, but aerial copulation is apparently rare and has only been reported on several occasions.
7601. An abandoned mud ovenbird nest. Martins may occasionally take over occupied nests.
7602. No. Glaucous is derived from the Latin *glaucus* meaning "blue-gray," in reference to the slightly grayish cast of the plumage of the wings and back.
7603. No.

Answers

7604. The primary feathers.
7605. Scarlet, but the crest is generally concealed.
7606. Red, as it is during the breeding season.
7607. No. In the past several decades, the population declined by as much as 36 percent in the east, and 19 percent in the midwest.
7608. The Bobwhite, followed by the Ring-necked (Common) Pheasant.
7609. Most, if not all, exhibit little or no sexual dimorphism.
7610. No, some snakes can climb cholla.
7611. The Ovenbird.
7612. Tits, generally Varied Tits, draw and open fortune cards.
7613. The Half-moon Conure.
7614. Yes.
7615. A large, 2-pound frog (.9 kg) from Dominica.
7616. The Black-capped Petrel. Some authorities suggest the Cahow may merely be a subspecies.
7617. The jay does not have a crest.
7618. Yes. The five toucanets in the genus *Selenidera* are dimorphic, such as the Spot-billed Toucanet, and at least one aracari (Green Aracari), in which the male has a black head, but the female has a chestnut-brown head.
7619. Unlike vultures, storks generally all soar in the same direction.
7620. Just over 31 years.
7621. Yes.
7622. The name is a corruption in English of a Portuguese corruption of a Tupi Indian word for the macaw palm tree, the fruits of which the huge parrots eat.
7623. Yes.
7624. Yes, some will not fly until nearly stepped upon.
7625. The name is Greek for *insect-eater*, implying that anis are insect-eaters.
7626. Orange-yellow.
7627. Estimates range from 1,200 to 10,000 rails.
7628. The raven's black coat may actually be cooler than a white one because dark colors trap heat at the surface, preventing it from penetrating to the skin. The surface feathers become very hot, but the heat is carried away by the wind. White remains cooler at the surface, but allows the sun's rays to penetrate deeper, bringing heat close to the skin. Black is only advantageous to large birds because a thick feather layer is necessary to separate the hot surface from the skin. As a result, many large desert birds, such as the Golden Eagle and Turkey Vulture, are dark colored, but small desert birds tend to be pale colored.
7629. Black.
7630. Yes, the shorebirds are highly gregarious.
7631. About 2.5 grams (0.09 oz).
7632. Harcourt's (Band-rumped) Storm-Petrel, that weighs about 1.5 ounces (42.5 gms).
7633. Generally 2-3 days prior to hatching, at which time the fetus commences breathing air.
7634. Yes, although not commonly.
7635. Swift.
7636. No, the majority are gregarious.
7637. None.
7638. Yes, but only very rarely in southernmost Texas.

7639. The sucked udder would wither, and the goat would go blind.
7640. Yellow.
7641. Crested Francolins cock their longish tails up like bantam chickens.
7642. No, its shorter bill is only slightly curved.
7643. Yes.
7644. The two African ground-hornbills.
7645. No, quite the opposite. The obligate brood parasitic-nesting birds have fledging periods as long as 38-40 days. This is a long time for little birds only weighing 10-55 grams (0.35-1.94 oz).
7646. Yes.
7647. Yes, even though the grouse is a much larger bird.
7648. The Double-crested Cormorant.
7649. Louis Pantaleon Costa de Beauregard, a Sardinian patriot, statesman, military commander, historian and accomplished amateur archaeologist.
7650. *Mother Goose.*
7651. The Lesser Goldfinch.
7652. The American Kestrel.
7653. The long, loose feathers are distinctly parted down the middle of the back.
7654. A crane or heron, a stork and an owl.
7655. No. Nest sites can sometimes be located because of the bird's long tail hanging out of nesting cavities.
7656. Gambel's Quail.
7657. No, most mannikins are somber colored, but many neotropical manakins are brightly colored.
7658. An eagle.
7659. No, its bill is smaller with a less distinct hook.
7660. No, it is a resident of sagebrush habitat.
7661. The bird is fond of mistletoe berries.
7662. A gamebird biologist's term describing a location male ducks and their mates use to find isolation and freedom from molestation by other drakes. The nest may be a mile (1.6 km) or more away, and the hen may visit the drake daily (or more often) during the egg-laying period.
7663. No, the berries are sour.
7664. Slightly more than three feet (0.9 m).
7665. Yes.
7666. Yes, all have pale eye-rings.
7667. No.
7668. Red.
7669. Nestlings.
7670. White, but the closely-related Western Grebe has gray chicks.
7671. The birds are seized by one wing and beaten on the ground.
7672. Only three, all of which are found south of the Sahara.
7673. Anhinga chicks have buffy-yellow skin with pale pinkish heads, whereas darter chicks have brownish to blackish skin. Within a day or two of hatching, the nestlings are covered with down.
7674. Yes.
7675. A bridge.

Answers

7676. Rocky Mountain females generally have mostly yellow bills, but eastern Barrow's Goldeneye hens have dark bills, sometimes with yellow tips.
7677. A jay.
7678. The petrels.
7679. Yes.
7680. About 64 million, or 1.3 birds per acre (3.2 per hectare).
7681. About 60-70 days of age. Development of wings and flight muscles are delayed.
7682. In 1990, about 2,600 breeding pairs, with an equal number of non-breeding birds, for a total of about 10,400 eagles.
7683. An owl.
7684. The American Avocet.
7685. The Pelagic Cormorant.
7686. Yes. Their wingbeat is slow, but the upstroke is accomplished with a powerful flick or jerk.
7687. A 25-50 percent reduction in individual heat loss.
7688. Yes.
7689. Horned Larks and Snow Buntings.
7690. No, it preys chiefly on insects captured at upland sites.
7691. Yes, and their plumage is more iridescent.
7692. The beautiful blue plumage is shiny and enamel-like.
7693. Yes.
7694. No, the Canyon Towhee (formerly conspecific with the California Towhee) also has a rufous crown.
7695. The extinct Reunion Solitaire.
7696. No. The Osprey's egg is white to pink or cinnamon, heavily blotched and spotted with dark browns, and is among the most handsome of hawk eggs.
7697. Gray, but the naked head of an adult is yellow to orange or pink, with a purple-red patch on the lower side of the neck.
7698. Yes, in those species that migrate.
7699. The Black-throated Sparrow.
7700. At least 69 species, of which 27 definitely became established, with 11 others probable.
7701. Cranes.
7702. Honeysuckers.
7703. Estrildid finches.
7704. Yellow.
7705. No, they seldom dive, but when alarmed they submerge to the neck.
7706. No, but the long-legged raptors do have frog-like croaks.
7707. The Papuan (Blyth's) Hornbill, a New Guinea, Moluccan and Solomon Islands species.
7708. Yes.
7709. A live aspen.
7710. The Trinidad Piping-Guan.
7711. Yes, all breed during the northern winter, the equivalent of the austral summer.
7712. The display is centered on the female, and wherever she goes, her mate goes to drive away other males.
7713. No.
7714. Both sexes have yellow bills, but the base of the male's bill is bluish, whereas that of the female is pinkish.

Answers

7715. Not normally, but in 1963, a large wreck of Horned Puffins, including two live ones, turned up in the northwest Hawaiian Islands.
7716. The silky appearance is due to the structure of the feathers, which is unique among birds, with only one in every two or three barbules of the inner barb clasping the next barb.
7717. A wood-warbler.
7718. Not normally, but it is a vagrant in Egypt.
7719. The Tree Swallow, that occurs as far north as New York during the winter.
7720. Yes, inland water species may prey on turtles, and occasionally small birds and mammals.
7721. No, the tern also nests in Mexico.
7722. No, all are relatively solitary.
7723. Yes.
7724. Yes, but the parrot's flight muscles are degenerate, although it can glide short distances.
7725. Between 7 and 12 eggs.
7726. The only physical difference appears to be the minute webbing between the bases of each of the three toes in the Semipalmated Plover, and only between two toes in the Ringed Plover.
7727. Yes, at least in Africa. Snakes are immobilized by beating them vigorously against a low-hanging branch.
7728. As the smallest penguin, it is the shallowest diver and a crepuscular feeder, probably because a greater proportion of its prey is near the surface at that time.
7729. No, but males do.
7730. Caressing with the beak (bill to bill).
7731. The duck-billed platypus. Its original scientific name *(Platypus anatinus)* meant just that. The generic name *Platypus* was subsequently dropped when it was discovered a group of beetles already had the name, but platypus has survived in popular usage.
7732. Males have red eyes, whereas females have dark-brown eyes.
7733. Sixteen.
7734. The naked, helpless chicks have shiny black skin.
7735. Probably as stowaways on four wooden whaling ships that sailed from Montevideo, Uruguay, in 1919.
7736. Yes, the Ceylon Frogmouth inhabits India and Sri Lanka.
7737. No. Ostriches are numerous in South Africa, but the only pure birds of the South African race are probably the several hundred surviving within the protected confines of Kruger National Park, plus a few in Namibia and the northern parts of its original range. The remaining Ostriches in South Africa were cross-bred to birds imported from North Africa during Ostrich-farming days.
7738. Nancy Howe was the first female artist to win the contest in the 40 years that it had been held. Her entry was of a pair of King Eiders.
7739. No, but flying birds are caught with their beaks.
7740. No, they feed primarily on wood-boring larvae and ants, not winged insects.
7741. Eagless is an old name for a female eagle.
7742. No, the storks lacked desirable plumes, and were also considered unfit for the pot.
7743. Three. South Dakota (Ring-necked Pheasant), Rhode Island (Rhode Island red chicken), and Delaware (blue hen chicken).
7744. The muskrat.
7745. The Hermit (Northern Bald) Ibis or Waldrapp. It formerly bred as far north as Switzerland, but the ibis disappeared in Europe in the 17th century.

Answers

7746. The Sulphur-bellied Flycatcher.
7747. Yes, in the Philippines, the Ruddy Kingfisher uses such stones to smash large land snails.
7748. Yes.
7749. No, the parrot may be extinct. No documented sightings have occurred since November of 1927, in southern Queensland.
7750. The Old World vultures.
7751. Leucistic describes plumage aberration of genetic origin, typically of faded or washed-out coloration, not to be confused with albinism.
7752. A walloon, because of the loon's vocalizations.
7753. Cherry trees.
7754. California and Oregon.
7755. No, most have long tails.
7756. No, it is very thin-billed for a vulture.
7757. Both sexes have double crests.
7758. Inside the barn, but not always. Cliff Swallows typically nest outside under the eaves.
7759. The Eskimo Curlew (if it still exists).
7760. Yes, the tip of its tail is pale colored.
7761. Yes. From below, the tail is pale, but it may transmit a hint of red.
7762. Waterfowl are too heavy to accommodate a gradual molt. Gaps in the flight feathers would result in inefficient flight.
7763. No, it weighs up to 1.7 ounces (47.4 gms).
7764. Red.
7765. No.
7766. The average nest weight, at least on Kangaroo Island, is 7.5 tons (6,800 kg). The megapode mound consists of about 448 cubic feet (12.7 cubic meters) of material.
7767. Juveniles have dark-colored bills, but the bills become yellow within three months of fledging. In breeding birds, the bill turns reddish with a yellow tip.
7768. The American Avocet.
7769. No.
7770. No, its bill is strongly hooked.
7771. Yes.
7772. The pencil-like protrusion behind the bill (snood).
7773. As a rule, they are duller and have paler underparts.
7774. Bewick's Wren. The insular race has not been recorded since 1897.
7775. Thief, alluding to the frigatebird's habit of stealing fish from other birds.
7776. No.
7777. No, young snowcocks feed primarily on legumes. However, the chicks of most pheasants, quail and partridges feed extensively on insects and other invertebrates.
7778. No, its tail is square.
7779. In 1913, James Chapin of the American Museum of Natural History acquired a strange feather from a native's hat when he was in Zaïre (formerly the Belgium Congo). Years later, while visiting the Congo Museum near Brussels, he discovered two birds labeled imported juvenile blue peafowl, and the brown feather he obtained 25 years earlier matched feathers on the two birds.
7780. No.
7781. *The Sea Hawk.*
7782. Yes, the first record was of a pelican seen in the summer of 1981.

7783. No, but several species are vulnerable.
7784. The Rock Sandpiper.
7785. Yes, as a rule.
7786. Cliff faces.
7787. The Killdeer.
7788. Yes, flocks of up to 50 birds may form.
7789. Yes.
7790. The Little Spotted Kiwi. Numbering about 1,000 birds, it no longer occurs on the mainland, and inhabits only a few offshore islands.
7791. Yes.
7792. Generally the male.
7793. No, it is generally inferior, but there are exceptions, such as the owls.
7794. The Rock Wren. Its flattened body enables it to creep far into fissures where the wren retreats to escape the midday desert heat.
7795. The Violet-green Swallow.
7796. No, but nearly so. It had long, powerful legs and ran its prey down. It disappeared in 1914.
7797. Yes.
7798. Terns.
7799. Yes, even if the intruders are human.
7800. No, females have higher-pitched voices.
7801. Yes.
7802. Yes. Copulating pairs cling together in a twittering, tumbling free-fall and may strike the ground before separating.
7803. Orange.
7804. No.
7805. About 3.0-3.2 ounces (86-90 gms).
7806. Troupial is the Anglicized version of the French *troupiale*, which is derived from *troupe*, meaning "flock." Troupials are flocking neotropical icterids.
7807. In 1544, William Turner, an Englishman residing in Germany, published *Avium Praecipuarum Quarum apud Plinium et Aristotelem Mentio est Brevis & Succincta Historia*. A few copies of this priceless volume still exist. Because of his Protestant beliefs, combined with the fact that two of his friends were burned at the stake for heresy, Turner was compelled to seek refuge on the Continent from religious persecution.
7808. Yes, and prey is generally struck against a branch prior to swallowing.
7809. The grouse.
7810. Yes.
7811. A hawk fly.
7812. Yes, albinistic kiwis are known.
7813. No, most hornbills lay white, oval-shaped eggs.
7814. Yes, some species do.
7815. Yes. The flesh-colored, tapering bill is about 2.75 inches (7 cm) long in females and 2.5 inches (6.4 cm) long in males.
7816. Honeyguides are obligate brood parasitic nesters, and the white feathers may be used to lure the host species away from its nest so eggs can be laid in the unprotected nests. Only the Yellow-rumped Honeyguide lacks white outer tail feathers in those species whose breeding habits are known, and it is the only one that may not be a parasitic nester.

Answers

7817. No, the shorebirds have a juvenile plumage.
7818. White feathers lack pigment and reflect all wavelengths of visible light.
7819. The Mew Gull.
7820. To crow.
7821. Possibly because migrating across deserts is a problem and many warblers perish.
7822. In most birds, only the left ovary is functional, but both are functional in kiwis. Thus, as the bird lays its huge egg, a second can be forming.
7823. The House (English) Sparrow.
7824. Yes, it ranges from Mexico and the West Indies south to northern Argentina.
7825. A single female. The island was subsequently stocked with several owls from New Zealand (a different subspecies), and hybrid chicks were produced.
7826. No, all are apparently exclusively insectivorous.
7827. Up to 1.25 pounds (567 gms).
7828. Yes.
7829. Peter *Finch*.
7830. Yes.
7831. No, Virginia Rails are more insectivorous.
7832. No, the female's head is brown with white stripes.
7833. Semi-tame Brown Pelicans loafing on piers begging fish from fishermen.
7834. All races have conspicuous white rumps.
7835. Yes.
7836. Nine breed regularly within the city limits of Canberra.
7837. The Peterson bird field guides.
7838. Yes, some do. At least six were collected in 1984.
7839. Yes, the auk's single egg is white or greenish with dark spots and scrolls.
7840. No, but one of the Old World warblers is called the Wood Warbler.
7841. Yes.
7842. Brownish-gray.
7843. The endemic alpine parrot has a red rump.
7844. Parents regurgitate food.
7845. Yes, but in nidifugous species where chicks depart the nest shortly after hatching, shells are ignored. In both groups, eggs failing to hatch are left in the nest.
7846. Yes, the New Zealand kiwis may utilize hollow logs.
7847. At least nine of the 19 species.
7848. At least four million pairs of Sooty Shearwaters breed on the islands.
7849. Yes.
7850. About 75 percent.
7851. The Cape Cormorant. Duiker is the Afrikaaner name for cormorant.
7852. Harris' Hawk.
7853. No, feeding flocks are quite noisy.
7854. Yes, but the gland is just barely functional.
7855. Procne was transformed into a swallow, hence the generic name *Progne* for the martin, the Latin equivalent of Procne, and the former generic name *Iridoprocne* for the Tree Swallow.
7856. The glands may be very large in the eyes of seabirds, and the oily secretion protects the eye surface from the effects of salt water.
7857. Generally fur or feathers.

Answers

7858. African Fish-Eagles.
7859. The Willet.
7860. No, both sexes feed the young.
7861. Only the Stewart Island race of the Brown Kiwi.
7862. Yes, all have hooked beaks.
7863. The Nicobar Pigeon of the Indo-Australian region.
7864. Woodrow Wilson (July 3, 1918).
7865. No, western birds are distinctly smaller.
7866. The House Sparrow and Java Sparrow.
7867. Unlike typical raptors that lay roundish eggs, Secretary-bird eggs are rather long, but still oval.
7868. No, it was named after George Newbold Lawrence, a 17th-century American ornithologist.
7869. The American Woodcock.
7870. No, a new cavity is generally excavated annually.
7871. Bright red or orange, but the crest is generally partially concealed.
7872. The Green-cheeked Amazon.
7873. Most, if not all, have a distinct musky odor.
7874. Beam feathers.
7875. About 500,000.
7876. None.
7877. Yes, the feathers are elevated if the birds are excited.
7878. Five to eight pounds (2.3-3.6 kg).
7879. The Fiscal Shrike.
7880. No, males call up to three times as often, although duets are common.
7881. Franklin's Gull.
7882. Yes.
7883. Yes.
7884. The Two-barred Crossbill.
7885. Only 2.5 million. The main reason for the decline was habitat destruction, but the drought of the late 1980's was also a negative factor. Blue-winged Teal were 25 percent below the average of the last 30 years, down to 3.6 million from 4.6 million in 1977. Mallards were roughly half as numerous as in 1958, when 12.9 million were counted. The decline for many other ducks was equally appalling.
7886. The sound of evergreen pine cone seeds being cracked by feeding crossbills.
7887. The Brown Eared-Pheasant.
7888. The Scarlet Ibis.
7889. No, for unknown reasons, female mortality is higher.
7890. Yes.
7891. No, the color is a result of the pigment cotingin.
7892. As a rule, yes, but hatching can take place at any time of day or night.
7893. Yes, as opposed to an enclosed, jug-like mud nest.
7894. No, the birds climb up a tree.
7895. The Pied-billed Grebe.
7896. Wilson's Phalarope.
7897. No, it is a thrush, as is a Whinchat.
7898. Introduced to control introduced rats, it was easier for the owls to catch birds, and in some areas the White (Fairy) Tern was essentially eliminated. The white-plumaged terns are

apparently highly visible at night. A bounty was subsequently placed on the owls.
7899. Its large bill is red and blue.
7900. Some authorities indicate only 19 percent, but thereafter mortality is relatively low, possibly only five percent annually. A molting Emperor Penguin's flippers swell up proportionally greater than other penguins, and possibly banded penguins perished as a result of constricting bands. If so, first-year mortality might be less extensive than believed.
7901. No, their heads are flatter.
7902. No, but that of the Thick-billed Murre does.
7903. Not normally, but in birds that hibernate or become torpid, body temperatures may be considerably lower.
7904. The Ashy Storm-Petrel.
7905. No, caracaras are rather long-legged.
7906. Probably the White-headed Woodpecker, possibly because of the high proportion of dry matter in its diet.
7907. The Cornell Laboratory of Ornithology, Cornell University, Ithaca, New York.
7908. Yes, many do.
7909. The Philippine Cockatoo, also known as the Red-vented Cockatoo.
7910. Yes.
7911. About 250 heartbeats per minute.
7912. No, just the opposite. It has the head of a Red-shafted Flicker (red mustache) with the body of a Yellow-shafted Flicker (wing and tail linings usually yellow).
7913. No, the plover has a Y-shaped bar across its chest.
7914. Three to seven feet (1.0-2.1 m) long.
7915. At least 350.
7916. The endangered Hawaiian Crow.
7917. Yes, catbird chicks are often fed nestlings, but adults eat mostly fruit.
7918. The neotropical potoos.
7919. The Double-striped Thick-knee. The pet birds function as nocturnal "watchdogs."
7920. Four: Chukar, Ring-necked (Common) Pheasant, Wild Turkey and White-tailed Ptarmigan.
7921. Dusky colored, but adults generally have orange to salmon-colored bills.
7922. Yes, primarily Baja California.
7923. The Australian Pilotbird often keeps company with lyrebirds because the much larger birds scratch up forest-floor litter, exposing insects and grubs. Early observers believed that humans were led or piloted to lyrebirds by Pilotbirds, but the name possibly originated because Pilotbirds lead lyrebirds to food. The White-browed Scrubwren also associates with lyrebirds.
7924. Saturn. According to the myth, Picus was transformed into a woodpecker by Circe, whose love he had rejected.
7925. Yes.
7926. Based on average territory size and potential habitat, as many as 12 million kiwis could have inhabited the country. Currently, between 20,000 and 50,000 remain, but they are declining in some areas.
7927. Yes.
7928. Yes.
7929. No, it is generally the first.

Answers

7930. The Bearded Tit (Reedling), because of its unusually long tail.
7931. Costa's Hummingbird, some of which live year-round in the Colorado Desert.
7932. Between 11.9 and 14.2 inches (30-36 cm).
7933. A fungus.
7934. The Yellow-eyed Penguin. Hoiho is a Maori name meaning *noise-shouter*.
7935. Yes, as opposed to four years in larger gulls.
7936. No, prey is carried in their feet. Prey is picked up with the bill, but as the shrike launches into the air, the prey is transferred to the feet.
7937. Yes, in loose flocks.
7938. No.
7939. No, females are responsible for nocturnal incubation.
7940. No, their eyes are open.
7941. The Eastern Bluebird.
7942. The Surf Scoter. A gibbous bill has a pronounced hump.
7943. MacGregor's Bird-of-paradise occurs up to 13,000 feet (3,962 m).
7944. The three jaegers.
7945. Yes, some kingfishers occasionally catch and consume small birds, such as white-eyes.
7946. California with 11.
7947. *The Flight of the Phoenix.*
7948. Yes.
7949. The wren may puncture and eat the bittern's eggs.
7950. Yes, slightly so. Most males have black throats, whereas female throats are brown.
7951. The Purple Swamphen.
7952. Yes.
7953. Yes, but with their very long bills and tiny vestigial wings, the birds appear a bit ridiculous.
7954. Yes, as far north as the tree limit.
7955. Just inside the nest.
7956. Yes, hybrids are frequent where their ranges overlap.
7957. No, it is restricted to Southeast Island (Chatham Islands). In 1990, the breeding population consisted only of about 44 pairs, with a total population of about 130 birds.
7958. None.
7959. African shrikes.
7960. Yes.
7961. Mergansers take their ducklings to deeper water, and shelducks are not adapted to diving.
7962. No.
7963. McCown's Longspur.
7964. The virtually extinct Buller's Shearwater. No other breeding sites are known, and by 1984 at least 100,000 occupied burrows covered the islands.
7965. No.
7966. Yes.
7967. Lewis' Woodpecker.
7968. The *dead-leaf* pattern.
7969. Yes.
7970. The South Polar Skua has been sighted at the Russian Antarctic station of Vostok, 750 miles (1,207 km) from the coast, where in July of 1983 the planetary low temperature of -129.9^{o}F (-89.6^{o}C) was recorded.

Answers

7971. Shortly after midnight.
7972. Herons, egrets, American White Ibises and Double-crested Cormorants.
7973. An African warbler.
7974. No, their bills are thin.
7975. Yes.
7976. Yes.
7977. The Yellow-billed Oxpecker.
7978. No.
7979. None. Old World vulture distribution is confined to more or less open country.
7980. No, it is one of the most common.
7881. Small parrots, because crossbills often dangle while feeding.
7982. Yes.
7983. The Gray-headed Chickadee.
7984. The Curve-billed Thrasher.
7985. Yes, but more typically, waxwings feed on berries.
7986. The Japanese name for the Yokohama chicken or Phoenix fowl, the ornamental chicken with the longest feathers of any bird.
7987. The Ibisbill of highland central Asia.
7988. Skins were prized for furry slippers and hats, and also for rugs and bed spreads.
7989. Yes, although a number of snipe inhabit South America.
7990. Chestnut.
7991. No, all four are restricted to Asia and New Guinea.
7992. Only the Yellow-eyed Penguin of New Zealand.
7993. Yes.
7994. Restricted to Chichi Shima, a small island south of Japan, the thrush has not been seen since four specimens were collected in 1828. The bird was probably wiped out by rats that escaped from whaling ships. At least five Bonin Islands birds have disappeared.
7995. Males and newly-hatched young have dark eyes, but adult females have cream-colored eyes. Males of southwest New Mexico and west Texas have black or black-flecked cheeks, and were formerly considered a separate species.
7996. Yes, but only rarely. Fish are almost invariably swallowed prior to flight.
7997. The American Robin.
7998. About twice as great.
7999. Burmeister's (Black-legged) Seriema.
8000. Yes, because of its powerful broad wings, but becoming airborne is cumbersome, requiring a hefty leap.
8001. The endangered, flightless gallinules defend an acre (0.4 hectare) or more.
8002. Yes.
8003. Yes.
8004. Yes, especially in central California.
8005. The Buff-faced Pygmy-Parrot weighs only a half-ounce (14.2 gms).
8006. Approximately 35 days. Following territory establishment and courtship, males are responsible for the first two-week incubation stretch, during which time they may lose up to 45 percent of their weight.
8007. A faint pink, but females have brownish breasts.
8008. Some of the African bulbuls.
8009. Bonaparte's Gull.

Answers

8010. Most species have longer legs.
8011. The Kerguelen fur seal and the leopard seal.
8012. No, their wing tips are white.
8013. At least 15.
8014. An owl.
8015. Jamaica (*Laterallus jamaicensis*).
8016. Yes, in addition to a range of twittering sounds.
8017. The hollow base of a baobab tree. Ground-hornbills also excavate their own holes in earth-banks, or even use old stick nests of other birds. Unlike other hornbills, they are generally not arboreal nesters, and the nest entrance is not sealed up.
8018. Seven.
8019. Yes.
8020. No, but there are exceptions.
8021. No, the 5-6 eggs are yellowish with fine yellowish-brown speckles.
8022. Breeding drake Common and Spectacled Eiders.
8023. No, female eye color is typically paler.
8024. No, but most do. A few, such as the Yellow-breasted and D'Arnaud's Barbet of Africa, nest in holes in banks or earthen cliffs, or in walls of houses, or even the sides of wells.
8025. Black.
8026. Yes.
8027. Boldly colored with rich blues, but some species are patterned with brilliant red as well.
8028. The polar bear.
8029. The shorebird may lie flat on the ground with outstretched head and neck until the danger has passed.
8030. Ostriches and parrots.
8031. Yes, nesting density may exceed 200 nests per acre (0.4 hectare), although some White-winged Doves nest in solitude.
8032. Blue.
8033. The Jungle Nightjar. It does not breed in Japan, but is a common summer visitor throughout the country.
8034. Yes, adult plumage is not attained until seven years of age, during which time the distinctive raptor passes through three intermediate phases that are recognizable in the field.
8035. The Toronto Blue Jays.
8036. Frank Chapman, along with 27 bird-watchers throughout the country, started the counts to protest the side hunt. The side hunt was a widespread tradition in which teams of hunters competed to see which side could kill the most wild birds and mammals in the course of an afternoon. In the first count, 5,000 birds were recorded, ushering in a new era for the sport of birding.
8037. The Rocky Mountain grouse lack a tail-band.
8038. Boxing day. This tradition probably conformed with some half-forgotten folklore decreeing that on the day after Christmas, certain small birds, notably the wren, were ritually hunted.
8039. It is likely a consequence of the kiwi's low body temperature. Biologists formerly believed that in excess of 80 days was necessary because the huge egg could not be completely covered by the incubating bird.

Answers

8040. A special pouch of skin behind the tongue enables a nutcracker to carry as many as 150 seeds. During a single season, the corvid may bury more than 30,000 seeds in many hundreds of separate holes.
8041. The Montezuma (Mearns') Quail ranges from the southwestern U.S. to Oaxaca, Mexico.
8042. No, mating Emus are typically rather silent.
8043. The Humboldt Penguin.
8044. Belgium, in about 1880, approximately 40 years after Budgies were initially imported from Australia.
8045. The Yellow Bittern of southern Asia, New Guinea and western Oceania.
8046. Yes.
8047. The frigatebirds.
8048. From the Latin *aureolus* meaning "golden" or "yellow."
8049. The skimmers.
8050. The Canary.
8051. The first two neck vertebrae are fused.
8052. It generally exceeds 100 days.
8053. No, but giant-petrels become paler with age.
8054. No, its temperature tends to fall.
8055. Yes, generally so.
8056. No, most are in North America.
8057. The American Coot.
8058. No, it averages about 3.0^{O}F (1.8^{O}C) below that of an adult, but by 10 days of age, it is the same.
8059. The Common Nighthawk. It is one of the few nightjars to do so, and unlike its close relatives, seldom calls when sitting.
8060. It twisted its bill on the nails of the cross of Christ.
8061. Rarely, if ever.
8062. No, females are green, whereas males are a brilliant turquoise-blue.
8063. The upper mandible is not longer. The lower mandible is decidedly longer.
8064. The bright scarlet skin is featherless.
8065. A giant saguaro cactus. The woodpecker may also excavate a nesting cavity in mesquite, hackberry, cottonwood, oak, walnut or other trees, or even in fence posts or utility poles.
8066. Yes, but only partially so at the base of their toes.
8067. Yes, at least in those species still surviving.
8068. Florida.
8069. Probably the aggressive Snowy Egret.
8070. A peculiar feature of the family is the exceptional variability in the position of the nostril and its shape that ranges from circular to slit-shaped.
8071. The Siberian White Crane.
8072. No, the Purple Finch is the state bird.
8073. Yes.
8074. Barbets and woodpeckers, but numerous other birds serve as hosts as well.
8075. No.
8076. The American Goldfinch, because it eats lettuce.
8077. At one time, the Heath Hen was found in some abundance in all New England states, and as far south as Virginia. While protected as early as 1791, the grouse was still hunted, and by the early 19th century was confined to Martha's Vineyard, Massachusetts. In 1908, the

population numbered less than 200, and a reserve of 1,600 acres (648 ha) was created to preserve them, and by 1915 the population had increased to 2,000. Then disaster struck, commencing with a grass fire in the reserve in the summer of 1916, followed by an especially severe winter, an irruption of Northern Goshawks that preyed heavily on the grouse, and finally an epidemic of blackhead, a disease of domestic poultry. The population was reduced to 13 in the mid-1920's, and the last Heath Hen succumbed in 1932.

8078. Yes.
8079. The time varies, ranging from a month up to a year, an unusual situation for passerines.
8080. No.
8081. Yes.
8082. The woodcocks and snipe.
8083. Pure white.
8084. Japan.
8085. Probably to avoid matting its plumage with sticky juice and nectar on which it feeds.
8086. No, but pairs may nest in loose association with one another.
8087. Tyrone Power and Maureen O'Hara.
8088. Both sexes incubated.
8089. Yes.
8090. Unlike most African ungulates, hartebeest are intolerant of oxpeckers climbing about on them.
8091. About 1,500,000.
8092. The peculiar African passerines dwell in dark forest caves and forage during twilight hours.
8093. Yes.
8094. About 80 percent.
8095. No.
8096. Large nests may reach nearly six feet (1.8 m) in length.
8097. No.
8098. The fleshy and bony parts of a bird's tail, where the tail feathers are attached. The name refers to the shape that recalls the headgear of the Pope. The term is generally applied to chickens and turkeys.
8099. Yes, and it often takes some time to drain the water prior to swallowing the catch. A gallon of water weighs 8.3 pounds (3.8 kg) and up to three gallons (11.4 liters) may be engulfed, weighing 24.9 pounds (11.3 kg).
8100. No.
8101. In one study, the average time for a European (Common) Cuckoo was a mere nine seconds. The host species consisted primarily of Meadow Pipits.
8102. Yes, but the neotropical ducks are domesticated.
8103. No, it is one of the first.
8104. At least 17.
8105. The cassowary.
8106. The Old World Common (River) Kingfisher.
8107. No, they are primarily vegetarian and frugivorous.
8108. Yes, some or most apparently mate for life.
8109. The raptor population of southern Spain at the time was more numerous than elsewhere in Europe, and the birds caught infected rabbits so quickly that the disease never spread.

Answers

8110. None, but both the Fork-tailed and Spine-tailed Swift are occasional stragglers to New Zealand.
8111. At least 10.
8112. The Lesser Seed-Finch.
8113. Yes.
8114. The *crowing area.*
8115. When a platypus was relaxing on the surface of the water, the hunter shot just below it, stunning it long enough for a dog to retrieve it. Thus, skins were not damaged by bullet holes.
8116. Thermal dependence is not necessary for a raptor feeding primarily on palm nuts.
8117. The owls.
8118. Cleaning, feather maintenance and parasite removal.
8119. Nearly five feet (1.5 m).
8120. Yes.
8121. Mallards readily hybridized with the native Gray Duck. Possibly no more than 2,000 genetically pure Gray Ducks remain.
8122. The six crested penguins. The first-laid egg is 20 to 60 percent smaller than the second.
8123. While appearing on Sesame Street during Bird Quest '89, Dr. James F. Clements was asked by a student the scientific name of Big Bird. He coined the name *Megagallus humongous* on the spot.
8124. Not more than 7,500 cranes, and possibly as few as 6,000.
8125. The Semipalmated Sandpiper, that lacks spotting on its flanks; the Western Sandpiper has spotted flanks.
8126. Yes.
8127. Yes, even Antarctica.
8128. The spinifex bush. The pigeon is so named because of its preference for nesting in a scrape under a spinifex bush, and is locally common in spinifex-covered rocks and sand ridges, usually near water. Outside Australia, it is better known as the Plumed Pigeon.
8129. Either sex.
8130. Yes, it lives year-round in some of the hottest, driest regions of North America.
8131. Yes.
8132. More than 6.5 million erythrocytes.
8133. Japan. The woodpecker is an uncommon resident of northern Okinawa Island.
8134. Yes.
8135. Only one. Mammals have two, providing them with a greater degree of head movement.
8136. American Coots have a white patch under the tail, whereas the under tail of the Eurasian Coot is slate-gray. The frontal shield of the Old World birds is entirely white, but the shield of the American Coot is dark at the top.
8137. Yes, but generally not at other times of the year.
8138. The Red-faced Warbler.
8139. Flags.
8140. The Pomarine Jaeger.
8141. A Rhinoceros Hornbill.
8142. *Otus*, and *Bubo* as well, are Latin for "horned owl," suggesting that the owl is horned.
8143. Yes.
8144. No, birds should be darker in color in humid climates.
8145. A lek display site.

Answers

8146. Griffons.
8147. Yes, a laterally compressed bill is higher than it is wide for much of its length.
8148. No, cardinals sing year-round.
8149. Mice.
8150. *Eating like a bird.* The expression is inaccurate because many birds eat more than 50 percent of their weight daily.
8151. *The Ostrich.*
8152. Approximately 66 percent.
8153. No.
8154. Diablotin means *little devil,* referring to the petrel's strange nocturnal calls. In Trinidad, the Oilbird is also known as a diablotin for the same reason.
8155. Yes.
8156. March, 1981. Its status is not known, but the rail is believed to be rare, and may not exceed 500 birds.
8157. Hawking.
8158. Yes, but this is certainly not a typical petrel habit. Some species, such as the Mottled Petrel, may nest in forests, and the birds scramble up trees, no doubt making take-off easier as the petrels can launch themselves directly into the air.
8159. Yes.
8160. Yes, but most often in juveniles and nestlings. The behavior is less common in adults because it precludes a rapid take-off in an emergency.
8161. Yes.
8162. Yes.
8163. Turaco feathers are largely deficient in barbules.
8164. Yes, 0.32-0.42 ounces (9.2-11.8 gms) versus 0.37-0.42 ounces (10.5-12.0 gms).
8165. The mythical Phoenix.
8166. The 1890's.
8167. The cuckoo.
8168. Yes.
8169. African mousebirds typically perch with their feet and shoulders at about the same level, with the belly hanging down exposed between the legs. When roosting, the birds are often touching one another.
8170. Yes.
8171. The crows and ravens.
8172. A gopher.
8173. Yes, some species do.
8174. Yes, and of the 26 specimens, 22 were collected in November of 1879.
8175. No, the birds float high.
8176. The breeding-plumaged male Scarlet Tanager.
8177. Yes, as did all surviving flightless birds.
8178. No, the western birds are drab gray.
8179. About 3.9 ounces (110 gms), or more then twice as much.
8180. Up to 98 percent of its diet consists of insects.
8181. The call of Gambel's Quail is four-noted; that of the California Quail is three-noted.
8182. Yes.
8183. Despite measuring only one-tenth of the total body length, the wings aid in thermoregulation. In very hot weather, the wings are raised, exposing the complex

structure of surface veins underneath, thus facilitating heat loss.

8184. Yes.
8185. Yes, some do.
8186. An owl, no doubt because of her nocturnal activities.
8187. Yes.
8188. The Chimango Caracara.
8189. Alexander von Humboldt.
8190. Salmonella.
8191. Eggs have been recorded year-round.
8192. No, some, such as the Swift Parrot, have elongated and pointed wings.
8193. No, the same nest site of previous seasons is often used.
8194. Hovering flight is strenuous, and a fanned tail enables the bird to gain lift from any wind.
8195. The Mute Swan.
8196. Both sexes cock and spread their tails, and males erect their crests.
8197. Yes, rollers of many species readily fly at humans and raptors.
8198. No, the lower mandible does not start to grow longer than the upper until the young are almost full-grown and ready to fly.
8199. Orange.
8200. Yes.
8201. 1828.
8202. Demoiselle is French and means *damsel.*
8203. Yes.
8204. Generally 8-12 days after hatching.
8205. No, it typically nests on steep, inaccessible sea cliffs.
8206. Duckweed (*Lemma minor*).
8207. No.
8208. Orange.
8209. Cock-fighting.
8210. White.
8211. The Spotted Sandpiper.
8212. Yes.
8213. Eighty, but 13 to 30 eggs is more typical.
8214. No.
8215. Yes.
8216. When fish are trapped in drying pools, and thus easier to catch.
8217. About one-third.
8218. No, females greatly outnumber males.
8219. The American Bittern, that utters sounds resembling resonant pounding.
8220. The Peregrine Falcon.
8221. Five: Great, Highland, Little, Thicket and Slaty-breasted Tinamou.
8222. Between 1,500 and 2,700.
8223. Yes.
8224. Yes, many do.
8225. Yes, some species occasionally do.
8226. Japan.
8227. Bright red.
8228. Seven.

Answers

8229. The bird nests high on the sides of inaccessible cliffs, thus its nest is difficult to reach.
8230. The Red-footed and Brown Booby.
8231. Through the use of satellite telemetry.
8232. Only seven, but these are the only ones that regularly fish in open water.
8233. No, but the Ruddy Turnstone does.
8234. Only two: Eurasian and Red-knobbed Coot.
8235. Its chorus of laughter every morning was a signal for the sky people to light the great fire that illuminates and warms the earth by day. To imitate the laugh was taboo because the sky people might take offense and plunge the earth back into eternal darkness.
8236. The Hermit Thrush.
8237. Yes.
8238. On cliff ledges or sheltered tree limbs, where very few currently nest.
8239. At least 125.
8240. Yes.
8241. Yes, the Australian megapode lays up to 34 eggs a year, and 8-11 young may fledge annually. However, due to habitat destruction and the introduction of foxes, the bird is endangered.
8242. No.
8243. An Old World flycatcher endemic to the Hawaiian Islands.
8244. Goldfinches have an undulating flight.
8245. About 50 pounds (22.7 kg).
8246. No, its bill is held slightly open.
8247. The claws are suitable for grasping reeds and grasses.
8248. Feather-brained or bird-brained.
8249. The Old World white-eyes.
8250. The white-plumaged juvenile Little Blue Heron.
8251. A duck-billed dinosaur from the late Cretaceous.
8252. Apparently squid.
8253. Yes, the birds only mingle during the breeding season.
8254. Chestnut.
8255. Yes.
8256. The Northern Mockingbird.
8257. The Magnificent Hummingbird.
8258. Yes, reportedly so.
8259. Cuckoos drop their eggs from some distance into the open cup-shaped nests of the host, and a thicker shell provides some protection. A thicker shell also prevents puncturing of the egg by the host bird as it attempts to remove the egg with its bill.
8260. Ospreys tend to dive less often when skies are cloud-covered. Cloud cover apparently interferes with their visual perception.
8261. None.
8262. The guillemots, especially the Black Guillemot. The birds readily approach small boats and settle on the water nearby, often in large numbers.
8263. Yes.
8264. Yes, but the young are attended by their parents for some days thereafter.
8265. The Hermit Thrush.
8266. Aldabra ibises have China-blue eyes, whereas mainland birds have brownish-red eyes.
8267. The African oxpeckers.

Answers

8268. Yes.
8269. Both commonly nest on utility poles, much to the distress of power company officials. In a census by the state electric company in Rio Grande do Sul, Brazil, in 1987, there were 580 Rufous Hornero nests on company structures, with 265 at risk of provoking short-circuits.
8270. Yes.
8271. The Black-bellied Plover.
8272. Yes.
8273. The inner hind-toe of the Three-toed Jacamar has been lost.
8274. Yes, generally so.
8275. Northeastern Brazil.
8276. The Summer Tanager has a robin-like song, not the hoarse call typical of other northern tanagers.
8277. No, the birds occasionally eat berries.
8278. No, but it has essentially doubled.
8279. Wax.
8280. Eyton's Whistling-Duck, more commonly known in Australia as the Grass Whistle-Duck.
8281. The Cyprus Pied Wheatear and Cyprus Warbler.
8282. Yes.
8283. The Great Crested Flycatcher.
8284. No, some, especially shags, have pink, red, or even yellow feet.
8285. Rich violet.
8286. An estrildid finch, named for the stripe of red feathers across its throat, resembling a slashed throat.
8287. Yes.
8288. About 1833.
8289. The bill is used to slit the stems of banana-like plants.
8290. Yes.
8291. No. The pale-color morph predominates in the Atlantic, whereas dark-color morph birds are more common in the Pacific.
8292. Crouching obscures or reduces the tell-tale shadow.
8293. To make tobacco pouches and sheaths.
8294. No, the birds defend no territory except for the nest.
8295. Yes, at times.
8296. Drunkenness. To give a child an owl's egg would insure he would never become a drunkard. If the soup were taken while the moon was waning, epilepsy would be cured.
8297. None, but some bones are in museum collections.
8298. Degenerate, hair-like plumage.
8299. Slightly more than four feet (1.2 m).
8300. Birds with four toes. Most birds are anisodactylous.
8301. No, their burrows are very shallow, or eggs are merely concealed under a log or in a rock crevice.
8302. Yes, but not always.
8303. Yes, mollusks are struck against a rock or root until the shell breaks. Such sites are often marked by quantities of broken shells. Limpkins can also extract snails without breaking the shells.
8304. The cuckoo feeds extensively on fruit, unlike most other cuckoos, which are insectivorous.
8305. The Black-necked Stilt.

8306. No, it is serrated.
8307. Cranes, especially when the birds are dancing.
8308. Red.
8309. No. The fledging period is probably 49-56 days, but some authors indicate flight may not be possible until about 75 days of age.
8310. About 228,500 times as large.
8311. The Old World Little Grebe, in which the gland accounts for 0.61 percent of the bird's weight.
8312. Yes, especially in diurnal birds.
8313. No, they are striped.
8314. Yes.
8315. The three kiwis.
8316. The Barn Owl, because of its heart-shaped facial disc.
8317. Some of the filter-feeding prions. The petrels rest lightly on the water and use their feet to skim swiftly over the surface, with their wings outstretched, and the bill or even the whole head submerged.
8318. Yes and no. Only adult male American Anhingas have crests.
8319. Grit.
8320. No, but Greater Flamingos do.
8321. No, its legs are longer.
8322. No.
8323. Yes.
8324. No, females incubate at night.
8325. Between 4,000 and 5,000 pairs.
8326. The Horned Lark.
8327. The Brown-headed Cowbird.
8328. About $350 million, as of 1989.
8329. Nowhere.
8330. All are colonial nesters.
8331. Quite possibly. It is certainly the most common wild pigeon of Britain.
8332. A serrated tongue enables vultures to rapidly swallow slippery food.
8333. No, the birds are uniformly blue-black.
8334. The parakeet does not nest in vegetation, but rather under it in burrows that may be more than three feet (1 m) deep.
8335. Yes.
8336. Ivory-white, often with a greenish tinge, and highly glazed.
8337. Yes, and only warriors of outstanding valor were entitled to wear one.
8338. The glands greatly enlarge.
8339. A human soul was being summoned to the grave.
8340. No, the raptor was named for its jackal-like call.
8341. The little birds can walk surprisingly fast, and hop and run as well.
8342. No, turacos have short, abruptly curved claws.
8343. Brother Matthias Newell, a respected teaching missionary from Prussia, was known as the *rattlesnake catcher*, a tribute to his remarkable zeal in natural history.
8344. Mono Lake, in east-central California. Up to 750,000 grebes gather on the lake in the fall, accounting for possibly a third of the world's population. The lake is also critical for migrating Wilson's Phalaropes, but less important than Great Salt Lake in Utah.

8345. A loud, snuffling noise is produced by expelling air forcibly through its nostrils, which are located at the tip of its bill. Nasal valves prevent soil from entering when the bill is buried.
8346. Not until 1969.
8347. Yes, very much so, whereas adults are relatively non-vocal.
8348. *Eat crow.*
8349. With the bird's own guano.
8350. The British Ornithologists' Union.
8351. No, most are rather dull colored.
8352. Helmeted Guineafowl were possibly brought to the West Indies as early as 1508, and would no doubt be more numerous on some islands were it not for the introduced Indian mongoose. Even so, guineafowl are important gamebirds in Cuba and Hispaniola.
8353. Yes.
8354. Feathers from about 80,000 Mamos were required to complete the cloak. Each honeycreeper yielded only 6-7 of the highly-sought yellow feathers and about 450,000 feathers were necessary. The Mamo is extinct, and the great king died before the cloak was completed.
8355. It fans or spreads its tail.
8356. The Black-legged Kittiwake.
8357. Indonesia, followed by Thailand. In a single year alone, more than 3.5 million cave swiftlet nests were exported from Borneo to China.
8358. Yes.
8359. Yes.
8360. The Sora rail.
8361. No, a single-egg clutch is typical.
8362. Clark's and Eurasian Nutcracker.
8363. The Pin-tailed Whydah. The African finch escaped in Hawaii, and probably bred in the mid-1970's, but failed to become established.
8364. Leach's Storm-Petrel. In other storm-petrels, both sexes dig the burrow.
8365. On the average, 4-12 feedings per hour. This is highly variable from species to species and even from individual to individual.
8366. The axillars.
8367. They use a distraction display.
8368. Not normally, but some have lived longer than 20 years.
8369. Yes.
8370. Never; the storks are solitary.
8371. In 1898, Sylvester D. Judd, a biologist from the U.S. Biological Survey, recorded with a gramophone and trumpet speaker the songs of his pet Brown Thrasher named *Rustler*.
8372. Yes.
8373. Yes.
8374. No, its long hindtoe is at the same level as the front three.
8375. Five to six females, although rarely up to 14.
8376. The flightless Weka rail. Eggs are even taken from under incubating birds.
8377. Ducks on the surface are far more vulnerable to turtles and are taken twice as frequently.
8378. The Chimango Caracara, in 1928.
8379. Juveniles of all three species are barred ventrally.
8380. No, it is smaller.
8381. Chickens.

8382. Yes, as a rule.
8383. Yes.
8384. The nipple.
8385. A leaf-gleaner.
8386. No.
8387. *Old Abe.* The famous eagle was named after President Abraham Lincoln and ultimately became the Wisconsin State mascot.
8388. Yes.
8389. Yes, cocks do.
8390. The ovenbird has a preference for sewage effluent.
8391. The kingfisher is gregarious, and up to five adult helpers may assist with a nest.
8392. The Winter Wren.
8393. No, their gobble is lower-pitched.
8394. Yes.
8395. Yes.
8396. No. Except for Trinidad and Tobago, antbirds are absent from the region.
8397. Dipper voices must be loud enough to be heard over the clamor of mountain torrents.
8398. No, their flight is undulating.
8399. Bright red.
8400. Yes.
8401. No. When the sharp edge of the lower mandible strikes prey, the head of a skimmer doubles back under its body and the jaws snap shut. Prey is drawn out of the water while the bird's head faces back or down.
8402. Ten: rheas, screamers, trumpeters, Sunbittern, seriemas, Magellanic Plover, anis, Hoatzin, seedsnipe and Oilbird.
8403. The Yellow-rumped Warbler.
8404. No, young birds are somewhat darker.
8405. No, young often depart when their parents are absent.
8406. The average time is about 40 days.
8407. The Philadelphia Vireo.
8408. The honeyguides, but there may be exceptions. Of the seven species whose breeding habits are known, six are definitely parasitic, but a Himalayan species may not be.
8409. Yes, in some species, but only the entrance. The burrow branches and both pairs maintain separate nesting chambers. Only the nesting chamber is defended, not the entire burrow.
8410. Possibly because of the regular dipping of cattle to control ticks.
8411. The Ground Woodpecker.
8412. An owl; *And yesterday the bird of night did sit, even at noonday, upon the market place, hooting and shrieking.*
8413. Helmetshrikes are characterized by intense sociability. Groups of 6-12 birds of all ages may feed and roost together, and share nest-building and chick-feeding. Most shrikes are not gregarious.
8414. No, only during the breeding season.
8415. No. While the overall color of the endemic bird is slate-gray, there is a conspicuous pattern of white, reddish and black bars on its broad, rounded wings that is visible when the wings are spread.
8416. Yes.
8417. No, but Yellow-billed Magpies are common in Sacramento.

Answers

8418. Lewis and Clark.
8419. No, this behavior was not documented until 1901.
8420. The Night Parrot. It had not been seen since September 21, 1912, but the mummified remains of the nocturnal bird were discovered on October 17, 1990, alongside a road in southwestern Queensland, apparently a road casualty. Two birds were seen in June of 1993 near Cluncurry, northwest Queensland.
8421. Parents regurgitate food for them.
8422. A breed of wire-haired European dog, usually grayish in color, used in the hunting of gamebirds.
8423. The Yellow-billed Cuckoo.
8424. Mules.
8425. The shorebirds.
8426. No.
8427. The Demoiselle Crane.
8428. Western Gulls require four years to assume adult plumage, but adult plumage is attained in three years in the Yellow-footed Gull. Until relatively recently, the Yellow-footed Gull of western Mexico was considered a race of the Western Gull.
8429. Yes, but some of the smaller toucans are content with dead trees.
8430. About nine years.
8431. The Indigo Bunting.
8432. The ulna, the larger of the two bones of the forearm.
8433. No. Breeding generally does not commence until at least the second year, possibly because young males are dominated by older gobblers and psychologically prevented from copulating. Domestic toms do breed in their first year.
8434. Yes, at which time they resemble females.
8435. Yes, but its status is accidental.
8436. The Central and South American tinamous have hearts accounting for only 1.6-3.1 percent of their overall weight, compared to 12 percent in domestic chickens.
8437. No, only females have brood-patches.
8438. No, their nests are generally flimsy.
8439. Ringworm is not caused by a worm, but rather by a fungus.
8440. Yes.
8441. The White Stork; it meant *bringer of good fortune*.
8442. No.
8443. Yes.
8444. The call that arboreal, cavity-nesting female ducks, such as the Wood Duck or Bufflehead, use to encourage their ducklings to leap to the ground.
8445. Yes, they have been known to chase off intruders as large as Mallards.
8446. The Oregon, Gray-headed, Slate-colored and White-winged Junco.
8447. The Razorbill.
8448. During food exchanges, the eyes of both adults and chicks are closed, probably to protect the eyes.
8449. Yes, many do.
8450. Yes, and the tropical flowers are also called canary-bird flowers.
8451. Yes.
8452. Yes.
8453. A lark.

Answers

8454. Yes, occasionally.
8455. None.
8456. No, the starling was intentionally introduced in 1903.
8457. Stephen's Lory.
8458. Yes, their flight consists of alternate bursts of flapping and gliding.
8459. Yes, kiwis do, but such behavior has rarely been observed.
8460. The attachment of the upper mandible to the skull in most birds is quite flexible, enabling the bird to move the mandible slightly, thus increasing its gape. It is most pronounced in parrots because of a well-developed hinge between the bill and skull.
8461. Yes, but only slightly.
8462. No, an owl's skin is relatively thin.
8463. Yes, but some are primarily nectarivorous, whereas others are fruit eaters.
8464. The White-throated Swift.
8465. The Common Quail.
8466. No, the terns are gregarious.
8467. No, its song sounds rather like a buzzing insect.
8468. The Long-tailed Jaeger.
8469. Africa. Some swans winter in Africa, and a feral population of Mute Swans is established in South Africa. Black-necked Swans have strayed to Antarctica.
8470. No, their eggs are remarkably spherical in shape.
8471. Yes.
8472. Yes.
8473. No, but mud is used.
8474. The Tufted Titmouse.
8475. Yes, but in captivity the cracid is also partially diurnal.
8476. Yes, the population may have increased as much as tenfold, and is currently in excess of 700,000 Emus.
8477. The largest of ovenbird nests, the huge, thorny structure is up to five feet (1.5 m) in diameter. The sturdy nest is strong enough to support the weight of a man without sustaining damage.
8478. Yes, most species are.
8479. Under the eves of the American Museum of Natural History in New York City, in 1890, the same year the pest was introduced.
8480. The Black-capped Chickadee.
8481. The larger nostrils of a Turkey Vulture possibly reflect its superior olfactory sense.
8482. To combat the previously introduced rabbit.
8483. Spiny lobster larvae.
8484. A screech-owl.
8485. *Jay*hawkers.
8486. The Marbled and Kittlitz's Murrelet.
8487. Primarily African, the Red-knobbed Coot also inhabits southern Spain. During the breeding season, two red cherry-like knobs appear on the coot's crown.
8488. No, the bird is terrestrial.
8489. Yes.
8490. Yes.
8491. The 14-inch-long (36 cm) Magellanic Woodpecker.
8492. Scouts.

Answers

8493. Mexico.
8494. The name came from the Betsimasaraka and Sakalava tribes of northern Madagascar for the indigenous drongos. The name has since passed into universal use for the family.
8495. Yes, but only a few.
8496. Yes. The coniferous forest of the far northern regions of the Old World is called the taiga.
8497. No, grebes are primarily temperate. When found in the tropics, most species inhabit montane lakes.
8498. From the liver-red color of the male.
8499. At backyard feeders in winter.
8500. The cissas turn blue because of cartenoid-poor diets.
8501. Yes, colonies may consist of thousands of nests.
8502. Yes.
8503. Red, but the loral spot is yellow at other times.
8504. The birds nervously flick their wings.
8505. The globular nest of a Verdin.
8506. Yes. Construction may be initiated by the male, but the female is responsible for most of it.
8507. Yes.
8508. Yes, the temperature of some passerines may vary as much as 10°F (5.6^{O}C) during the course of a day.
8509. The Crested Caracara is holding a snake.
8510. Yes, many species are.
8511. A termite nest.
8512. Possibly as a defense against cuckoldry. Males do not collect nesting material.
8513. The kiwi. Glow worms are apparently one of its favored foods. Maoris attracted kiwis with pieces of smoldering bark; the sparks presumably resembled glow worms.
8514. 1961.
8515. The tropicbirds.
8516. The 5 by 3-inch-egg (12.7x7.6 cm) generally exceeds a pound (0.45 kg).
8517. No, their wings are short and rounded, and are small for the bird's size.
8518. Yellow.
8519. The Nicobar Pigeon of the Indo-Australasian region.
8520. No, generally only the males of the two brown pelicans.
8521. At least 330 species in 60 families.
8522. The owls.
8523. Grayish, but the wingtips of other white pelicans are black.
8524. Nests are not really constructed because the finch nests in a cavity, but a thin-walled nest of grass stems might be built.
8525. Yes, but wattles are generally restricted to males.
8526. Yes.
8527. About 20 years, and a few old-timers *might* survive for a half-century.
8528. The Golden and Great Argus Pheasant. Females of both species reportedly incubate continuously without departing the nest, not even to eat or drink.
8529. Yes, some, such as the Spotted Sandpiper, do.
8530. Yes, but breeding at such an early age is exceptional. Nesting does not generally commence until the second year.
8531. The white-eyes.
8532. The Yellow-billed Loon.

Answers

8533. The mating, treading or coupling time.
8534. Chile.
8535. The Black-bellied (Gray) Plover. Other Arctic-breeding plovers also breed south of the Arctic.
8536. Yes.
8537. No, but trained raptors are somewhat successful as deterrents.
8538. Yes.
8539. The South American Black-headed Duck.
8540. The Prothonotary Warbler.
8541. The tropicbirds and frigatebirds.
8542. Yes, but there are exceptions, especially during courtship.
8543. No, males are among the first spring migrants, generally several weeks in advance of their mates.
8544. Yes, and the chicks are often of different ages and sizes. Broods from nests as far as 7.5 miles (12 km) apart may merge. The largest recorded brood aggregation consisted of 380 young, and large groups are generally accompanied by immatures.
8545. The Warbling Vireo.
8546. No.
8547. No, the disease is viral.
8548. Yes.
8549. Aftershafts (afterfeathers) may increase the thermal insulating property of the feathers.
8550. Up to 2.75 pounds (1.2 kg).
8551. No, woodcock eggs are paler in color.
8552. No.
8553. The megapode deposits its eggs in black volcanic soil along certain beaches that readily absorb solar radiation.
8554. Yes, the birds twitter continually on long flights.
8555. Most Wrybills winter on North Island, but the entire population breeds on South Island.
8556. About 47.
8557. Hutton's Vireo, that ranges from British Columbia to Guatemala.
8558. The birds remove ectoparasites and probably function as sentinels for their nearsighted hosts.
8559. Adolphus Lewis Heermann of Heermann's Gull fame.
8560. Yes.
8561. Yellow.
8562. An eagle.
8563. No, it is a Rocky Mountain species.
8564. The birds tend to withhold their excreta until they are at the breeding site.
8565. Yes.
8566. The American Coot.
8567. The long detour around the water is undertaken to take advantage of thermals over land. The storks cross the Mediterranean via two unusually narrow fronts: the Straits of Gibraltar and the narrow corridor of Asia Minor.
8568. Yes.
8569. Yes. The phenomenon of nonbreeding birds attending the nests of breeding members has been recorded, at least casually, for at least 32 families and subfamilies of passerines, but also in such non-passerines as rails, kingfishers, bee-eaters, woodpeckers and swifts.

Answers

8570. No, these Old World flycatchers have only a nasal call note and a rather indifferent warbling song.
8571. Yes.
8572. By panting.
8573. A large rail.
8574. Somalia.
8575. The Satin Bowerbird, from which it steals green berries, leaves and cicada remains.
8576. In flight, when they call out with characteristic whistles and clucks.
8577. Roving peccaries, that flush insects for the birds.
8578. Great Britain has hosted the I.O.C. three times: 1905 in London, and 1934 and 1966 in Oxford.
8579. No.
8580. Ten.
8581. No, both species are insectivorous.
8582. In one extreme case, three House Sparrows became trapped in a coal mine in Yorkshire, England, 2,100 feet (640 m) below the surface, from the summer of 1975 until the spring of 1978. In 1977, one pair nested and reared three young, but these ultimately died.
8583. Papyrus marshes.
8584. The study of the interactions between birds and humans.
8585. Yes.
8586. Yes. Grandry's corpuscles are a group of nerve endings that may be receptors of touch.
8587. *A wild goose chase.*
8588. No, but some are serious pests to small-scale farmers.
8589. The Humboldt Current.
8590. Possibly because the bird is able to detect vibrations produced by an insect, thus is aware that food is near.
8591. Yes.
8592. No, the noise in a nesting cave can be deafening.
8593. Egg-bound.
8594. Yes, essentially so, but a few animals may be taken, such as insects, lizards or small tortoises.
8595. The shelducks and sheldgeese. Both groups are extremely aggressive during the breeding season.
8596. Yes.
8597. The tail feathers of a Rhinoceros Hornbill.
8598. No, it has increased dramatically.
8599. Only the Maleo.
8600. The Australian Magpie Goose. Ganders undertake twice as much of the nest building.
8601. Yes.
8602. No.
8603. Yes and no. Larger species may, but in smaller species, prey is brought back to the perch and beaten into immobility prior swallowing.
8604. *Fratercula arctica* means "little brother of the north."
8605. Probably, but this has not been confirmed. Broken carpal spurs have been found embedded deep in the breasts of screamers.
8606. Yes.
8607. No, the precocial chicks develop slowly and depend on the male for up to four months.

Answers

8608. Possibly because the feathers protect the chicks from the sun. The eagles nest in open, unshaded nests.
8609. The Rough-faced Shag, a species endemic to the Cook Strait area in New Zealand. The population was never numerous and the 300 shags are restricted to five breeding locations.
8610. A tiny transmitter, but bells were formerly used.
8611. Yes.
8612. No, the bill of James' Flamingo is designed for smaller food items.
8613. This may function as visual orientation, acquainting one bird with another's presence. Rails also flick their tails when no other birds are present, which might be an automatic response to release nervous energy.
8614. Generally widely spread.
8615. Possibly. The crepuscular Hook-billed Kingfisher of New Guinea takes most of its prey on the ground, and also apparently feeds at night.
8616. No, but Black-billed Magpies do.
8617. The Red Sea Swallow is known from a single specimen, and the White-eyed River-Martin from Thailand is known only from a few individuals.
8618. No. Based on zoo records, other types of animals apparently suffer tumors twice as frequently as birds.
8619. No.
8620. Yes, as a rule. Pigeons, for example, have a single pair of syringeal muscles and do not have complex vocalizations, but most songbirds have seven to nine pairs.
8621. No, it is oval-shaped.
8622. Yes.
8623. Adult males of this bird-of-paradise from the northern Moluccas have a pair of incredibly long white pennants or plumes extending from the bend of the wing.
8624. The Bobolink, that breeds in North America and winters in South America.
8625. Its bill is well suited to digging for underground bulbs, a favored food during dry periods.
8626. No, the shorebird generally occurs above 9,000 feet (2,743 m). The snipe was recently rediscovered in Peru, after having been considered extinct for 100 years, and was known only from two specimens.
8627. Yes.
8628. Yes.
8629. White.
8630. The Fiordland Crested Penguin (New Zealand).
8631. Yes, more or less.
8632. About 55 days.
8633. Fat: about 15 percent.
8634. Yes.
8635. Woodswallows.
8636. No, young fledge at 70-80 percent of an adult's weight.
8637. Yes, but the huge, tropical forest eagles are extremely rare in Mexico.
8638. No, the eastern jays are much bolder and more confiding.
8639. Some of the penguins, especially the King Penguin.
8640. No, cotingas lay unusually large eggs for passerines.
8641. The Common Redshank.
8642. Insects. Like many Old World cuckoos, cuckoo-bees foist their broods upon the care of others.

Answers

8643. The Java Sparrow.
8644. No, figbird nests are very flimsy, and the three pale-green eggs are often clearly visible from below.
8645. Yes, but only during the breeding season when the female's head and neck are more densely covered in black feathers, and the bare parts are more intensely blue than in the male.
8646. Yes, some call nocturnally during the breeding season.
8647. Yes, especially males.
8648. The wattlebirds.
8649. A *jay*walker.
8650. Yes, but turacos are almost exclusively vegetarian.
8651. No, the syrinx of a rhea is more highly developed.
8652. Cocoons.
8653. The King Eider. The name is of Indian origin from *ouarnicouti*, meaning "head filled out full."
8654. Two to three weeks.
8655. Juveniles have blue or grayish-green legs, whereas adults have yellow or reddish-orange legs.
8656. Yes, but there are a few notable exceptions, such as the petrels and kiwis.
8657. Escaped Ring-necked Parakeets have gained a foothold in several localities.
8658. No, the best mimics are typically ground dwellers.
8659. No, its bill is twice as long.
8660. Yes. The nominate race of the Winter Wren is *Troglodytes troglodytes troglodytes,* and the lengthy name for such a little bird became something of a joke.
8661. Wilson's Phalarope.
8662. None.
8663. The endangered Hawaiian birds have more extensive black on the sides of their neck and face.
8664. No, todies capture most insects from the undersides of leaves and twigs. The birds perch with their bills pointing upward, scanning the lower surfaces of the leaves above them. Less than 10 percent of the time do the little birds sally out in flycatcher fashion to capture insects on the wing.
8665. Yes, but the volume of his song is lower than when the grosbeak is away from the nest.
8666. Yes, in terms of the total birds killed. In 1953, two hailstorms in Alberta, Canada killed an estimated 148,000 waterfowl alone. On another occasion, 1,000 Sandhill Cranes were killed by hailstones in just 30 minutes. Even cattle are killed by large hailstones.
8667. After molting, the new feathers are almost black, but with time and exposure to the sun, the feathers start to lose melanin, and take on a grayish-brown shade.
8668. Maximilian's Jay.
8669. No.
8670. The Large-billed Tern.
8671. In wet or humid areas, the guano becomes sticky, coating both eggs and chicks.
8672. The Boat-billed Flycatcher.
8673. Yes, some guans have a wing-whirring or drumming display performed in flight, although this is unique in the family.
8674. No, Oilbirds do not have crops.
8675. No.

Answers

8676. Yes, as do some Horned Grebes.
8677. A claw.
8678. Yes, but coastal breeding is more typical.
8679. The Least Storm-Petrel.
8680. Yes, and the populations of the same species on different islands also sing different songs.
8681. Yes.
8682. No, its enormous black bill is slightly turned up.
8683. As recently as 1913, robins were sold in eastern U.S. markets for about 60 cents a dozen.
8684. *To grouse.*
8685. Yes.
8686. Lacking a cooling mechanism, the temperature would be over 750°F (399°C). About 155,000 calories would be burned daily.
8687. Yes.
8688. Yes, the frequency of fly-catching is related to tail length. The Gray Wagtail, with the longest tail, does the most. The Yellow Wagtail, with the shortest tail, does the least.
8689. The piccolo.
8690. Up to 2.2 pounds. (1 km). Some pebbles measure may be an inch (2.5 cm) in diameter.
8691. Up to 50 percent.
8692. The Yellow-billed Stork.
8693. Koel is a Hindustani name for cuckoo.
8694. Distinctly triangular-shaped.
8695. A Nightingale.
8696. Many chicks perish when their nest becomes heavily infested with blood-sucking larvae of the bluebottle fly.
8697. Yes, some species so.
8698. The island formerly had thriving colonies, but the approximately 2,200 feral cats eliminated every one. The cats accounted for 455,000 seabirds a year, with diving-petrels a major prey species. Nearby Prince Edward Island has no introduced predators, and diving-petrels flourish there.
8699. Woodhoopoes are cavity nesters, but unlike most hole-nesting species, they do not lay white eggs. Their eggs are blue, green or gray.
8700. Lady Sarah Amherst (1762-1838), the wife of William Pitt Amherst, a British diplomat (1773-1857) who was in China in 1816, and later became the Governor General of India.
8701. Probably not.
8702. The Houbara Bustard.
8703. About 90 percent.
8704. Charles Darwin (Darwin's Tanager), currently known as the Blue-and-yellow Tanager.
8705. Bushshrikes with soft and elongated rump feathers.
8706. Yes, some species do.
8707. No.
8708. Some of the ovenbirds (*Cinclodes*), such as the Tussockbird of the Falkland Islands.
8709. No, but they nest in association with gulls and terns.
8710. The hooked or curved tip of the upper mandible, as in some plovers and pigeons.
8711. No, mimicking has only been noted in captivity, generally with solitary birds.
8712. The storm-petrels. The name reflected the mariner's belief that the birds were invisible until the onset of a storm or tempest.
8713. Yes, so far as is known.

Answers

8714. No, as pairs.
8715. The International Union for the Conservation of Nature and Natural Resources (IUCN). The books list rare and endangered species.
8716. A musket.
8717. Yes. The metabolic disorder probably occurs in all birds, but it is most commonly seen in domestic poultry and captive raptors, including owls.
8718. Rooks often mob the loudspeaker prior to dispersing.
8719. Yes, a flying grebe has a hunch-backed look.
8720. About three percent.
8721. Yes, probably so.
8722. Yes, but as adults the bill color is bluish.
8723. Usually aggressive, threatening, or attacking behavior, but the term is also used to describe fleeing or submissive behavior. Agonistic behavior is often used as synonymous with aggressive behavior.
8724. It possibly inhibits attack by adult kittiwakes during the breeding season.
8725. 1983.
8726. The Gray Catbird.
8727. About 27.
8728. In the few days before laying, female kiwis may stand in water. This might relieve their legs of the weight of the enormous egg, and may soften the skin for the big event. Females about to lay must walk with their legs apart, and when they pause, their stomachs may touch the ground.
8729. Black.
8730. It was believed beneficial in treating rheumatism.
8731. Yes.
8732. At night.
8733. Yes, many species do.
8734. Blue.
8735. Yes, in most species.
8736. Yes.
8737. Yes, well beyond its short, forked tail.
8738. A pack (typically a group larger than a covey).
8739. If power lines are present, woodswallows readily perch on them.
8740. The Great Spotted Kiwi.
8741. A male Northern Goshawk greater than one year of age that was obtained as a nestling.
8742. No, its bill is slightly upturned.
8743. About 578.
8744. No, their eggs are invariably spotted.
8745. Acorns, and the jay's distribution closely coincides with acorn-producing oaks.
8746. Yes.
8747. Yes, and some taxonomists include them in the same order.
8748. Yellow-billed and Marabou Storks and Pink-backed Pelicans.
8749. The terrestrial raptors may capitalize on animals killed in a fire, but generally do not feed on carrion.
8750. No. Members of a flock often call at the same time, but not necessarily in unison.
8751. Christchurch, New Zealand.
8752. Yes.

8753. The migrating swallows feed en route.
8754. An owl.
8755. Yes, under normal circumstances.
8756. Yes.
8757. No, it moves backward and forward.
8758. The Yellow Warbler.
8759. Probably Song Thrushes and Nightingales in England, about 1900. A captive Shama Thrush had its voice recorded even earlier in Germany, on an Edison wax cylinder.
8760. The Southern Screamers.
8761. The skin, tendons and other coarse pieces of tissue. The vulture tends not to consume large amounts of muscle tissue.
8762. The watershrews. Their body mass is only about one-fifth that of a dipper.
8763. 1969.
8764. The pelican.
8765. Yes.
8766. Yes. Sometimes as many as a dozen oxpeckers forage at the same time on a large animal, such as a giraffe, but feeding groups are generally smaller.
8767. Continuous hovering.
8768. The cassowary, and cassowary bones often adorn their ears. Bare primary quills were made into nose-pins or earrings, and the strong leg bones fashioned into long daggers, spoons or scrapes.
8769. Yes.
8770. Possibly as a response to changes in farming practice. Progressively earlier hay cutting may have selected increasingly against godwits laying late in the season.
8771. Yes, but the breeding population is not large.
8772. Nests may be 80 feet (24 m) up in cypress trees.
8773. Yes.
8774. Apparently to isolate the nest in case of fire.
8775. The northern populations of Long-eared and Short-eared Owls.
8776. The entire population does not exceed 100 condors, possibly no more than 30.
8777. No, but this is typical of other sulids.
8778. Lyrates. The tail consists of the two lyrates, two medians and 12 filamentaries.
8779. Yes.
8780. Hair, fur, or feathers.
8781. Yes. Campo is the Latin American name for grassy plains with scattered bushes and low trees.
8782. The endemic New Zealand Scaup.
8783. The Trumpet Manucode (Trumpetbird), because the aggressive butcherbird keeps predators away.
8784. The potoos.
8785. No, all are grassland species.
8786. The Northern Cardinal.
8787. The Long-tailed Widowbird.
8788. The Australasian Bittern.
8789. A small enclosure where cock-fights are staged is called a cockpit.
8790. Some of the ovenbirds, specifically the spinetails.
8791. Yes, and males may defend temporary sites in the club where they advertise for females.

Answers

8792. White.
8793. This makes the bird more difficult to see.
8794. Traditionally at Skylarks.
8795. Yes.
8796. No, habitat destruction was partially responsible, but more likely commercial feather-hunters sealed the doom of the large wattlebird, that has been presumed extinct since December 28, 1907. The Maoris regarded the bird as sacred, and the greatest chiefly rank was to wear its white-tipped tail feathers. When King George V visited New Zealand in 1901, a Huia tail feather was placed in his hat by a high-ranking Maori woman, signifying that the king was a great chief. Unfortunately, many English subjects wanted to follow royal fashion, and the price for a single feather soared to as much as £5. Under such relentless commercial pressure, the Huia was doomed.
8797. Apparently so.
8798. Yes.
8799. Yes.
8800. The geese nest in Iceland, the only location where lava grit could be obtained.
8801. At least six weeks.
8802. Tropical and southern Africa.
8803. When bathing and preening, penguins often roll about flailing the water. To observers unfamiliar with such behavior, it might appear as if the birds were in distress.
8804. Ithaca, New York, in 1962.
8805. Emu oil was used to fuel lamps.
8806. Growing feathers will become red.
8807. Yes.
8808. Both are aggressive birds that drive off obligate brood parasitic-nesting Brown-headed Cowbirds.
8809. Its bulky nest consists predominately of regurgitated fruit of paste-like consistency. This is applied directly from the side of the bill. Disgorged seeds and, to a lesser extent, the bird's own guano, also contribute to the structure.
8810. The Helmeted Guineafowl.
8811. Yes, as a rule.
8812. Typically not before three years of age.
8813. No, mating takes place at the nest site.
8814. The mousebirds (colies).
8815. Females have longer, more convoluted tracheas than males.
8816. To house raptors during the molt. Mews are currently used to house falconer's birds year-round.
8817. Yes, but there are some exceptions.
8818. The endangered Yellow-eyed Penguin.
8819. A man-made shelter for doves or pigeons.
8820. The bird is alarmed or is in an aggressive state.
8821. No, the paired coracoids of the avian pectoral girdle are absent in mammals.
8822. No, the thin tail is often difficult to see.
8823. Petrels are among the few birds with a superior olfactory sense, and the tubes may enhance their ability to smell. Some petrels can smell food, and burrows may be located by smell.
8824. Yes.
8825. About 16 days.

8826. Parents feed their chicks bill-to-bill.
8827. No, in the few species studied, jacamar eggs were tended almost constantly.
8828. Yes.
8829. The late 1930's.
8830. The Pygmy Kingfisher.
8831. One to three eggs, generally two.
8832. The Superb Parrot.
8833. The hummingbirds.
8834. The Saddle-billed and Black-necked Stork, in which the eye color of the sexes differs. The red at the base of the neck of a male neotropical Jabiru may be brighter than that of the female.
8835. The male incubates first for an average of 18.5 days, after which the sexes alternate.
8836. No, but some authors suggest otherwise.
8837. No.
8838. A band or bar, usually white or whitish, in the wing. It is concealed in some species when the wing is folded, but revealed in flight.
8839. During displays.
8840. No, males outnumber females by about six to one.
8841. The bird was overlooked because it bears a striking resemblance to the Black-and-white Warbler, a common migrant in Puerto Rico.
8842. The King Bird-of-paradise.
8843. It jerks its tail.
8844. No. All moas were apparently entirely herbivorous, demonstrating a preference for the twigs from a variety of woody plants, as well as some seeds, fruits and leaves.
8845. Both are aviculturist terms describing a duck, goose, or swan where one or both wings are drooped over. This generally results when a young bird is fed a protein-rich diet, and the weight of the growing blood feathers (primaries) causes the wing to fold over. If the drooping wing(s) is not immediately taped, the unsightly condition will be permanent.
8846. Sungrebe toes are gaudily banded black and yellow *zebra-striped,* as opposed to the solid red of the African birds and pea-green yellow of the Asian species. Their habitat consists of muddy waters, so it is unlikely that colored feet or patterned toes are related to flushing or attracting prey.
8847. The Yellow-headed Blackbird.
8848. Yes, but the tiny perching-ducks more typically feed on the surface.
8849. None.
8850. Yes, very much so.
8851. Yes.
8852. To *duck* an issue.
8853. Only one, the Palestine Sunbird.
8854. Yes.
8855. Duck plague. The disease is an acute, contagious, often fatal herpesvirus infection, primarily of domestic waterfowl, but it is easily transmitted to wild populations.
8856. No.
8857. The Magdalena Tinamou.
8858. Yes. The Pied Kingfisher of Africa and Asia is a hovering species that can catch two fish during a foraging flight, and is capable of swallowing a fish without having to return to a perch. Such behavior is not typical of most kingfishers.

8859. The quail may not nest at all.
8860. A single chick, but its clutch consists of two eggs.
8861. Yes, the strikingly-patterned Snow Pigeon, for example.
8862. No.
8863. No, pigeons typically lay early in the afternoon.
8864. Yes, based on the fossil record.
8865. No, it is an African bushshrike.
8866. During the winter.
8867. The Dodo has traditionally been regarded as a very slow and clumsy bird, but some researchers have recently concluded that it could probably run rather swiftly.
8868. White.
8869. Up to 25 percent more.
8870. Yes.
8871. All three.
8872. Yes.
8873. Probably because the eggs are buried and adults seldom touch them, thus a thicker shell is not required.
8874. Yes.
8875. No, most are tropical.
8876. No, the huge bird had three toes.
8877. Blood hemoglobin and bile pigments.
8878. The Winter Wren. The gland accounts for 0.58 percent of the bird's weight.
8879. Yes.
8880. Yes. The Maoris found the flightless parrots very good eating, and warm, fragrant feather capes were fashioned from the sweet-smelling, moss-green skins.
8881. No, but unoccupied males are fair game, and a female may copulate with all her mates in less than a half hour.
8882. The paired, elongated kidneys each consist of three main lobes in most birds, and four lobes in some, but kiwi kidneys have five lobes.
8883. A Rook.
8884. Yes, as a rule.
8885. About 15,000 pairs, in Botswana.
8886. Emperor Penguins weigh up to 90 pounds (41 kg), and possibly even 100 pounds (45 kg). A captive bird at Sea World, San Diego, weighed more than 120 pounds (54 kg).
8887. The Zoological Society of London, published annually since 1864.
8888. Down is lacking in megapode chicks. The natal plumage consists of contour-like feathers, enabling chicks to fly almost at once.
8889. The name is an anagram of pintado, the Portuguese name for the petrel.
8890. The Emu.
8891. No, it is the largest. The American Sungrebe is smallest.
8892. Up to four or five months.
8893. One Barred Owl had 9,206 feathers.
8894. No, it is an Australian warbler.
8895. No, mating occurs ashore, generally at the nest site.
8896. Nothing.
8897. White, and the tail usually has a broad, distal black band.
8898. The Laysan Finch, a Hawaiian honeycreeper.

Answers

8899. Flanges.
8900. Yes, as a rule.
8901. No, flying rollers often vocalize.
8902. The swifts.
8903. The three cassowaries.
8904. The largest, strongest central pair are molted after all the other tail feathers have grown in. This molting sequence renders the tail maximally effective as a brace throughout the molt.
8905. About 18 inches (46 cm).
8906. About three weeks of age. The young leave the burrow and are probably completely independent, but may still be tolerated in its parent's territory for the first year.
8907. Black.
8908. No, the birds are generally quiet and inconspicuous.
8909. An arboreal cavity nester, it favors tree hollows.
8910. Only two: Vermilion and Galapagos Large-billed Flycatcher.
8911. Flying by young birds well before full adult size is attained, such as in megapodes, cracids and some pheasants.
8912. No.
8913. No, but most do. The few exceptions include the bright blue eggs of the Black-faced Spinetail, and a few other species that lay slightly bluish or off-white eggs.
8914. The rising sun.
8915. Squat.
8916. Yes.
8917. No, their tails are wedge-shaped.
8918. Sometimes, but not always.
8919. An explosive hiss, like that of a snake, specifically a copperhead.
8920. No one knows. When *Incas,* the last male died at the Cincinnati Zoo, arrangements were made to have his carcass shipped to the Smithsonian Institution. However, the body never arrived, and its fate is unknown. In excess of 700 Carolina Parakeet skins are deposited in collections around the world, with 76 in Washington, D.C. alone.
8921. No, the vultures are timid.
8922. Yes.
8923. The marten.
8924. Norway.
8925. Yes.
8926. The shorebird often nests on burnt-over ground, and the color provides a degree of camouflage.
8927. Yes.
8928. The light orange wood with a straight, coarse grain from several trees in the genus *Persea.*
8929. Yes.
8930. No, but mud is used.
8931. Yes.
8932. Sacred Ibis, herons, monitor lizards and even crocodiles.
8933. No.
8934. The wild type green.
8935. Gens.
8936. Twelve species.
8937. The Burrowing Owl. It is also known as a prairie dog owl or gopher owl.

Answers

8938. Most are, but Black-faced Antthrush chicks are densely covered with dark down.
8939. Robert Cushman Murphy.
8940. Yes.
8941. Yes.
8942. Yes, generally so.
8943. The Tui (Parsonbird).
8944. No, green is essentially absent in African barbets, with black, yellow and red predominating.
8945. The Rose-breasted Grosbeak, but the Evening Grosbeak ranges nearly as far north. The Pine Grosbeak inhabits most of Alaska, but it is not a true grosbeak.
8946. The Mexican birds are all dark, lacking the white rump.
8947. No, submerging is difficult for the buoyant phalaropes.
8948. Yes. Nearly two-thirds of all passerine species fledge at below adult weight, compared to only about 28 percent in non-passerines.
8949. A sakret.
8950. The geese are winter nesters.
8951. Yes, as a rule.
8952. Bats.
8953. The Straw-necked Ibis, with some colonies consisting of up to 80,000 pairs.
8954. *Dacelo* is simply an anagram of the Latin name for kingfisher (*Alcedo*).
8955. Abdim's Stork.
8956. No, the birds may occur as high as 10,000 feet (3,048 m).
8957. Charles Darwin.
8958. The shell is weakened because of the loss of mineral substances dissolved out. The minerals are transported by the blood, and are used to prepare the first ossification of the skeleton of the developing embryo.
8959. No, only slightly so.
8960. Yes, during the non-breeding season.
8961. No, the gulls are more inclined to nest on a cliff.
8962. No, it may be evident for weeks in some species.
8963. Yes.
8964. Yes, some species do.
8965. No. Its wattles are generally held close in to the throat, overlapping in adult males and meeting below the bill. The wattles are never pendulous as is often illustrated.
8966. The propane gas gun, a device in which gas from a cylinder is periodically released into a firing chamber where it is ignited. The interval between explosions is adjustable.
8967. No.
8968. Yes, but they only occur locally and are dependent on the presence of suitable caves. Most colonies are in mountainous country, although some Trinidad colonies are in sea-caves.
8969. Yellow, but the cere of a juvenile is pale blue-green.
8970. Typical shorebirds don't construct nests, but the two aberrant painted-snipes construct a woven plant-material nest, generally concealed in dense vegetation. The male alone is thought to be responsible for nest construction.
8971. Yes, as a rule.
8972. Both are obsolete or dialect names for the Gray Heron in Britain.
8973. About three days.
8974. Yes.

8975. Yes.
8976. About 400 birds are restricted to Narcondam Island in the Bay of Bengal, India.
8977. Yes.
8978. Yes, partially so.
8979. Approximately 100,000, maintaining perhaps two million pigeons in their lofts.
8980. The *stud post,* where females come to be bred.
8981. The Helmeted Hornbill.
8982. Marine mammals, bats and large terrestrial mammals, such as caribou and African wildebeest (gnu).
8983. It holds the prey down with its feet and bites off pieces.
8984. Yes.
8985. Yes, and for some time the dove was considered extinct.
8986. Yes.
8987. No, just behind, and balance is maintained by the toes.
8988. No, generally well before daylight.
8989. No, hundreds winter along the Alaska Peninsula.
8990. A puffin.
8991. Over-hunting, even though the pheasant is protected.
8992. The cornea of a penguin's eye is relatively flat, rather than rounded.
8983. Thorny twigs that provide much protection. The large, untidy, domed nest has a side entrance.
8994. The trogon may throw its head back with the bill pointing almost straight up.
8995. The Straw-necked Ibis.
8996. 1954.
8997. Two mammalian ribs. One rib has saw-teeth, across which the other bone is rhythmically drawn to reproduce the Corn Crake's rasping "song."
8998. Yes.
8999. The margin of the comb is serrated, rather like a comb.
9000. The Snowy Owl.
9001. No, the African stork nests in trees.
9002. The Black Storm-Petrel.
9003. Yes, especially during the winter.
9004. No, their dirty-white eggs have rather rough shells.
9005. A small, downy feather.
9006. With specially trained dogs, typically Labrador retrievers, fitted with muzzles and a bell.
9007. The Black-crowned Night-Heron.
9008. The Thrush Nightingale.
9009. The Pinyon Jay.
9010. Yes.
9011. The gas emitted by drying green leaves may be beneficial in controlling parasites. The leaves might also serve to supply moisture to the eggs.
9012. Yes.
9013. Yes.
9014. No, all three species are in separate genera.
9015. A falcon.
9016. No, generally in pairs.
9017. Yes.

Answers

9018. When Gray (Common) Partridges gather and nestle together at night, it is called *jugging*.
9019. Yes, females average 5.7 ounces (161 gms), whereas the much smaller males average only 3.2 ounces (91 gms).
9020. The Green Avadavat.
9021. A secondary.
9022. Yes.
9023. Feeding Puna and Andean Flamingos average 10-15 and 20-30 steps a minute respectively, but Chilean Flamingos move at a more rapid 40-60 steps a minute, stirring up more invertebrates from the lake bottom than the other two species.
9024. In excess of six tons (5,443 kg).
9025. The raccoon.
9026. No, only the major female. She lays about 10 eggs in the nest before incubation commences, and the minor hens may lay 10-30 or more eggs among them. At the start of incubation, the major hen pushes out a number of the surplus eggs, and these doomed eggs form a ring around the nest. The major hen rarely pushes out any of her own eggs, and these are retained with the 20 or so that she can effectively incubate.
9027. Three eggs.
9028. Cankerworms.
9029. White.
9030. The three New Zealand kiwis.
9031. Hair, wool and feathers.
9032. The birds are frugivorous. Adults, and chicks as well, feed predominantly or wholly on fruit.
9033. A hawk.
9034. No.
9035. The winds the birds require blow almost continually.
9036. Yes.
9037. The Egyptian Vulture.
9038. Yes, avian vocalizations were recorded in Antarctica in 1934, but not until the 1940's in South America.
9039. The Rufous Hornero.
9040. The Piping Plover.
9041. A jack.
9042. Yes.
9043. Yes, but immaculate eggs are unmarked.
9044. Four.
9045. Yes.
9046. Abbott's Booby incubates for 57 days.
9047. Yes.
9048. Nothing.
9049. Yes, some species do.
9050. The New Zealand Fiordland Crested Penguin begins egg laying as early as August.
9051. No, nestlings develop very slowly. Their eyes do not open until they are over two weeks of age, at which time the chicks are still naked.
9052. Sunbirds, but males of some species remain brightly colored year-round.
9053. The Long-tailed Tit and the Dunnock.

Answers

9054. The woodpecker became so tipsy on birch juice that it was unable to attend the official distribution of colors, and it acquired its many hues as a consolatory gift from other birds.
9055. White.
9056. Gannets and boobies lack brood-patches and incubate by actually standing on the eggs, thus the egg must be strong enough to withstand the weight of the bird.
9057. In flight.
9058. No, but there is some evidence that night-roosting Turkey Vultures can lower their body temperature.
9059. Yes.
9060. Australia.
9061. Yellow.
9062. Aves.
9063. Yes, and the birds lean on their wings while digging.
9064. The African Openbill stork. Hippos expose snails on which the stork feeds almost exclusively.
9065. This is a necessity for clinging to a moving mammal.
9066. Yes.
9067. In 1991, more than 76 million people, of which 60 million were birders. Approximately one million were active, or *committed*, birders.
9068. The Ascension Island Frigatebird.
9069. Yes.
9070. Robert H. *Finch*.
9071. Crayfish.
9072. Yes, but in migratory birds, the lek location may change yearly.
9073. No. Most birds run relatively slowly, but there are numerous exceptions, such as roadrunners and ratites.
9074. No, they can be surprisingly confiding.
9075. No, it is proportionally longer.
9076. The Little Pied Cormorant.
9077. Yes, the Willow Tit, for example.
9078. No, Magnificent and Ascension Island Frigatebirds, and some races of the Great Frigatebird, do not.
9079. Despite large incubated clutches of 21-25 eggs, heavy egg and chick losses typically result in low breeding success. Depredation by Egyptian Vultures, jackals and hyenas can be severe, and typically only about 10 percent of all eggs laid hatch. Combining all adults involved at an incubated nest, typically 4-6 birds, the fledging rate is surprisingly low; in one study, only 0.15 young per adult per year.
9080. The shorebirds stand in one spot and spin using their feet to excavate shallow holes in areas of decaying vegetation where fly larva are common.
9081. The Keel-billed Toucan.
9082. Yes.
9083. They have both precocial and altricial traits. Sunbittern chicks hatch with thick, short down like precocial birds and resemble plover or snipe chicks. Like altricial birds, they remain in the nest for up to three weeks where they are fed by their parents.
9084. A landing stork is especially noisy because of air rushing through its large wings.
9085. No, it is shy and retiring.
9086. No, it has a few quiet calls of short, seldom-heard, melodious phrases.

Answers

9087. The condition of having feathers on the tarsus and toes.
9088. The five Caribbean todies have incubation periods of 21-22 days.
9089. The Maleo. Its wings and tail are black, and it has pinkish underparts.
9090. No, they occasionally nest on the ground.
9091. Yes.
9092. An average of about 76 birds, with a range of 63-84.
9093. No, although a moa bone was found in Australia, but it was probably taken there by humans.
9094. No.
9095. The secondary shaft.
9096. The Crimson-backed Tanager was introduced from Panama, sometime prior to 1940.
9097. Kentucky, in 1926.
9098. No.
9099. To avoid predators and presumably also to reduce daily energy requirements.
9100. No, a fossil motmot is described from the lower Oligocene in Switzerland.
9101. No, the tern is confined mainly to freshwater habitats.
9102. Yes.
9103. The cave-dwelling Oilbird of South America.
9104. The terns.
9105. The two turnstones, that seldom venture far from the coast.
9106. Yes, but only slightly.
9107. Yes.
9108. No, the endemic honeyeater was named for its explosive call note of *tzit* or *stitch.*
9109. The Cerulean and Black-throated Blue Warbler.
9110. A robin or grackle, but old nests of other species are used as well.
9111. Yes, but only the broad, flat nail of the middle toe.
9112. Aspergillosis.
9113. The American writer and naturalist Henry David Thoreau (1817-1862).
9114. No.
9115. The cuckoo.
9116. Red.
9117. Based on one California study, 320 acres (130 hectares). Territory size no doubt varies depending on the terrain and population density.
9118. The distance between the forward edge of a bird's nostril and the tip of its bill.
9119. *Apteryx* is derived from the Greek *a*, a prefix meaning "not," or "there is not," and *pterus*, meaning "a wing." *Apteryx* implies that the kiwi is wingless.
9120. No, but adults do.
9121. Their rich, melodious warbles are ear-pleasing.
9122. Yes.
9123. The Ladder-backed Woodpecker.
9124. The Western and Clark's Grebe.
9125. Thirty pounds (13.6 kg), but 15-17 pounds (6.8-7.7 kg) is more typical.
9126. Yes.
9127. A Hawaiian honeycreeper
9128. No, quite the opposite.
9129. No, but the White-faced Whistling-Duck, Southern Pochard and Comb Duck occur on both continents.

Answers

9130. The Black Vulture.
9131. Hairy caterpillars.
9132. Yes.
9133. The kiwi.
9134. Yes.
9135. All introduction attempts failed.
9136. No, their eyes are red.
9137. The wren.
9138. No, its tail is tipped with black.
9139. No, the bowers of most species are open at the top.
9140. As of 1990, a total of 108 in 18 collections.
9141. Seventy-two.
9142. The male Mute Swan, but only when *busking*. Angry or excited cobs paddle vigorously with both feet in unison, resulting in jerky forward movements, causing water to "boil" or foam up against the breast.
9143. No, their mouths are usually closed.
9144. Probably because of increased cultivation.
9145. Yes, the daily rate of water intake is inversely related to body size.
9146. The Benguela Current.
9147. Yes, and nests increase in height as material is added.
9148. The Australian White Ibis.
9149. New Zealand. In areas where kiwis are common, the birds are run over on remote country roads.
9150. A pheasant.
9151. The Atoll Fruit-Dove inhabits the low islands in the Tuamotu Archipelago where there are no fruit trees. It is the only insectivorous member of the *Ptilinopus* genus.
9152. Yes, but the Emperor Penguin does not. Both sexes of King Penguins alternately incubate, and a territory enables the returning partner to rapidly locate its mate.
9153. The Ruff.
9154. Yes.
9155. Hackle feathers dyed red to match the blood spilled by the regiment at Fontenoy.
9156. The bills of Ostriches and rheas are rather duck-like in shape.
9157. Bamboo.
9158. Yes.
9159. No, generally not until their second or third year.
9160. As ground nesters, the dark color possibly is less conspicuous to predators.
9161. No, not as a rule. The mammalian resting heart rate is typically about one-and-a-half times as fast.
9162. Yes, the screamer incubation period of 43-45 days is slightly longer even than that of the Torrent Duck (40-44 days).
9163. Defended feeding territories are centered around favored food flowers. When the food supply is concentrated, territories tend to be smaller. Territory size can vary by several hundredfold, although each contains approximately the same number of flowers. Apparently the nectar production of the flowers is just sufficient to support an individual's daily energy requirements.
9164. No, a Prairie Falcon is more maneuverable.
9165. The Reddish Egret, because it is often observed on reefs.

Answers

9166. Yes, salamanders, lizards and frogs are taken.
9167. The Irish Wildbird Conservancy.
9168. Grebes.
9169. Thirteen.
9170. Fidelity to the home area, used especially for the tendency of migrants to return to a previous breeding or wintering area. The term is widely used in English publications.
9171. The Prothonotary and Lucy's Warblers.
9172. Yes, it generally flew only when traversing an opening in the forest, or to cross from one tree to another.
9173. Each foot is shaken, presumably to shake off water.
9174. The Gray Vireo of the southwest, but only about five percent of its feeding bouts take place on the ground.
9175. Also known as the Colombo Crow, the birds sometimes arrive on trading vessels, apparently boarding the ships in India or Colombo, Sri Lanka, and disembarking at the next port of call, usually Freemantle, Australia. The crow apparently became established in Natal from passing ships whose crew members kept the birds as pets.
9176. The Butterfly Bird, because of its butterfly-like flight.
9177. None.
9178. Red.
9179. Yes, the wing shape varies from short and rounded to long and somewhat pointed.
9180. The Rose-breasted Grosbeak.
9181. Azure.
9182. Yes.
9183. Yes.
9184. Based on the fossil record, Old World vultures once inhabited the New World.
9185. Swallow.
9186. The Varied Lorikeet. Males are brighter in color with a wine-colored chest.
9187. The two American yellowlegs. The name conforms to the names for the Old World Greenshank and Redshank.
9188. Yes, the Carolina Chickadee, for example, but stink bugs are avoided by most birds.
9189. The Lugger Falcon.
9190. No, only rarely.
9191. No, at the beginning of the breeding season, the starling loses the feathering on the head and long wattles grow, but afterwards, the wattles are re-absorbed and feathers reappear.
9192. No, it seldom gathers in large flocks.
9193. During the winter.
9194. Mandibulating.
9195. The American Painted-snipe.
9196. The flying fox (a huge fruit bat).
9197. The Carolina Wren.
9198. The Cornell Laboratory of Ornithology at Ithaca, New York. The Library of Natural Sounds was established in 1931.
9199. Yes, but they also skim. A young bird flying with its parents probably quickly learns the best places to forage.
9200. The Gray-cheeked Thrush.
9201. The Great Tit.
9202. Yes, almost without exception.

Answers

9203. The Three-toed Parrotbill is remarkable among passerines in only having three toes, the outer one being reduced to a clawless stump attached to the middle toe.
9204. The steppes and deserts of the southern Palearctic region.
9205. The Gull-billed Tern.
9206. In contrast to the true swifts, treeswifts are sexually dimorphic.
9207. Yes.
9208. Both crop and gizzard are reduced in size in those species feeding extensively on pollen, nectar and fruit, but most parents have well-developed crops and gizzards.
9209. The American Woodcock.
9210. Corvids, especially the Black-billed Magpie, but also the Rook, Common Raven, Eurasian Jay and Azure-winged Magpie.
9211. Most jacanas lay beautiful, scrolled and highly polished eggs, but the rich bronze-colored eggs of the Pheasant-tailed Jacana are unmarked.
9212. The waxbill's breeding range is largely outside that of the cowbird.
9213. The See-see Partridge.
9214. Nowhere.
9215. Figs.
9216. No.
9217. Yes, but courting Old World quail become vocal.
9218. Yes.
9219. Corn husks.
9220. No, all have short tails.
9221. Not as a young bird, but aged individuals may have almost entirely white heads and bellies.
9222. The Lesser Seed-Finch. The birds are locally known as curío, and are highly valued because of their superior singing ability. There are numerous curío clubs, and even a National Federation of Breeders. Contests are often held, and the death of a champion curío is announced in the newspapers like the passing of an important person.
9223. No, the little rail is quite inquisitive.
9224. Yes, there were a number of failed attempts.
9225. Nests consist of little more than shallow scrapes.
9226. Yes.
9227. No, their diet may be augmented with fruit, but most babblers are insectivorous.
9228. Between 11 and 15 weeks.
9229. Yes, it swims frequently.
9230. No, mussels are swallowed whole.
9231. No, egg color is highly variable.
9232. No. This Old World warbler was named for its insect-like song.
9233. They crow loudly when they take off.
9234. 1947.
9235. Jamaica, where the white ants nest in trees.
9236. The Japanese Murrelet has a black crest, whereas the Ancient Murrelet lacks a crest.
9237. The Peruvian Diving-petrel.
9238. The Great Argus or Indian Peacock. Some authorities suggest that the Golden Pheasant was the original of the Phoenix itself.
9239. The Red (Gray) Phalarope.
9240. The Fish Crow.

Answers

9241. Probably the Atlantic race of the Brent goose. Some move considerable distances east or west from their breeding grounds to reach the continental shores along which they migrate. Many breeding in the Canadian Arctic move east across Greenland to Iceland, and then southwards to their wintering grounds in Ireland.
9242. No, dozing hummingbirds squat on a twig with body feathers fluffed, the head tilted back slightly, and the bill pointing up at an angle.
9243. The Salton Sea, California.
9244. Yes.
9245. The omnivorous Nihoa Finch includes seabird eggs in its varied diet.
9246. Yes.
9247. Loss of suitable habitat.
9248. One who alters the color or form of living or stuffed birds by artificial processes, usually with intent to deceive.
9249. A primary feather.
9250. Nestlings may regurgitate food on which the vultures pounce.
9251. No, their plumage is soft and lax.
9252. Yes.
9253. Yes, but effectiveness may diminish with prolonged use.
9254. The New Zealand wrens, with the Rifleman listed first. Of the four species, the Stephen Island Wren is extinct, and the Bush Wren is probably extinct.
9255. The Meadow Pipit.
9256. Doves, because of their frequent mating.
9257. Between 47 and 49 days.
9258. About three times as great.
9259. Ship's surgeon and mineralogist.
9260. Morocco.
9261. Yes, although not as numerous, its range is much more extensive, but much of the expansion is due to introductions.
9262. No, it is a hummingbird (Black-breasted Plovercrest) of the savannas of southeastern Brazil to northeast Argentina.
9263. The plumage streamlines the body.
9264. Yes, even during the non-breeding season, the little birds feed constantly from dawn to dusk.
9265. No, cave swiftlets probably have developed echo-location to a higher degree than any other bird.
9266. The Rosy Pastor (Rosy Starling).
9267. Yes, but the term is not very suitable either because females are by no means promiscuous, but rather are selective in their choice of mates. Males are relatively undiscriminating. Polybrachygamy might be more a appropriate term as it implies many brief matings.
9268. No, nunbirds are more far more gregarious.
9269. No, but it is prone to wander during the winter.
9270. All three.
9271. Glacier Bay, Alaska.
9272. The Bronze-winged Jacana of Asia.
9273. An Australian logrunner (rail-babbler family).
9274. No, unlike Ostriches that run with their heads held high, the necks of running Lesser Rheas may be stretched out almost parallel to the ground.

Answers

9275. Yes, as a general rule.
9276. The Asian Openbill stork.
9277. A bird-call fashioned from bamboo is used.
9278. A hanging bunch of Spanish moss.
9279. The powder is considered a powerful remedy against venomous snake bites.
9280. Yes, but nests vary from flimsy to compact.
9281. It is included in a special group of animals some conservationists call charismatic macrovertebrates (CMVs). Condors, like some other animals, such as cetaceans, wolves and eagles, tend to evoke strong human emotions.
9282. Kiskadees, Monk Parakeets, Chopi Blackbirds and the Thrush-like Wren. These birds all help solidify the nest by bringing their own nesting material, especially the wren that brings mud to provide a filling of mortar. Additional eyes also increase the probability of spotting predators.
9283. While this is often stated, the feat requires confirmation.
9284. *Anastomus* is Greek for "an opening," in reference to the gap or opening in the bill of an openbill stork..
9285. From an old German word originally meaning *a sleeping house for fowls*.
9286. Robert Frost.
9287. Yes, as a rule.
9288. Their tails are rapidly quivered.
9289. Yes, males have ruddy breasts that are lacking in females.
9290. The Black-capped Chickadee.
9291. Yes.
9292. No, the bills are flexible.
9293. Communally. As many as 50 wrens may gather at dusk, cramming themselves into suitable cavities.
9294. No. The little estrildid finches are remarkable for their very weak voices, emitting hissing and bubbling sounds, sometimes inaudible only a few feet (0.6 m) away.
9295. Early sealers, to provide a food source. In late 1990, the last Weka was finally removed because the predatory rails posed a threat to burrow-nesting petrels.
9296. Several dozen dippers may roost under a bridge, apparently taking advantage of the shelter.
9297. About 20 percent.
9298. Hawaii is currently in the Oceanic Zoogeographical region, but was formerly included in the Australasian biogeographical region.
9299. Yes.
9300. The Australian Gannet. One of the largest colonies is on Cape Kidnappers (North Island), a "mainland" site. In 1980, the gannetry contained approximately 5,200 breeding pairs.
9301. No, the hawks soar with a dihedral flight profile.
9302. Up to 3.5 pounds (1.6 kg).
9303. A long, slender feather on the neck of some gallinaceous birds, especially domestic roosters.
9304. Yes, screamers are larger than all ducks, although some flightless steamer-ducks weigh more. The Horned Screamer is the largest species, weighing up to 10 pounds (4.5 kg).
9305. Red and orange, but there is no scientific evidence to support this belief.
9306. Migrating finches occur as high as 21,000 feet (6,400 m) in the Karakorum Range of northern India.
9307. The Sacred Ibis, with which it is often regarded as a subspecies.

Answers

9308. A swift.
9309. Not according to some authorities.
9310. Yes, many do.
9311. No, most bee-eaters have a line through the eye, but it is black, not blue.
9312. They have more elongated bodies and longer necks.
9313. An African corvid.
9314. The Wild Birds Protection Act of 1954.
9315. The Rufous and Noisy Scrub-bird.
9316. Twice as much. Males weigh more than 15 pounds (6.8 kg).
9317. The Rock Ptarmigan. The Willow Ptarmigan is absent from Greenland and Svalbard (Spitzbergen).
9318. Yes, the calls of both sexes are loud and penetrating.
9319. There is no difference; the sexes of winter-plumaged phalaropes is identical.
9320. Yes.
9321. No, the toe is well developed.
9322. No, it is primarily Mexican.
9323. Lance-shaped.
9324. Yes, but the Arctic raptors are typically cliff nesters.
9325. The Australian Brolga crane, but the introduction failed. The cranes were introduced sometime prior to 1926 by a planter on Mango, an island near Taveuni.
9326. The King-of-Saxony Bird-of-paradise.
9327. Probably the Tricolored (Louisiana) Heron.
9328. No, shrike nostrils are oval-shaped.
9329. No, the abundant white down is long.
9330. Long, drawn-out, plaintive whistles, which help to keep them together in their groups, particularly when danger threatens.
9331. All three guillemots have bright red legs and feet.
9332. The Vesper Sparrow.
9333. No, the resultant progeny are sterile.
9334. The Java Sparrow and Javan Munia.
9335. About 9.5 ounces (269 gms).
9336. The Common Tern.
9337. A brilliant rose-pink, and its bill is coral-red.
9338. An estimated seven million pairs.
9339. The Dolphin (Scoresby's) Gull, and less commonly, the Kelp Gull.
9340. Yes, males have brownish eyes, whereas females have garnet-red eyes.
9341. The flightless, endemic Takahe (*Notornis mantelli*).
9342. Yes.
9343. The Diamond and Emerald Dove.
9344. Yes, the crest is especially prominent in males.
9345. Yes.
9346. The Australian Gannet. In 1980, the population was estimated at 56,664 pairs, of which 46,000 pairs were in New Zealand.
9347. Coffee plantations.
9348. A single egg, unlike most cranes. If two eggs are laid, the pair generally departs the nest after the first chick hatches, abandoning the second egg.
9349. Fighting, as in cock-fighting.

Answers

9350. About 45 degrees.
9351. Not as a rule, but a few bird groups, such as the parrots, are exceptions.
9352. Red, but the interior of the casque is golden.
9353. The Eurasian Wryneck.
9354. Iowa.
9355. Yes, some do, and winter territories may be vigorously defended.
9356. Pensile nests.
9357. A distance 40 times as great.
9358. The Common Tern. For many years ornithologists did not recognize the exclusively North American Forster's Tern as a separate species. The calls of the two terns differ, and the wingtips of the Common Tern are noticeably darker.
9359. The African Social Weaver. A single communal nest structure may be in excess of 27 feet (8.2 m) in length and six feet (1.8 m) high, and contains more than 100 individual nest chambers.
9360. A pear tree.
9361. I don't know, and I have never seen a reference relating to this. Moreover, I have been closely associated with birds all my life, both in captivity and in the wild, and I have never heard a bird's stomach growl.
9362. Yes, the Shaft-tailed Whydah, for example, only lays in the nest of a Violet-eared Waxbill.
9363. Food is regurgitated for the nestlings, and second-year birds may help with feeding.
9364. The bittern's feathers were used in the making of trout flies.
9365. No, its tail is long and broad.
9366. The Gila Woodpecker.
9367. A nuchal crest.
9368. The thrush.
9369. The White-tailed Tropicbird.
9370. No, it is one of the most numerous.
9371. The Red-necked Grebe, but its earlier name is no longer used.
9372. No, in the Ethiopian highlands, it may occur up to 13,000 feet (4,000 m).
9373. Yes.
9374. No, they are generally encountered in large, active flocks.
9375. The Oriental Darter. The diving birds were used in the same manner that Chinese and Japanese used cormorants.
9376. Not really, but there is some webbing at the base of the toes.
9377. Chest-bands.
9378. Aberrant Australian flowerpeckers, also known as diamond-birds. The name "pardalote" is derived from a Greek word meaning *spotted.*
9379. The birds feed on the ground.
9380. No.
9381. 16 to 32 days.
9382. Antshrikes attack small birds caught in the nets.
9383. The tail is bobbed continuously.
9384. No, their nails are sharp.
9385. Yes. In the center of each wing there is a large round shield of deep chestnut-orange set in an area of pale orange-buff, and the wings are boldly patterned. A displaying Sunbittern spreads its wings, with the richly-colored upper surface tilted forward, and fans out its raised tail to fill the gap between them, thereby forming a semicircle of plumage, in the

midst of which the head stands.

9386. Not as a rule, but a few, such as female Plum-throated and Turquoise Cotingas, are larger than their mates.
9387. Two or more species similar in form, habits and breeding that are separated geographically, such as the three species of gannets.
9388. Twelve, including two endemics and six endemic races.
9389. The Ruff.
9390. In 1874, an Australian Darter skin was discovered nailed to a shed wall on the west coast of New Zealand.
9391. The chicks are not completely naked at hatching, as is typical of other kingfishers.
9392. The wood-warblers.
9393. Verdin is from the French, meaning yellowhammer.
9394. No, and soaked chicks may die of exposure as a result.
9395. The female's tail is generally deflected laterally.
9396. Yes.
9397. The Ostrich.
9398. A cartwheel. In the U.S., cartwheels are used as Osprey nesting platforms.
9399. Christopher Columbus noted their behavior in his log on his first voyage to America in the late 15th century.
9400. The two color morphs occur in about equal numbers.
9401. Yes.
9402. Yes, terns do not swim underwater.
9403. No, more taste buds are in the palate and oral mucosa of the sides of the lower jaw.
9404. No, all have soft tails.
9405. No.
9406. No, the petite Old World merganser is largely unfit for the table.
9407. Yes.
9408. No.
9409. Yes, but the reverse is true in herons.
9410. The European Nightjar and Whimbrel, but exactly why is unknown.
9411. Only eight.
9412. The Yellow-shafted Flicker, with no less than 132.
9413. Yes.
9414. Yes.
9415. A night owl.
9416. Yes, and its nest can be rather bulky.
9417. No, all have rather slender necks.
9418. Hispaniola (Haiti and the Dominican Republic) with two species.
9419. Loons, but their heads and necks have soft, velvety feathers.
9420. Yes.
9421. Feral dogs can kill a kiwi by simply picking it up because the chest lacks the protection of flight muscles. In August and September of 1987, a female German shepherd ran amok in Waitangi State forest, killing 500 of the 900 kiwis living there, but few were consumed. The dog was finally shot and the slaughter stopped.
9422. No, but it is twice as long.
9423. Two: Peruvian and Chinese Crested-Tern.
9424. No, Neddicky is a South African name for the Piping Cisticola, an Old World warbler.

Answers

9425. Yes.
9426. Yes.
9427. No, the birds are prone to gather in flocks.
9428. The Prairie Falcon. All are desert falcons.
9429. The Phainopepla.
9430. Yes, for short distances.
9431. The rolling, trilling notes of its exquisite song.
9432. The martin. The mammal spelling differs (marten), but the pronunciation is the same.
9433. No, it is pointed.
9434. Yes, in many species.
9435. No, the rare finch is a high-altitude, pine forest resident of the Canary Islands.
9436. Probably the Whiskered Auklet and Kittlitz's Murrelet. Their limited distribution and tendency to concentrate in large numbers makes them vulnerable.
9437. White.
9438. Yes, this is not atypical behavior at waterholes.
9439. Black, but the male has a red bill.
9440. Yes, some may migrate.
9441. The Gray Junglefowl *(Gallus sonneratii).*
9442. No.
9443. No, the large wattlebird often descended to the ground where it normally moved rapidly with a series of bounds or jumps.
9444. Five to seven inches (12.7-17.9 cm).
9445. No, but it formerly did.
9446. Scarlet Ibis feathers. The sale of such flowers is a thriving business, much to the detriment of the ibis.
9447. Between 6 and 12 months.
9448. A bird-trapping technique where tall folding nets are clapped together to capture birds as they fly out when disturbed from their roosts at night.
9449. Wilson's Storm-Petrel.
9450. None.
9451. The Siberian Tit of northern Eurasia and Alaska.
9452. Yes.
9453. Liza Minelli.
9454. Yes.
9455. No, they have relatively long tails for rails. The flightless birds resemble small bantam chickens.
9456. White.
9457. The Gray Vireo, that prefers the hot, dry, chaparral and pinion-juniper woodlands of the American southwest.
9458. A retort (a laboratory glass container in which substances are distilled).
9459. Yes, even though ibises do not have webbed feet.
9460. The eagle.
9461. No.
9462. Yes.
9463. The Spot-billed Pelican.
9464. Yes, a male will allow chicks from other broods to join his group, as long as they are smaller than his own offspring.

Answers

9465. Winter. The yellow bill becomes a beautiful light green in the spring.
9466. The cutworms, eggs and larvae of the Rocky Mountain grasshopper.
9467. No, the grouse of the interior do, but males inhabiting the coast have orange neck-sacs.
9468. The western population of the Siberian Crane migrates over 5,000 miles (8,047 km).
9469. The Ovenbird, a North American warbler.
9470. No.
9471. No, but it formerly did.
9472. *Night singer.*
9473. The juvenile Bald Eagle, that he believed was a separate species.
9474. Yes.
9475. Probably to minimize the probability of encountering just-exploited feeding areas. African Cape Cormorants that normally feed on large shoals of small fish near the sea surface apparently do the same thing as other species of cormorants.
9476. The Gouldian Finch.
9477. The endemic plover is restricted to the island, and during his exile, Napoleon was also confined to St. Helena.
9478. Hawaii: Hawaiian Goose (Nene).
9479. No, feeding catbirds typically scratch with both feet simultaneously, using a hopping motion.
9480. The Red-fronted Parakeet from New Zealand and adjacent islands. Named by the Maoris, the birds are highly prized by aviculturists.
9481. Yes, northern Wrentits are reddish-brown.
9482. Yes.
9483. The Chinese Goose.
9484. No, penguins immediately follow grebes.
9485. Calls are generally short and simple, produced by both sexes at any time. Songs are long, complex and produced by males only during the breeding season. There are considerable difficulties in exactly defining the two terms.
9486. The *Pelican*, but the ship was later renamed the *Golden Hind.*
9487. Wine-colored.
9488. Yes.
9489. Yes.
9490. The noted American ornithologist Robert Cushman Murphy reportedly suggested either the Blue Petrel or the prion. When these seabirds twist in flight, exposing their gray dorsal surfaces, they practically disappear against the gray sea.
9491. No, the 5-6 remiges are hollow.
9492. No, it is a lowland resident.
9493. Some will readily attack, even if an intruder is human.
9494. The Yellow-nosed Albatross.
9495. The Arctic fox and Glaucous Gull.
9496. In some parts of its range, it uses cotton in its nest.
9497. Yes.
9498. Probably polychlorinated biphenyls (PCB's) and heavy metal pollutants, such as mercury.
9499. A gull.
9500. Yes, when foraging.
9501. Yes, it resembles juvenile plumage, but lacks barring on the underwing coverts, and the long tail feathers may be absent.

Answers

9502. Yes.
9503. Yes.
9504. A dove.
9505. The Lammergeier, even though its feet are relatively small.
9506. Predominately red.
9507. Yes.
9508. Possibly because such plumage is less likely to be damaged when the huge bird crashes through dense, thorny vegetation.
9509. No, they are more robust.
9510. No, but many are spotted.
9511. The female throws the egg out, probably as protection from the obligate brood parasitic-nesting Dideric Cuckoo.
9512. Yes.
9513. Yes.
9514. Near Cabo San Lucas, Baja California, Mexico.
9515. Yellow, or yellowish-green, but its legs are black.
9516. No, the shorebirds become especially vocal.
9517. Two.
9518. No, it is slightly larger.
9519. Its tail is twitched from side to side.
9520. Slightly more than 2.2 pounds (1 kg).
9521. Cigana means *gypsy*, referring to the extravagant dress of the South American Hoatzin.
9522. Yes.
9523. The American Kestrel.
9524. No, it is Celtic in origin.
9525. The Saker Falcon.
9526. About 5-6 days of age.
9527. The 2-foot-long (0.6 m) Pel's Fishing-Owl.
9528. The Black-faced Spoonbill breeds only on a few small rocky islands off North Korea. The endangered spoonbill's population is estimated at about 288 birds.
9529. In parts of their range they frequent tobacco fields.
9530. Yes, the stork habitually wades in shallow water, catching fish and amphibians like a large heron.
9531. Audubon showed Alexander Wilson the cranes in March of 1810, near Louisville, Kentucky.
9532. The Greater Pewee, formerly Coues' Flycatcher, was named for its call.
9533. No, its nape feathers are white only at the base. Unless the feathers are ruffled in flight or by the wind, the white is not visible.
9534. The Eastern Bluebird.
9535. An Australian warbler.
9536. Yes.
9537. Yes, as a rule.
9538. Another name for the alula or bastard wing.
9539. Yes.
9540. Yes, the raptors dive less frequently, thus fewer fish are caught.
9541. Three large digital claws.
9542. No, wagtails are infrequent visitors.

Answers

9543. The nearly flightless bird is extinct.
9544. The warbler inhabits the bare coral atolls in the Tuamotu Archipelango where there are no reeds or swamps.
9545. No. Adult Glossy Ibises have brown eyes, but the brownish eyes of juvenile White-faced Ibises become red as adults.
9546. The crow.
9547. No, only six of the 25 species are Eurasian. With the exception of an Australian species, all others are African.
9548. The Black-collared Hawk, because it preys on fish and crabs.
9549. The Northern Fulmar.
9550. Down feathers usually cling to the entrance hole of the tree cavity, spread by the female as she departs the nest.
9551. Bright chestnut.
9552. No, white-eyes have short bills.
9553. Scheepmaker's Crowned-Pigeon.
9554. The tiny Rifleman.
9555. Abdim's and European White Storks, because they control potentially devastating swarms of locusts. Swarming locusts can be very destructive to crops. In 1784, a South African swarm was estimated to encompass 2,000 square miles (5,180 sq km).
9556. No, but this is often the case.
9557. The neotropical tinamous have translucent flesh, caused by poor circulation.
9558. When begging for food, chicks keep their bills closed, but when water is desired, this is signaled with open mouths. The parent then flies to nearby water and returns to regurgitate water into the mouths of the offspring.
9559. Its feet do not project beyond the tail at all.
9560. The necks of the two combatants may be twisted around one another.
9561. Yes, dense flocks mob aerial and, less commonly, ground predators.
9562. Xantus' Hummingbird, Gray Thrasher and Peninsular (Belding's) Yellowthroat.
9563. No, the kinglet is astonishingly fearless, sometimes allowing itself to be stroked, or even picked up.
9564. Yes, in the southern part of their range.
9565. Yes, but only several species, such as the White-headed Kingfisher of New Guinea.
9566. Yes.
9567. A gullery.
9568. Bird-eating tarantula spiders are sometimes known as jayhawkers.
9569. An arrow with a broad, blunt end used for killing birds without damaging the plumage.
9570. The Hamerkop. Regarded as a bird of ill omen, if one flies over a hut and croaks, the inhabitants depart.
9571. Yes, in some species.
9572. The Parasitic Jaeger.
9573. Bruce Willis.
9574. Four: Malaysian, Madagascar and Javan Plover, and Senegal Lapwing. The Crocodile-bird is called the Egyptian Plover, but it is a courser, not a plover.
9575. Yes.
9576. The Galapagos Islands.
9577. No, but the Rock Ptarmigan does.
9578. The Eared Trogon.

Answers

9579. The Red-crowned (Japanese or Manchurian) Crane.
9580. Yellow.
9581. It was accidentally introduced into the Miami area in 1949, where the oriole is currently a relatively common resident.
9582. The guineafowl.
9583. Alexander Wilson, in 1811.
9584. No, but Common Mergansers do.
9585. The Brazilian Red-crested Cardinal.
9586. Yes, except when the rollers are migrating.
9587. The Olivaceous Flycatcher.
9588. A spectator at the first game remarked that the pins, on being struck, *flew like ducks*.
9589. No.
9590. The Nankeen Night-Heron.
9591. Cory's Shearwater.
9592. The White-winged Guan, previously known from only three skins.
9593. All 11 species carry nesting material in their rump feathers.
9594. The Pekin Robin.
9595. Up to three months.
9596. Bright red.
9597. A Hottentot girl from South Africa, between the years 1765 and 1782. Very little is known about her, but she evidently died young.
9598. The bizarre male bowerbirds of New Guinea and Australia.
9599. The Gyrfalcon.
9600. *Robin* Wood.
9601. No, its upper tail is white, but not the rump.
9602. No. Soaked birds prefer to climb out on overhanging vegetation prior to taking flight, apparently to dry off.
9603. Condores.
9604. Yes, generally in trees.
9605. The White-throated and Black-throated Magpie-Jay.
9606. No, each is in a separate family.
9607. The Red-tailed Tropicbird.
9608. Yes.
9609. The White (Fairy) Tern.
9610. Spider webs (silk).
9611. On August 22, 1927, Fuertes died at age 54 when his automobile was struck by a train.
9612. Yes.
9613. In 1978, the name was officially changed to the Philippine Eagle. Then-President Marcos felt the former name denigrated the qualities of a bird *in whose rarity and confident bearing the Philippines can take pride*. It is currently known as the Great Philippine Eagle.
9614. The gulls, such as Herring and Western Gulls.
9615. The habitat available in a single rainforest tree is much more diverse than the entire Antarctic ecosystem. Far fewer species, but many more individuals, inhabit higher latitudes than the tropics. While the habitat and available food is less diverse in polar regions, the abundance of food is such that enormous populations can be supported.
9616. No.

Answers

9617. Yes, based on midden remains, Little Blue (Fairy) Penguins were eaten.
9618. Rufous, but the color is generally only evident in flight.
9619. Northern Ravens, Steller's Jays and possibly Great Horned Owls. Ten of the 23 tree nests found up to 1993 in North America were depredated, and all three of these birds have been seen at nest sites. The murrets themselves are possibly preyed upon as well.
9620. Black.
9621. Harlan's Hawk. Once considered a distinct species, it is currently a race of the Red-tailed Hawk.
9622. The White-crowned Sparrow.
9623. Yes, on two occasions condor nests were discovered high up in the hollow of a sequoia (redwood) tree, the latest in 1984.
9624. No, their tails are extremely short.
9625. Yes.
9626. No, just the opposite. The length of the primaries, secondaries and tail feathers of a year-old Bald Eagle exceed the length of adult feathers by as much as 8, 13 and 23 percent respectively.
9627. The Wood Stork.
9628. More than 40 illustrated books.
9629. Yes.
9630. The bird is a female. Males have blue faces.
9631. Kittiwakes.
9632. The Common (Ring-necked) and Reeve's Pheasants.
9633. A frigatebird.
9634. The Lesser Yellowlegs.
9635. The two scavenging giant-petrels.
9636. Between 15 and 20 percent.
9637. No.
9638. Yes, except in terrestrial species.
9639. Alligators lie in wait among a hunter's decoys and as ducks are shot, the reptiles immediately claim them. Retriever dogs have been consumed by gaters as well.
9640. No, it is dark colored.
9641. A rabbit burrow.
9642. The Cockatiel.
9643. Yes.
9644. The mule was named after Lind, the *Swedish Nightingale,* a famous singer of the times, "...on account of her musical bray."
9645. A light shotgun for shooting birds.
9646. The introduced European (Common) Starling usurps its nesting holes, and may even kill woodpeckers.
9647. The zoo purchased four Michigan pairs in 1878 or 1879. At the time, the price for live pigeons was high, $2.50 a pair, whereas dead ones sold for 35-50 cents each in Chicago markets. The zoo colony did not prosper, and died out in 1914 when *Martha*, the last of her kind, expired.
9648. Using border collies, because dogs are far more efficient at tracking geese on steep terrain than biologists.
9649. The curlew berry.
9650. John *Jay* (1745-1829).

Answers

9651. The Spectacled Owl, because its call has been compared to the prolonged rapid tapping of a woodpecker.
9652. The Pacific Gull of Australia.
9653. Yes.
9654. The Jamaican Becard.
9655. No, in some areas (mainly Alaska), the eagles commonly nest on ridges, cliffs and sea-stacks.
9656. In the latest taxonomic revision, the little sparrow-like birds are placed between the pipits and weavers. It was formerly believed that their nearest relatives were the thrushes.
9657. The Spectacled Guillemot.
9658. The Skylark.
9659. Yes.
9660. Bright lemon-yellow.
9661. Coral-red.
9662. The Brown-eared Bulbul.
9663. Theodore A. Parker III.
9664. Blue Bird-of-paradise skins sold for as much as £40.
9665. Yes.
9666. The four diving-petrels (family Pelecanoidae; genus *Pelecanoides*). The little petrels do not resemble pelicans, but like them, they have an extensible throat pouch.
9667. The Old World pratincoles.
9668. The Sharpbill.
9669. Brewer's Sparrow.
9670. Dig out nesting burrows seeking the tiny eggs to eat.
9671. No, the birds can go for days without water. Most of their water is obtained from succulent plants, and the Ostrich has no problem surviving in regions where the annual rainfall is under eight inches (20 cm).
9672. The Phoenix Cardinals.
9673. The Merlin.
9674. The buttercup.
9675. No. The Two-banded Plover is a South American and Falkland Islands species, whereas the Double-banded Dotterel inhabits Australia and New Zealand.
9676. Yes.
9677. Yes, sometimes all night long.
9678. Prince Maximilian of Weid, a German naturalist and traveler in early 19th century America. The bird is currently called the Brown-crested Flycatcher.
9679. No, they are quite aggressive.
9680. Small hard seeds, hence their need to drink every day.
9681. Essentially diurnal, the stork rarely feeds at night, and even then only during extensive moonlight, or in pools illuminated by fishermen's fires.
9682. The Short-eared Owl. The name alludes to its flight style and irregular wingbeat, comparable to that of a giant butterfly.
9683. Yes.
9684. The guans.
9685. Yes, in years of a double rainy season, Hoatzins may produce two broods, but in dry years, none at all.
9686. At least a year.

Answers

9687. About 18 percent.
9688. No.
9689. A melanistic or erythristic color form of the Least Bittern, formerly regarded as a distinct species.
9690. No.
9691. An average of about 120 paces.
9692. Some ducks have crossed the English Channel and have become established in continental Europe where there is fear of hybridization with the endangered White-headed Duck. At least one hybrid has been documented.
9693. Surfaced sea turtles.
9694. Between 6,000 and 8,000, as of 1990.
9695. Coues' (Washington Island) Gadwall, a dwarf form of the Common Gadwall, known only from a single pair (apparently juveniles), both of which are deposited in the National Museum, Washington, D.C. The two ducks were collected in January, 1874 by Thomas H. Streets, an assistant U.S. Navy surgeon. The obscure race was probably limited to the lake and peat bogs of Washington and New York Islands of the mid-Pacific Fanning Island group. Migrant waterfowl were regularly hunted by the island inhabitants during the late 1800's, and if Coues' Gadwall was a non-migratory resident, it could easily have been extirpated by hunters in a few years.
9696. At about nine days of age, but by 16 days, the bill shape resembles that of the adult.
9697. No, the 23-day incubation period is long for a passerine.
9698. A ravenry.
9699. A widespread Philippine parrot.
9700. The New Zealand shorebird is extinct, having disappeared in 1964.
9701. A nightingale.
9702. The neotropical birds were named after their calls.
9703. Bird thistle.
9704. Four: Magellanic, Humboldt, Rockhopper and Macaroni Penguins.
9705. The Wood Stork.
9706. *Beccafico* is a familiar Italian name meaning fig-eater. It is applied almost indiscriminately to small passerines frequenting gardens during the autumn, many of which are caught for the table.
9707. Pure white, but this is often obscured due to a powdering of iron oxide, acquired by cosmetic means, either by dusting on ledges, or by bathing in springs or rivulets of iron-rich water. The rufous color persists most of the time, but it may wash off in wet weather.
9708. On the ground.
9709. Yes. In 1960, Red-whiskered Bulbuls were released in south Miami, where the exotic birds are currently locally common.
9710. *Quitting cold turkey.*
9711. Yes, although hens tend to be somewhat duller.
9712. Yes.
9713. The Ruddy Duck.
9714. The Common Grackle.
9715. A contrasting color only on the shaft of a feather, not extending onto the web.
9716. Yes.
9717. The clutch size is 1-2 eggs, but apparently only a single chick is reared. The smaller chick is not killed as a result of sibling rivalry, but usually starves.

Answers

9718. Females were especially vulnerable to collectors when trapped in their sealed-up nesting cavities.
9719. The cranes.
9720. The Black Kite, because it has adapted to living in close association with humans.
9721. The Rosy Finch.
9722. On April 18, 1901, along the east coast of Florida. Another was reportedly collected in Kansas in August, 1904.
9723. No, the vulture was introduced from Cuba during the late 1800's, and is currently well established.
9724. Yes, in Alaska and Siberia.
9725. The Least Seedsnipe. The other three are named for their belly color: Rufous-bellied, White-bellied and Gray-breasted Seedsnipe.
9726. No, the continental geese are significantly brighter.
9727. The Western Gull.
9728. The Red-throated Loon.
9729. The red bill is tipped yellow.
9730. A weaver.
9731. The Burrowing Owl may use an armadillo burrow.
9732. A creature with the supernatural power of taking on at will the form of a swan, or of a beautiful maiden.
9733. No, but it winters there.
9734. He could recall the vocalizations of over 3,000 species of birds.
9735. No, the little rails are inhabitants of rough grasslands and hay fields.
9736. Yes, up to four nests may be constructed on top of one another.
9737. Bright yellow.
9738. No.
9739. The subspecies is in serious trouble. During the early 1950's, the population was estimated at 200, but in 1990, only 54 cranes reporetedly remained.
9740. The raptor preys on hatchlings.
9741. The Hamerkop stork.
9742. It is migratory and only seen in North America in the summer.
9743. Mutton-birders.
9744. A huge, large-bore gun, generally mounted on a special shallow-draft boat, in which a hunter lies prone and stalks waterfowl. At times, hundreds of ducks were dropped with a single blast. In most areas, punt-gunning is illegal.
9745. As the bird of Satan.
9746. The Mistletoe-bird.
9747. Yes.
9748. Yes.
9749. The American Ornithologists' Union. The award was for work published in the previous 6-10 calendar years that is judged to be the most important work relating in whole or in part to the birds of the western hemisphere. In recent years, the geographical restriction has been dropped.
9750. The name is derived from the Greek *scopus*, meaning "see."
9751. The kiwis. The attachment to their territory is so strong that even if their forest is chopped down, they are liable to remain there for several weeks, until forced to move away because of scarcity of food arising from loss or drying out of the surface layer of soil.

9752. The Dickcissel, because of its song.
9753. No, the huge aircraft is constructed of birch wood.
9754. At least 14 species, of which five definitely became established, with three others probable.
9755. An old sportsman's term generally applied to two birds of the same species bagged.
9756. The Ruddy Duck.
9757. It varies from a minimum of 38 to a maximum of 176 heartbeats per minute.
9758. Yes, condors have succumbed to lead poisoning.
9759. The Squirrel Cuckoo.
9760. No, they have cackling calls.
9761. The feathers are stained during preening when secretions from the oil gland are spread on the wing.
9762. Brown.
9763. No, northern Bald Eagles are larger.
9764. No, only cocks have the colorful combs.
9765. Yes.
9766. The Pileated Woodpecker.
9767. Bannerman's Turaco of the montane forests of southwestern Cameroon is endangered, and Prince Ruspoli's Turaco from southern Ethiopia is rare and restricted to about 10 square miles (26 sq km) of highland juniper woods.
9768. No, a few tinamous are not polyandrous, such as the Ornate Tinamou that forms stable pair bonds, or the Spotted Tinamou that is initially monogamous, but becomes polyandrous when older.
9769. No, nests are typically located 3-20 feet (0.9-6.1 m) above the ground.
9770. The harbinger of ill omens. This belief supposedly came about because on foggy evenings cormorants coming in to roost become disoriented and were occasionally attracted to campfires, sometimes crashing into the flames, spreading burning embers among the thatched houses, and setting them on fire.
9771. Some have lived over two decades.
9772. The Water Thick-knee of Africa.
9773. No, most have rounded tails.
9774. Eight days, with a range of 5-12 days.
9775. No, while common in Jamaica, the finch was introduced from South America.
9776. The Crozet Islands (Indian Ocean).
9777. None.
9778. Once restricted to the rocky shores of the North Atlantic during the winter, the birds now winter as far south as central Florida. Large stone breakwaters provide ideal habitat for the marine life on which the sandpipers depend, and the birds simply followed breakwater construction south.
9779. The worm.
9780. Yes.
9781. Yes.
9782. Guatemala.
9783. Yes.
9784. Approximately 50.
9785. No, the terns tend not to fly higher than 10 feet (3 m).
9786. A soup made of a capon boiled with leeks.
9787. Eighteen.

Answers

9788. No, but it does have a neck-ruff of long, pointed feathers that wave in the wind, giving the ibis a somewhat shaggy appearance. It is also known as the Northern Bald Ibis, because of its bare head.
9789. The Northern Goshawk, and some have even attempted to carry decoys off.
9790. Yes, the eggs of some species are so glossy that reflections can be seen in them.
9791. A loony, because a loon is featured on one side.
9792. *Bone-swallower,* alluding to its scavenging habits.
9793. Yes.
9794. The Zebra Dove.
9795. No, the bird feeds primarily on beetles, caterpillars, wasps and spiders.
9796. Yes, but the birds more typically probe for insects on the ground.
9797. The Wattled-crow (Kokako), because of its rich melodious song.
9798. The Cape Sparrow.
9799. Yes, but the herons are essentially nocturnal.
9800. The American Ornithologists' Union.
9801. The woodcreepers.
9802. Yes.
9803. White or gray.
9804. The Asian Pheasant-tailed Jacana. It is the only migrating jacana.
9805. Williamson's Sapsucker. Dr. Newberry, a surgeon assigned to the Pacific Railroad Survey, named the woodpecker in honor of his commanding officer, Lt. R. S. Williamson.
9806. Dr. Jean Delacour, the well-known French ornithologist and aviculturist.
9807. The Northern Shoveler, because of its guttural calls.
9808. When passerines land on a perch, their weight causes the leg tendons to tighten, clamping the toes tightly shut.
9809. The nests of tree-dwelling termites (termitaria).
9810. No.
9811. Probably one of the races of Crested Serpent-Eagle, such as those from Nias Island (Indonesia), or the Nicobar Islands.
9812. Yes.
9813. The Platte River (Nebraska).
9814. The Lesser Frigatebird. The natives were aware that the birds often heralded the approach of a storm by coming ashore to rest on mangroves with outspread wings, disappearing again as soon as the storm abated.
9815. Yes, on fence posts, for example.
9816. Up to 7.25 inches long (18.4 cm).
9817. Alaska.
9818. Cattle ranchers of Isla Santa Cruz (Indefatigable Island) introduced the Smooth-billed Ani during the mid-1970's because they erroneously believed the birds (also known as tickbirds) would feed on the ticks plaguing cattle. Following the El Niño event of 1982/83, the population dramatically increased and anis currently inhabit many of the islands.
9819. Yes.
9820. No, adult plumage is not acquired until its third year.
9821. The average dive length is about 33 seconds, but the little alcids can remain underwater for more than a minute.
9822. No.
9823. The Slender-billed Prion.

Answers

9824. The European Honey-buzzard, Eleonora's and Red-footed Falcon.
9825. A beautiful emerald-green.
9826. Unlike many herons, they tend to nest alone and do not require the stimulus of a colony to breed successfully.
9827. The Anhinga and darters.
9828. Mistletoe berries.
9829. No, the Andean Avocet is a high-altitude resident, generally occurring above 10,500 feet (3,217 m).
9830. A young chicken, especially one only a few months old, used for broiling or frying. In slang usage, a *spring chicken* applies to a young person.
9831. The Ruffed Grouse.
9832. Immortality.
9833. Grayish-yellow.
9834. The Dolphin Gull of South America and the Falkland Islands.
9835. Swing their tails from side to side, and up and down.
9836. *Robin* Hood, a 12th century outlaw.
9837. Hop, but the two African ground-hornbills walk.
9838. The Macaroni Penguin beta egg is up to 71 percent larger than the alpha egg.
9839. Based on 1991 counts, about 3,900.
9840. Richard Burton, Roger Moore and Richard Harris.
9841. A duck squack.
9842. The Antipodes Island Green Parakeet inhabits the subantarctic islands of Antipodes.
9843. Yes.
9844. The White-tailed Tropicbird.
9845. Dr. Thomas S. Traill (1781-1862) was a Scottish doctor and professor of medicine at Edinburgh University. He was also the editor of the 8th edition of *The Encyclopedia Britannica.*
9846. In 1967, a pair commenced breeding in the Shetland Islands where the owls were rigorously protected. The pair fledged 21 owlets over the next nine years, but in 1975 the male disappeared, and young have not been produced since.
9847. Not normally, but Eastern Meadowlarks can cause extensive damage to sprouting corn in North and South Carolina in the early spring.
9848. The Hawaiian Goose (Nene).
9849. The burying of chicks and adults in their burrows by earth slides caused by people or grazing sheep.
9850. No, junglefowl nest on the ground.
9851. *Parrot of the dark*, because the nocturnal, terrestrial parrot shunned daylight.
9852. No.
9853. At about 7-8 days, the young leave the nest. At 12-13 days chicks can fly a short distance, and at 26-34 days are completely independent.
9854. The American Kestrel.
9855. Four: Laysan, Black-footed, Short-tailed and Waved Albatross.
9856. No, chicks are not able to stand until about 7-10 days.
9857. No.
9858. For firewood, one of the most valuable commodities in that impoverished nation.
9859. Snakes do not occur in New Zealand.

Answers

9860. The owl has evolved a strategy of partial migration where adult males are resident, but females and young migrate.
9861. No, males are more numerous.
9862. Greater Rhea eggs are yellowish in color, but fade to off-white. Lesser Rhea eggs are greenish.
9863. Hair.
9864. The Yellow-rumped Warbler.
9865. Yes, the loose neck skin can be greatly distended when inflated.
9866. France.
9867. Probably the Great Knot and Sanderling.
9868. The White (Fairy) Tern.
9869. The Algerian Nuthatch, the most recently (1980) described Western Palearctic species.
9870. The four Marx brothers.
9871. None, but a number were introduced.
9872. Six.
9873. Yes, but only males.
9874. No, not compared to other storks.
9875. Yes, as a general rule.
9876. A gull.
9877. When gathered at watering holes.
9878. No, but there is a record of a Eurasian Spoonbill surviving at least 28 years.
9879. Yes.
9880. Black, bordered by two broad white patches.
9881. No, it is a band of usually dark feathering on the wing between the wrist (carpel joint) and elbow, particularly noticeable in some species of juvenile terns.
9882. No, the large, spectacular wattlebird measured up to 19 inches (48 cm) in length.
9883. Trainbearers are long-tailed hummingbirds of the Andes of northern South America.
9884. The neotropical Masked Duck.
9885. *Sharming.*
9886. Red.
9887. Yes, nearly so.
9888. Yes, the endemic bird is capable of gliding down slopes, and can run rapidly.
9889. Yes.
9890. Yes, he thought the huge bird was from Madagascar.
9891. Yes, but cotingas also feed from the perch.
9892. Yes.
9893. On August 17, 1978, the *Double Eagle II* with three Americans aboard crossed the Atlantic, landing near Paris, France.
9894. None has primary feathers.
9895. John James Audubon, who thought of the Bald Eagle as a coward.
9896. Yes.
9897. The birds snap up small animals attempting to escape the hunting carnivores.
9898. Over 66 percent. Of the 12 resident land birds present in 1788, only four survive.
9899. In the late 1700's and early 1800's, Bewick was regarded as the foremost bird illustrator of the day, and he developed wood engraving into an art form.
9900. The serrations possibly produce a whistling sound in flight that might be related to courtship.

Answers

9901. It is an American award for animals that served with conspicuous "gallantry" during wartime. The famous carrier pigeon *G.I Joe*, for example, was awarded the medal for saving an allied-occupied Italian village from bombing.
9902. Boo birds.
9903. The Amur River forming the far eastern border between Russia and China. Five cranes breed in the region: Red-crowned, White-naped, Hooded, Common and Demoiselle Crane. In addition, the Siberian White Crane migrates through the area.
9904. Juveniles have naked, pale yellow or whitish faces and throat pouches, as opposed to the red and blue of the adults.
9905. Yes, so far as is known.
9906. The Western Kingbird.
9907. Quetzaltenango, named after the national bird, the Resplendent Quetzal.
9908. Yes.
9909. A canary: *He sang like a canary.*
9910. Not as a rule.
9911. A trainer of short-winged hawks, mainly Northern Goshawks.
9912. A harrier.
9913. Yes, although this is not typical foraging behavior.
9914. Yes.
9915. The Dunlin.
9916. The Common Raven.
9917. The Willet.
9918. Alabama selected the flicker.
9919. The White-chinned Petrel.
9920. Yes.
9921. No, not as a rule.
9922. Yes.
9923. The flightless parrot's clutch consists of but a single egg.
9924. *The Ibis* was initially published in 1859.
9925. Possibly 50-70 years.
9926. The Red-throated Loon. The short, frequently repeated cackle is typically uttered when flying over its own or a neighboring territory.
9927. No, quite the contrary. It is known from only seven specimens from submontane Borneo.
9928. Yes.
9929. Yes.
9930. No, the nostrils are round.
9931. The Bare-crowned Antbird of Central America has a bare blue crown, forehead and face.
9932. The Philippines (Tawi-Tawi Island).
9933. Yes.
9934. Yes.
9935. *Two in the bush.*
9936. Yes.
9937. About one percent.
9938. The mosquito.
9939. No, only infrequently, and the first hybrid was not recorded until 1988.
9940. Mistletoe, the fruit of which the cotinga feeds on almost exclusively. The bird regurgitates seeds onto branches, where the seeds germinate. The mistletoe is dependent on the cotinga

for seed dispersal because no other seed-eater occurs at that elevation. Without the cotinga, the seeds would simply fall to the ground, where they cannot survive.

9941. The Greater Roadrunner.
9942. Yes, but the stork is more at ease on the ground near water.
9943. While it is claimed that rheas feed on poisonous snails, such behavior has not been documented.
9944. A bird that characteristically scratches the ground while foraging, such as a chicken.
9945. Swan feathers (down) are placed over a seal's snow-covered breathing hole, and when a hunter sees the feathers move, he is alerted that a seal is rising to breathe.
9946. Males are smaller. The female is substantially larger, an unusual passerine trait.
9947. The Sandhill Crane, but the meadowlark prevailed.
9948. No, its breast is streaked, not spotted.
9949. Red.
9950. No, not since the recent taxonomic revision. The two African ground-hornbills are now in a separate family, the Bucorvidae.
9951. The dippers.
9952. No, but generally so.
9953. The Mute Swan.
9954. Dr. Elliott Coues.
9955. About nine inches (22.9 cm).
9956. No, the turaco was named after Prince Eugenio Ruspoli (1866-1893), an Italian zoologist who was in northeastern Africa from 1891 to 1893, where he was killed by an elephant.
9957. Yes, they are among the few American birds that do.
9958. The Mallard.
9959. No, the site is invariably changed.
9960. No, the White-headed Vulture kills adults as well as young; flamingo eggs are also eaten
9961. Yes.
9962. Melanin, the same as for the color black.
9963. Peter Ilyich Tchaikovsky (1840-1893), the famous Russian composer.
9964. Males have reddish streaked breasts, but streaking is lacking in females.
9965. France.
9966. Yes.
9967. Dr. W. J. Burchell (1782-1863), a naturalist who explored South Africa, in 1811.
9968. No, the waxwing breeds in southeastern Siberia and northern Manchuria.
9969. No, jacamars favor lowland habitat.
9970. The House (English) Sparrow.
9971. No, the birds are rare in Tasmania.
9972. Campbell Island. European Blackbirds, Song Thrushes, Common Starlings, Redpolls, Chaffinches, white-eyes and Dunnocks all breed on the island. The birds reached Campbell from New Zealand on their own, aided by the wind. Most are exotics first introduced into New Zealand from Europe.
9973. St. Lucia in the West Indies. There have only been five records since the 1920's, and the endemic warbler is probably extinct.
9974. Yes, some species do.
9975. Goose grass.
9976. Alexander Wilson discovered a torpid Ruby-throated Hummingbird in 1810. Torpid hummingbirds can lower their temperature 30-50°F (17-28°C) below active levels. While

normally considered a nightly hibernation, Chimborazo Hillstars at 16,000 feet (5,400 m) on Mt. Cotopaxi in Ecuador went into a torpor between each feeding session, and were easily approached and picked up by hand during these frequent diurnal "hibernations."

9977. About 109,000 pelicans in 55 colonies, as of 1992.
9978. Yes.
9979. A common European bunting.
9980. Yes, but only in domestic breeds. The crested or top-knotted duck has a large globular crest.
9981. The barn would never be struck by lightening.
9982. No, all three crowned-pigeons nest in trees.
9983. Yes, gannets dive straight down, whereas boobies are more inclined to plunge-dive at an angle.
9984. No, avian toes are moved by tendons, not muscles.
9985. Their bills are brownish, whereas those of adults are red.
9986. No, it has a single entrance.
9987. Yes, generally so.
9988. The Cape Gannet.
9989. Not until 1903. The warbler was first discovered in 1851, in Ohio.
9990. Usually the reigning ornithological hierarchy in each country, such as the American Ornithologists' Union, British Ornithologists' Union, Royal Australian Ornithologists' Union, etc. To many people, the changes often appear arbitrary and unwarranted. Historical precedent is often thrown to the winds, especially with such name changes as Double-wattled Cassowary to Southern Cassowary, Baltimore Oriole to Northern Oriole, and Myrtle Warbler to Yellow-rumped Warbler. Some, like the latter two, are caused by lumping, but many are put in effect by arrogance on the part of the ornithologists. The proposed name change of the Sharp-tailed Streamcreeper to the Streamside Lochmias (voted down, fortunately) is a good example of the short-sightedness of many of these "eminent" ornithologists. Many regional authors still favor historical precedent, which is why names still vary from one field guide to another and standardization of common names worldwide is still a long way off.
9991. Yes.
9992. Yes, but they are poor fliers.
9993. Pink.
9994. Yes, with ibises and other herons.
9995. No, rainforest areas are avoided, even when extensively cleared and cultivated.
9996. Yes, some species may rear five broods in succession.
9997. Yes, and this might be regarded as "play" behavior.
9998. No, it is an abundant bird.
9999. No, but because of the alternate bands of light and dark colors, the long feathers *appear* to be notched.
10,000. Yes, in the western highlands, but less numerous to the east.
10,001. *Ibis* Publishing Company.

Records

In order to make the index more functional, certain key categories appear in **boldface**. Not every bird or item has been cross-referenced in the index, as this would often give away the answer.

This page lists major categories for records, and the following page lists the boldface headings for other categories.

Major Index Categories

Index

Index

Index

Index

Index

Index

Index

Index

Index

Index

Index